**208 Representations of Elementary Abelian $p$-Groups and Vector Bundles**

CAMBRIDGE TRACTS IN MATHEMATICS



A complete list of books in the series can be found at www.cambridge.org/mathematics. Recent titles include the following:

173. Enumeration of Finite Groups. By S. R. BLACKBURN, P. M. NEUMANN, and G. VENKATARAMAN
174. Forcing Idealized. By J. ZAPLETAL
175. The Large Sieve and its Applications. By E. KOWALSKI
176. The Monster Group and Majorana Involutions. By A. A. IVANOV
177. A Higher-Dimensional Sieve Method. By H. G. DIAMOND, H. HALBERSTAM, and W. F. GALWAY
178. Analysis in Positive Characteristic. By A. N. KOCHUBEI
179. Dynamics of Linear Operators. By F. BAYART and É. MATHERON
180. Synthetic Geometry of Manifolds. By A. KOCK
181. Totally Positive Matrices. By A. PINKUS
182. Nonlinear Markov Processes and Kinetic Equations. By V. N. KOLOKOLTSOV
183. Period Domains over Finite and $p$-adic Fields. By J.-F. DAT, S. ORLIK, and M. RAPOPORT
184. Algebraic Theories. By J. ADÁMEK, J. ROSICKÝ, and E. M. VITALE
185. Rigidity in Higher Rank Abelian Group Actions I: Introduction and Cocycle Problem. By A. KATOK and V. NIŢICĂ
186. Dimensions, Embeddings, and Attractors. By J. C. ROBINSON
187. Convexity: An Analytic Viewpoint. By B. SIMON
188. Modern Approaches to the Invariant Subspace Problem. By I. CHALENDAR and J. R. PARTINGTON
189. Nonlinear Perron–Frobenius Theory. By B. LEMMENS and R. NUSSBAUM
190. Jordan Structures in Geometry and Analysis. By C.-H. CHU
191. Malliavin Calculus for Lévy Processes and Infinite-Dimensional Brownian Motion. By H. OSSWALD
192. Normal Approximations with Malliavin Calculus. By I. NOURDIN and G. PECCATI
193. Distribution Modulo One and Diophantine Approximation. By Y. BUGEAUD
194. Mathematics of Two-Dimensional Turbulence. By S. KUKSIN and A. SHIRIKYAN
195. A Universal Construction for Groups Acting Freely on Real Trees. By I. CHISWELL and T. MÜLLER
196. The Theory of Hardy's $Z$-Function. By A. IVIĆ
197. Induced Representations of Locally Compact Groups. By E. KANIUTH and K. F. TAYLOR
198. Topics in Critical Point Theory. By K. PERERA and M. SCHECHTER
199. Combinatorics of Minuscule Representations. By R. M. GREEN
200. Singularities of the Minimal Model Program. By J. KOLLÁR
201. Coherence in Three-Dimensional Category Theory. By N. GURSKI
202. Canonical Ramsey Theory on Polish Spaces. By V. KANOVEI, M. SABOK, and J. ZAPLETAL
203. A Primer on the Dirichlet Space. By O. EL-FALLAH, K. KELLAY, J. MASHREGHI, and T. RANSFORD
204. Group Cohomology and Algebraic Cycles. By B. TOTARO
205. Ridge Functions. By A. PINKUS
206. Probability on Real Lie Algebras. By U. FRANZ and N. PRIVAULT
207. Auxiliary Polynomials in Number Theory. By D. MASSER
208. Representations of Elementary Abelian $p$-Groups and Vector Bundles. By D. J. BENSON

# Representations of Elementary Abelian $p$-Groups and Vector Bundles

DAVID J. BENSON
*University of Aberdeen*

CAMBRIDGE
UNIVERSITY PRESS

**CAMBRIDGE**
**UNIVERSITY PRESS**

University Printing House, Cambridge CB2 8BS, United Kingdom

One Liberty Plaza, 20th Floor, New York, NY 10006, USA

477 Williamstown Road, Port Melbourne, VIC 3207, Australia

4843/24, 2nd Floor, Ansari Road, Daryaganj, Delhi – 110002, India

79 Anson Road, #06–04/06, Singapore 079906

Cambridge University Press is part of the University of Cambridge.

It furthers the University's mission by disseminating knowledge in the pursuit of education, learning, and research at the highest international levels of excellence.

www.cambridge.org
Information on this title: www.cambridge.org/9781107174177
DOI: 10.1017/9781316795699



First published 2017

*A catalogue record for this publication is available from the British Library.*

*Library of Congress Cataloging-in-Publication Data*
Names: Benson, D. J. (David J.), 1955–
Title: Representations of elementary abelian p-groups and vector bundles /
David J. Benson, University of Aberdeen.
Description: Cambridge : Cambridge University Press, [2017] |
Series: Cambridge tracts in mathematics ; 208 |
Includes bibliographical references and index.
Identifiers: LCCN 2016046127 | ISBN 9781107174177 (alk. paper)
Subjects: LCSH: Abelian p-groups. | Abelian groups. | Vector bundles.
Classification: LCC QA180 .B46 2017 | DDC 512/.25–dc23
LC record available at https://lccn.loc.gov/2016046127

ISBN 978-1-107-17417-7 Hardback

# Contents

# Preface

The origins of this book lie in an extended visit that I made in the Spring of 2008 to MSRI in Berkeley, California as the Simons Professor for the programme on Representation Theory of Finite Groups and Related Topics. Jon Carlson and Julia Pevtsova were there for a large part of this time, and Eric Friedlander passed through town several times. Through numerous conversations and lectures, they ignited my interest in the theory of modules of constant Jordan type. At first, I was reluctant to be drawn in. But then I managed to prove one of the conjectures from a paper of Carlson, Friedlander and Pevtsova [81], and I was hooked. This work has been published [42] and appears here as Section 5.12.

After that initial success, I started working with Julia Pevtsova on the vector bundles on projective space associated to modules of constant Jordan type. We refined a previous definition of Friedlander and Pevtsova, and proved a realisation theorem. This work has also been published [53] and appears here as Theorem 8.1.1, Sections 8.2–8.9 and Section 10.2.

I wrote several other papers soon after that, and soon it was getting to the point where I had so much material that it made sense to make a book based on the resulting series of papers as well as a great deal of other unpublished work. This is that book.

My thanks go to MSRI for its hospitality in 2008 and 2013 while parts of this work were being written; David Eisenbud for sharing his knowledge of vector bundles on projective spaces; Jon Carlson, Eric Friedlander and Julia Pevtsova for numerous conversations; Serge Bouc, Radu Stancu and Jon Carlson for their extensive feedback on earlier versions of the text; Jeremy Rickard for various interesting comments and questions, and particularly for formulating Conjecture 5.13.1; Andrew Granville for helping me with some number theoretic questions, and especially for formulating and supplying a proof of a statement similar to Lemma 12.10.1. I'd also like to thank Jesse Burke, Jon Carlson, Claudia Miller, Julia Pevtsova and Greg Stevenson for enlightening conversations and feedback related to Chapter 11, and Mark Walker for giving a talk at a conference in Seattle that stimulated my interest in Orlov's correspondence.

# Introduction

Let us begin with an example. Consider the following $30 \times 30$ matrices with entries in an algebraically closed field $k$ of characteristic two.

$$
\left(\begin{array}{cccccccccccccccccccccccccccccc}
&&&&&&&&&&&&&&&&&&&&&&&&&&&&& \\
&&&&&&&&&&&&&&&&&&&&&&&&&&&&& \\
a&&&&&&&&&&&&&&&&&&&&&&&&&&&&& \\
b&&&&&&&&&&&&&&&&&&&&&&&&&&&&& \\
c&&&&&&&&&&&&&&&&&&&&&&&&&&&&& \\
d&&&&&&&&&&&&&&&&&&&&&&&&&&&&& \\
e&&&&&&&&&&&&&&&&&&&&&&&&&&&&& \\
f&&&&&&&&&&&&&&&&&&&&&&&&&&&&& \\
&a&&&&&f&e&&&&&&&&&&&&&&&&&&&&&& \\
&b&&&&f&&d&&&&&&&&&&&&&&&&&&&&&& \\
&c&&&&e&d&&&&&&&&&&&&&&&&&&&&&&& \\
&d&&c&b&&&&&&&&&&&&&&&&&&&&&&&&& \\
&e&c&&a&&&&&&&&&&&&&&&&&&&&&&&&& \\
&f&b&a&&&&&&&&&&&&&&&&&&&&&&&&&& \\
&&d&&f&a&&c&&&&&&&&&&&&&&&&&&&&&& \\
&&e&&&&a&&&&&&&&&&&&&&&&&&&&&&& \\
&&f&&&&&a&&&&&&&&&&&&&&&&&&&&&& \\
&&&d&&b&&&&&&&&&&&&&&&&&&&&&&&& \\
&&&e&f&&b&c&&&&&&&&&&&&&&&&&&&&&& \\
&&&f&&&&b&&&&&&&&&&&&&&&&&&&&&& \\
&&&&d&c&&&&&&&&&&&&&&&&&&&&&&&& \\
&&&&e&&c&&&&&&&&&&&&&&&&&&&&&&& \\
&&&&&&&&&&&&f&e&&b&c&&a&&&&&&&&&&& \\
&&&&&&&&&&&f&&d&b&&&a&&c&&&&&&&&&& \\
&&&&&&&&&&&e&d&&c&&&&c&&a&b&&&&&&&& \\
&&&&&&&&&c&b&&&&&&&e&d&&f&&&&&&&&& \\
&&&&&&&&c&&a&&&&e&d&&&&&&f&&&&&&&& \\
&&&&&&&&b&a&&&&&f&&d&&f&e&&&&&&&&&& \\
&&&&&&&&d&e&f&a&b&c&&&&&&&&&&&&&&&& \\
&&&&&&&&&&&&&&&&&&&&&&d&e&f&a&b&c&&
\end{array}\right)
$$

Blank entries are taken to be zero (there are two blank rows at the top and two blank columns on the right). These matrices, for different values of $a, b, c, d, e, f \in k$, all commute and square to zero. Furthermore, as long as $a, b, c, d, e, f$ are not all zero, the matrix has rank exactly 14. So the kernel modulo the image is a two-dimensional vector space. This means that we have assigned to each point $(a : b : c : d : e : f) \in \mathbb{P}^5$ a vector space of dimension two. This gives us a rank two *algebraic vector bundle* on projective 5-space, which happens to be the indecomposable *Tango bundle*.

On the other hand, if we add the identity to the six matrices obtained by setting one of the variables equal to 1 and the rest equal to 0, we obtain six commuting matrices which square to the identity. In other words, we have a $kE$-module, where $E \cong (\mathbb{Z}/2)^6$ is an elementary abelian 2-group of rank six. This module has *constant Jordan type* $[2]^{14}[1]^2$, meaning that the matrices above all have the same Jordan canonical form with 14 blocks of length two and two blocks of length one.

This illustrates the connection between modules of constant Jordan type for elementary abelian $p$-groups and vector bundles on projective space in characteristic $p$ investigated in this book. Because the people studying these two subjects are almost disjoint, I have tried to include plenty of introductory material. The reader should feel free to skip this if appropriate.

So what are modules of constant Jordan type? Among all modules, they are analogous to the vector bundles among the sheaves.

As Dade put it in [97], "There are just too many modules over $p$-groups!" More explicitly, the group algebra of a finite $p$-group in characteristic $p$ usually has *wild representation type*, as we explain in Section 1.2. For a more general finite group, the Sylow $p$-subgroup controls the representation type. It follows that in general we do not hope to classify all the finite-dimensional indecomposable representations of a finite group.

The theory of varieties for modules was developed by Carlson and others [24, 73, 74, 75] as a way of getting at module structure without making such a classification. Many aspects of this theory are controlled by the elementary abelian $p$-subgroups. So it makes sense to study modules for an elementary abelian $p$-group as a subject in its own right. We concentrate on modules of constant Jordan type. For these modules the variety gives essentially no information, so the theory supplements the now well-established variety theory. Modules of constant Jordan type are still wild whenever the representation type of the group is wild, as we show in Section 5.5. These modules are much more rigidly behaved than the general module, but the theory is nonetheless surprisingly rich.

The formal definition of constant Jordan type is as follows. Let $E = \langle g_1, \ldots, g_r \rangle \cong (\mathbb{Z}/p)^r$ be an elementary abelian $p$-group and let $k$ be an algebraically closed field of characteristic $p$. We set

$$X_i = g_i - 1 \in kE$$

for $1 \le i \le r$, so that $X_i^p = 0$. If $\alpha = (\lambda_1, \ldots, \lambda_r) \in \mathbb{A}^r(k)$, we define

$$X_\alpha = \lambda_1 X_1 + \cdots + \lambda_r X_r \in kE.$$

If $\alpha \neq 0$ then $g_\alpha = 1 + X_\alpha$ is a unit of order $p$ in $kE$.

A *cyclic shifted subgroup* of $E$ is a subgroup of the group algebra $kE$ of the form $E_\alpha = \langle g_\alpha \rangle$ for $0 \neq \alpha \in \mathbb{A}^r(k)$. A finitely generated $kE$-module is said to have *constant Jordan type* if the Jordan canonical form of $X_\alpha$ on $M$ is independent of $\alpha$ for $0 \neq \alpha \in \mathbb{A}^r(k)$.

We are far from understanding what Jordan types occur for a module $M$ of constant Jordan type. Of course if $r = 1$ then the problem is trivial. So let us assume that $r \geq 2$. We write $[a_1]\dots[a_t]$ for a Jordan type with blocks of lengths $a_1, \dots, a_t$, each of which is an integer between 1 and $p$. We often wish to ignore Jordan blocks of length $p$, and the *stable Jordan type* is the same list with the length $p$ blocks omitted. The first important theorem in the subject is Dade's lemma, from his 1978 paper [98], which states that if the stable Jordan type is empty, in other words if $M$ has constant Jordan type $[p]^n$ for some $n \geq 0$, then $M$ is projective. In particular, $n$ is divisible by $p^{r-1}$. Using this, it is not hard to show that a module $M$ of stable constant Jordan type [1] or $[p-1]$ is *endotrivial*, in the sense that $M \otimes_k M^*$ is trivial plus projective. Dade's classification of endotrivial modules for an elementary abelian $p$-group then implies that $M$ is isomorphic to $\Omega^n(k)$ plus a projective for some $n \in \mathbb{Z}$.

In the paper of Carlson, Friedlander and Pevtsova [81], it is conjectured that for $r \geq 2$ and $p \geq 5$ there is no module of stable constant Jordan type [2]. In other words, there is no module with the property that every $X_\alpha$ acts with Jordan blocks all of length $p$ except for a single block of length two. In Section 5.12 we prove the more general statement that for $r \geq 2$ and $2 \leq a \leq p-2$ there is no $kE$-module of stable constant Jordan type $[a]$. This completes the analysis of modules of constant Jordan type with one non-projective Jordan block. For larger stable Jordan types, our knowledge is much more limited. The following conjectures appear in Sections 5.13 and 5.15. We continue to assume that $E$ has rank $r \geq 2$.

**Conjecture** (Rickard) If a $kE$-module of constant Jordan type has no Jordan blocks of length $j$ then the total number of Jordan blocks of length larger than $j$ (including the blocks of length $p$) is divisible by $p$.

In Section 5.13 we prove the cases $j = 1$ and $j = p - 1$ of Rickard's conjecture. The proof involves the notion of *generic kernel* for modules over rank two elementary abelian groups, developed by Carlson, Friedlander and Suslin [82].

**Conjecture** (Suslin) If a $kE$-module of constant Jordan type has Jordan blocks of length $j$ then it also has to have Jordan blocks of length either $j + 1$ or $j - 1$. In other words, there are no isolated lengths.

**Conjecture** (Carlson, Friedlander and Pevtsova) If $M$ is a $kE$-module of stable constant Jordan type $[2][1]^j$ then $j \geq r - 1$.

In Section 10.4 we prove a weak version of the conjecture of Carlson, Friedlander and Pevtsova, on modules of stable constant Jordan type $[2][1]^j$. Namely we prove that $j \geq r-2$ if $p$ is large enough. The proof of this uses the theory of Chern classes for vector bundles on projective space, a subject which we discuss in detail in Chapter 7.

When we talk of vector bundles, we are referring to *algebraic vector bundles*, namely locally free sheaves of modules over the structure sheaf. In the case of projective space $\mathbb{P}^{r-1}$, the only rank one vector bundles (line bundles) are twists of the structure sheaf $\mathcal{O}(a)$. There are plenty of indecomposable vector bundles of every rank at least $r - 2$ if $r \geq 3$, but very little is known about vector bundles whose rank $s$ is in the range $2 \leq s \leq r - 3$. The only values of $r$ and $s$ in this range for which we know of indecomposable vector bundles are $r = 5$, $s = 2$ (Horrocks–Mumford), $r = 6$, $s = 3$ (Horrocks), and in characteristic two $r = 6$, $s = 2$ (Tango). A vector bundle in this range with other values of $r$ and $s$ will be referred to as a *new low rank vector bundle* on projective space.

Part of the point of Chapters 6 and 7 is to give an account of the theory of vector bundles on projective space, leading quickly and efficiently to the definition of Chern classes and a proof of the Hirzebruch–Riemann–Roch theorem in this case. This will greatly facilitate the discussion of restrictions on Jordan type coming from Chern classes.

If $M$ is a $kE$-module of constant Jordan type, then we associate to $M$ vector bundles $\mathcal{F}_1(M), \dots, \mathcal{F}_p(M)$ on projective space $\mathbb{P}^{r-1}$ in such a way that the rank of $\mathcal{F}_i(M)$ is equal to the number of Jordan blocks of length $i$ on a cyclic shifted subgroup of $kE$. The way that projective space enters the game is that points in $\mathbb{P}^{r-1}$ correspond to cyclic shifted subgroups of $kE$ (up to a scalar). In some sense, $\mathcal{F}_i(M)$ associates to each point of projective space the socle of the sum of the Jordan blocks of length $i$ of the restriction of $M$ to the corresponding cyclic shifted subgroup. The twists $\mathcal{F}_i(M), \mathcal{F}_i(M)(1), \dots, \mathcal{F}_i(M)(i - 1)$ associate to each point on projective space the successive socle layers of the length $i$ blocks of the restriction.

At this point, an interesting question arises. What vector bundles occur this way? The answer to this question is quite different for $p = 2$ and $p$ odd, as we shall see in Section 8.9. For $p = 2$, given any vector bundle $\mathcal{F}$ of rank $s$ on $\mathbb{P}^{r-1}$, there exists a $kE$-module $M$ of stable constant Jordan type $[1]^s$ such that $\mathcal{F}_1(M) \cong \mathcal{F}$. The same construction with $p$ odd only shows that given $\mathcal{F}$, there is a $kE$-module $M$ of stable constant Jordan type $[1]^s$ such that $\mathcal{F}_1(M) \cong F^*(\mathcal{F})$, the pullback of $\mathcal{F}$ through the Frobenius map $F$ on $\mathbb{P}^{r-1}$. And indeed, it turns out that without pulling back through the Frobenius map, there are restrictions coming from Chern classes. In Section 10.3 we show that if $M$ has stable constant Jordan type $[1]^s$ then the Chern numbers $c_1(\mathcal{F}_1(M)), \dots, c_{p-2}(\mathcal{F}_1(M))$ are divisible by $p$.

Further congruences on Chern numbers come from the Hirzebruch–Riemann–Roch theorem. For example, in Section 10.8 we prove the following. Let $E \cong (\mathbb{Z}/2)^r$. If $M$ is a $kE$-module of constant Jordan type $[2]^n[1]^m$ and $n$ is not congruent to one of the integers $0, -1, \dots, -m$ modulo $2^{r-1}$ then $\mathcal{F}_1(M)$ is a vector bundle on $\mathbb{P}^{r-1}$ which is not a sum of line bundles. In particular, if $r > 6$ and $m \leq r - 3$ then $\mathcal{F}_1(M)$ is a new low rank vector bundle on projective space. The Tango bundle example from the beginning of the introduction shows why the restriction $r > 6$ is necessary.

Finally, the last chapter is a bit more speculative. We investigate the general question of how to construct small modules with interesting properties. The constructions are basically the same as those used to construct modules of constant Jordan type, but we see that the applicability of the methods is much wider. Again, $p = 2$ behaves quite differently from $p$ odd. So, for example, we shall see that a module of Loewy length $p$ can have an arbitrary hypersurface as its variety if $p = 2$ but only finite unions of hyperplanes can be realised when $p$ is odd.

# 1

# Modular Representations and Elementary Abelian Groups

## 1.1 Introduction

This chapter introduces background material on the relevant aspects of modular representation theory of finite elementary abelian $p$-groups. We work over an algebraically closed field $k$ of characteristic $p$, and we only consider finitely generated modules, even when some statements are true more generally.

Since the group algebra of an elementary abelian 2-group is just an exterior algebra, many of the methods we describe apply equally well to modules over exterior algebras in any characteristic. So to some extent we develop the theory of exterior algebras in parallel to that of elementary abelian $p$-groups.

In the representation theory of elementary abelian $p$-groups, we often have to describe separately the cases $p = 2$ and $p$ odd. In the case of exterior algebras, usually no such separation of cases is necessary.

## 1.2 Representation Type

The *trichotomy theorem* (Drozd [107], Crawley–Boevey [96]) partitions finite-dimensional algebras over an algebraically closed field into three mutually disjoint classes:

(i) Finite representation type: in this case there are only a finite number of isomorphism classes of finitely generated indecomposable modules.

(ii) Tame representation type: in this case there are infinitely many isomorphism classes of finitely generated indecomposables, but in any given dimension they come in one-parameter families, with finitely many exceptions. For algebras of tame representation type, one usually hopes to write down a complete classification of the finitely generated indecomposable modules.

(iii) Wild representation type: an algebra $A$ has wild representation type if there is a finitely generated $A$-$k\langle X, Y\rangle$-bimodule $B$, free as a right $k\langle X, Y\rangle$-module, such that the functor $B \otimes_{k\langle X,Y\rangle} -$ from finite-dimensional $k\langle X, Y\rangle$-modules to finite-dimensional $A$-modules preserves indecomposability and reflects isomorphisms.

Broadly speaking, this means that a classification of finite-dimensional indecomposable $A$-modules would entail the classification of pairs of square matrices under simultaneous conjugation. This problem is thought to be unsolvable.

Group algebras of finite groups are almost always of wild representation type.

**Theorem 1.2.1** (Bondarenko and Drozd [57]; see also Ringel [220]) *Let $G$ be a finite group and $k$ have characteristic $p$.*

(i) *If the Sylow $p$-subgroups of $G$ are cyclic then $kG$ has finite representation type.*

(ii) *If $p = 2$ and the Sylow $p$-subgroups of $G$ are dihedral, semidihedral or generalised quaternion, then $kG$ has tame representation type.*

(iii) *In all other cases $kG$ has wild representation type.*

Looking in particular at an elementary abelian $p$-group $E$, this says that finite representation type happens when $E$ has rank one, tame representation type happens only for $(\mathbb{Z}/2)^2$, and otherwise the representation type is wild.

In the rank one case, the classification follows easily from the theory of the Jordan canonical form. There is one indecomposable module for each size of Jordan block between 1 and $p$. In the case of $(\mathbb{Z}/2)^2$, the classification follows from Kronecker's classification of matrix pencils, see for example Section 4.3 of [39].

Next we discuss exterior algebras. Let $\Lambda = \Lambda(X_1, \dots, X_r)$ be an exterior algebra over a field $k$ on generators $X_1, \dots, X_r$. We can either consider this as an ungraded algebra, and look at ungraded modules over it, or we can consider it as a graded algebra, and look at graded modules. In the latter case, we put the generators $X_i$ in degree one. The relations satisfied by the $X_i$ are $X_i^2 = 0$ $(1 \le i \le r)$ and $X_i X_j = -X_j X_i$ $(1 \le i < j \le r)$.

**Theorem 1.2.2** *If $\Lambda = \Lambda(X_1, \dots, X_r)$ is an exterior algebra over a field $k$ on generators $X_1, \dots, X_r$, then the representation type of $\Lambda$ either as an ungraded or as a graded algebra is as follows:*

(i) *finite if $r = 1$,*
(ii) *tame if $r = 2$, and*
(iii) *wild if $r \ge 3$.*

In case (i), if $\Lambda$ is graded, we really mean that there are finitely many isomorphism classes of indecomposable modules up to shift in degree. In fact, there are only two isomorphism classes of indecomposables, of dimensions one and two, and shifts of them in the graded situation. If $r = 2$ then again the classification of the indecomposables follows from Kronecker's classification. For $r \ge 3$, wildness follows for example from Ringel [220], or from the wildness of the quiver $\bullet \substack{\longrightarrow\\ \longrightarrow\\ \longrightarrow} \bullet$.

Given that we usually do not hope to classify the indecomposable $kE$-modules or the indecomposable $\Lambda$-modules, there are several approaches to making progress. One approach is to make classifications that are less refined. For example, the classification of thick subcategories of the stable module category is achieved in [49]. Another is to find general properties of modules, short of their classification. Examples of theorems in this direction are Dade's lemma 1.9.5 and Carlson's theory of rank varieties, discussed in Section 1.9, both of which we shall make use of in these notes. A third way to make progress is to restrict the class of modules under consideration. That is the approach taken in these notes. We shall concentrate largely on modules of constant Jordan type. There are many other interesting subclasses of the class of all finitely generated $kE$-modules, and we are in no way suggesting that this is the most important class.

## 1.3 Shifted Subgroups

We begin with a discussion of shifted subgroups of an elementary abelian $p$-group $E$. These are certain subgroups of the group algebra $kE$.

The notation that we use throughout this book is as follows. A finite group $E$ is said to be an *elementary abelian p-group* if it is abelian and has exponent $p$. Equivalently, $E$ is isomorphic to a direct product of $r$ copies of a cyclic group of order $p$ for some $r \geq 0$:

$$E = \langle g_1, \dots, g_r \rangle \cong (\mathbb{Z}/p)^r$$

where $[g_i, g_j] = 1$ for $1 \leq i, j \leq r$ and $g_i^p = 1$ for $1 \leq i \leq r$.

We begin with the definition of shifted subgroup. Let

$$X_i = g_i - 1 \in kE$$

for $1 \leq i \leq r$. Since we are in characteristic $p$ we have

$$X_i^p = (g_i - 1)^p = g_i^p - 1^p = 0$$

in $kE$. The images of $X_1, \dots, X_r$ form a basis for $J(kE)/J^2(kE)$, where $J(kE)$ denotes the Jacobson radical of $kE$. If $\alpha = (\lambda_1, \dots, \lambda_r)$ is an element of affine space $\mathbb{A}^r(k)$, we define

$$X_\alpha = \lambda_1 X_1 + \dots + \lambda_r X_r \in kE.$$

This is an element of $J(kE)$ which again satisfies $X_\alpha^p = 0$. It follows that $(1 + X_\alpha)^p = 1$. So if $\alpha \neq 0$ then $g_\alpha = 1 + X_\alpha$ is an element of order $p$ in the group of units of the group algebra $kE$.

**Definition 1.3.1** A *cyclic shifted subgroup* of $E$ is a subgroup $E_\alpha$ of the group of units of $kE$ generated by such an element $g_\alpha = 1 + X_\alpha$ with $\alpha \neq 0$. More generally, a *shifted subgroup* is a subgroup $E'$ of the group of units in $kE$ generated by elements $g_{\alpha_1}, \dots, g_{\alpha_s}$ where $\alpha_1, \dots, \alpha_s$ are linearly independent elements of

$\mathbb{A}^r(k)$. This linear independence condition is equivalent to the statement that the induced map $\rho\colon kE' \to kE$ is injective. If this is the case then $kE$ is free as a module over the image of $\rho$.

A *maximal shifted subgroup* is a shifted subgroup $E'$ of the same rank as $E$. The induced map $\rho\colon kE' \to kE$ is then an isomorphism. Choosing a basis of $E'$ allows us to identify it with $E$ and obtain an automorphism $\rho$ of $kE$.

This notion of a cyclic shifted subgroup is dependent on the choice of generators for $E$. For a different choice of generators, if we make the corresponding linear transformation on $\mathbb{A}^r(k)$, the resulting element $X_\alpha$ with respect to the new basis differs from the old one by an element of $J^2(kE)$.

Let us illustrate this by example. If $E = \langle g_1, g_2\rangle \cong (\mathbb{Z}/p)^2$ and $X_1 = g_1 - 1$, $X_2 = g_2 - 1$, then a multiplicative basis change on $E$ using the matrix

$$\begin{pmatrix} 1 & 0 \\ 1 & 1 \end{pmatrix}$$

gives new generators $E = \langle g_1 g_2, g_2\rangle$. Now we have

$$g_1 g_2 - 1 = X_1 + X_2 + X_1 X_2.$$

So modulo the square of the radical, we have the additive base change on $X_1$ and $X_2$ given by the same matrix. However, the linear span of $g_1g_2 - 1$ and $g_2 - 1$ is not the same as the linear span of $X_1$ and $X_2$, so the definition of a shifted subgroup is basis dependent. In Chapter 4, we shall make a systematic study of the effect of adding an element of $J^2(kE)$.

The following is immediate from the definitions.

**Theorem 1.3.2** *The group algebra $kE$ is a truncated polynomial ring:*

$$kE = k[X_1, \ldots, X_r]/(X_1^p, \ldots, X_r^p).$$

We define the *rank variety* of $E$ to be the affine space $V_E^{\mathrm{rk}} \cong \mathbb{A}^r(k)$ obtained by taking the linear span in $J(kE)$ of the elements

$$X_1 = g_1 - 1, \ldots, X_r = g_r - 1.$$

Thus $V_E^{\mathrm{rk}}$ is a complement in $J(kE)$ of $J^2(kE)$. This is the home for the theory of rank varieties, as described in Section 1.9.

**Warning 1.3.3** It is easy fall into the trap of thinking that the definitions

$$\begin{aligned} X_i &= \log(g_i) = (g_i - 1) - \tfrac{1}{2}(g_i - 1)^2 + \cdots \\ X_\alpha &= \lambda_1 X_1 + \cdots + \lambda_r X_r \\ g_\alpha &= \exp(X_\alpha) = 1 + X_\alpha + \tfrac{1}{2!}X_\alpha^2 + \cdots \end{aligned}$$

give a more invariant definition of shifted subgroups. The problem is that it is *not true* that

$$\log(gh) = \log(g) + \log(h).$$

For example, in characteristic two we have $\log(g) = 1+g$, $\log(g)+\log(h) = g+h$ and $\log(gh) = 1 + gh$.

The existence of an invariant definition of shifted subgroups would imply the existence of a non-zero $GL(r, \mathbb{F}_p)$-invariant homomorphism $\phi\colon E \to J(kE)$. Such a homomorphism exists if and only if $E$ is either cyclic or isomorphic to $(\mathbb{Z}/2)^2$.

To see this, we use the following argument, due to Serge Bouc. Let $\phi$ be such a homomorphism and write

$$\phi(g) = \sum_{\substack{h\in E\\ h\neq 1}} \phi_h(g)(h-1).$$

Invariance amounts to the statement that for $\alpha \in GL(r, \mathbb{F}_p)$ we have $\phi_h(\alpha(g)) = \phi_{\alpha^{-1}(h)}(g)$. The group $GL(r, \mathbb{F}_p)$ is transitive on pairs of elements $g, h$ such that $g$ is not in $\langle h\rangle$. So $\phi_h(g)$ is a constant $a$, independent of $h$ and $g$, provided $g \notin \langle h\rangle$. If $E$ is neither cyclic nor isomorphic to $(\mathbb{Z}/2)^2$ then we can find linearly independent elements $g$ and $h$ such that $\langle g\rangle$, $\langle h\rangle$ and $\langle gh\rangle$ do not exhaust $E$. Then the statement that $\phi(gh) = \phi(g) + \phi(h)$ implies first that $2a = a$, so that $a = 0$, and then that $\phi = 0$ since $\phi(gh)$, $\phi(g)$ and $\phi(h)$ have supports intersecting in the identity.

The discussion of rank varieties for exterior algebras is essentially the same. Let $\Lambda = \Lambda(X_1, \dots, X_r)$ be an exterior algebra on generators $X_1, \dots, X_r$. If $\alpha = (\lambda_1, \dots, \lambda_r) \in \mathbb{A}^r(k)$, we define

$$X_\alpha = \lambda_1 X_1 + \cdots + \lambda_r X_r \in \Lambda.$$

This is an element of $J(\Lambda)$ which satisfies $X_\alpha^2 = 0$. The *rank variety* of $\Lambda$ is the affine space $V_\Lambda^{\mathrm{rk}} \cong \mathbb{A}^r(k)$ obtained by taking the linear span of $X_1, \dots, X_r$ in $J(\Lambda)$. It is a complement to $J^2(\Lambda)$ in $J(\Lambda)$. Details can be found in the paper of Aramova, Avramov and Herzog [12].

## 1.4 The Language of $\pi$-Points

Much of the modern literature on Jordan type for finite group schemes [80, 81, 82, 118, 124, 125, 126, 127, 128] is written in the language of $\pi$-points. It is worth making a few remarks on the translation between this and the language of cyclic shifted subgroups, in the case of an elementary abelian $p$-group. Understanding this section is not logically necessary for the rest of the book, but will help reconcile it with the rest of the literature.

Giving a finite group scheme $G$ over $k$ is equivalent to giving a finite-dimensional cocommutative Hopf algebra $kG$ over $k$. For example, if $G$ is a finite group then $kG$ is just its group algebra in the normal sense. Other examples include the restricted universal enveloping algebra of a finite-dimensional $p$-restricted Lie algebra. In this example the underlying variety of the group scheme has only one point, but the local ring at that point contains all the information – it is just the dual Hopf algebra.

If $G$ is a finite group scheme then a $\pi$-point of $G$ is defined to be a *flat* homomorphism of algebras $K[t]/(t^p) \to KG$ over some extension field $K$ of $k$, which factors through some unipotent abelian subgroup scheme of $KG$.

To say that a homomorphism $K[t]/(t^p) \to KG$ is flat means that $KG$ is flat as a module over $K[t]/(t^p)$, which is equivalent to each one of the following adjectives: *free, projective, injective.*

An equivalence relation is put on $\pi$-points as follows. Two $\pi$-points are said to be equivalent if for every finitely generated $kG$-module $M$, the restriction of $K \otimes_k M$ along one is free if and only if the restriction along the other is free. This condition is difficult to work with in practice, since it is expressed in terms of all finitely generated modules, but can be reduced to an easier cohomological condition. See for example Theorem 3.6 of [125].

In the case where $G = E$ is a finite elementary abelian $p$-group, $KE$ is unipotent, so the second condition is automatically satisfied. The flatness condition is equivalent to the statement that the image of $t$ lies in $J(KE)$ but not in $J^2(KE)$. The equivalence relation in this case is as follows. By Lemma 6.4 of Carlson [76], two such homomorphisms give equivalent $\pi$-points if and only if the differences between the images of $t$ lies in $J^2(KE)$. Thus the set of equivalence classes of $\pi$-points over $K$ is in bijection with $\mathbb{A}^r(K) \setminus \{0\}$. With respect to a given choice of generators of $E$, in each equivalence class there is a unique representative which is given by a cyclic shifted subgroup.

Note that an exterior algebra in characteristic not equal to two is not an example of a finite group scheme. Nonetheless, the notion of $\pi$-point makes perfect sense in this context, if defined as a flat embedding $K[t]/(t^2) \to K \otimes_k \Lambda$.

## 1.5 The Stable Module Category

In general, if $G$ is a finite group the cohomology ring $H^*(G, k)$ is defined to be $\mathrm{Ext}^*_{kG}(k, k)$. By a theorem of Evens [115] and Venkov [242] (or an earlier theorem of Golod [136] in the case of finite $p$-groups), this is a finitely generated graded commutative algebra over $k$.

The *module category* $\mathsf{mod}(kG)$ is the category of finitely generated $kG$-modules and module homomorphisms. The *stable module category* $\mathsf{stmod}(kG)$ has the same objects as $\mathsf{mod}(kG)$ but its arrows are given by

$$\underline{\mathrm{Hom}}_{kG}(M, N) = \mathrm{Hom}_{kG}(M, N)/P\mathrm{Hom}_{kG}(M, N),$$

where $P\mathrm{Hom}_{kG}(M,N)$ is the linear subspace of $\mathrm{Hom}_{kG}(M,N)$ consisting of homomorphisms that factor through some projective $kG$-module. Whereas $\mathsf{mod}(kG)$ is an abelian category, $\mathsf{stmod}(kG)$ is a triangulated category.

If $M$ is a finitely generated $kG$-module, we define $\Omega(M)$ to be the kernel of the projective cover of $M$ and $\Omega^{-1}(M)$ to be the cokernel of the injective hull of $M$. It follows from the fact that $kG$ is self-injective that $\Omega(\Omega^{-1}(M))$ and $\Omega^{-1}(\Omega(M))$ are naturally isomorphic to $M$ in the stable module category $\mathsf{stmod}(kG)$. If $M$ has no projective summands, this implies that they are isomorphic to $M$ in the module category $\mathsf{mod}(kG)$, but the isomorphism is not natural. If $n > 0$ we have

$$\mathrm{Ext}^n_{kG}(M,N) \cong \underline{\mathrm{Hom}}_{kG}(\Omega^n(M),N), \tag{1.1}$$

and in particular since $P\mathrm{Hom}_{kG}(\Omega^n(k),k) = 0$ we have

$$H^n(G,k) \cong \mathrm{Ext}^n_{kG}(k,k) \cong \mathrm{Hom}_{kG}(\Omega^n(k),k).$$

If $\zeta \in H^n(G,k)$, we write $\hat{\zeta}$ for the corresponding homomorphism $\Omega^n(k) \to k$. If $\zeta \neq 0$ then $\hat{\zeta}$ is surjective, and the *Carlson module* $L_\zeta$ is defined to be the kernel of $\hat{\zeta}$. We shall study these modules further in Sections 1.10 and 12.4.

Tate cohomology is defined by extending (1.1) from positive degrees to all degrees:

$$\widehat{\mathrm{Ext}}^n_{kG}(M,N) \cong \underline{\mathrm{Hom}}_{kG}(\Omega^n(M),N) \qquad (n \in \mathbb{Z}).$$

The Tate cohomology ring of $G$ is defined to be

$$\hat{H}^*(G,k) = \widehat{\mathrm{Ext}}^*_{kG}(k,k)$$

with multiplication given by shifting and composing. This makes it into a graded commutative $k$-algebra with the ordinary cohomology $H^*(G,k)$ as the subalgebra of non-negative degree elements. The $k$-algebra $\hat{H}^*(G,k)$ is finitely generated if and only if it is periodic, which in turn happens if and only if all abelian $p$-subgroups of $G$ are cyclic.

Tate duality (see for example Cartan and Eilenberg [83], section XII.6) states that for $n \in \mathbb{Z}$ there are natural vector space isomorphisms

$$(\hat{H}^{-n-1}(G,M))^* \cong \widehat{\mathrm{Ext}}^n_{kG}(M,k).$$

In particular, $\hat{H}^{-n-1}(G,k)$ is the vector space dual of $\hat{H}^n(G,k)$.

If $H$ is a subgroup of $G$, then transfer in Tate cohomology from $H$ to $G$ is Tate dual to restriction from $G$ to $H$.

Exactly the same definitions make sense in the case of an exterior algebra. If $\Lambda = \Lambda(X_1,\dots,X_r)$ is an exterior algebra over $k$ then the cohomology ring $H^*(\Lambda,k) = \mathrm{Ext}^*_\Lambda(k,k)$ is a polynomial ring $k[Y_1,\dots,Y_r]$. If $\Lambda$ is regarded here as an *ungraded* algebra, then the $Y_i$ are in cohomological degree one. So if $k$ does not have characteristic two and $r > 1$ then $H^*(\Lambda,k)$ is not a graded commutative ring, because the $Y_i$ do not commute with each other in the graded sense. On the other hand, if $\Lambda$ is regarded as a *graded* algebra with the $X_i$ in degree one, then

$H^*(\Lambda, k)$ is bigraded, with the $Y_i$ in degree $(1, 1)$. The total degree is two, so the cohomology ring is graded commutative in this case.

Tate cohomology and Tate duality for exterior algebras works in the same way as for group algebras.

## 1.6 The Derived Category

If $\mathcal{A}$ is an abelian category, we write $\mathsf{D}^b(\mathcal{A})$ for the *bounded derived category* of $\mathcal{A}$. Its objects are the bounded chain complexes of objects and arrows in $\mathcal{A}$. The morphisms are the homotopy classes of degree preserving chain maps, with the quasi-isomorphisms inverted; a *quasi-isomorphism* is a morphism that induces an isomorphism in homology. Thus a morphism $\mathbf{C} \to \mathbf{D}$ in $\mathsf{D}^b(\mathcal{A})$ is given by an equivalence class of diagrams of chain complexes

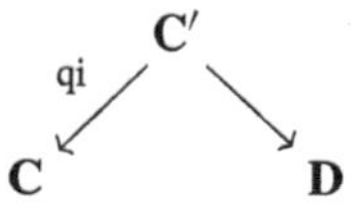

where "qi" denotes a quasi-isomorphism. Two such diagrams $\mathbf{C} \overset{\text{qi}}{\leftarrow} \mathbf{C}' \to \mathbf{D}$ and $\mathbf{C} \overset{\text{qi}}{\leftarrow} \mathbf{C}'' \to \mathbf{D}$ are equivalent if there is a third $\mathbf{C} \overset{\text{qi}}{\leftarrow} \mathbf{C}''' \to \mathbf{D}$ that maps to both of them to make a commutative diagram

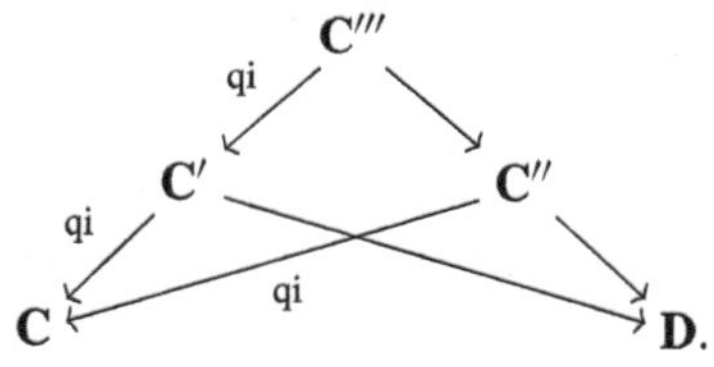

To compose arrows we form a pullback:

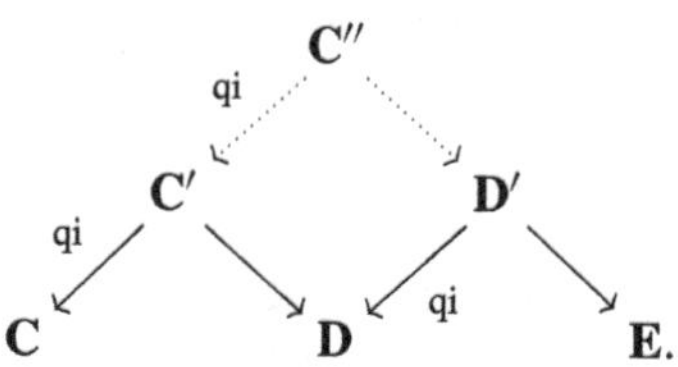

The distinguished triangles in $\mathsf{D}^b(\mathcal{A})$ are the triangles isomorphic to

$$\mathbf{C} \xrightarrow{f} \mathbf{D} \to M_f \to \mathbf{C}[1]$$

where $M_f$ is the mapping cone of $f$.

The derived categories we shall be interested in are $\mathsf{D}^b(kG) = \mathsf{D}^b(\mathsf{mod}(kG))$, the derived category of $kG$-modules, and $\mathsf{D}^b(\mathsf{Coh}(X))$, the derived category of coherent $\mathcal{O}_V$-modules for a variety $V$.

Next we describe the construction of a functor

$$\mathcal{R}\colon \mathsf{D}^b(kG) \to \mathsf{stmod}(kG).$$

This functor appeared in the late 1980s in the work of several people, see Buchweitz [66], Keller and Vossieck [176], Rickard [219, Theorem 2.1]. We shall use this construction in Section 8.9 as part of the proof of the realisation theorem.

**Definition 1.6.1** A *perfect complex* is an object in $\mathsf{D}^b(kG)$ which is isomorphic to a bounded complex of projective modules. We write $\mathsf{perf}(kG)$ for the full subcategory of $\mathsf{D}^b(kG)$ whose objects are the perfect complexes.

The quotient category $\mathsf{D}^b(kG)/\mathsf{perf}(kG)$ has the same objects as $\mathsf{D}^b(kG)$. The arrows are formed by inverting arrows whose mapping cone is in $\mathsf{perf}(kG)$.

**Theorem 1.6.2** *The quotient category* $\mathsf{D}^b(kG)/\mathsf{perf}(kG)$ *is equivalent to the stable module category* $\mathsf{stmod}(kG)$.

*Proof (sketch).* We begin with the functor $\mathsf{mod}(kG) \to \mathsf{D}^b(kG)$ which takes a module $M$ to the complex which consists of $M$ in degree zero and the zero module in all other degrees. The composite of this with the quotient functor

$$\mathsf{D}^b(\mathsf{mod}(kG)) \to \mathsf{D}^b(kG)/\mathsf{perf}(kG)$$

takes projective modules to zero, and so it factors through $\mathsf{mod}(kG) \to \mathsf{stmod}(kG)$ to give a well-defined functor

$$\mathsf{stmod}(kG) \to \mathsf{D}^b(kG)/\mathsf{perf}(kG).$$

One can check that this functor takes distinguished triangles to distinguished triangles. It is also full, and takes non-zero objects to non-zero objects. To prove that it is an equivalence, it therefore remains to prove that every object in $\mathsf{D}^b(kG)/\mathsf{perf}(kG)$ is isomorphic to an object in the image of this functor.

Given a bounded chain complex of finitely generated $kG$-modules $\mathbf{C}$, we can resolve it to give a quasi-isomorphism

$$\begin{array}{ccccccccccc} 0 & \to & K & \to & P_s & \to & \cdots & \to & P_t & \to & 0 \\ & & & & \downarrow & & & & \downarrow & & \\ & & 0 & \to & M_s & \to & \cdots & \to & M_t & \to & 0 \end{array}$$

where the $P_i$ are projective modules, and without loss of generality we can take $s > 0$. Now take an injective resolution of $K$. Then there is a quasi-isomorphism

$$\begin{array}{ccccccccccccc} 0 & \to & K & \to & Q_s & \to & \cdots & \to & Q_1 & \to & Q_0 & \to & 0 \\ & & \downarrow & & \downarrow & & & & \downarrow & & \downarrow & & \\ 0 & \to & 0 & \to & 0 & \to & \cdots & \to & 0 & \to & \Omega^{-s-1}(K) & \to & 0. \end{array} \tag{1.2}$$

Then modulo $\mathsf{perf}(kG)$ the original complex is isomorphic to $K$ in degree $s+1$, which is in turn isomorphic to the top row and then the bottom row of (1.2). The latter is in the image of $\mathsf{stmod}(kG)$. □

**Definition 1.6.3** We write $\mathcal{R}$ for the functor

$$\mathcal{R}\colon \mathsf{D}^b(kG) \to \mathsf{stmod}(kG)$$

whose existence is guaranteed by the theorem. This functor takes distinguished triangles in $\mathsf{D}^b(kG)$ to distinguished triangles in $\mathsf{stmod}(kG)$.

The proof of the theorem shows that the image of a complex under this functor is obtained by resolving, taking the kernel, and then shifting it back to degree zero. It is possible, but tedious to check directly that this process is functorial.

It follows from the theorem that an object in $\mathsf{D}^b(kG)$ goes to zero in $\mathsf{stmod}(kG)$ if and only if it is a perfect complex.

In the case of the exterior algebra $\Lambda$, there are two versions of the derived category and of the stable module category, according to whether we are talking about ungraded modules or graded modules. In both cases, the discussion of the derived category and its relationship with the stable module category works in the same way as above. More generally, it works for any finite-dimensional ungraded or graded self-injective algebra.

## 1.7 Singularity Categories

As a generalisation of the equivalence $\mathsf{stmod}(kG) \simeq \mathsf{D}^b(kG)/\mathsf{perf}(kG)$ of Theorem 1.6.2, we have the following definition, which will become important in Chapter 11.

**Definition 1.7.1** Let $R$ be a ring. Then the *singularity category* of $R$ is the Verdier quotient

$$\mathsf{D}_{\mathrm{sg}}(R) = \mathsf{D}^b(R)/\mathsf{perf}(R).$$

Likewise, if $R$ is a graded ring, we denote by $\mathsf{D}^b(R)$ the bounded derived category of finitely generated graded $R$-modules and $\mathsf{D}_{\mathrm{sg}}(R)$ the quotient by the perfect complexes of graded modules.

**Warning 1.7.2** In commutative algebra, this definition is much better behaved for Gorenstein rings than for more general commutative Noetherian rings. For a Gorenstein ring, the singularity category is equivalent to the stable category of maximal Cohen–Macaulay modules (Buchweitz [66]), but the following example is typical of the behaviour for non-Gorenstein rings.

**Example 1.7.3** Let $R$ be the ring

$$k[X, Y]/(X^2, XY, Y^2).$$

Then the radical of $R$ is isomorphic to $k \oplus k$, so there is a short exact sequence of $R$-modules

$$0 \to k \oplus k \to R \to k \to 0.$$

This means that in $\mathsf{D}_{\mathrm{sg}}(R)$ the connecting homomorphism of this short exact sequence gives an isomorphism $k \cong k[1] \oplus k[1]$. We have

$$k \cong k[1]^{\oplus 2} \cong k[2]^{\oplus 4} \cong k[3]^{\oplus 8} \cong \cdots$$

and so $k$ is an infinitely divisible module. Its endomorphism ring $\mathrm{End}_{\mathsf{D}_{\mathrm{sg}}(R)}(k)$ is the colimit of

$$k \to \mathrm{Mat}_2(k) \to \mathrm{Mat}_4(k) \to \mathrm{Mat}_8(k) \to \cdots$$

where each matrix ring is embedded diagonally into a product of two copies, sitting in the next matrix ring. In fact, this endomorphism ring is an example of a von Neumann regular ring. For a generalisation of this example to finite-dimensional algebras with radical square zero, see Chen [88].

The reason why $\mathsf{D}_{\mathrm{sg}}(R)$ is called the "singularity category" is that it only "sees" the singular locus of $R$.

**Example 1.7.4** If $R$ is a regular ring then $R$ has finite global dimension. So we have $\mathsf{D}^b(R) = \mathsf{perf}(R)$ and hence $\mathsf{D}_{\mathrm{sg}}(R) = 0$.

More generally, we have the following definition and theorem.

**Definition 1.7.5** Let $R$ be a [graded] Noetherian commutative ring. Then the *singular locus* of $R$ is the set of [homogeneous] prime ideals $\mathfrak{p}$ of $R$ such that the [homogeneous] localisation $R_{\mathfrak{p}}$ is not regular.

**Remark 1.7.6** Provided that $R$ satisfies a mild technical condition known as "excellence," the singular locus is a Zariski closed set, so that it is of the form $V(I)$ for some [homogeneous] radical ideal $I$ of $R$. Thus $a \in I$ if and only if $R[a^{-1}]$ is regular. Quotients of polynomial rings, for example, are excellent.

**Theorem 1.7.7** *Let $R$ be a [graded] Noetherian commutative ring of finite Krull dimension whose singular locus is a Zariski closed set. Then $\mathsf{D}_{\mathrm{sg}}(R)$ is generated by [the graded shifts of] the modules $R/\mathfrak{p}$ where $\mathfrak{p}$ is a [homogeneous] prime ideal in the singular locus of $R$.*

**Remark 1.7.8** In the ungraded case, the theorem of Schoutens [223] implies the above theorem, but it is stronger, and the proof is more complicated. Schoutens' theorem also holds in the graded case, with minor adjustments to the proof.

I'd like to thank Srikanth Iyengar for suggesting the simple proof presented here. The idea behind this argument also appears in Lemma 2.2 of Herzog and Popescu [151], and in Iyengar and Takahashi [168]. This theorem will be used in the proof of Proposition 11.9.1.

*Proof of Theorem 1.7.7.* Let $d = \dim R$ and let $V(I)$ be the singular locus of $R$. It suffices to show that if $M$ is a finitely generated [graded] $R$-module then $M$ is in the thick subcategory of $\mathsf{D}^b(R)$ generated by [graded shifts of] $R$ and of $R/\mathfrak{p}$ with $\mathfrak{p} \in V(I)$, i.e., with $\mathfrak{p} \supseteq I$.

The first step is to replace $M$ by its $d$th syzygy $\Omega^d(M)$, i.e., the $d$th kernel in any [graded] resolution of $M$ by finitely generated free [graded] $R$-modules (we allow free graded modules to be sums of *shifts* of $R$). Thus we may assume that $M$ is a $d$th syzygy.

We claim that if $a$ is a [homogeneous] element of $I$ then for some $n > 0$, $a^n$ annihilates $\mathrm{Ext}^1_R(M, \Omega(M))$. This is because $R[a^{-1}]$ has global dimension at most $d$, so the fact that $M$ is a $d$th syzygy implies that $M[a^{-1}]$ is also a $d$th syzygy and is hence projective as an $R[a^{-1}]$-module. So $\mathrm{Ext}^1_R(M, \Omega(M))[a^{-1}] = \mathrm{Ext}^1_{R[a^{-1}]}(M[a^{-1}], \Omega(M[a^{-1}])) = 0$.

Apply this to the extension $0 \to \Omega(M) \to F \to M \to 0$, with $F$ a finitely generated free [graded] $R$-module. Multiplying this extension by $a^n$ amounts to forming the pullback $X$ in the following diagram:

$$\begin{array}{ccccccccc}
0 & \longrightarrow & \Omega(M) & \longrightarrow & X & \longrightarrow & M & \longrightarrow & 0 \\
 & & \Vert & & \downarrow & & \downarrow{\scriptstyle a^n} & & \\
0 & \longrightarrow & \Omega(M) & \longrightarrow & F & \longrightarrow & M & \longrightarrow & 0
\end{array}$$

The resulting extension splits, so $X \cong M \oplus \Omega(M)$. The snake lemma implies that the middle vertical arrow gives rise to an exact sequence

$$0 \to \mathsf{Ker}\,(a^n, M) \to X \to F \to M/a^n M \to 0.$$

Now $\mathsf{Ker}\,(a^n, M)$ and $M/a^n M$ are annihilated by $a^n$. So $X$, and hence $M$, is in the thick subcategory of $\mathsf{D}^b(R)$ generated by $R$ and modules supported on $R/aR$. Now inducting on a finite set of generators for the ideal $I$, we see that $M$ is in the thick subcategory of $\mathsf{D}^b(R)$ generated by $R$ and modules supported on $V(I)$. The latter are in turn generated by the $R/\mathfrak{p}$ with $\mathfrak{p} \in V(I)$. □

## 1.8 Cohomology of Elementary Abelian $p$-Groups

The ring structure of the cohomology of $E \cong (\mathbb{Z}/p)^r$ depends on whether $p = 2$ or $p$ is odd.

**Case 1.** If $p = 2$ then the cohomology is a polynomial ring

$$H^*(E, k) = k[y_1, \ldots, y_r]$$

where the generators $y_i$ are in degree one.

**Case 2.** If $p$ is odd then the cohomology is the tensor product of an exterior algebra and a polynomial ring

$$H^*(E, k) = \Lambda(y_1, \ldots, y_r) \otimes_k k[x_1, \ldots, x_r]$$

where the exterior generators $y_i$ are in degree one and the polynomial generators $x_i$ are in degree two. The Bockstein map $\beta\colon H^1(E, k) \to H^2(E, k)$ is the semilinear map given by

$$\beta(\sum_{i=1}^{r} \lambda_i y_i) = \sum_i \lambda_i^p x_i. \tag{1.3}$$

It is semilinear with respect to the Frobenius map, in the sense that for $y \in H^1(E, k)$ and $\lambda \in k$ we have $\beta(\lambda y) = \lambda^p \beta(y)$. The Bockstein map is related to the $p$-fold Massey product by the formula

$$\beta(y) = -\langle y, y, \ldots, y\rangle \qquad (p \text{ terms})$$

for $y \in H^1(E, k)$ (Kraines [180], Theorem 14). The Frobenius twist involved in the Bockstein map will manifest itself in Theorem 8.1.1, where it explains the difference in behaviour between $p = 2$ and $p$ odd for the realisation of vector bundles from modules of constant Jordan type.

In the proof of Theorem 8.1.1, we shall need to use the nature of explicit maps representing the elements $y_i$ ($p = 2$), $x_i$ ($p$ odd). We begin by discussing the case of a cyclic group of order $p$. Let $C = \langle g\rangle \cong \mathbb{Z}/p$, and let $\mathbf{C}$ be the periodic projective resolution of $k$ for $kC$ given by

$$\cdots kC \xrightarrow{1-g} kC \xrightarrow{1+g+\cdots+g^{p-1}} kC \xrightarrow{1-g} kC \to 0.$$

If $p = 2$ this is periodic of period one, and there is a periodicity map $\hat{y}\colon \mathbf{C} \to \mathbf{C}$ of degree one representing $y \in H^1(C, k)$. If $p$ is odd it has period two, and there is a periodicity map $\hat{x}\colon \mathbf{C} \to \mathbf{C}$ of degree two representing $x \in H^2(C, k)$.

Now if $E = \langle g_1, \ldots, g_r\rangle \cong (\mathbb{Z}/p)^r$, we let $\mathbf{P} = \mathbf{C} \otimes \cdots \otimes \mathbf{C}$ ($r$ factors). Then $\mathbf{P}$ is a minimal resolution of $k$ as a $kE$-module. Let

$$\hat{y}_i = 1 \otimes \cdots \otimes \hat{y} \otimes \cdots \otimes 1\colon \mathbf{P} \to \mathbf{P} \qquad (p = 2)$$
$$\hat{x}_i = 1 \otimes \cdots \otimes \hat{x} \otimes \cdots \otimes 1\colon \mathbf{P} \to \mathbf{P} \qquad (p \text{ odd}).$$

Here, all terms in the tensor product of maps are equal to 1 except the $i$th term, which is $\hat{y}$, respectively $\hat{x}$. The elements $\hat{y}_i$ and $\hat{x}_i$ represent the elements $y_i \in H^1(E, k)$ ($p = 2$), respectively $x_i \in H^2(E, k)$ ($p$ odd). These maps commute, and so we get a ring homomorphism

$$k[y_1, \ldots, y_r] \to \operatorname{Hom}^*_{kE}(\mathbf{P}, \mathbf{P}) \qquad (p = 2)$$

$$k[x_1, \dots, x_r] \to \operatorname{Hom}^*_{kE}(\mathbf{P}, \mathbf{P}) \qquad (p \text{ odd}).$$

Letting $\Omega^n(k)$ be the $n$th kernel in the complex $\mathbf{P}$, we get induced maps which by abuse of notation we write as

$$y_i : \Omega^{n+1}(k) \to \Omega^n(k) \qquad (p = 2)$$
$$x_i : \Omega^{n+2}(k) \to \Omega^n(k) \qquad (p \text{ odd})$$

for all $n \geq 0$. They strictly commute, in the sense that $y_i y_j$ and $y_j y_i$ give the same map $\Omega^{n+2}(k) \to \Omega^n(k)$ $(p = 2)$; and $x_i x_j$ and $x_j x_i$ give the same map $\Omega^{n+4}(k) \to \Omega^n(k)$ $(p$ odd$)$.

**Example 1.8.1** It is easy to compute explicit minimal resolutions over elementary abelian $p$-groups, either by hand or in larger cases using computer algebra packages such as Magma or Macaulay2. For example, if $E = (\mathbb{Z}/p)^2$, with $X_1 = g_1 - 1$ and $X_2 = g_2 - 1$, here is the beginning of the minimal resolution of $k$ as a module over $kE$

$$6kE \xrightarrow{\begin{pmatrix} X_1 & X_2 & 0 & 0 & 0 & 0 \\ 0 & X_1^{p-1} & X_2^{p-1} & 0 & 0 & 0 \\ 0 & 0 & X_1 & X_2 & 0 & 0 \\ 0 & 0 & 0 & X_1^{p-1} & X_2^{p-1} & 0 \\ 0 & 0 & 0 & 0 & X_1 & X_2 \end{pmatrix}} 5kE \xrightarrow{\begin{pmatrix} X_1^{p-1} & -X_2 & 0 & 0 & 0 \\ 0 & X_1 & -X_2^{p-1} & 0 & 0 \\ 0 & 0 & X_1^{p-1} & -X_2 & 0 \\ 0 & 0 & 0 & X_1 & -X_2^{p-1} \end{pmatrix}} 4kE$$

$$\xrightarrow{\begin{pmatrix} X_1 & X_2 & 0 & 0 \\ 0 & X_1^{p-1} & X_2^{p-1} & 0 \\ 0 & 0 & X_1 & X_2 \end{pmatrix}} 3kE \xrightarrow{\begin{pmatrix} X_1^{p-1} & -X_2 & 0 \\ 0 & X_1 & -X_2^{p-1} \end{pmatrix}} 2kE \xrightarrow{(X_1\ X_2)} kE \longrightarrow k$$

and here is the beginning of the minimal resolution of the three-dimensional $kE$-module $kE/J^2(kE)$

$$9kE \xrightarrow{\begin{pmatrix} X_1^2 & X_1X_2 & X_2^2 & 0 & 0 & 0 & 0 & 0 & 0 \\ 0 & X_1^{p-1} & X_1^{p-2}X_2 & X_2^{p-1} & 0 & 0 & 0 & 0 & 0 \\ 0 & 0 & X_1^{p-1} & X_1X_2^{p-2} & X_2^{p-1} & 0 & 0 & 0 & 0 \\ 0 & 0 & 0 & X_1^2 & X_1X_2 & X_2^2 & 0 & 0 & 0 \\ 0 & 0 & 0 & 0 & X_1^{p-1} & X_1^{p-2}X_2 & X_2^{p-1} & 0 & 0 \\ 0 & 0 & 0 & 0 & 0 & X_1^{p-1} & X_1X_2^{p-2} & X_2^{p-1} & 0 \\ 0 & 0 & 0 & 0 & 0 & 0 & X_1^2 & X_1X_2 & X_2^2 \end{pmatrix}} 7kE$$

$$\xrightarrow{\begin{pmatrix} X_1^{p-2} & -X_2 & 0 & 0 & 0 & 0 & 0 \\ 0 & X_1 & -X_2 & 0 & 0 & 0 & 0 \\ 0 & 0 & X_1 & -X_2^{p-2} & 0 & 0 & 0 \\ 0 & 0 & 0 & X_1^{p-2} & -X_2 & 0 & 0 \\ 0 & 0 & 0 & 0 & X_1 & -X_2 & 0 \\ 0 & 0 & 0 & 0 & 0 & X_1 & -X_2^{p-1} \end{pmatrix}} 6kE \xrightarrow{\begin{pmatrix} X_1^2 & X_1X_2 & X_2^2 & 0 & 0 & 0 \\ 0 & X_1^{p-1} & X_1^{p-2}X_2 & X_2^{p-1} & 0 & 0 \\ 0 & 0 & X_1^{p-1} & X_1X_2^{p-2} & X_2^{p-1} & 0 \\ 0 & 0 & 0 & X_1^2 & X_1X_2 & X_2^2 \end{pmatrix}} 4kE$$

$$\xrightarrow{\begin{pmatrix} X_1^{p-2} & -X_2 & 0 & 0 \\ 0 & X_1 & -X_2 & 0 \\ 0 & 0 & X_1 & -X_2^{p-2} \end{pmatrix}} 3kE \xrightarrow{\left(X_1^2\ X_1X_2\ X_2^2\right)} kE \longrightarrow kE/J^2(kE).$$

The number of copies of $kE$ at the $n$th stage of the minimal resolution of a module $M$ is $\dim_k \operatorname{Ext}^n_{kE}(M, k)$. The Poincaré series

$$\sum_{n=0}^{\infty} \dim_k \operatorname{Ext}^n_{kE}(M, k) t^n$$

is a rational function of $t$ of the form (polynomial)/$(1 - t^2)^c$ where $c$ is the complexity of $M$. The value of the polynomial in the numerator at $t = 1$ is a positive integer which we shall denote $d(M)$. David Eisenbud asks whether it is always true that $d(M) \geq 2^r$ if $c = r$. The example of $kE/J^2(kE)$ above shows that it does not have to be an integer multiple of $2^r$ (in this case $c = r = 2$ and the degree is 6).

## 1.9 Chouinard's Theorem, Dade's Lemma and Rank Varieties

Part of the importance of elementary abelian groups in the modular representation theory of finite groups is explained by the following theorem of Chouinard. First recall that if $G$ is a finite group and $k$ is a field then the conditions that a $kG$-module is projective, injective, and flat are all equivalent, and that if $G$ is a finite $p$-group then these conditions are also equivalent to the $kG$-module being free.

**Theorem 1.9.1** (Chouinard [89]) *Let $G$ be a finite group. A $kG$-module $M$ is projective if and only if the restriction of $M$ to every elementary abelian $p$-subgroup of $G$ is projective.*

**Remark 1.9.2** This theorem was proved by Chouinard in the generality of an arbitrary commutative ring $k$ of coefficients and arbitrary modules without the restriction of finite generation. In this context one needs to use the elementary abelian $p$-groups for all $p$ dividing the group order.

**Example 1.9.3** To see how remarkable this theorem is, let us look at the example of the generalised quaternion group of order $2^n$ ($n \geq 3$),

$$G = Q_8 = \langle g, h \mid g^{2^{n-1}} = 1, h^2 = g^{2^{n-2}}, hg = g^{-1}h^{-1} \rangle$$

with $k$ a field of characteristic two. Set $z = g^{2^{n-2}}$, so that $Z(G) = \{1, z\}$. The subgroup $Z(G)$ is the only non-trivial elementary abelian subgroup of $G$, so the theorem states that a $kG$-module $M$ is projective if and only if $M{\downarrow}_{Z(G)}$ is projective. Since the only indecomposable modules for $Z(G)$ are the trivial module of dimension one and the regular representation of dimension two, it follows that if $M$ is finite dimensional then it is projective if and only if the rank of the matrix representing $1 + z \in kG$ is as large as it can be, namely one half of the dimension of $M$.

**Remark 1.9.4** There are many other related aspects of group cohomology and modular representation theory where the elementary abelian $p$-subgroups play a crucial role. See the papers of Quillen [206, 207, 210], Carlson [72, 73, 74, 76, 77, 79], Alperin and Evens [7, 8], Avrunin and Scott [24], Jackowski and McClure [169], Kroll [181], Linckelmann [188], Symonds [232], Webb [243], etc.

Chouinard's theorem suggests that it is important to be able to decide whether a $kE$-module is projective. The following test only works for finitely generated $kE$-modules with $k$ algebraically closed.

**Lemma 1.9.5** (Dade's lemma; [98], lemma 11.8) *A finitely generated $kE$-module $M$ is projective if and only if the restriction of $M$ to each cyclic shifted subgroup (with respect to a chosen system of generators for $E$) is projective.*

**Remark 1.9.6** Dade's lemma in the form stated is not true for infinitely generated modules, nor for finitely generated modules over a field that is not algebraically closed. For further information see Benson, Carlson and Rickard [48].

Dade's lemma motivates the following definition, introduced by Carlson.

**Definition 1.9.7** Recall that $V_E^{\mathrm{rk}}$ denotes the affine space $\mathbb{A}^r(k)$ of dimension equal to the rank of $E$. The *rank variety* $V_E^{\mathrm{rk}}(M)$ of a finitely generated $kE$-module $M$ is defined to be

$$V_E^{\mathrm{rk}}(M) = \{0\} \cup \{0 \neq \alpha \in V_E^{\mathrm{rk}} \mid M{\downarrow_{\langle g_\alpha \rangle}} \text{ is not projective}\}.$$

To test whether $M{\downarrow_{\langle g_\alpha \rangle}}$ is projective, look at the rank of the linear transformation effected by $X_\alpha$ on $M$. This is at most $(p-1)/p$ times the dimension of $M$, with equality if and only if the restriction is projective.

**Lemma 1.9.8** (Carlson [74, §6]) *Let $x, y$ be elements of $J(kE) \setminus J^2(kE)$ with $x - y \in J^2(kE)$ and let $M$ be a finitely generated $kE$-module. Then $M{\downarrow_{\langle 1+x \rangle}}$ is projective if and only if $M{\downarrow_{\langle 1+y \rangle}}$ is projective.*

This may be seen as a consequence of the following, which appears as Proposition 2.2 in Friedlander and Pevtsova [124].

**Proposition 1.9.9** *Let $M$ be a $k$-vector space (recall that $k$ has characteristic $p$), and let $\alpha, \beta, \gamma$ be pairwise commuting endomorphisms of $M$ with $\alpha^p = 0$, $\beta^p = 0$ and $\gamma$ nilpotent. Then $M$ is free as a $k[t]/(t^p)$-module with $t$ acting via $\alpha$ if and only if it is free with $t$ acting via $\alpha + \beta\gamma$.*

*Proof.* A $k[t]/(t^p)$-module is free if and only if $\mathsf{Ker}\,(t^{p-1}) = \mathsf{Im}\,(t)$, or equivalently $\mathsf{Ker}\,(t) = \mathsf{Im}\,(t^{p-1})$. So we need to check that if $\mathsf{Ker}\,((\alpha + \beta\gamma)^{p-1}, M) = \mathsf{Im}\,(\alpha + \beta\gamma, M)$ then $\mathsf{Ker}\,(\alpha^{p-1}, M) = \mathsf{Im}\,(\alpha, M)$. Let $V = \mathsf{Ker}\,(\alpha^{p-1}, M)/\mathsf{Im}\,(\alpha, M)$. Since $\gamma$ is nilpotent, it suffices to show that the map

$V \to V$ induced by $\gamma$ is injective. Suppose that $x \in \mathsf{Ker}\,(\alpha^{p-1}, M)$ with $\gamma x \in \mathsf{Im}\,(\alpha, M)$. Then $\alpha^{p-1}x = 0$ and there exists $y \in M$ with $\gamma x = \alpha y$. Let

$$\delta = \sum_{i=0}^{p-2}(-1)^i \alpha^{p-2-i}\beta^{i+1}\gamma^i = \alpha^{p-2}\beta - \alpha^{p-3}\beta^2\gamma + \cdots - \beta^{p-1}\gamma^{p-2},$$

so that

$$\alpha\delta = (\alpha + \beta\gamma)^{p-1}\beta, \qquad \gamma\delta = \alpha^{p-1} - (\alpha + \beta\gamma)^{p-1}.$$

Then we have

$$(\alpha + \beta\gamma)^{p-1}x = (\alpha^{p-1} - \gamma\delta)x = -\alpha\delta y = -(\alpha + \beta\gamma)^{p-1}\beta y,$$

so $x + \beta y$ is in the kernel of $(\alpha + \beta\gamma)^{p-1}$. Therefore there is an element $z \in M$ with

$$(\alpha + \beta\gamma)z = x + \beta y.$$

Then we have

$$(\alpha + \beta\gamma)y = \gamma x + \beta\gamma y = \gamma(\alpha + \beta\gamma)z$$

and therefore there is an element $u \in M$ with

$$(\alpha + \beta\gamma)^{p-1}u = y - \gamma z.$$

Finally, we have

$$\alpha(z - \delta u) = (x + \beta y - \beta\gamma z) - (\alpha + \beta\gamma)^{p-1}\beta u = (x + \beta y - \beta\gamma z) - \beta(y - \gamma z) = x,$$

and so $x \in \mathsf{Im}\,(\alpha, M)$. □

**Remark 1.9.10** Lemma 1.9.8 is a special case of Theorem 4.5.3. Proposition 1.9.9 is closely related to the discussion in Chapter 4.

The following summarises the basic properties of rank varieties that we shall be using. Proofs can be found in Carlson [74, 75] or in Benson [40].

**Proposition 1.9.11** *Let $M$ and $N$ be $kE$-modules.*

(1) *$V_E^{\mathrm{rk}}(M)$ is a closed homogeneous subvariety of $V_E^{\mathrm{rk}} = \mathbb{A}^r(k)$.*
(2) *$\dim V_E^{\mathrm{rk}}(M)$ is equal to the complexity of $M$.*
(3) *$V_E^{\mathrm{rk}}(M \oplus N) = V_E^{\mathrm{rk}}(M) \cup V_E^{\mathrm{rk}}(N)$.*
(4) *$V_E^{\mathrm{rk}}(M \otimes_k N) = V_E^{\mathrm{rk}}(M) \cap V_E^{\mathrm{rk}}(N)$.*
(5) *If $0 \to M_1 \to M_2 \to M_3 \to 0$ is a short exact sequence of $kE$-modules then*
$$V_E^{\mathrm{rk}}(M_2) \subseteq V_E^{\mathrm{rk}}(M_1) \cup V_E^{\mathrm{rk}}(M_3).$$
(6) *(Carlson's connectedness theorem) If $V_E^{\mathrm{rk}}(M) = V_1 \cup V_2$ where $V_1$ and $V_2$ are closed homogeneous subsets of $V_E^{\mathrm{rk}}$ with $V_1 \cap V_2 = \{0\}$ then $M$ decomposes as $M_1 \oplus M_2$ with $V_E^{\mathrm{rk}}(M_1) = V_1$ and $V_E^{\mathrm{rk}}(M_2) = V_2$.*

An exactly parallel discussion holds for exterior algebras.

**Lemma 1.9.12** (Dade's lemma for exterior algebras) *Let $\Lambda = \Lambda(X_1, \dots, X_r)$ be an exterior algebra over an algebraically closed field $k$. Then a finitely generated $\Lambda$-module $M$ is projective if and only if it is projective on restriction to $X_\alpha$ for every $0 \neq \alpha \in \mathbb{A}^r(k)$.*

*Proof.* See Theorem 3.1 of Aramova, Avramov and Herzog [12]. □

**Definition 1.9.13** Recall that $V_\Lambda^{\mathrm{rk}}$ denotes the affine space $\mathbb{A}^r(k)$. The *rank variety* $V_\Lambda^{\mathrm{rk}}(M)$ of a finitely generated $\Lambda$-module $M$ is defined to be

$$V_\Lambda^{\mathrm{rk}}(M) = \{0\} \cup \{0 \neq \alpha \in V_\Lambda^{\mathrm{rk}} \mid M{\downarrow_{\langle X_\alpha \rangle}} \text{ is not projective}\}.$$

Again, the test for projectivity on restriction to $X_\alpha$ is easy. The matrix for the action of $X_\alpha$ on $M$ has rank at most half the dimension of $M$, with equality if and only if the restriction of $M$ to $X_\alpha$ is projective.

**Proposition 1.9.14** *The rank varieties of $\Lambda$-modules satisfy the properties listed in Proposition 1.9.11 for rank varieties of $kE$-modules.*

*Proof.* Again, see Theorem 3.1 of Aramova, Avramov and Herzog [12]. □

## 1.10 Carlson's $L_\zeta$ Modules, and a Matrix Version

In this section, we describe Carlson's $L_\zeta$ modules, and variations on them defined using matrices of elements of cohomology. This section will be crucial to the discussions in Chapter 12, where we also explore variations using relative cohomology.

Let $0 \neq \alpha = (\lambda_1, \dots, \lambda_r) \in \mathbb{A}^r(k)$. Recall that $X_\alpha = \sum_{i=1}^r \lambda_i X_i \in kE$ and $g_\alpha = 1 + X_\alpha$, so that $\langle g_\alpha \rangle \cong \mathbb{Z}/p$. Let $H^*(\langle g_\alpha \rangle, k) = k[y_\alpha]$ ($p = 2$) or $k[y_\alpha, x_\alpha]/(y_\alpha^2)$ ($p$ odd), where $y_\alpha$ and $x_\alpha$ are the elements of degrees one and two described in Section 1.8.

The following theorem appears as Proposition 2.20 in Carlson [74].

**Theorem 1.10.1** *The restriction of $y_i$ to $\langle g_\alpha \rangle$ is equal to $\lambda_i y_\alpha$. If $p$ is odd, then the restriction of $x_i$ to $\langle g_\alpha \rangle$ is equal to $\lambda_i^p x_\alpha$.*

*Proof.* For the $y_i$ this is easy, using the duality between $H^1(E, k)$ and $V_E^{\mathrm{rk}}$. If $p$ is odd, the formula for the restriction of $x_i$ comes from the fact that restriction commutes with the Bockstein map, together with the formula (1.3). □

**Corollary 1.10.2** *Let $f$ be a polynomial in $r$ variables. If $p = 2$ then the restriction of $f(y_1, \dots, y_r)$ to $\langle g_\alpha \rangle$ is equal to $\sum f(\lambda_1, \dots, \lambda_r) y_\alpha$. If $p$ is odd then the restriction of $f(x_1, \dots, x_r)$ to $\langle g_\alpha \rangle$ is equal to $f(\lambda_1^p, \dots, \lambda_r^p) x_\alpha$.* □

This allows us to compute the rank varieties of Carlson's $L_\zeta$ modules. Let $\zeta \in H^n(E, k)$ be represented by a cocycle $\hat{\zeta} : \Omega^n(k) \to k$. As long as $\zeta$ is non-zero, $\hat{\zeta}$ is surjective, and we write $L_\zeta$ for the kernel. Thus we have a short exact sequence

$$0 \to L_\zeta \to \Omega^n(k) \xrightarrow{\hat{\zeta}} k \to 0.$$

If $\zeta = 0$, we define $L_\zeta$ to be $\Omega^n(k) \oplus \Omega(k)$. The reasoning for this is that in both cases, we are really completing $\hat{\zeta}$ to a triangle in $\mathsf{stmod}(kE)$:

$$L_\zeta \to \Omega^n(k) \to k.$$

Since restriction to $\langle g_\alpha \rangle$ gives a well-defined functor from $\mathsf{stmod}(kE)$ to $\mathsf{stmod}(k\langle g_\alpha \rangle)$, we have the following.

**Proposition 1.10.3** *For $p = 2$, let $\zeta = f(y_1, \dots, y_r)$. Then $V_E^{\mathrm{rk}}(L_\zeta)$ is the hypersurface defined by the equation $f(Y_1, \dots, Y_r) = 0$.*

*For $p$ odd, let $\zeta = f(x_1, \dots, x_r)$. Then $V_E^{\mathrm{rk}}(L_\zeta)$ is the hypersurface defined by the equation $f^{(p)}(Y_1, \dots, Y_r) = 0$, where $f^{(p)}$ is obtained from $f$ by raising all the coefficients to the $p$th power.*

*Proof.* A point $\alpha$ is in $V_E^{\mathrm{rk}}(L_\zeta)$ precisely when the restriction of $\hat{\zeta}$ is not an isomorphism in $\mathsf{stmod}(k\langle g_\alpha \rangle)$, namely when the restriction of $\zeta$ is zero. Now use Corollary 1.10.2. □

**Corollary 1.10.4** (Carlson) *Every closed homogeneous subvariety of $V_E^{\mathrm{rk}}$ is the rank variety of some finitely generated $kE$-module.*

*Proof.* A closed homogeneous subvariety of affine space can always be expressed as an intersection of a finite set of closed homogeneous hypersurfaces. For such a hypersurface, we can apply Proposition 1.10.3. For the intersection, we tensor together these modules and use Proposition 1.9.11 (4). □

A variation on this is given by using matrices of cohomology elements. Let $(f_{ij})$ be an $n \times n$ matrix of homogeneous polynomials in $r$ variables. Let us suppose that these satisfy the following condition.

**Condition 1.10.5** There are integers $a_1, \dots, a_n$ and $b_1, \dots, b_n$ such that $f_{ij}(y_1, \dots, y_r)$ ($p = 2$), respectively $f_{ij}(x_1, \dots, x_r)$ ($p$ odd), is an element of $H^{a_i - b_j}(E, k)$.

Then we can represent it by a cocycle $\hat{f}_{ij}\colon \Omega^{b_j}(k) \to \Omega^{a_i}(k)$. Thus the matrix $(f_{ij})$ gives us a map

$$(\hat{f}_{ij})\colon \bigoplus_{j=1}^{n} \Omega^{b_j}(k) \to \bigoplus_{i=1}^{n} \Omega^{a_i}(k).$$

Complete this to a triangle in $\mathsf{stmod}(kE)$:

$$\bigoplus_{i=1}^{n} \Omega^{a_i+1}(k) \to L_{(f_{ij})} \to \bigoplus_{j=1}^{n} \Omega^{b_j}(k) \to \bigoplus_{i=1}^{n} \Omega^{a_i}(k).$$

**Theorem 1.10.6** *If* $p = 2$ *then* $V_E^{\mathrm{rk}}(L_{(f_{ij})})$ *is the hypersurface defined by the equation* $\det(f_{ij}) = 0$. *If* $p$ *is odd, then* $V_E^{\mathrm{rk}}(L_{(f_{ij})})$ *is the hypersurface defined by the equation* $\det(f_{ij})^{(p)} = 0$.

*Proof.* Again, a point $\alpha$ is in $V_E^{\mathrm{rk}}(L_{(f_{ij})})$ precisely when the restriction of $(\hat{f}_{ij})$ is not an isomorphism in $\mathsf{stmod}(\langle g_\alpha \rangle)$, namely when the determinant restricts to zero. Again use Corollary 1.10.2. □

## 1.11 Diagrams for Modules

We shall sometimes describe $kE$-modules by drawing diagrams. The meaning of these diagrams is that vertices represent basis elements and edges represent the action of the elements $X_i$. Not all modules have such diagrams, but when they do, the diagram is very convenient for understanding the module. More formal details of diagrams for modules can be found in Benson and Carlson [46]. We shall illustrate the ideas with some examples without going into formalities. It will be instructive to compare and contrast these diagrams with the ones occurring in Section 9.2.

For our first example, we let $E = (\mathbb{Z}/3)^2$, and we draw the following diagram for the group algebra $kE$.

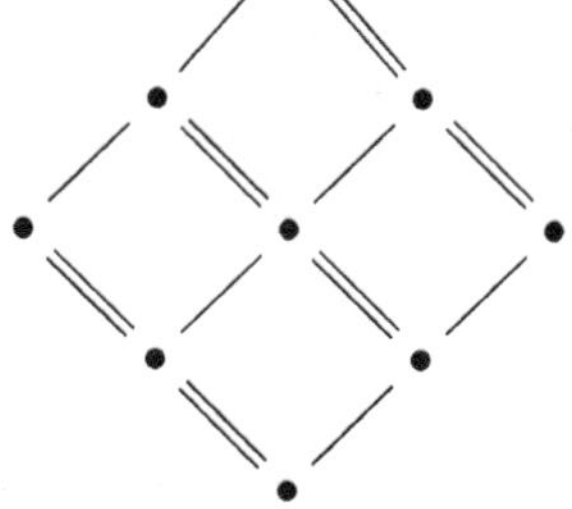

Going downwards, single edges represent multiplication by $X_1$ and double edges by $X_2$. If there is no edge of a particular kind leading downwards from a basis element, then the corresponding $X_i$ acts as zero.

The top basis element, labelled "1", represents the identity element of $kE$. So the two basis elements just below that are $X_1$ and $X_2$, the three below that are $X_1^2$, $X_1X_2$ and $X_2^2$, the two below that are $X_1^2X_2$ and $X_1X_2^2$, and finally the bottom basis element is $X_1^2X_2^2$, which is annihilated by both $X_1$ and $X_2$.

The augmentation map $kE \to k$ takes the top basis element to 1 and the remaining basis elements to zero, and the kernel of this map is $\Omega(k)$. Similarly, there is a map $k \to kE$ sending 1 to $X_1^2X_2^2$, whose cokernel is $\Omega^{-1}(k)$. It follows that the diagrams for these modules are as follows.

$\Omega(k) =$ 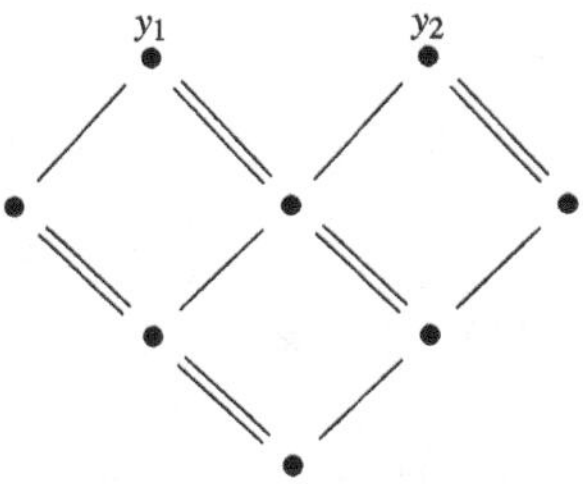


$\Omega^{-1}(k) =$ 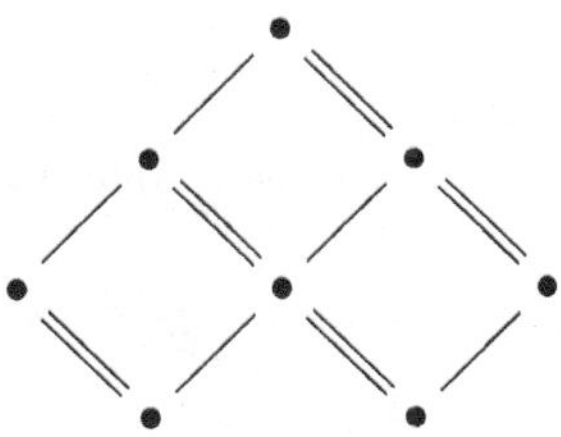

The top vertices of $\Omega(k)$ have been labelled with the exterior generators

$$y_1, y_2 \in H^1(E, k) \cong \mathrm{Ext}^1_{kE}(k, k) \cong \mathrm{Hom}_{kE}(\Omega(k), k).$$

The homomorphism $\hat{y}_1 : \Omega(k) \to k$ corresponding to $y_1 \in H^1(E, k)$, for example, sends the top left-hand vertex to 1 and the remaining vertices to zero.

The projective cover of $\Omega(k)$ is two copies of $kE$, in such a way that the identity elements map to the two top vertices of $\Omega(k)$. The four vertices in the middle get hit twice, so there are diagonal elements in the kernel. It is not hard to see that this gives the following diagram for $\Omega^2(k)$.

$\Omega^2(k) =$ 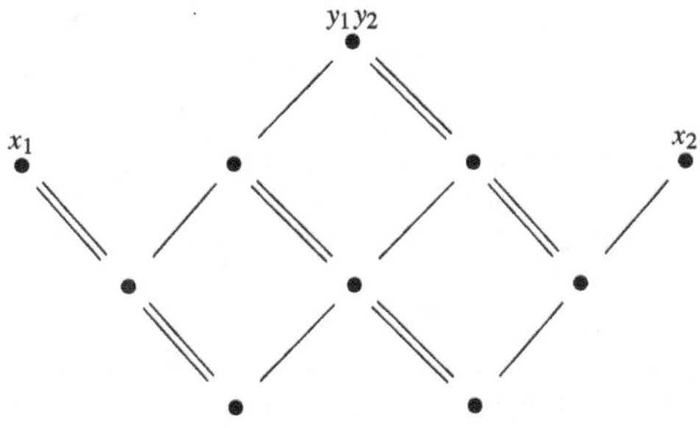

Similarly, the following is the diagram for $\Omega^3(k)$.

$\Omega^3(k) =$

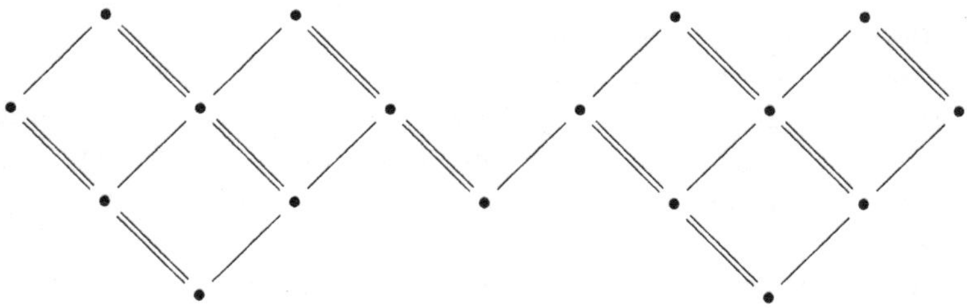

From the diagram for $\Omega^2(k)$, we can see immediately for example that there is a short exact sequence of the form

$$0 \to \Omega^{-1}(k) \to \Omega^2(k) \to k \oplus k \to 0.$$

Under the isomorphism

$$H^2(E,k) \cong \mathrm{Ext}^2_{kE}(k,k) \cong \mathrm{Hom}_{kE}(\Omega^2(k),k),$$

let $\hat{x}_1 \colon \Omega^2(k) \to k$ correspond to $x_1 \in H^2(E,k)$. Then $\hat{x}_1$ is the homomorphism that takes the top left-hand vertex of the diagram for $\Omega^2(k)$ to 1 and the remaining basis elements to zero. So we have a Carlson module

$L_{x_1} = \mathsf{Ker}\,(\hat{x}_1) =$

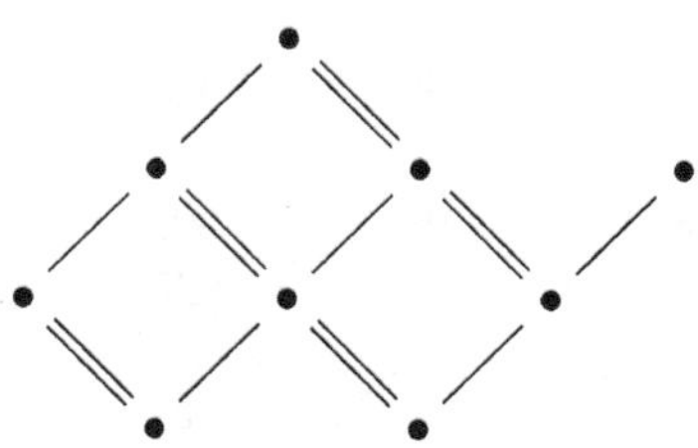

Similarly we have

$L_{x_2} = \mathsf{Ker}\,(\hat{x}_2) =$

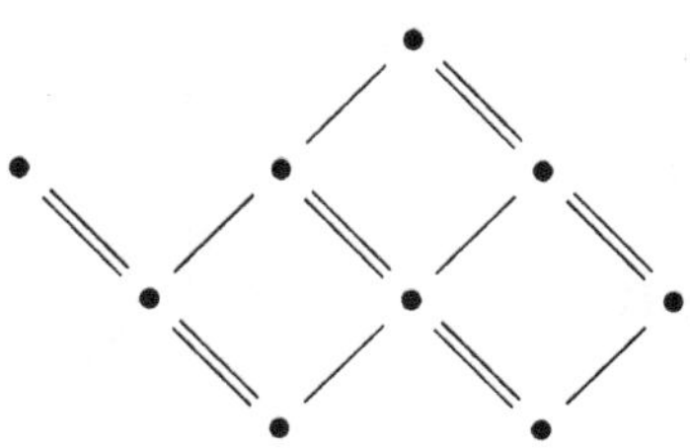

These are periodic modules with period two, i.e., $\Omega^2(L_{x_i}) \cong L_{x_i}$. We also have $L^*_{x_i} \cong \Omega(L_{x_i})$.

For $p = 2$ the syzygies of the trivial module are easier to compute, and are as follows:

$\Omega(k) =$ 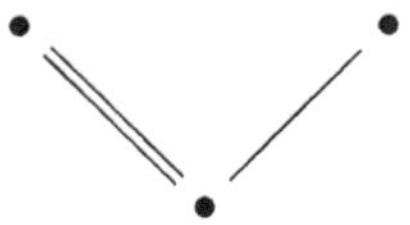 $\Omega^2(k) =$ 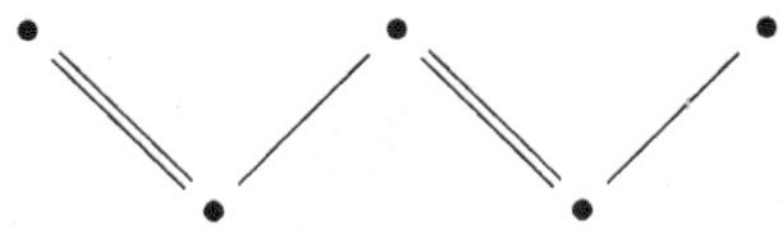

$\Omega^{-1}(k) =$ 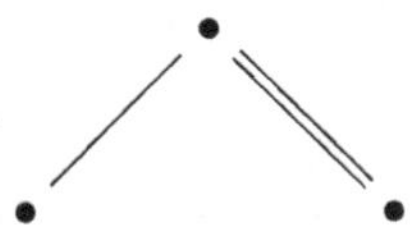 $\Omega^{-2}(k) =$ 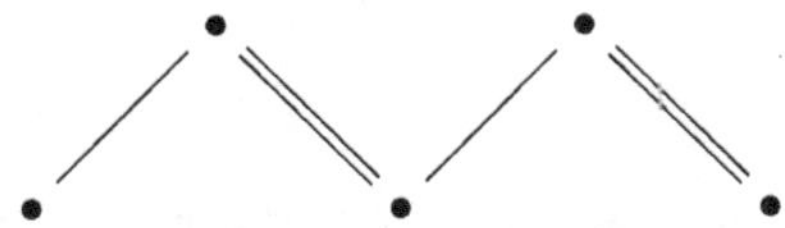

For elementary abelian groups of rank three and higher, it is harder to draw diagrams except in very restricted circumstances, because the page is two dimensional. We will see just one rank four example in Section 12.2. But it is worth demonstrating how to draw diagrams for $(\mathbb{Z}/2)^3$. For this purpose, we use a bird's eye view, looking from the "top" of the module downwards. So for example $\Omega(k)$ and $\Omega^{-1}(k)$ are represented by a cube with the top, respectively bottom vertex removed:

$\Omega(k) =$ 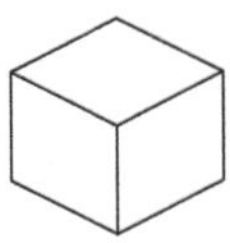 $\Omega^{-1}(k) =$ 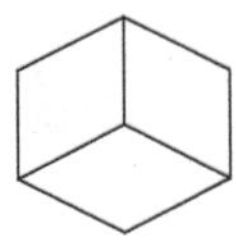

To find $\Omega^2(k)$, we map three copies of $kE$ onto the three top vertices of $\Omega(k)$, which in the diagram are top, bottom left and bottom right. The kernel then has the following diagram.

$\Omega^2(k) =$ 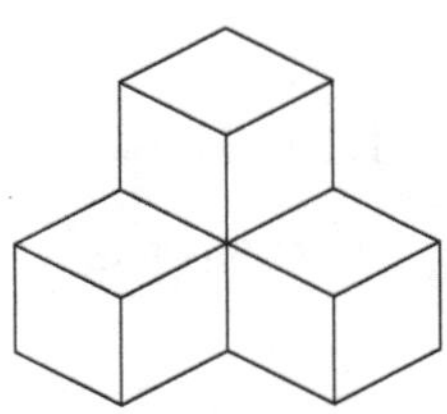

The only thing to watch in this diagram is that the centre vertex really represents a vector space of dimension two, with three incoming maps and three outgoing maps corresponding to the three subspaces of dimension one defined over $\mathbb{F}_2$. Thus $\Omega^2(k)$ has dimension 17.

It is easy to continue this way, so that we have

$\Omega^3(k) =$

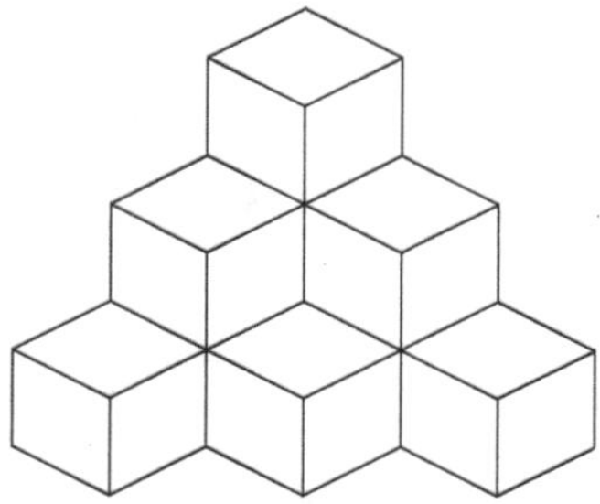

There are now three vertices in this diagram that represent vector spaces of dimension two, while the rest are one dimensional. Thus $\Omega^3(k)$ is 31 dimensional.

Carlson modules can be described as before. So for example if we let $x_1$ be the element of $H^1(E, k)$ corresponding to the uppermost vertex in the diagram for $\Omega(k)$, then

$L_{x_1} =$

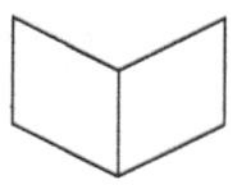

$L_{x_1^2} =$

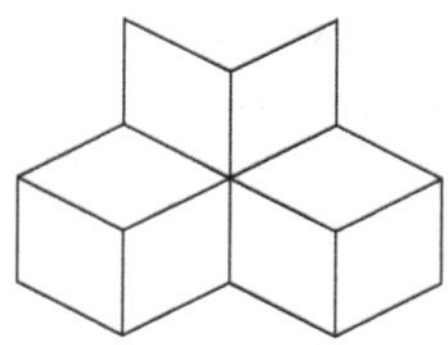

$L_{x_1^3} =$

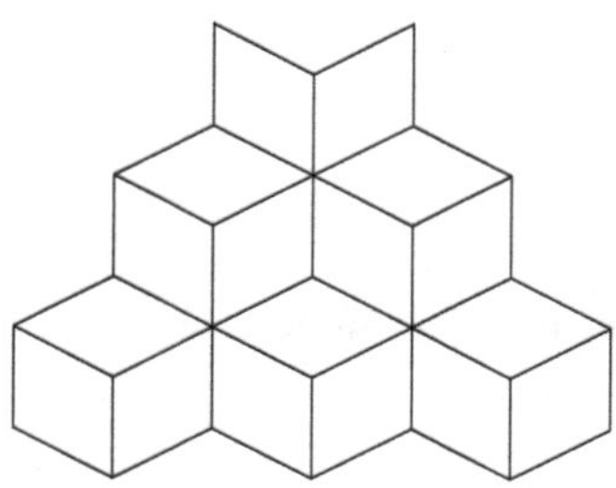

## 1.12 Tensor Products

If $G$ is any finite group, and $M$ and $N$ are $kG$-modules, we make $M \otimes_k N$ into a $kG$-module via

$$g(m \otimes n) = gm \otimes gn$$

for $g \in G$, $m \in M$ and $n \in N$. Note, however, that for a linear combination of group elements the action is given by

$$(\sum_i \lambda_i g_i)(m \otimes n) = \sum_i \lambda_i (g_i m \otimes g_i n)$$

and not by

$$(\sum_i \lambda_i g_i)(m \otimes n) = (\sum_i \lambda_i g_i m) \otimes (\sum_i \lambda_i g_i n).$$

Indeed, the latter formula does not even define an action, because it is not linear.

This can be seen in terms of the diagonal map $\Delta\colon kG \to kG \otimes_k kG$ given by

$$\Delta(\sum_i \lambda_i g_i) = \sum_i \lambda_i (g_i \otimes g_i).$$

Then any element $x \in kG$ acts on $M \otimes_k N$ by taking $m \otimes n$ to $\Delta(x)(m \otimes n)$. The elements $x \in kG$ satisfying $\Delta(x) = x \otimes x$ are precisely the elements of $G$.

**Warning 1.12.1** As a result of the way the action on tensor products is defined, if $E$ is an elementary abelian $p$-group and $M$ and $N$ are $kE$-modules, then the restriction of $M \otimes_k N$ to a cyclic shifted subgroup is not necessarily isomorphic to the tensor product of the restrictions.

**Example 1.12.2** Let $E = \langle g_1, g_2 \rangle \cong \mathbb{Z}/2 \times \mathbb{Z}/2$ and let $M$ and $N$ both be the representation of dimension two given by

$$g_1 \mapsto \begin{pmatrix} 1 & 0 \\ 1 & 1 \end{pmatrix} \qquad g_2 \mapsto \begin{pmatrix} 1 & 0 \\ \lambda & 1 \end{pmatrix}$$

with $\lambda \in k \setminus \{0, 1\}$. Let $\alpha = (\lambda, 1) \in \mathbb{A}^2(k)$, so that

$$X_\alpha = \lambda(g_1 - 1) + (g_2 - 1), \qquad g_\alpha = 1 + X_\alpha.$$

Then $M$ and $N$ each restricts to $\langle g_\alpha \rangle$ to give a sum of two copies of the trivial module, while $M \otimes_k N$ restricts to give a sum of two copies of the trivial module and an indecomposable module of dimension two.

There is a way to remedy this problem by changing the diagonal map. Setting

$$E = \langle g_1, \dots, g_r \rangle, \qquad X_i = g_i - 1 \in kE$$

as usual, the map $\Delta$ defined above sends $J(kE)$ into

$$J(kE) \otimes kE \oplus kE \otimes J(kE),$$

and

$$\Delta(X_i) \equiv X_i \otimes 1 + 1 \otimes X_i \pmod{J(kE) \otimes J(kE)}.$$

If instead of $\Delta$ we use a different diagonal map $\tilde{\Delta}$ given by

$$\tilde{\Delta}(X_i) = X_i \otimes 1 + 1 \otimes X_i$$

then the tensor product defined using this new diagonal map does indeed commute with restriction to shifted subgroups. To distinguish between these two module structures on the tensor product, we write $M \otimes_k N$ for the *group theoretic tensor product* defined using $\Delta$ and $M \tilde{\otimes}_k N$ for the *restricted Lie tensor product* defined using $\tilde{\Delta}$. This name comes from the fact that the diagonal map $\tilde{\Delta}$ is the one obtained by viewing $kE$ as the restricted enveloping algebra of the restricted Lie algebra given by the linear span of $X_1, \dots, X_r$ with trivial Lie bracket and trivial $p$th power map. The reason why restriction to shifted subgroups commutes with this tensor product is that a shifted subgroup is the restricted enveloping algebra of a restricted Lie subalgebra.

We also remark that the difference between the two diagonal maps $\Delta$ and $\tilde{\Delta}$ consists of "second order terms". This observation will be useful in Chapter 4 when we want to show that the shift from one tensor product to the other does not alter the generic Jordan type.

Next, we discuss tensor products of modules for exterior algebras. Let $\Lambda = \Lambda(X_1, \dots, X_r)$, and $M$ and $N$ be graded $\Lambda$-modules. Then $M \otimes_k N$ is a $\Lambda$-module via

$$X_i(m \otimes n) = X_i m \otimes n + (-1)^{|m|} m \otimes X_i n.$$

The corresponding diagonal map $\Delta \colon \Lambda \to \Lambda \otimes_k \Lambda$ is given by

$$\Delta(\sum_i \lambda_i X_i) = \sum_i \lambda_i (X_i \otimes 1 + 1 \otimes X_i).$$

**Warning 1.12.3** As long as we are not in characteristic two, the formula above for $\Delta$ is only an algebra homomorphism if we treat $\Lambda$ as a graded algebra. This is because if we try to use this formula for the ungraded algebra then

$$\Delta(X_i)^2 = 2(X_i \otimes X_i) \neq 0.$$

The signs introduced by the grading make sure that $\Delta(X_i)^2 = 0$.

The consequence of this is that if we wish to form tensor products of $\Lambda$-modules then we must work in the graded context.

## 1.13 Duality

Just as with tensor products, there are two ways to define duality, namely the *group theoretic dual* and the *restricted Lie dual*, corresponding to the two different *antipodes* on $kE$. The antipode $\eta \colon kE \to kE$ for the group theoretic dual is the algebra (anti-)automorphism defined by $\eta(g) = g^{-1}$ for $g \in E$. Thus

$$\eta(X_i) = -X_i + X_i^2 - \cdots + X_i^{p-1}.$$

The action of $kE$ on the dual $M^* = \mathrm{Hom}_k(M, k)$ of a $kE$-module $M$ is given by $(xf)(m) = f(\eta(x)m)$ for $x \in kE$, $m \in M$ and $f \in M^*$.

On the other hand, the antipode $\tilde{\eta} \colon kE \to kE$ for the restricted Lie dual is the algebra (anti-)automorphism defined by $\tilde{\eta}(X_i) = -X_i$. The action of $kE$ on the restricted Lie dual $M^{\tilde{*}}$ is given by $(xf)(m) = f(\tilde{\eta}(x)m)$. Restriction to a shifted subgroup commutes with the restricted Lie dual, but not with the group theoretic dual.

The maps $\Delta$ and $\eta$ make $kE$ into a Hopf algebra, as do $\tilde{\Delta}$ and $\tilde{\eta}$. The behaviours of these two different Hopf structures on $kE$ are quite different.

**Example 1.13.1** Let $k$ be an algebraically closed field of characteristic $p \geq 3$ and let $E = \langle g_1, g_2 \rangle \cong \mathbb{Z}/p \times \mathbb{Z}/p$. Let $M_{\lambda,\mu}$ be the $kE$-module

$$X_1 \mapsto \begin{pmatrix} 0 & 0 & 0 \\ 1 & 0 & 0 \\ 0 & 1 & 0 \end{pmatrix} \qquad X_2 \mapsto \begin{pmatrix} 0 & 0 & 0 \\ \lambda & 0 & 0 \\ \mu & \lambda & 0 \end{pmatrix}.$$

It is easy to check that for different pairs $(\lambda, \mu)$ these modules are non-isomorphic. The group theoretic dual $M^*_{\lambda,\mu}$ is given by the matrices

$$X_1 \mapsto \begin{pmatrix} 0 & -1 & 1 \\ 0 & 0 & -1 \\ 0 & 0 & 0 \end{pmatrix} \qquad X_2 \mapsto \begin{pmatrix} 0 & -\lambda & \lambda^2 - \mu \\ 0 & 0 & -\lambda \\ 0 & 0 & 0 \end{pmatrix}$$

so that it is isomorphic to $M_{\lambda,\lambda^2-\lambda-\mu}$, whereas the Lie theoretic dual $M^{\tilde{*}}_{\lambda,\mu}$ is given by the matrices

$$X_1 \mapsto \begin{pmatrix} 0 & -1 & 0 \\ 0 & 0 & -1 \\ 0 & 0 & 0 \end{pmatrix} \qquad X_2 \mapsto \begin{pmatrix} 0 & -\lambda & -\mu \\ 0 & 0 & -\lambda \\ 0 & 0 & 0 \end{pmatrix}$$

so that it is isomorphic to $M_{\lambda,-\mu}$.

Now let $\lambda \notin \{0, 1\}$ and $M = M_{\lambda,0}$. Set $\alpha = (\lambda, -1)$, so that $X_\alpha = \lambda X_1 - X_2$. Then the restriction to $X_\alpha$ of $M^*$ has Jordan canonical form [2] [1] while the restriction of $M^{\tilde{*}}$ is trivial, with Jordan type $[1]^3$. So for the group theoretic dual, the restriction of the dual to $X_\alpha$ can have a different Jordan type than the restriction of the module.

Furthermore, in this example the kernel of the map $kE \to \mathrm{End}_k(M^*)$ is not equal to the kernel of $kE \to \mathrm{End}_k(M)$.

**Example 1.13.2** Given a diagram for a module in the sense of Section 1.11, a diagram for the restricted Lie dual may be obtained by turning it upside down and negating the edges, in the following sense. Let $m_1, \ldots, m_n$ is a basis for $M$ such that $X_i$ takes $m_j$ to $\lambda m_{j'}$. Then with respect to the dual basis $m_1^*, \ldots, m_n^*$, $X_i$ takes $m_{j'}^*$ to $-\lambda m_j^*$. Often one can replace the elements of the dual basis by multiples in order to reduce the number of minus signs.

For example, if $M$ is the module for $\mathbb{Z}/p \times \mathbb{Z}/p$ with diagram

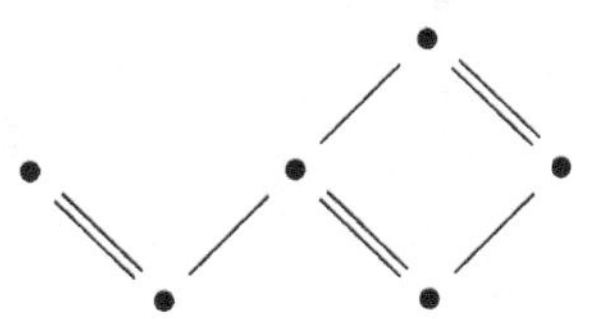

then $M^{\tilde{*}}$ has diagram

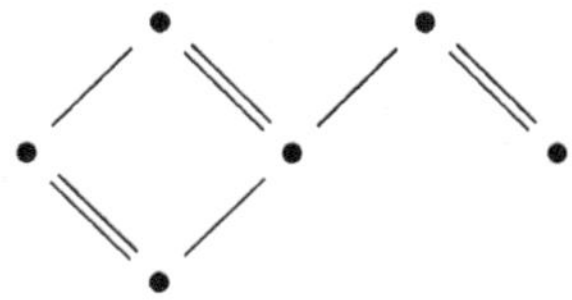

The sign problem is solved in this case by negating the basis elements in the middle row.

More care is needed with the group theoretic dual, and as an exercise the reader may wish to verify that in the example above, the group theoretic dual $M^*$ has the same diagram after a suitable change of basis.

Duality for exterior algebras works in a similar way. Let $\Lambda = \Lambda(X_1, \dots, X_r)$. The antipode $\eta\colon \Lambda \to \Lambda$ is given by

$$\eta(X_i) = -X_i.$$

So the action of $\Lambda$ on the dual $M^* = \mathrm{Hom}_k(M, k)$ of a graded module $M$ is given by $X_i f(m) = -f(X_i m)$.

The maps $\Delta$ and $\eta$ make $\Lambda$ into a graded Hopf algebra.

## 1.14 Symmetric and Exterior Powers

We define the $d$th symmetric power $S^d(M)$ of a module $M$ to be the quotient of the $d$-fold tensor product $M^{\otimes d}$ by the linear span of the elements

$$m_1 \otimes \cdots \otimes m_d \;-\; m_{\pi(1)} \otimes \cdots \otimes m_{\pi(d)}$$

with $m_1, \dots, m_d \in M$ and $\pi \in \mathfrak{S}_d$, the symmetric group of degree $d$.[1] We write $m_1 \dots m_d$ for the image of $m_1 \otimes \cdots \otimes m_d$ in $S^d(M)$. The following lemma is clear from the definition.

**Lemma 1.14.1** *If $M$ and $N$ are modules then*

$$S^d(M \oplus N) \cong \bigoplus_{i=0}^{d} S^i(M) \otimes S^{d-i}(N).$$

Similarly, we define the $d$th exterior power $\Lambda^d(M)$ as the quotient of $M^{\otimes d}$ by the elements

$$m_1 \otimes \cdots \otimes m_d$$

[1] Note that the natural action of $\mathfrak{S}_d$ on $M^{\otimes d}$ is given by

$$\pi(m_1 \otimes \cdots \otimes m_d) = m_{\pi^{-1}(1)} \otimes \cdots \otimes m_{\pi^{-1}(d)}.$$

with $m_1, \ldots, m_d \in M$, and such that there exist $i \neq j$ with $m_i = m_j$. As long as $p > 2$, this is the same as the quotient by the elements

$$m_1 \otimes \cdots \otimes m_d - \operatorname{sgn}(\pi) m_{\pi(1)} \otimes \cdots \otimes m_{\pi(d)}$$

with $m_1, \ldots, m_d \in M$ and $\pi \in \mathfrak{S}_d$. We write $m_1 \wedge \cdots \wedge m_d$ for the image of $m_1 \otimes \ldots m_d$ in $\Lambda^d(M)$.

**Remark 1.14.2** The antisymmetrisation map

$$m_1 \wedge \cdots \wedge m_d \mapsto \sum_{\pi \in \mathfrak{S}_d} \operatorname{sgn}(\pi) m_{\pi(1)} \otimes \cdots \otimes m_{\pi(d)}$$

gives a natural embedding of $\Lambda^d(M)$ in $M^{\otimes d}$.

Beware, though, that $S^d(M)$ can only be naturally embedded in $M^{\otimes d}$ when $d!$ is invertible. The fixed points of $\mathfrak{S}_d$ on $M^{\otimes d}$ give the $d$-fold *divided power* $D^d(M)$, which is related to the $d$-fold symmetric power by $D^d(M)^* \cong S^d(M^*)$. The divided power $D^d(M)$ is naturally isomorphic to $S^d(M)$, provided $d!$ is invertible; see Lemma 1.14.4.

**Lemma 1.14.3** *If M and N are modules then*

$$\Lambda^d(M \oplus N) \cong \bigoplus_{i=0}^{d} \Lambda^i(M) \otimes \Lambda^{d-i}(N).$$

**Lemma 1.14.4** *If $d < p$, then $S^d(M)$ and $\Lambda^d(M)$ are direct summands of $M^{\otimes d}$.*

*Proof.* The natural surjections $M^{\otimes d} \to S^d(M)$ and $M^{\otimes d} \to \Lambda^d(M)$ are split by the maps

$$m_1 \ldots m_d \mapsto \frac{1}{d!} \sum_{\pi \in \mathfrak{S}_d} m_{\pi(1)} \otimes \cdots \otimes m_{\pi(d)},$$

respectively

$$m_1 \wedge \cdots \wedge m_d \mapsto \frac{1}{d!} \sum_{\pi \in \mathfrak{S}_d} \operatorname{sgn}(\pi) m_{\pi(1)} \otimes \cdots \otimes m_{\pi(d)}. \qquad \square$$

## 1.15 Schur Functions

We begin with some standard definitions concerning partitions. If

$$\lambda_1 \geq \lambda_2 \geq \cdots \geq \lambda_t \geq 0$$

with $\sum_{i=1}^{t} \lambda_i = d$, we write $[\lambda_1] \ldots [\lambda_t]$ for the corresponding partition of $d$. We usually ignore trailing zeros, so that $[\lambda_1] \ldots [\lambda_t]$ represents the same partition as

$[\lambda_1]\dots[\lambda_t][0]\dots[0]$. This way we can assume that if we are working with several partitions, the values of $t$ are equal. Repeated parts are denoted with indices, so that $[2]^3[1]$ means $[2][2][2][1]$, for example.[2]

**Definition 1.15.1** If $\lambda = [\lambda_1]\dots[\lambda_t]$ is a partition, then the *conjugate partition* $\lambda' = [\lambda'_1][\lambda'_2]\dots$ is defined by

$$\lambda'_i = |\{j \mid \lambda_j \geq i\}|.$$

**Lemma 1.15.2** *For every partition $\lambda$ we have $\lambda'' = \lambda$. In other words, the conjugate of the conjugate is the original partition.*

**Definition 1.15.3** Let $x_1, \dots, x_n$ be variables and let $\lambda = [\lambda_1][\lambda_2]\dots$ be a partition of $d$. The *Schur function* $s_\lambda$ is defined by

$$s_\lambda = s_\lambda(x_1, \dots, x_n) = \frac{\det(x_i^{\lambda_j+n-j})_{1\leq i,j\leq n}}{\det(x_i^{n-j})_{1\leq i,j\leq n}}.$$

The denominator is the Vandermonde determinant

$$\prod_{i<j}(x_i - x_j),$$

and it divides the numerator because setting any two of the variables equal makes the numerator vanish. It follows that $s_\lambda$ is a symmetric polynomial in $x_1, \dots, x_n$ with integer coefficients.

For example if $\lambda = [2]$ then

$$s_\lambda = \frac{\begin{vmatrix} x_1^{n+1} & \cdots & x_n^{n+1} \\ x_1^{n-2} & & x_n^{n-2} \\ \vdots & & \vdots \\ 1 & \cdots & 1 \end{vmatrix}}{\begin{vmatrix} x_1^{n-1} & \cdots & x_n^{n-1} \\ x_1^{n-2} & & x_n^{n-2} \\ \vdots & & \vdots \\ 1 & \cdots & 1 \end{vmatrix}} = x_1^2 + \cdots + x_n^2 + x_1x_2 + x_1x_3 + \cdots + x_{n-1}x_n. \quad (1.4)$$

More generally, $s_{[d]}$ is the *complete symmetric function* $h_d$ of degree $d$.

The following formula, variously known as *Jacobi–Trudi identity*, the *first Gambelli formula* or the *determinantal form*, expresses $s_\lambda$ in terms of the $h_j$ (see for example Macdonald [189], Section I.3):

$$s_\lambda = \det(h_{\lambda_i-i+j})_{1\leq i,j\leq r} \quad (1.5)$$

[2] It is more common to write $[2^3, 1]$, but this notation is ambiguous, because for example it does not distinguish between the partitions $[p^2]$ and $[p]^2$ in our notation.

($r$ at least the number of parts in $\lambda$). For example we have

$$s_{[2][1]} = \begin{vmatrix} h_2 & h_3 \\ h_0 & h_1 \end{vmatrix} = h_2 h_1 - h_3.$$

The easiest way to write down this determinant is to put the $h_{\lambda_i}$ down the diagonal and then fill in horizontally, decreasing the subscript to the left and increasing it to the right. Note that $h_0 = 1$ and that if $j < 0$ then $h_j = 0$.

Let $e_d$ be the elementary symmetric function of degree $d$, so that for example

$$e_2 = x_1 x_2 + x_1 x_3 + \cdots + x_{n-1} x_n.$$

Then $e_d$ is equal to the Schur function $s_{[1]^d}$. The *second Gambelli formula* expresses $s_\lambda$ in terms of the $e_j$ (again see Macdonald [189], Section I.3):

$$s_\lambda = \det(e_{\lambda'_i - i + j})_{1 \le i, j \le r}$$

($r$ at least the number of parts of $\lambda'$, the conjugate of the partition $\lambda$).

For use in Section 2.4, we also record the obvious recurrence relations satisfied by the complete and elementary symmetric functions $h_d$ and $e_d$.

**Proposition 1.15.4** *We have*

(1) $e_d(x_1, \ldots, x_n) = e_d(x_1, \ldots, x_{n-1}) + x_n e_{d-1}(x_1, \ldots, x_{n-1})$,
(2) $h_d(x_1, \ldots, x_n) = h_d(x_1, \ldots, x_{n-1}) + x_n h_{d-1}(x_1, \ldots, x_n)$. □

## 1.16 Schur Functors

If $M$ is a module, we saw in Section 1.14 that there is a natural embedding of $\Lambda^d(M)$ in $M^{\otimes d}$ and a natural surjection of $M^{\otimes d}$ onto $S^d(M)$. If $\lambda$ is a partition of $d$, with conjugate partition $\lambda'$, we look at the composite map

$$\Lambda^{\lambda'_1}(M) \otimes \Lambda^{\lambda'_2}(M) \otimes \cdots \to M^{\otimes d} \to S^{\lambda_1}(M) \otimes S^{\lambda_2}(M) \otimes \cdots .$$

Here, the tensor factors of $M^{\otimes d}$ should be thought of as being indexed by the boxes in the partition $\lambda$. The exterior powers refer to the columns and the symmetric powers refer to the rows of the partition.

**Definition 1.16.1** The Schur functor $S^\lambda$ takes $M$ to the image of the map described above.

**Example 1.16.2** We have

$$S^{[d]}(M) = S^d(M), \qquad S^{[1]^d}(M) = \Lambda^d(M).$$

**Example 1.16.3** Let $\lambda = [3][1]$, $\lambda' = [2][1]^2$:

| $m_1$ | $m_2$ | $m_3$ |
|---|---|---|
| $m_4$ | | |

The maps

$$\Lambda^2(M) \otimes M \otimes M \to M^{\otimes 4} \to S^3(M) \otimes M$$

are given by

$$(m_1 \wedge m_4) \otimes m_2 \otimes m_3 \to m_1 \otimes m_2 \otimes m_3 \otimes m_4 - m_4 \otimes m_2 \otimes m_3 \otimes m_1$$
$$\to m_1 m_2 m_3 \otimes m_4 - m_2 m_3 m_4 \otimes m_1.$$

This definition was introduced by Akin, Buchsbaum and Weyman [1] as a characteristic free definition of the Schur functor. Their definition associates this functor to the conjugate partition $\lambda'$ rather than $\lambda$. Our choice of notation is designed to fit better with the labelling of Schur functions in Section 1.15.

If we take for $M$ the natural module of dimension $n$ for the general linear group $GL(n,k)$ then the modules $S^\lambda(M)$ are the contragredient duals of the Weyl modules $W^\lambda$. In characteristic zero we have $S^\lambda(M) \cong W^\lambda$, and these give a set of representatives for the isomorphism classes of simple polynomial representations of $GL(n,k)$ as $\lambda$ runs over the set of partitions with at most $n$ parts. But in characteristic $p$ they are not in general simple. In this case the simple module $L^\lambda(M)$ is isomorphic to the unique simple submodule of $S^\lambda(M)$ and the unique simple quotient of $W^\lambda$, and the modules $L^\lambda(M)$ give a set of representatives for the isomorphism classes of simple polynomial representations as $\lambda$ runs over the set of partitions with at most $n$ parts.

If $k$ has characteristic zero or characteristic $p > d$ then there is another description of $S^\lambda(M)$ as a direct summand of $M^{\otimes d}$. In this case, the group algebra of the symmetric group $k\mathfrak{S}_d$ is semisimple, and its irreducible representations are indexed by partitions of $d$. If $M$ is a representation of a group $G$ then $M^{\otimes d}$ is a representation of $G \times \mathfrak{S}_d$, where $\mathfrak{S}_d$ permutes the tensor factors. So if $e_\lambda \in k\mathfrak{S}_d$ is a primitive idempotent corresponding to a partition $\lambda$ of $d$ then

$$S^\lambda(M) \cong e_\lambda . M^{\otimes d}.$$

All choices of $e_\lambda$ are conjugate in $k\mathfrak{S}_d$, so the right-hand side is independent of this choice up to isomorphism.

In this semisimple case, the Schur functors are simple functors. On the other hand, if $p \le d$ then the Schur functor is often not simple. For example, even the $p$th symmetric power functor $S^p$ has a subfunctor $F^p$ (Frobenius $p$th power functor) where $F^p(M)$ consists of the $p$th powers of elements of $M$. In general $S^\lambda$ has a unique simple subfunctor. It is denoted $L^\lambda$. Every simple polynomial functor is isomorphic to a unique $L^\lambda$. The functors $L^\lambda$ will appear again in Section 1.17.

The relationship between Schur functors and Schur functions is as follows. If we choose a basis $v_1, \dots, v_n$ of $M$ consisting of eigenvectors for a semisimple element $g \in G$ with eigenvalues $x_1, \dots, x_n$ then the eigenvalues of $g$ on $S^\lambda(M)$ are precisely the monomials appearing in $s_\lambda(x_1, \dots, x_n)$. Thus the Schur function computes the character of the Schur functor.

**Theorem 1.16.4** *Suppose that $\lambda$ is a partition of $d$ and the characteristic of $k$ is either zero or $p > d$. Then we have a determinantal form*

$$S^\lambda(M) = \det(S^{\lambda_i - i + j}(M))_{1 \le i,j \le r}$$

*($r$ at least the number of parts of $\lambda$).* □

For example we have

$$S^{[2][1]}(M) = \begin{vmatrix} S^2(M) & S^3(M) \\ S^0(M) & S^1(M) \end{vmatrix}.$$

Since $S^0(M) = k$ and $S^1(M) = M$, the meaning of this equation is that

$$S^2(M) \otimes M \cong S^{[2][1]}(M) \oplus S^3(M).$$

## 1.17 Radical Layers of $kE$

In this section we describe the radical layers of $kE$ in terms of Schur functors. Recall from Section 1.3 that $kE$ is a truncated polynomial ring

$$kE = k[X_1, \dots, X_r]/(X_1^p, \dots, X_r^p).$$

The $n$th radical layer $J^n(kE)/J^{n+1}(kE)$ is isomorphic to the image of the space of homogeneous polynomials of degree $n$ in $X_1, \dots, X_r$ in this quotient ring. It has a basis consisting of monomials in which each $X_i$ appears to at most the $(p-1)$st power. Thus we have

$$J^{(p-1)r+1}(kE) = 0.$$

We let $GL(r, k)$ act on $kE$ by linear substitutions on the linear span $V$ of $X_1, \dots, X_r$. Hence $GL(r, k)$ acts on the radical layers $J^n(kE)/J^{n+1}(kE)$ for $0 \le n \le (p-1)r$, making it a quotient of the $n$th symmetric power $S^n(V)$:

$$J^n(kE)/J^{n+1}(kE) \cong S^n(V)/S^n(V) \cap (X_1^p, \dots, X_r^p).$$

Thus for $n \ge p$ there is an exact sequence

$$F^p(V) \otimes S^{n-p}(V) \to S^n(V) \to J^n(kE)/J^{n+1}(kE) \to 0$$

where $F^p(V)$ is the Frobenius $p$th power of $V$, spanned by the monomials $X_1^p, \dots, X_r^p$.

The submodule structure of $S^n(V)$ was determined by Doty [105]. It follows from the calculations there that $J^n(kE)/J^{n+1}(kE)$ is a simple $GL(r, k)$-module for $0 \le n \le (p-1)r$. The corresponding partition is given as follows. Let $n = (p-1)m + s$ with $0 \le s \le p-1$ and let $L^\lambda(V)$ be the simple $GL(r, k)$-module corresponding to a partition $\lambda$. Thus $L^\lambda$ is the simple functor discussed in Section 1.16. Then

$$J^n(kE)/J^{n+1}(kE) \cong L^{[p-1]^m[s]}(V).$$

See Section 9.3 for how this structure is reflected in the vector bundles coming from the regular representation of $kE$.

For example, the radical layers of $k(\mathbb{Z}/3)^3$ are

$$k,\quad V,\quad S^2(V),\quad L^{[2][1]}(V),\quad L^{[2]^2}(V),\quad L^{[2]^2[1]}(V),\quad L^{[2]^3}(V)$$

of dimensions, 1, 3, 6, 7, 6, 3, 1. Note that for $SL(3,k)$ we have $L^{[2]^2}(V) \cong S^2(V^*)$, $L^{[2]^2[1]}(V) \cong V^*$ and $L^{[2]^3}$ is the trivial representation. For $GL(3,k)$ one has to tensor with the determinant representation.

More generally for any $p$ and $r$, we have

$$L^{[p-1]^m[s]}(V) \cong L^{[p-1]^{r-m-1}[p-s]}(V^*) \otimes \det.$$

**Theorem 1.17.1** *If $n < p$ then the ideal of $kE$ generated by the elements $X_\alpha^n$, $\alpha \in \mathbb{A}^r$, is equal to $J^n(kE)$.*

*Proof.* We offer two proofs: one is at the level of elements and the other is at the level of representations.

The first proof goes as follows. Let $I$ be the ideal generated by the $X_\alpha^n$. Clearly $I \subseteq J^n(kE)$. To prove that $J^n(kE) \subseteq I$, we will show that every monomial of degree $n$ in $X_1, \ldots, X_r$ is in $I$.

We begin with the case $r = 2$, as this is easiest. For $\lambda \in k$ we have

$$(X_1 + \lambda X_2)^n = X_1^n + n\lambda X_1^{n-1}X_2 + \binom{n}{2}\lambda^2 X_1^{n-2}X_2^2 + \cdots + \lambda^n X_2^n \in I.$$

Choose $n+1$ distinct values $a_1, \ldots, a_{n+1}$ for $\lambda$. Since the matrix

$$A = \begin{pmatrix} 1 & na_1 & \binom{n}{2}a_1^2 & \cdots & a_1^n \\ 1 & na_2 & \binom{n}{2}a_2^2 & \cdots & a_2^n \\ \vdots & & & & \vdots \\ 1 & na_{n+1} & \binom{n}{2}a_{n+1}^2 & \cdots & a_{n+1}^n \end{pmatrix}$$

is obtained by multiplying the columns of a Vandermonde matrix by binomial coefficients, which are non-zero because $n < p$, we have

$$\det A = \prod_{i=0}^{n}\binom{n}{i} \cdot \prod_{1 \le i < j \le n+1} (a_j - a_i) \neq 0.$$

It follows that $A$ is invertible, and so each monomial $X_1^n, X_1^{n-1}X_2, \ldots, X_2^n$ is in $I$.

For larger values of $r$, the argument is similar except that the matrix is more complicated. For $\lambda_2, \ldots, \lambda_r \in k$ we have

$$(X_1 + \lambda_2 X_2 + \cdots + \lambda_r X_r)^n = \sum_{\substack{0 \le i_2 \le n \\ \cdots \\ 0 \le i_r \le n}} \binom{n}{i_2 \ldots i_r} \lambda_2^{i_2} \ldots \lambda_r^{i_r} X_1^{n-i_2-\cdots-i_r} X_2^{i_2} \ldots X_r^{i_r}.$$

This sum has $(n+1)^{r-1}$ terms, but the ones with $i_2+\cdots+i_r > n$ are zero. Choose $n+1$ distinct elements $a_1, \ldots, a_{n+1} \in k$ and consider this identity for each choice of values $a_1, \ldots, a_{n+1}$ for each of $\lambda_2, \ldots, \lambda_r$, giving a total of $(n+1)^{r-1}$ such choices.

Now consider the matrix $B$ obtained by taking a Kronecker tensor product of $r-1$ copies of the Vandermonde determinant

$$\begin{pmatrix} 1 & a_1 & \cdots & a_1^n \\ \vdots & & & \vdots \\ 1 & a_{n+1} & \cdots & a_{n+1}^n \end{pmatrix}.$$

Then

$$\det B = \left( \prod_{1 \le i < j \le n+1} (a_j - a_i) \right)^{r-1}$$

and so $B$ is non-singular. Deleting the columns where the total degree is larger than $n$ gives a rectangular matrix with full column rank. It follows that there exists a choice of rows giving a non-singular square matrix. Since $n < p$, multiplying the columns by the multinomial coefficients $\binom{n}{i_1 \ldots i_r}$ again gives a nonsingular matrix, and so each monomial of degree $n$ in $X_1, \ldots, X_r$ is in $I$.

The second proof using representation theory is much shorter. It consists of the observation that $J^n(kE)/J^{n+1}(kE)$ is irreducible as a module for $GL(r,k)$ (see above). The image in this space of the linear span of the $X_\alpha^n$ is a non-zero invariant subspace, and is hence the whole of $J^n(kE)/J^{n+1}(kE)$. Now use Nakayama's lemma. □

## 1.18 Twisted Versions of $kE$

Let $p$ be an odd prime, and let $k$ be a field of characteristic $p$. Consider the group $G = (\mathbb{Z}/p)^2 \rtimes Q_8$ where $Q_8 = \langle i, j, k \mid i^2 = j^2 = k^2 = ijk \rangle$ is the quaternion group of order eight, and where the action of $Q_8$ on $(\mathbb{Z}/p)^2$ is given by letting the central involution $z = i^2 = j^2 = k^2$ act trivially, and the quotient $Q_8/Z(Q_8) \cong (\mathbb{Z}/2)^2$ act by inverting the two generators of $(\mathbb{Z}/p)^2$. Thus the quotient $G/\langle z \rangle \cong D_{2p} \times D_{2p}$, a direct product of two copies of the dihedral group of order $2p$.

The principal block idempotent is $e_0 = \frac{1+z}{2}$, and we have $ekG \cong k(G/\langle z \rangle)$. The remaining block $e_1 kG$, where $e_1 = 1 - e_0 = \frac{1-z}{2}$ has just one simple module, which has dimension two. We have $e_1 kG \cong \mathrm{Mat}_2(C)$, where

$$C = k\langle X_1, X_2 \rangle / (X_1^p, X_2^p, X_1X_2 + X_2X_1).$$

The algebra $C$ can be regarded as a quantised version of $k(\mathbb{Z}/p)^2$, and the module categories of $C$ and $e_1 kG$ are equivalent.

Generalisations of the example above are studied in Benson and Green [50], Holloway and Kessar [158], Benson and Kessar [52]. The general setup is as follows. Let $k$ be an algebraically closed field of characteristic $p$. Let $\mathbf{q} = (q_{ij}) \in \mathrm{Mat}_r(k)$ be such that $q_{ij}q_{ji} = 1$ for $1 \le i, j \le r$, and $q_{ii} = 1$ for $1 \le i \le r$. Let $\mathbf{n} = (n_1, \dots, n_r)$ be an $r$-tuple of strictly positive integers. Let $C_{\mathbf{q},\mathbf{n}}$ (or just $C_{\mathbf{q}}$ if each $n_i = 1$) be the $k$-algebra with generators $X_1, \dots, X_r$ and relations

$$X_i X_j = q_{ij} X_j X_i \quad (1 \le i, j \le r), \qquad X_i^{p^{n_i}} = 0.$$

Such an algebra might reasonably be called a *quantum complete intersection*, cf. §2 of Avramov, Gasharov and Peeva [23]. The following statements are proved in [50, 158]:

(i) If $q_{ij}^{p-1} = 1$ for $1 \le i, j \le r$ then there exists a finite group $G$ with a normal Sylow subgroup $P \cong \mathbb{Z}/p^{n_1} \times \cdots \times \mathbb{Z}/p^{n_r}$, a block idempotent $e \in Z(kG)$ with defect group $P$, abelian inertial quotient and a single isomorphism class of simple modules, such that $ekG \cong \mathrm{Mat}_d(C_{\mathbf{q},\mathbf{n}})$. Here, $d$ is the dimension of the simple $ekG$-module.

(ii) If $G$ is a finite group and $e \in Z(kG)$ is a block idempotent with normal defect group isomorphic to $\mathbb{Z}/p^{n_1} \times \cdots \times \mathbb{Z}/p^{n_r}$, abelian inertial quotient, and a single isomorphism class of simple modules, then there exists a matrix $\mathbf{q} = (q_{ij}) \in \mathrm{Mat}_r(k)$ satisfying $q_{ij}q_{ji} = 1$ for $1 \le i, j \le r$ and $q_{ii} = 1$ for $1 \le i \le r$ such that $ekG \cong \mathrm{Mat}_d(C_{\mathbf{q},\mathbf{n}})$. Again, $d$ is the dimension of the simple $ekG$-module.

Notice that there is a gap between statements (i) and (ii): in statement (ii) it is not necessarily true that $q_{ij}^{p-1} = 1$.

The Ext algebra of the algebra $C_{\mathbf{q},\mathbf{n}}$ is also calculated in [158], under the assumption that each $p^{n_i} > 2$. The algebra $\mathrm{Ext}^*_{C_{\mathbf{q},\mathbf{n}}}(k, k)$ is the $k$-algebra with generators $y_1, \dots, y_r$ in degree one and $x_1, \dots, x_r$ in degree two[3] satisfying the following relations:

$$x_i x_j = q_{ji}^{p^{n_i+n_j}} x_j x_i, \quad y_i x_j = q_{ji}^{p^{n_j}} x_j y_i, \quad y_i y_j = -q_{ji} y_j y_i, \quad y_i^2 = 0.$$

We shall be particularly interested in the case where all the $n_i$ are equal to one, corresponding to the case where $P$ is elementary abelian. This case is further analysed in [52], with the following conclusion. In case the algebra $C_{\mathbf{q}}$ arises from a block $ekG$ as in (ii) above, with $P$ elementary abelian, there exists a permutation $\sigma$ of $\{1, \dots, r\}$ such that for all $1 \le i, j \le p$ we have $q_{\sigma(i),j} = q_{i,\sigma(j)} = q_{ij}^p$. In particular, it follows that $q_{\sigma(i),\sigma(j)} = q_{ij}^{p^2}$. Since $\sigma$ has to be of finite order, it follows that the $q_{ij}$ lie in some finite subfield $\mathbb{F}_{p^s}$ of $k$.

Many theorems about representation theory of elementary abelian $p$-groups generalise naturally to this context, but there are some differences, and in general

[3] Beware that despite what is written in [158], the algebra $\mathrm{Ext}^*_{C_{\mathbf{q},\mathbf{n}}}(k, k)$ is not graded commutative.

the representation theory of the algebras $C_{\mathbf{q}}$ discussed in this section is harder to analyse than that of elementary abelian $p$-groups.

Representations of $C_{\mathbf{q}}$ and of $kE$ are both examples of nilvarieties of exponent $p$ in the sense defined in Section 4.1. In particular, when we prove theorems about nilvarieties, we are also proving theorems about $C_{\mathbf{q}}$-modules.

# 2

# Cyclic Groups of Order $p$

## 2.1 Introduction

In this chapter, we discuss representations of $\mathbb{Z}/p$ in characteristic $p$. This will be applied when we examine tensor products, exterior and symmetric powers, and other Schur functors of modules of constant Jordan type. As a bonus, we also get information about the representations of $SL(2, \mathbb{C})$ in characteristic zero.

Tensor products, exterior powers and symmetric powers of representations of cyclic $p$-groups have been studied by many authors over the years, including Almkvist [2, 3, 4, 5], Almkvist and Fossum [6], Barry [30, 31], Bryant and Johnson [65], Carlson [71], Fossum [122, 123], Glasby, Praeger and Xia [135], Green [138], Hannula, Ralley and Reiner [143], Himstedt and Symonds [155], Hughes and Kemper [162], Kouwenhoven [179], McFall [192, 193], Nam and Oh [196, 197], Norman [199, 200], Ralley [211], Renaud [215, 216, 217, 218], Srinivasan [228], Symonds [233], Wehlau [244].

We restrict our attention to cyclic groups of order $p$. In Section 2.9 we give a recipe for computing Schur functors $S^\lambda$ on indecomposable modules in this case, where $\lambda$ is a partition of $n$ and where $n < p$. The case of exterior and symmetric powers was addressed by Almkvist and Fossum [6], in terms of Gaussian polynomials $g_{n,d}(q)$. The formula for a more general Schur functor $S^\lambda$ generalises theirs, and involves generalised Gaussian polynomials $g_{n,\lambda}(q)$. The computation of these polynomials is given by a hook formula. The same formula computes the Schur functors on the representations of $SL(2, \mathbb{C})$ in characteristic zero.

## 2.2 Modules for $\mathbb{Z}/p$

If $E \cong \mathbb{Z}/p = \langle g \rangle$ is a cyclic group of order $p$, set $X = g - 1$ so that $kE = k[X]/(X^p)$. If $M$ is an indecomposable $kE$-module then the action of $X$ on $M$ is a single Jordan block of length between 1 and $p$ with eigenvalue zero. We write $J_i$ $(1 \leq i \leq p)$ for the indecomposable module of dimension $i$, and by abuse of notation we also write $J_i$ for a Jordan block of length $i$ with eigenvalue zero. So a module is projective (or equivalently injective or flat) if and only if it is isomorphic to a direct sum of copies of $J_p$.

Let us begin by discussing subquotients of modules for a cyclic group of order $p$. Consider the forgetful functor from $kE$-modules to vector spaces. We can define functors $F_{i,j}$ for $0 \le j < i \le p$ by taking the following subquotients of the forgetful functor:

$$F_{i,j}(M) = \frac{\mathsf{Ker}\,(X^{j+1}) \cap \mathsf{Im}\,(X^{i-j-1})}{(\mathsf{Ker}\,(X^{j+1}) \cap \mathsf{Im}\,(X^{i-j})) + (\mathsf{Ker}\,(X^{j}) \cap \mathsf{Im}\,(X^{i-j-1}))}.$$

Using the second isomorphism theorem, we can see that $M$ has a filtration where the $F_{i,j}(M)$ are the filtered subquotients. In fact, each $F_{i,j}$ is a simple functor (it has no non-trivial proper subfunctors) and the isomorphism types of these simple functors only depends on $i$ and not on $j$. The effect of the functor $F_{i,j}$ is to pick out the $(j+1)$st socle layers of all the Jordan blocks of length $i$.

The lattice of subfunctors of the forgetful functor is distributive, and the poset of join irreducible elements in this lattice looks as follows, with each one labelled by the corresponding composition factor of the forgetful functor.

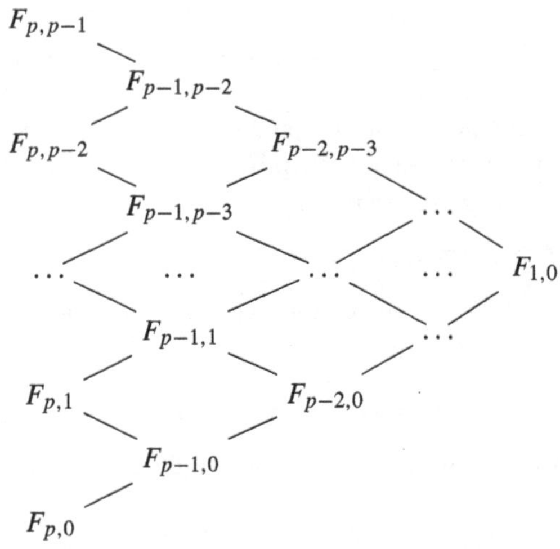


In particular, the composition length of the forgetful functor is $p(p+1)/2$. We shall see this diagram again when we describe certain vector bundles coming from modules of constant Jordan type in Section 8.2.

## 2.3 Tensor Products

In this section we shall compute the tensor products of the modules $J_i$ for $\mathbb{Z}/p$. We shall see that the group theoretic tensor product $J_i \otimes_k J_j$ is always isomorphic to the restricted Lie tensor product $J_i \tilde{\otimes}_k J_j$.

**Lemma 2.3.1** *If* $1 \le i < p$ *then*

$$J_2 \otimes_k J_i \cong J_2 \tilde{\otimes}_k J_i \cong J_{i-1} \oplus J_{i+1}.$$

*The case $i = p$ is given by the formula*

$$J_2 \otimes_k J_p \cong J_2 \tilde{\otimes}_k J_p \cong J_p \oplus J_p.$$

*Proof.* This is straightforward, by computing the ranks of powers of $X$ on these tensor products. In the case of $J_2 \otimes_k J_i$ the matrix representing $X$ is $\begin{pmatrix} J_i & I + J_i \\ 0 & J_i \end{pmatrix}$, and so the matrix representing $X^n$ is $\begin{pmatrix} J_i^n & n\left(J_i^{n-1} + J_i^n\right) \\ 0 & J_i^n \end{pmatrix}$. Thus we have

$$\operatorname{rank}(X^n) = \begin{cases} 2i - 2n & n < i \\ 1 & n = i < p \\ 0 & n = i = p. \end{cases}$$

In the case of $J_2 \tilde{\otimes}_k J_i$ the matrix representing $X$ is $\begin{pmatrix} J_i & I \\ 0 & J_i \end{pmatrix}$, and so the matrix representing $X^n$ is $\begin{pmatrix} J_i^n & nJ_i^{n-1} \\ 0 & J_i^n \end{pmatrix}$. The rank of $X^n$ is the same as before. □

It is possible to compute all tensor products of $kE$-modules inductively from the lemma, using associativity and the Krull–Schmidt theorem. For example, if we want to compute $J_3 \otimes_k J_i$ we use the identity

$$(J_3 \otimes_k J_i) \oplus J_i \cong (J_3 \oplus J_1) \otimes_k J_i \cong (J_2 \otimes_k J_2) \otimes_k J_i \cong J_2 \otimes_k (J_2 \otimes_k J_i).$$

The same holds for $J_3 \tilde{\otimes}_k J_i$, and so we have $J_3 \otimes_k J_i \cong J_3 \tilde{\otimes}_k J_i$. For example, provided $i < p - 1$ this gives the formula

$$J_3 \otimes_k J_i \cong J_3 \tilde{\otimes}_k J_i \cong J_{i-2} \oplus J_i \oplus J_{i+2}.$$

Continuing this way, an easy inductive argument gives the following theorem.

**Theorem 2.3.2** *If $M$ and $N$ are $kE$-modules with $E$ cyclic of order $p$ then*

$$M \otimes_k N \cong M \tilde{\otimes}_k N.$$

*Proof.* By the Krull–Schmidt theorem, it suffices to prove this with $M$ and $N$ indecomposable. In this case it follows inductively from Lemma 2.3.1 and the associativity of the tensor product $J_2 \otimes_k J_i \otimes_k J_j$. □

In fact, it is not hard to write down an explicit formula for $J_i \otimes_k J_j$. The following formulas are special cases of Feit [119, Theorem VIII.2.7]. If $i \le j$ and $i + j \le p$ we have

$$J_i \otimes_k J_j \cong J_i \tilde{\otimes}_k J_j \cong \bigoplus_{s=1}^{i} J_{j-i+2s-1}$$

and

$$J_{p-i} \otimes_k J_{p-j} \cong J_{p-i} \tilde{\otimes}_k J_{p-j} \cong (p - i - j)J_p \oplus (J_i \otimes_k J_j).$$

Using the commutativity of tensor product, all cases are covered by these two statements, since either $i + j \le p$ or $(p - i) + (p - j) \le p$.

Note that for $1 \le \ell \le p-1$ the number of copies of $J_\ell$ as a summand of a tensor product of Jordan blocks is at most one, and there is such a summand precisely in the cases

$$J_{i+j} \otimes_k J_{i+\ell-1-j}, \qquad 1 \le i \le p - \ell, \qquad 0 \le j \le \ell - 1. \tag{2.1}$$

Tables of tensor products of modules for $\mathbb{Z}/p$ can be found in Appendix A. For cyclic groups of order a power of a prime, Green [138] has described all tensor products of indecomposable modules, but the formulas are not quite so direct.

## 2.4 Gaussian Polynomials

In preparation for a discussion of symmetric and exterior powers of modules for $\mathbb{Z}/p$, we introduce the Gaussian polynomials.

**Definition 2.4.1** The (homogeneous) *Gaussian polynomial* $G_{n,d}(X, Y)$ $(n \ge r \ge 0)$ is defined to be

$$G_{n,d}(X, Y) = \frac{(X^n - Y^n)(X^{n-1} - Y^{n-1}) \dots (X^{n-d+1} - Y^{n-d+1})}{(X^d - Y^d)(X^{d-1} - Y^{d-1}) \dots (X - Y)}.$$

We shall see in Corollary 2.4.3 that this is indeed a polynomial in $X$ and $Y$. The more usual (inhomogeneous) Gaussian polynomial is $G_{n,d}(X, 1)$.

An easy calculation gives the following properties of the Gaussian polynomials.

**Proposition 2.4.2** *We have*

$$G_{n,d}(X, Y) = G_{n,d}(Y, X) = G_{n,n-d}(X, Y) = X^{n-d} G_{n-1,d-1}(X, Y) + Y^d G_{n-1,d}(X, Y).$$ □

**Corollary 2.4.3** $G_{n,d}(X, Y) \in \mathbb{Z}[X, Y]$ *is a polynomial of degree* $d(n - d)$.

*Proof.* This follows from the recurrence relation in Proposition 2.4.2 by induction. □

The relationship between the Gaussian polynomials and the complete and elementary symmetric functions is given in the following proposition.

**Proposition 2.4.4** *We have*

(1) $e_d(X^{n-1}, X^{n-2}Y, \dots, XY^{n-2}, Y^{n-1}) = (XY)^{\binom{d}{2}} G_{n,d}(X, Y)$,
(2) $h_d(X^n, X^{n-1}Y, \dots, Y^n) = G_{n+d,d}(X, Y)$.

*Proof.* From Proposition 1.15.4 we have

$$\begin{aligned} e_d(X^{n-1}, X^{n-2}Y, \dots, Y^{n-1}) &= X^d e_d(X^{n-2}, X^{n-3}Y, \dots, Y^{n-2}) \\ &\quad + X^{d-1}Y^{n-1}e_{d-1}(X^{n-2}, X^{n-3}Y, \dots, Y^{n-2}) \end{aligned}$$

and

$$\begin{aligned} h_d(X^n, X^{n-1}Y, \dots, Y^n) &= X^d h_d(X^{n-1}, X^{n-2}Y, \dots, Y^{n-1}) \\ &\quad + Y^n h_{d-1}(X^n, X^{n-1}Y, \dots, Y^n). \end{aligned}$$

The proposition now follows using the recurrence formula in Proposition 2.4.2 and induction on $n$. □

**Definition 2.4.5** We define

$$g_{n,d}(q) = G_{n,d}(q, q^{-1}) = q^{-d(n-d)}G_{n,d}(q^2, 1) \in \mathbb{Z}[q, q^{-1}].$$

**Example 2.4.6** We have

$$\begin{aligned} g_{7,3}(q) &= \frac{(q^7 - q^{-7})(q^6 - q^{-6})(q^5 - q^{-5})}{(q^3 - q^{-3})(q^2 - q^{-2})(q - q^{-1})} \\ &= q^{12} + q^{10} + 2q^8 + 3q^6 + 4q^4 + 4q^2 + 5 \\ &\quad + 4q^{-2} + 4q^{-4} + 3q^{-6} + 2q^{-8} + q^{-10} + q^{-12}. \end{aligned} \tag{2.2}$$

**Proposition 2.4.7** *We have*

(1) $e_d(q^{n-1}, q^{n-3}, \dots, q^{-n+1}) = g_{n,d}(q)$,
(2) $h_d(q^n, q^{n-2}, \dots, q^{-n}) = g_{n+d,d}(q)$.

*Proof.* This follows directly from Proposition 2.4.4. □

## 2.5 Generalised Gaussian Polynomials and a Hook Formula

In this section we introduce the generalised Gaussian polynomials $g_{n,\lambda}(q)$ and prove a hook formula for calculating them. This section is taken from Example 1 in Section I.3 of Macdonald [189], and will be needed in in Section 2.9.

**Definition 2.5.1** Let $\lambda$ be a partition of $d$. We define the *generalised Gaussian polynomial* for $\lambda$ to be

$$g_{n,\lambda}(q) = s_\lambda(q^{n-1}, q^{n-3}, \dots, q^{-n+1})$$

where $s_\lambda$ is the Schur function, see Definition 1.15.3.

**Example 2.5.2** By Proposition 2.4.7, we have $g_{n,[1]^d}(q) = g_{n,d}(q)$ and $g_{n+1,[d]}(q) = g_{n+d,d}(q)$.

The following is the transpose of statement (1.7) in Chapter I of [189].

**Lemma 2.5.3** *Let $\lambda$ be a partition with conjugate partition $\lambda'$, and let $m \geq \lambda_1$, $n \geq \lambda'_1$. Then the $m+n$ numbers $\lambda'_i + m - i$ $(1 \leq i \leq m)$ and $m - \lambda_j - 1 + j$ $(1 \leq j \leq n)$ are a permutation of $\{0, 1, \ldots, m+n-1\}$.*

*Proof.* The diagram for $\lambda$ is contained in an $n \times m$ rectangle. Number the edges of the boundary between $\lambda$ and its complement as in the diagram below, starting with zero at the top right and ending with $m+n-1$ at the bottom left.

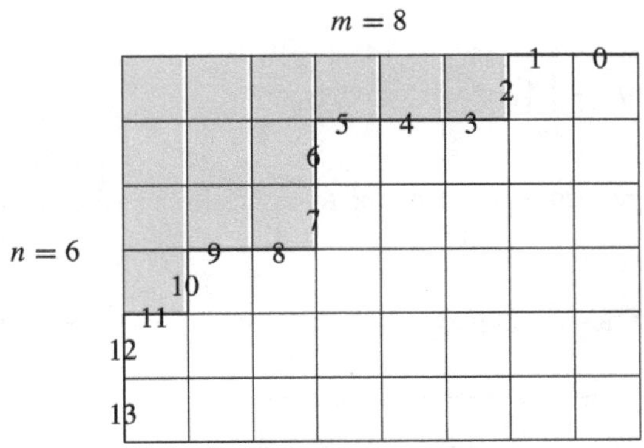


$\lambda = [6]\,[3]^2[1]$

$\lambda' = [4]\,[3]^2[1]^3$

The numbers on the horizontal edges are $\lambda'_i + m - i$ $(1 \leq i \leq m)$, while those on the vertical edges are $(m+n-1)-(\lambda_j+n-j) = m-\lambda_j-1+j$ $(1 \leq j \leq n)$. □

**Definition 2.5.4** The *hook length* of a partition $\lambda$ at a box $x = (i, j) \in \lambda$ is defined to be

$$h(x) = \lambda_i + \lambda'_j - i - j + 1.$$

This is the number of boxes directly below and directly to the right of $x$, including $x$ itself.

**Lemma 2.5.5** *Let $\lambda$ be a partition with at most n parts. Then we have*

$$\prod_{x\in\lambda}(q^{h(x)}-q^{-h(x)}) \prod_{1\leq i<j\leq n}(q^{\lambda_i-\lambda_j-i+j}-q^{-(\lambda_i-\lambda_j-i+j)}) = \prod_{i=1}^{n}\prod_{j=1}^{\lambda_i+n-i}(q^j-q^{-j}).$$

*Proof.* Let $m = \lambda_1$, and observe that $n \geq \lambda'_1$. It follows from Lemma 2.5.3 that

$$\prod_{i=1}^{m}(q^{\lambda'_i+m-i}-q^{-(\lambda'_i+m-i)}) \prod_{j=2}^{n}(q^{m-\lambda_j-1+j}-q^{-(m-\lambda_j-1+j)}) = \prod_{j=1}^{m+n-1}(q^j-q^{-j})$$

because the set of terms on the left is a permutation of the set of terms on the right. Note that we have omitted the term $(q^0-q^{-0})$ from both sides for obvious reasons, and that this explains why the second product on the left side omits the term $j=1$ and the product on the right side omits the term $j=0$.

The exponents $\lambda'_i + m - i$ in the first product on the left are the hook lengths along the first row of $\lambda$. Writing down this identity for the partitions $(\lambda_i, \lambda_{i+1}, \ldots, \lambda_n)$ for $1 \le i \le n$ and multiplying, we obtain the identity in the statement of the lemma. □

**Definition 2.5.6** The *content* of the box $x = (i, j)$ is defined to be $c(x) = j - i$.

**Proposition 2.5.7** *Let $\lambda$ be a partition with at most $n$ parts. Then we have the following hook formula for $g_{n,\lambda}(q)$:*

$$g_{n,\lambda}(q) = \prod_{x \in \lambda} \frac{q^{n+c(x)} - q^{-(n+c(x))}}{q^{h(x)} - q^{-h(x)}}.$$

**Remark 2.5.8** If $\lambda$ has more than $n$ parts then the formula still holds, but it says that zero is equal to zero.

**Example 2.5.9** For the partition [3][2] we have the following values for $n + c(x)$ and $h(x)$:

| $n$ | $n+1$ | $n+2$ |
|---|---|---|
| $n-1$ | $n$ | |

| 4 | 3 | 1 |
|---|---|---|
| 2 | 1 | |

and so

$$g_{n,[3][2]}(q) = \frac{(q^{n+2} - q^{-n-2})(q^{n+1} - q^{-n-1})(q^n - q^{-n})^2(q^{n-1} - q^{-n+1})}{(q^4 - q^{-4})(q^3 - q^{-3})(q^2 - q^{-2})(q - q^{-1})^2} \in \mathbb{Z}[q, q^{-1}].$$

*Proof of Proposition 2.5.7.* By Definitions 1.15.3 and 2.5.1, we have

$$\begin{aligned} g_{n,\lambda}(q) &= s_\lambda(q^{n-1}, q^{n-3}, \ldots, q^{-n+1}) \\ &= \frac{\det(q^{(n+1-2i)(\lambda_j+n-j)})_{1\le i,j\le n}}{\det(q^{(n+1-2i)(n-j)})_{1\le i,j\le n}}. \end{aligned}$$

The top of this fraction is some power of $q$ times a Vandermonde determinant in the variables $q^{2(\lambda_j+n-j)}$, and the bottom is some power of $q$ times the Vandermonde determinant in the variables $q^{2(n-j)}$. Using the usual formula for a Vandermonde determinant, we find that $g_{n,\lambda}(q)$ is some power of $q$ times

$$\frac{\prod_{1\le i<j\le n}(q^{2(\lambda_i+n-i)} - q^{2(\lambda_j+n-j)})}{\prod_{1\le i<j\le n}(q^{2(n-i)} - q^{2(n-j)})}.$$

We can either keep track directly of the power of $q$ in front of this expression, or we can just use the fact that the result has to be invariant under swapping $q$ and $q^{-1}$. Either way, we get

$$g_{n,\lambda}(q) = \frac{\prod_{1\le i<j\le n}(q^{\lambda_i-\lambda_j-i+j} - q^{-(\lambda_i-\lambda_j-i+j)})}{\prod_{1\le i<j\le n}(q^{j-i} - q^{-(j-i)})}.$$

The numerator on the right-hand side of this equation is equal to the second product on the left-hand side of Lemma 2.5.5, while the denominator is equal to $\prod_{i=1}^{n}\prod_{j=1}^{n-i}(q^j - q^{-j})$. Making these substitutions, we get

$$\begin{aligned} g_{n,\lambda}(q) &= \frac{\prod_{i=1}^{n}\prod_{j=1}^{\lambda_i+n-i}(q^j - q^{-j})}{\prod_{i=1}^{n}\prod_{j=1}^{n-i}(q^j - q^{-j})\prod_{x\in\lambda}(q^{h(x)} - q^{-h(x)})} \\ &= \frac{\prod_{i=1}^{n}\prod_{j=n-i+1}^{\lambda_i+n-i}(q^j - q^{-j})}{\prod_{x\in\lambda}(q^{h(x)} - q^{-h(x)})}. \end{aligned}$$

Finally, we observe that the numbers $n + c(x)$ for $x$ in the $i$th row of $\lambda$ are $n - i + 1, \dots, n - i + \lambda_i$. So the numerator in the right-hand side of the above expression for $g_{n,\lambda}(q)$ is equal to

$$\prod_{x\in\lambda}(q^{n+c(x)} - q^{-(n+c(x))}). \qquad \square$$

## 2.6 λ-Rings and Representations of $SL(2, \mathbb{C})$

It is convenient to introduce the notion of a λ-ring in order to make further computations with symmetric and exterior powers. For more information on λ-rings, see Atiyah and Tall [18].

**Definition 2.6.1** A *λ-ring* is a ring $R$ together with operations $\lambda^n : R \to R$ $(n \ge 0)$ such that for all $x, y \in R$ we have

(1) $\lambda^0(x) = 1$,
(2) $\lambda^1(x) = x$,
(3) $\lambda^n(x + y) = \sum_{i+j=n} \lambda^i(x)\lambda^j(y)$.

**Example 2.6.2** Lemma 1.14.3 shows that exterior powers on representation rings satisfy these properties.

In a λ-ring it is often convenient to introduce an indeterminate $t$ and define

$$\lambda_t : R \to R[[t]], \qquad \lambda_t(x) = \sum_{n\ge 0} \lambda^n(x)t^n.$$

In terms of this definition, axiom (3) states that

$$\lambda_t(x + y) = \lambda_t(x)\lambda_t(y).$$

Since we are working with representations in characteristic $p$, the exterior power operations $\Lambda^i$ are better behaved for $i < p$ than for $i \ge p$. We therefore make the following definition.

**Definition 2.6.3** A *p-λ-ring* is a ring $R$ together with operations $\lambda^n$ for $0 \le n < p$ satisfying axioms (1), (2) and (3).

In a $p$-λ-ring, we truncate $\lambda_t$ in the obvious way:

$$\lambda_t \colon R \to R[t]/(t^p), \qquad \lambda_t(x) = \sum_{0 \le n < p} \lambda^n(x) t^n.$$

A *special* λ-ring is a λ-ring with further axioms detailing the behaviour of the λ-operations on products, and the behaviour of compositions of λ-operations. We shall need a modification of this concept to deal with the fact that we are working with representations in characteristic $p$.

We begin by defining certain polynomials involved in the definitions. Let $x_1, \dots, x_s$ and $y_1, \dots, y_s$ be indeterminates, and let $e_i$ and $e'_i$ be the elementary symmetric functions in $x_1, \dots, x_s$, respectively $y_1, \dots, y_s$. Let

$$P_n(e_1, \dots, e_n; e'_1, \dots, e'_n)$$

be the coefficient of $t^n$ in $\prod_{j,k=1}^{s}(1 + x_j y_k t)$. As long as $s \ge n$, these polynomials are independent of $s$ and have integer coefficients. Similarly, we let

$$P_{m,n}(e_1, \dots, e_{mn})$$

be the coefficient of $t^m$ in $\prod_{i_1 < \cdots < i_n}(1 + x_{i_1} \dots x_{i_n} t)$. As long as $s \ge mn$, these polynomials are independent of $s$ and have integer coefficients.

**Definition 2.6.4** A *special λ-ring* is a λ-ring satisfying the following further axioms:

(4) $\lambda_t(xy) = P_n(\lambda^1(x), \dots, \lambda^n(x); \lambda_1(y), \dots, \lambda^n(y))$,
(5) $\lambda^m(\lambda^n(x)) = P_{m,n}(\lambda^1(x), \dots, \lambda^{mn}(x))$.

Note that the case $n = 0$ of (5) implies that $\lambda_t(1) = 1 + t$.

A *special p-λ-ring* is a $p$-λ-ring which satisfies axiom (4) for $n < p$ and axiom (5) for $mn < p$.

**Example 2.6.5** Using the exterior powers as the λ-operations, representation rings of characteristic zero representations are special λ-rings, while representation rings of characteristic $p$ representations are special $p$-λ-rings.

**Example 2.6.6** Let $\mathbb{T}$ be the 1-dimensional torus, or circle group $\{z \in \mathbb{C} \mid |z| = 1\}$. Every finite-dimensional continuous representation of $\mathbb{T}$ is a direct sum of representations of dimension one. There is a representation of dimension one for each integer $n$, namely $z \mapsto z^n$. Tensoring such representations adds the corresponding integer, and so the representation ring is the ring of Laurent polynomials $\mathbb{Z}[q, q^{-1}]$. The λ-ring structure is given by $\lambda_t(q^n) = 1 + q^n t$, and this makes $\mathbb{Z}[q, q^{-1}]$ into a special λ-ring.

**Example 2.6.7** Next, consider the group $SL(2,\mathbb{C})$ of $2 \times 2$ complex matrices of determinant one. Every finite-dimensional continuous representation is a direct sum of irreducible representations. There is one irreducible representation $V_n$ for each integer $n > 0$, and $V_n$ has dimension $n$. The restriction of $V_n$ to

$$\mathbb{T} = \left\{ \left(\begin{smallmatrix} z & 0 \\ 0 & z^{-1} \end{smallmatrix}\right) \mid |z| = 1 \right\} \subseteq SL(2,\mathbb{C})$$

corresponds to the Laurent polynomial $q^{n-1} + q^{n-3} + \cdots + q^{-n+1}$. This identifies the representation ring of $SL(2,\mathbb{C})$ with the fixed points of $\mathbb{Z}[q,q^{-1}]$ under the involution which swaps $q$ and $q^{-1}$. The special $\lambda$-ring structure is inherited from that of $\mathbb{Z}[q,q^{-1}]$.

**Proposition 2.6.8** *The representation $\Lambda^d(V_n)$ of $SL(2,\mathbb{C})$ corresponds to the element $g_{n,d}(q) \in \mathbb{Z}[q,q^{-1}]$ while $S^d(V_{n+1})$ corresponds to the element $g_{n+d,d}(q)$.*

*Proof.* This follows from Proposition 2.4.4. □

**Corollary 2.6.9** *As representations of $SL(2,\mathbb{C})$ we have*

$$\Lambda^d(V_{n+d}) \cong S^d(V_{n+1}) \cong S^n(V_{d+1}).$$ □

**Definition 2.6.10** A *$\lambda$-ideal* in a $\lambda$-ring $R$ is an ideal which is closed under the operations $\lambda^i$. A *$p$-$\lambda$-ideal* in a $p$-$\lambda$-ring is an ideal which is closed under the operations $\lambda^i$ for $i < p$.

**Lemma 2.6.11** *If $R$ is a special $\lambda$-ring (resp. special $p$-$\lambda$-ring) then an ideal $I$ is a $\lambda$-ideal (resp. a $p$-$\lambda$-ideal) if and only if it is generated by a set of elements whose images under $\lambda^n$ are in $I$ for all $n \geq 0$ (resp. for $0 \leq n < p$).*

*Proof.* This is proved in Lemma 6.4 of [18] for special $\lambda$-rings, and the proof for special $p$-$\lambda$-rings is identical. □

The following lemma is a version of the *splitting principle* for $\lambda$-rings.

**Lemma 2.6.12** *Let $x$ be an element of a special $\lambda$-ring (resp. special $p$-$\lambda$-ring) satisfying $\lambda^i(x) = 0$ for $i > n$ (resp. for $n < i < p$). Then $R[\xi]/(\xi^n - \lambda^1(x)\xi^{n-1} + \cdots + (-1)^n\lambda^n(x))$ with $\lambda_t(\xi) = 1 + \xi t$ is a special $\lambda$-ring (resp. special $p$-$\lambda$-ring).*

*Proof.* This is proved in Lemma 6.5 of [18] for special $\lambda$-rings, and the proof for special $p$-$\lambda$-rings is identical. □

## 2.7 The Representation Ring of $\mathbb{Z}/p$

Following Almkvist and Fossum [6], we examine the representation ring $\mathfrak{R}$ of the cyclic group of order $p$ in characteristic $p$. This has a basis consisting of the representations $J_n$ $(1 \le n \le p)$, and the multiplication is determined by Lemma 2.3.1. Using the exterior powers as the $\lambda$-operations, $\mathfrak{R}$ is a special $p$-$\lambda$-ring.

Let us apply Lemma 2.6.12 to the element $J_2$ of $\mathfrak{R}$. Since $\lambda_t(J_2) = 1 + J_2 t + t^2$, this gives us a ring

$$\hat{\mathfrak{R}} = \mathfrak{R}[q]/(q^2 - J_2 q + 1).$$

Rewriting the relation $q^2 - J_2 q + 1 = 0$, we obtain $J_2 = q + q^{-1}$.

**Lemma 2.7.1** *In $\hat{\mathfrak{R}}$, $J_n$ corresponds to the element*

$$q^{n-1} + q^{n-3} + \cdots + q^{-n+1}$$

*for* $1 \le n \le p$.

*Proof.* By Lemma 2.3.1, if $1 \le n < p$, we have the relation $J_2 \otimes_k J_n \cong J_{n-1} + J_{n+1}$. We also have

$$(q + q^{-1})(q^{n-1} + q^{n-3} + \cdots + q^{-n+1}) = (q^n + q^{n-2} + \cdots + q^{-n}) + (q^{n-2} + q^{n-4} + \cdots + q^{-n+2}).$$

Combining these proves the lemma by induction on $n$. □

**Proposition 2.7.2** *We have* $\hat{\mathfrak{R}} = \mathbb{Z}[q, q^{-1}]/(q-1)(q^{-1}-1)(q^{p-1}+q^{p-3}+\cdots+q^{-p+1})$. *This is a special $p$-$\lambda$-ring with $\lambda$-operations given by* $\lambda_t(q^n) = 1 + q^n t$.

*Proof.* By Lemma 2.7.1, the representation $J_p$ corresponds to the element $q^{p-1} + q^{p-3} + \cdots + q^{-p+1}$ of $\hat{\mathfrak{R}}$. By Lemma 2.3.1 we have

$$J_2 \otimes J_p \cong J_p \oplus J_p.$$

This gives the relation

$$(q + q^{-1})(q^{p-1} + q^{p-3} + \cdots + q^{-p+1}) = 2(q^{p-1} + q^{p-3} + \cdots + q^{-p+1})$$

or equivalently

$$(q - 1)(q^{-1} - 1)(q^{p-1} + q^{p-3} + \cdots + q^{-p+1}) = 0.$$

The quotient of $\mathbb{Z}[q, q^{-1}]$ by this relation is a free $\mathbb{Z}$-module of rank $2p$ with basis

$$q^p, q^{p-1}, \dots, q^{-p+1}.$$

Comparing $\mathbb{Z}$-ranks, this quotient must equal $\hat{\mathfrak{R}}$.

By Lemma 2.6.12, we have $\lambda_t(q) = 1 + qt$. Repeatedly applying axiom (4) then gives $\lambda_t(q^n) = 1 + q^n t$. □

**Remark 2.7.3** Proposition 2.7.2 allows us to identify $\mathfrak{R}$ with the fixed points of $\hat{\mathfrak{R}}$ under the automorphism interchanging $q$ and $q^{-1}$.

The following corollary gives a close relationship between the representation theory of $SL(2,\mathbb{C})$ in characteristic zero and of $\mathbb{Z}/p$ in characteristic $p$.

**Corollary 2.7.4** *The map $\mathbb{Z}[q,q^{-1}] \to \hat{\mathfrak{R}}$ is a homomorphism of special $p$-$\lambda$-rings.*

*Proof.* This follows by comparing Examples 2.6.6 and 2.6.7 with Proposition 2.7.2. □

**Remark 2.7.5** Referring to Lemma 2.6.11, the corollary shows that the ideal in $\mathbb{Z}[q,q^{-1}]$ generated by the element

$$(q-1)(q^{-1}-1)(q^{p-1}+q^{p-3}+\cdots+q^{-p+1})$$

is a $p$-$\lambda$-ideal. However, it is not a $\lambda$-ideal, as it is not closed under the operation $\lambda^p$. This explains why $\mathfrak{R}$ is not a special $\lambda$-ring.

**Remark 2.7.6** It is often computationally easier to work in the further quotient

$$\bar{\mathfrak{R}} = \mathbb{Z}[q,q^{-1}]/(q^{p-1}+q^{p-3}+\cdots+q^{-p+1}).$$

In this quotient, $q^p$ is identified with $q^{-p}$. The $2p-2$ elements $q^{p-1},q^{p-2},\ldots,q^{-p+2}$ form a free $\mathbb{Z}$-basis for $\bar{\mathfrak{R}}$. This may be regarded as the "stable" version of $\mathfrak{R}$ in which the projective module $J_p$ has been set equal to zero.

The map from the representation ring of $\mathbb{Z}/p$ to $\bar{\mathfrak{R}}$ is not injective, because $J_p$ goes to zero. However, the elements corresponding to $J_n$ for $1 \le n \le p-1$ are linearly independent. If we combine with the dimension homomorphism to $\mathbb{Z}$, we get an injective map from the representation ring of $\mathbb{Z}/p$ to $\mathbb{Z}\times\bar{\mathfrak{R}}$, sending $J_n$ to the pair $(n, q^{n-1}+q^{n-3}+\cdots+q^{-n+1})$ and in particular sending $J_p$ to $(p,0)$. The image is not quite the fixed points of the automorphism of $\mathbb{Z}\times\bar{\mathfrak{R}}$ interchanging $q$ and $q^{-1}$, but rather the elements of this of the form $(n, f(q))$ where $f(1)\equiv n \pmod{p}$.

As another alternative, less aesthetically pleasing but somehow more convenient for calculation, we can work in the quotient

$$\tilde{\mathfrak{R}} = \mathbb{Z}[q,q^{-1}]/(q-1)(q^{p-1}+q^{p-3}+\cdots+q^{-p+1}). \tag{2.3}$$

This quotient is a free $\mathbb{Z}$-module of rank $2p-1$ with basis $q^{p-1},q^{p-2},\ldots,q^{-p+1}$. Since

$$(q^{-1}+1)(q-1)(q^{p-1}+q^{p-3}+\cdots+q^{-p+1}) = q^p - q^{-p},$$

in this quotient ring $q^p$ is identified with $q^{-p}$. Furthermore, the elements

$$q^p+q^{p-2}+\cdots+q^{-p+2},\quad q^{p-1}+q^{p-3}+\cdots+q^{-p+1},\quad q^{p-2}+q^{p-4}+\cdots+q^{-p}$$

are all identified, and these represent the element $J_p$.

In the ring $\tilde{\mathfrak{R}}$, the elements representing $J_n$ $(1 \le n \le p)$ remain linearly independent, and the representation ring of $\mathbb{Z}/p$ is identified with the subring of $\tilde{\mathfrak{R}}$ consisting of the fixed points of the automorphism interchanging $q$ and $q^{-1}$.

**Example 2.7.7** Let us compute the tensor product $J_4 \otimes J_5$. Using Lemma 2.7.1, this corresponds to the element

$$(q^3 + q + q^{-1} + q^{-3})(q^4 + q^2 + 1 + q^{-2} + q^{-4}) = q^7 + 2q^5 + 3q^3 + 4q \\ + 4q^{-1} + 3q^{-3} + 2q^{-5} + q^{-7}.$$

If $p \ge 11$ it follows that

$$J_4 \otimes J_5 \cong J_8 \oplus J_6 \oplus J_4 \oplus J_2.$$

If $p = 7$ the first and last term in the polynomial combine to give $2q^7$, and so we get

$$J_4 \otimes J_5 \cong 2J_7 \oplus J_4 \oplus J_2.$$

Finally, if $p = 5$ then the term $q^7$ is replaced by $q^{-3}$, $q^{-7}$ is replaced by $q^3$ and $q^{-5}$ is replaced by $q^5$ to get

$$4q^5 + 4q^3 + 4q + 4q^{-1} + 4q^{-3}$$

and so

$$J_4 \otimes J_5 \cong 4J_5.$$

For tensor products this is not so exciting, but for symmetric and exterior powers and other Schur functors it becomes a very good method of computation.

## 2.8 Symmetric and Exterior Powers of Jordan Blocks

Symmetric and exterior powers of Jordan blocks for a cyclic group of order $p$ were investigated extensively in the work of Almkvist [2, 3, 4, 5], and Almkvist and Fossum [6]. We begin with the following observation, which appears as Proposition 2.10 from Chapter III of [6], and which will be needed in the proof of Theorem 5.12.1.

**Theorem 2.8.1** *As $k(\mathbb{Z}/p)$-modules, $S^d(J_i)$ is projective (i.e., a direct sum of copies of $J_p$) provided $d < p$ and $d + i > p$.*

*Proof.* Write $m$ for a non-zero element of $J_i$ spanning the fixed space of dimension one under $\mathbb{Z}/p$. Then there is an injective homomorphism $S^d(J_i) \to S^{d+1}(J_i)$ which sends $m_1 \dots m_d$ to $mm_1 \dots m_d$. The quotient module is isomorphic to $S^{d+1}(J_{i-1})$, where $J_{i-1}$ is identified with $J_i/(m)$. Thus there are exact sequences

$$0 \to S^d(J_i) \to S^{d+1}(J_i) \to S^{d+1}(J_{i-1}) \to 0.$$

If $i = p$ and $d < p - 1$ then both $S^d(J_i)$ and $S^{d+1}(J_i)$ are projective (and hence injective), so $S^{d+1}(J_{p-1})$ is projective. Repeat the argument using downward induction on $i$. □

The following formulas are proved in Corollary III.1.4 and Proposition III.2.6 of Almkvist and Fossum [6]. See also Hughes and Kemper [162].

**Theorem 2.8.2** *As elements of the ring $\hat{\mathfrak{R}}$ described in Section 2.7, we have*

$$\Lambda^d(J_n) = g_{n,d}(q), \qquad S^d(J_{n+1}) = g_{n+d,d}(q).$$

*for $d < p$.*

*Proof.* This follows from Proposition 2.6.8 and Corollary 2.7.4. □

**Corollary 2.8.3** *We have*

$$\begin{aligned} S^d(J_{n+1}) &\cong S^n(J_{d+1}) && d, n < p, \\ S^d(J_{n+1}) &\cong \Lambda^d(J_{n+d}) && n + d \le p. \end{aligned}$$ □

**Remark 2.8.4** The isomorphisms in Corollary 2.8.3 were discovered by Almkvist and Fossum [6]. We describe a generalisation to Schur functors in Section 2.9. Another interesting formula proved in the same paper is

$$S^{mp+d}(J_{n+1}) \cong S^d(J_{n+1}) \oplus \text{copies of } J_p \qquad 0 \le d < p.$$

**Example 2.8.5** Using equation (2.2) we calculate that for $p \ge 13$ we have

$$S^3(J_4) \cong J_{13} \oplus J_9 \oplus J_7 \oplus J_5 \oplus J_1.$$

If $p = 11$ we replace $q^{12}$ and $q^{-12}$ by $q^{-10}$ and $q^{10}$ in (2.2) to get

$$S^3(J_4) \cong J_{11} \oplus J_{11} \oplus J_7 \oplus J_5 \oplus J_1.$$

Using these methods, the following is an easy exercise.

**Proposition 2.8.6** *Let $J_a$, $J_b$ be the indecomposable representations of dimensions $a \le b \le p$ for a cyclic group of order $p$ over $k$. Then*

$$(1)\ J_a \otimes J_b \cong \begin{cases} J_{a+b-1} \oplus J_{a+b-3} \oplus J_{a+b-5} \oplus \cdots \oplus J_{b-a+1} & p \ge a+b-1 \\ (a+b-p)J_p \oplus J_{2p-a-b-1} & \\ \qquad \oplus J_{2p-a-b-3} \oplus \cdots \oplus J_{b-a+1} & p \le a+b-2 \end{cases}$$

$$(2)\ \Lambda^2(J_a) \cong \begin{cases} J_{2a-3} \oplus J_{2a-7} \oplus J_{2a-11} \oplus \dots & p \ge 2a-3 \\ (a - \frac{p+1}{2})J_p \oplus J_{2p-2a-1} \oplus J_{2p-2a-5} \oplus \dots & p \le 2a-5. \end{cases}$$

$$(3)\ S^2(J_a) \cong \Lambda^2(J_{a+1}).$$ □

Tables of symmetric and exterior powers of modules for $\mathbb{Z}/p$ can be found in Appendix A.

## 2.9 Schur Functors for $SL(2, \mathbb{C})$ and $\mathbb{Z}/p$

In this section we describe how to compute Schur functors on representations of $SL(2, \mathbb{C})$ in characteristic zero and of $\mathbb{Z}/p$ in characteristic $p$. Recall from Example 2.6.7 that the representation ring of $SL(2, \mathbb{C})$ is isomorphic to $\mathbb{Z}[q, q^{-1}]$ as a $\lambda$-ring. To compute in the representation ring of $\mathbb{Z}/p$ we work in the quotient ring $\hat{\mathfrak{R}}$ of $\mathbb{Z}[q, q^{-1}]$ described in Section 2.7. The following theorem generalises Theorem 2.8.2.

**Theorem 2.9.1** *Let $\lambda$ be a partition of $d$. Then we have the following.*

(1) *As a representation of $SL(2, \mathbb{C})$, the image of $S^\lambda(V_n)$ in $\mathbb{Z}[q, q^{-1}]$ is equal to the generalised Gaussian polynomial $g_{n,\lambda}(q)$.*
(2) *If $d < p$, as a representation of $\mathbb{Z}/p$ in characteristic $p$, the image of $S^\lambda(J_n)$ $(1 \le n \le p)$ in $\hat{\mathfrak{R}}$ is equal to $g_{n,\lambda}(q)$.*

*Proof.* From the determinantal form (1.5) we have

$$g_{n,\lambda}(q) = \det(g_{n,\lambda_i - i + j}(q))_{1 \le i,j \le r}$$

($r$ at least the number of parts of $\lambda$).

For the case of $SL(2, \mathbb{C})$, by Theorem 1.16.4 we have

$$S^\lambda(V_n) \cong \det(S^{\lambda_i - i + j}(V_n))_{1 \le i,j \le r}.$$

Part (1) of the theorem therefore follows from the special case $\lambda = [d]$, which is Proposition 2.6.8.

For the case of $\mathbb{Z}/p$, because of the condition $d < p$, by Theorem 1.16.4 we have

$$S^\lambda(J_n) \cong \det(S^{\lambda_i - i + j}(J_n))_{1 \le i,j \le r}.$$

Part (2) of the theorem therefore follows from the special case $\lambda = [d]$, which is Theorem 2.8.2. □

**Example 2.9.2** Let us calculate $S^{[3][2]}(V_4)$ for $SL(2, \mathbb{C})$. At the same time we compute $S^{[3][2]}(J_4)$ for $\mathbb{Z}/p$ with $p \ge 13$. Using the hook formula of Proposition 2.5.7, we have

$$\begin{aligned} g_{4,[3][2]}(q) &= \frac{(q^6 - q^{-6})(q^5 - q^{-5})(q^4 - q^{-4})^2(q^3 - q^{-3})}{(q^4 - q^{-4})(q^3 - q^{-3})(q^2 - q^{-2})(q - q^{-1})^2} \\ &= q^{11} + 2q^9 + 4q^7 + 6q^5 + 8q^3 + 9q + 9q^{-1} + 8q^{-3} + 6q^{-5} \\ &\quad + 4q^{-7} + 2q^{-9} + q^{-11} \end{aligned}$$

and so

$$S^{[3][2]}(V_4) \cong V_{12} \oplus V_{10} \oplus 2V_8 \oplus 2V_6 \oplus 2V_4 \oplus V_2 \qquad (\text{for } SL(2,\mathbb{C}))$$
$$S^{[3][2]}(J_4) \cong J_{12} \oplus J_{10} \oplus 2J_8 \oplus 2J_6 \oplus 2J_4 \oplus J_2 \qquad (\text{for } \mathbb{Z}/p,\ p \geq 13).$$

For smaller primes $p > 5$, we identify $q^p$ with $q^{-p}$ as usual, and we get

$$S^{[3][2]}(J_4) \cong \begin{cases} 2J_{11} \oplus 2J_8 \oplus 2J_6 \oplus 2J_4 \oplus J_2 & p = 11 \\ 8J_7 \oplus J_4 & p = 7. \end{cases}$$

Further results of such calculations may be found in Appendix A.

The following corollary generalises the isomorphism $S^d(J_{n+1}) \cong \Lambda^d(J_{n+d})$ of Almkvist and Fossum (see Corollary 2.8.3).

**Corollary 2.9.3** *Suppose that $\lambda$ be a partition of $d$ and that the characteristic of $k$ is either zero or $p > d$. Suppose further that there exists an integer $e$ such that for each diagonal box $x = (i, i)$ of $\lambda$ the quantity $\lambda_i - \lambda'_i$ is equal to $e$.*

(1) *As representations of $SL(2, \mathbb{C})$ we have*
$$S^{\lambda'}(V_{n+e}) \cong S^{\lambda}(V_n).$$

(2) *Provided that $d < p$, $n \leq p$ and $n + e \leq p$, as representations of $\mathbb{Z}/p$ in characteristic $p$ we have*
$$S^{\lambda'}(J_{n+e}) \cong S^{\lambda}(J_n).$$
□

**Example 2.9.4** One obvious class of partitions with the property described in the corollary is the hook partitions $[j][1]^{d-j}$ with $e = 2j - d + 1$. So we have

$$S^{[d-j+1][1]^{j-1}}(V_{n+2j-d+1}) \cong S^{[j][1]^{d-j}}(V_n) \qquad (\text{for } SL(2,\mathbb{C}))$$
$$S^{[d-j+1][1]^{j-1}}(J_{n+2j-d+1}) \cong S^{[j][1]^{d-j}}(J_n) \qquad (\text{for } \mathbb{Z}/p,\ p > d).$$

Another example is $[4][3][1]$ with $e = 2$, giving

$$S^{[3][2]^2[1]}(V_{n+2}) \cong S^{[4][3][1]}(V_n) \qquad (\text{for } SL(2,\mathbb{C}))$$
$$S^{[3][2]^2[1]}(J_{n+2}) \cong S^{[4][3][1]}(J_n) \qquad (\text{for } \mathbb{Z}/p,\ p \geq 11).$$

# 3

# Background from Algebraic Geometry

Many readers will wish to skip over this chapter, which is designed to provide background material from algebraic geometry in the form needed for our considerations. We make no attempt to be either general or comprehensive, but limit ourselves to the task in hand. More details can be found in EGA I and EGA II (Grothendieck [141, 142]), FAC (Serre [224]), and Chapters I–III of Hartshorne [147].

The first four sections of this chapter are all that is needed in order to read Chapter 4. The rest of this chapter consists of prerequisites for Chapter 6.

## 3.1 Affine Space and Affine Varieties

Let $k$ be an algebraically closed field. We write $\mathbb{A}^r = \mathbb{A}^r(k)$ for affine $r$-space over $k$, consisting of $r$-tuples $(a_1, \dots, a_r)$ with each $a_i \in k$. Let $X_i$ be the $r$-tuple $(0, \dots, 1, \dots, 0)$ with a 1 in the $i$th place and 0 in all the other places. Thus $X_1, \dots, X_r$ is the standard basis for $\mathbb{A}^r$.

Let $Y_1, \dots, Y_r$ be the linear functions on $\mathbb{A}^r$ defined by $Y_i(X_j) = \delta_{i,j}$ (Kronecker delta). Thus the coordinate ring of $\mathbb{A}^r$ is the ring $k[Y_1, \dots, Y_r]$ of polynomial functions on $\mathbb{A}^r$.

If $I$ is an ideal in $k[Y_1, \dots, Y_r]$ we write

$$V(I) = \{a \in \mathbb{A}^r \mid \forall f \in I,\ f(a) = 0\},$$

the set of simultaneous zeros of $I$. *Hilbert's Nullstellensatz* states that if $f \in k[Y_1, \dots, Y_r]$ vanishes at every point in $V(I)$ then some power of $f$ is in $I$. It follows that $V(I) = V(I')$ if and only if $\sqrt{I} = \sqrt{I'}$, where

$$\sqrt{I} = \{f \in k[Y_1, \dots, Y_r] \mid \exists n,\ f^n \in I\}.$$

An ideal I is said to be a *radical ideal* if $I = \sqrt{I}$.

We endow $\mathbb{A}^r$ with the *Zariski topology* in which the closed sets are the sets of the form $V(I)$ as $I$ runs over the ideals in $k[Y_1, \dots, Y_r]$, together with the empty set. Thus the non-empty closed sets are in one-to-one correspondence with the radical ideals. We refer to the closed subsets as *closed subvarieties*. Each point

$(a_1, \ldots, a_r)$ forms a closed set corresponding to the maximal ideal $\mathfrak{m} = (Y_1 - a_1, \ldots, Y_r - a_r)$.

An ideal is radical if and only if it is a finite intersection of prime ideals. Since $V(I \cap I') = V(I) \cup V(I')$, the *irreducible* closed sets, namely those which cannot be written as a union of two proper closed subsets, are precisely the closed sets corresponding to the *prime* ideals. Thus the closed irreducible subvarieties are those of the form $V(\mathfrak{p})$ with $\mathfrak{p}$ prime.

**Definition 3.1.1** An *affine variety* over $k$ is a closed subvariety $V$ of $\mathbb{A}^r(k)$ for some $r$. The Zariski topology on $V$ is inherited from that on $\mathbb{A}^r$ via the subspace topology.

**Example 3.1.2** $GL(n, k)$ is the closed subvariety of $\mathbb{A}^{n^2+1}$ defined by $n^2$ matrix entries $a_{i,j}$ together with another coordinate $a_0$ and the single equation $Y_0.\det(Y_{i,j}) = 1$. The same device can be used to make any open set in $\mathbb{A}^r$ defined by the non-vanishing of a single polynomial into a closed subvariety of $\mathbb{A}^{r+1}$.

## 3.2 Generic Points and Closed Points

If $V = V(\mathfrak{p})$ is a closed irreducible subvariety of $\mathbb{A}^r$ corresponding to a prime ideal $\mathfrak{p} \subseteq k[Y_1, \ldots, Y_r]$, we write $k[V]$ for the integral domain $k[Y_1, \ldots, Y_r]/\mathfrak{p}$. This is the ring of polynomial functions on $V$, and is called the *coordinate ring* of $V$. We write $k(V)$ for the field of fractions of $k[V]$. This is the field of *rational functions* on $V$, and is called the *function field* of $V$.

Let $\bar{Y}_1, \ldots, \bar{Y}_r$ be the images of $Y_1, \ldots, Y_r$ in $k[V] \subseteq k(V)$. Then the point

$$\bar{Y}_1 X_1 + \cdots + \bar{Y}_r X_r \in \mathbb{A}^r(k(V))$$

is called the *generic point* of $V$. It is a canonically defined point in $\mathbb{A}^r(k(V))$. It is independent of the choice of basis $X_1, \ldots, X_r$ of $\mathbb{A}^r(k)$. In fact it lies in the subvariety $V_{k(V)}$ obtained from $V$ by extending scalars from $k$ to $k(V)$, since the coordinates satisfy the equations defining $V$.

Let $K$ be algebraically closed and have transcendence degree at least $r$ over $k$. Then there is a $k$-embedding of $k(V)$ into $K$. Let $t_1, \ldots, t_r$ be the images of $\bar{Y}_1, \ldots, \bar{Y}_r$ under such an embedding. Then the point

$$t_1 X_1 + \cdots + t_r X_r \tag{3.1}$$

is called *a generic point* of $V$, and it is an element of $V_K$.

Conversely, given any extension field $K$ of $k$ and any point $\alpha$ in $\mathbb{A}^r(K)$, we write it in the form (3.1) with $t_1, \ldots, t_r \in K$. Letting $\mathfrak{p} \subseteq k[Y_1, \ldots, Y_r]$ be the ideal of polynomial relations satisfied by $t_1, \ldots, t_r$ over $k$, we see that $\mathfrak{p}$ is a prime ideal and so there is a corresponding closed irreducible subvariety $V \subseteq \mathbb{A}^r(k)$ with $k[V] = k[Y_1, \ldots, Y_r]/\mathfrak{p}$. This gives a uniquely determined field homomorphism $k(V) \to K$ such that the induced map $\mathbb{A}^r(k(V)) \to \mathbb{A}^r(K)$ sends the generic point of $V$ to $\alpha$, namely $\bar{Y}_i \mapsto t_i$. We then say that $\alpha$ is a generic point for $V$. In this way, every point of $\mathbb{A}^r(K)$ is a generic point for some closed irreducible subvariety of $\mathbb{A}^r(k)$. If the irreducible subvariety $V$ is a single point defined over $k$ then we say the generic point is a *closed point.*

**Lemma 3.2.1** *Suppose that we are given polynomials $p_1, \dots, p_n \in k[Y_1, \dots, Y_r]$ and a closed irreducible subvariety $V$ of $\mathbb{A}^r(k)$. Then the following are equivalent:*

- *$p_1, \dots, p_n$ vanish at the generic point of $V$ in $\mathbb{A}^r(k(V))$*
- *$p_1, \dots, p_n$ vanish at some generic point of $V$ defined over some extension field of $k$*
- *$p_1, \dots, p_n$ vanish on every point of $V$.*

*If these conditions fail, then the points of $V$ where $p_1, \dots, p_n$ do not vanish form a dense open subset of $V$ in the Zariski topology.* □

Because of the correspondence between generic points and prime ideals, it is customary to throw in abstract generic points to make the *spectrum* of $k[Y_1, \dots, Y_r]$. More generally, we have the following definition.

**Definition 3.2.2** If $R$ is a commutative ring, we write $\mathsf{Spec}\, R$ for the set of prime ideals of $R$. We endow this with the *Zariski topology* in which the closed sets are the sets of the form

$$V(I) = \{\mathfrak{p} \in \mathsf{Spec}\, R \mid I \subseteq \mathfrak{p}\},$$

where $I$ is an ideal in $R$. If $r \in R$, we write $V(r)$ for the closed set determined by the principal ideal $(r)$, and

$$D(r) = \{\mathfrak{p} \mid r \notin \mathfrak{p}\}$$

for its open complement. Every open set is a union of open sets of the form $D(r)$, so they form a base for the Zariski topology.

If $R \to R'$ is a ring homomorphism then $f^*\colon \mathsf{Spec}\, R' \to \mathsf{Spec}\, R$ takes a prime ideal in $R'$ to its inverse image in $R$.

So if we examine $\mathsf{Spec}\, k[Y_1, \dots, Y_r]$ we find that the closed points are the maximal ideals $(Y_1 - a_1, \dots, Y_r - a_r)$ corresponding to points $(a_1, \dots, a_r) \in \mathbb{A}^r(k)$. The remaining points are generic points for closed irreducible subvarieties of $\mathbb{A}^r(k)$. From now on, we shall write $\mathbb{A}^r$ or $\mathbb{A}^r(k)$ for $\mathsf{Spec}\, k[Y_1, \dots, Y_r]$ and talk of *closed points* when we wish to refer to elements of the form $(a_1, \dots, a_r)$.

## 3.3 Projective Space and Projective Varieties

Points in projective space correspond to lines through the origin in affine space. More precisely, we put an equivalence relation on the closed points of $\mathbb{A}^r(k) \setminus \{0\}$ by deeming that $(a_1, \ldots, a_r)$ and $(a'_1, \ldots, a'_r)$ are equivalent if and only if there exists $0 \neq \lambda \in k$ such that for $i = 1, \ldots, r$ we have $a'_i = \lambda a_i$. The set of equivalence classes is denoted $\mathbb{P}^{r-1}(k)$ or just $\mathbb{P}^{r-1}$, and called *projective* $(r-1)$*-space*. The point in $\mathbb{P}^{r-1}$ defined by a representative $(a_1, \ldots, a_r) \in \mathbb{A}^r \setminus \{0\}$ is denoted $(a_1 : \ldots : a_r)$.

The corresponding ring theoretic construction is to put a grading on $k[Y_1, \ldots, Y_r]$ in such a way that each $Y_i$ has degree one.

**Definition 3.3.1** We say that a polynomial $f(Y_1, \ldots, Y_r) \in k[Y_1, \ldots, Y_r]$ is *homogeneous* of degree $n$ if every monomial appearing with non-zero coefficient has degree exactly $n$. This is equivalent to the statement that for $\lambda \in k$ we have

$$f(\lambda Y_1, \ldots, \lambda Y_r) = \lambda^n f(Y_1, \ldots, Y_r).$$

For the grading on $k[Y_1, \ldots, Y_r]$, a homogeneous element of degree $n$ is a homogeneous polynomial of degree $n$ in $Y_1, \ldots, Y_r$. An ideal $I$ in $k[Y_1, \ldots, Y_r]$ is said to be *homogeneous* if it is generated by homogeneous elements.

The point of this is that a closed subvariety of $\mathbb{A}^r$ is a union of lines through the origin if and only if it is the set of zeros of a homogeneous ideal. So we define a closed subvariety of $\mathbb{P}^{r-1}$ to be the set of simultaneous zeros of a homogeneous ideal $I \subseteq k[Y_1, \ldots, Y_r]$. These form the closed sets in the *Zariski topology* on $\mathbb{P}^{r-1}$. Thus there is a one-to-one correspondence between the closed subsets of $\mathbb{P}^{r-1}$ and the homogeneous radical ideals of $k[Y_1, \ldots, Y_r]$. Under this correspondence, the closed irreducible subsets correspond to the homogeneous prime ideals.

The ideal $\mathfrak{m} = (Y_1, \ldots, Y_r)$ is the unique maximal homogeneous ideal, and corresponds to the empty subset of $\mathbb{P}^{r-1}$. This is called the *irrelevant ideal*, because it corresponds to the origin in $\mathbb{A}^r$, which was removed in order to construct $\mathbb{P}^{r-1}$.

The points in $\mathbb{P}^{r-1}$ correspond to the maximal homogeneous ideals among those properly contained in the irrelevant ideal. Namely, the point $(a_1 : \ldots : a_r) \in \mathbb{P}^{r-1}$ corresponds to the homogeneous ideal generated by the elements $a_i Y_j - a_j Y_i$ for $i \neq j$.

Again, we can throw in abstract generic points for closed irreducible subvarieties of $\mathbb{P}^{r-1}$ to form the *projective spectrum* of the graded ring $k[Y_1, \ldots, Y_r]$. More generally, we have the following definition, see EGA II [142, §2.3].

**Definition 3.3.2** Let $R = \bigoplus_{i \geq 0} R_i$ be a commutative graded ring. We write $\mathsf{Proj}\, R$ for the set of homogeneous prime ideals not containing the irrelevant ideal $\bigoplus_{i>0} R_i$.

We endow this with the *Zariski topology*, in which the closed sets are the sets of the form

$$V(I) = \{\mathfrak{p} \in \mathsf{Proj}\, R \mid I \subseteq \mathfrak{p}\}$$

where $I$ is a homogeneous ideal in $R$.

Thus $\mathsf{Proj}\, k[Y_1, \dots, Y_r]$ has as its closed points the ideals $(a_iY_j - a_jY_i,\ i \neq j)$ corresponding to the points $(a_1 : \dots : a_r) \in \mathbb{P}^{r-1}(k)$. The remaining points are generic points for other closed irreducible subvarieties of $\mathbb{P}^{r-1}(k)$. As in the affine case, we shall write $\mathbb{P}^{r-1}$ or $\mathbb{P}^{r-1}(k)$ for $\mathsf{Proj}\, k[Y_1, \dots, Y_r]$, and refer to *closed points* when we wish to refer only to elements of the form $(a_1 : \dots : a_r)$.

Projective space $\mathbb{P}^{r-1}$ has an open cover by $r$ copies of $\mathbb{A}^{r-1}$. At the level of closed points, the $i$th open set $U_i$ is the set of $(a_1 : \dots : a_r)$ such that $a_i \neq 0$. Such a point has a unique representative with $a_i = 1$, and the remaining coordinates give the correspondence with points in $\mathbb{A}^{r-1}$. At the level of $\mathsf{Proj}\, k[Y_1, \dots, Y_r]$, $U_i$ is the set of homogeneous primes not containing $Y_i$. This open set may be thought of as $\mathsf{Spec}\, k[Y_1/Y_i, \dots \overset{i}{\uparrow} \dots, Y_r/Y_i]$. These copies of $\mathbb{A}^{r-1}$ covering $\mathbb{P}^{r-1}$ are called the *affine patches*.

**Definition 3.3.3** A *projective variety* over $k$ is a closed subvariety of $\mathbb{P}^{r-1}(k)$ for some $r$. If $V$ is a projective variety then $V = \mathsf{Proj}\, R$ where $R = k[Y_1, \dots, Y_r]/I$ for some homogeneous ideal $I \subseteq k[Y_1, \dots, Y_r]$. A projective variety $V$ has an open cover consisting of the

$$V \cap U_i = \mathsf{Spec}\, R[Y_i^{-1}]_0$$

for which $Y_i$ is not nilpotent in $R$.

**Definition 3.3.4** A *quasi-projective variety* over $k$ is a *locally closed* subset of $\mathbb{P}^{r-1}$, i.e., the intersection of an open subset and a closed subset, for some $r$.

Thus affine varieties and projective varieties are examples of quasi-projective varieties. But, for example, $\mathbb{A}^2(k) \setminus \{0\}$ is a quasi-projective variety which is neither an affine variety nor a projective variety. On the other hand, every quasi-projective variety has a finite open cover by affine varieties.

**Definition 3.3.5** A *regular map* of quasi-projective varieties is one which is locally given by polynomials at every point. A *rational map*, on the other hand, is locally given by polynomials on some dense open subset, and not defined on the complement. Two rational maps are equal if and only if they agree where they are both defined.

Thus for example the regular maps $\mathbb{A}^r \to \mathbb{A}^1$ (or equivalently $\mathbb{A}^1 \to k$) are the elements of $k[Y_1, \dots, Y_r]$ while the rational maps are the elements of $k(Y_1, \ldots, Y_r)$. Regular or rational maps to $k$ are given the structure of a (commutative) $k$-algebra via the addition and multiplication in $k$.

The only regular maps $\mathbb{P}^{r-1} \to \mathbb{A}^1$ are the constant maps, whereas the rational functions are the elements of $k(Y_1/Y_r, \ldots, Y_{r-1}/Y_r)$, namely the rational functions of $Y_1, \ldots, Y_r$ whose numerator and denominator have the same degree.

## 3.4 Tangent Spaces

Let $V$ be a quasi-projective variety over $k$ and $a$ a (closed) point of $V$. We wish to define the tangent space to $V$ at $a$. This is supposed to be a local concept in the sense that if $V'$ is an open subset of $V$ containing $a$ then the tangent space to $V$ and to $V'$ at $a$ are naturally isomorphic. So by intersecting with an affine patch and taking the closure, we may assume that $V$ is a closed subset of $\mathbb{A}^r$ for some $r$.

The *linear space* $L_{V,a}$ of $V$ at $a = (a_1, \ldots, a_r)$ is defined to be the set of $b = (b_1, \ldots, b_r) \in \mathbb{A}^r$ with the property that for all $\phi \in k[Y_1, \ldots, Y_r]$ that vanish on $V$ we have

$$\sum_{i=1}^{r} \left( \frac{\partial \phi}{\partial Y_i} \right) (a)(b_i - a_i) = 0. \tag{3.2}$$

See for example Shafarevich [227, §II.1.3]. The notation is a little confusing here, so let me spell it out. What we are doing is evaluating the function $\frac{\partial \phi}{\partial Y_i}$ at $a$, multiplying by $b_i - a_i$ and then summing. This amounts to evaluating the element of the dual vector space of $\mathbb{A}^r$ with coordinates $\left( \frac{\partial \phi}{\partial Y_i} \right)(a)$ on the vector $b - a \in \mathbb{A}^r$.

The *tangent space* $T_{V,a}$ is defined to be the vector space whose elements are the vectors $b - a$ with $b \in L_{V,a}$. The linear space $L_{V,a}$ is the coset $a + T_{V,a}$. Thus the tangent space is the subspace defined by the equations

$$\sum_{i=1}^{r} \left( \frac{\partial \phi}{\partial Y_i} \right) (a) b_i = 0. \tag{3.3}$$

For $f \in k[Y_1, \ldots, Y_r]$ we write $d_a(f)(b)$ for $\sum_{i=1}^r \left( \frac{\partial f}{\partial Y_i} \right)(a) b_i$. So $d_a(f)$ is a linear form, and (3.3) says that $T_{V,a}$ is defined by the equations $d_a(\phi)(b) = 0$ for $\phi \in I$. We also have

$$\begin{aligned} d_a(f_1 + f_2) &= d_a(f_1) + d_a(f_2) \\ d_a(\lambda f) &= \lambda d_a(f) \quad (\lambda \in k) \\ d_a(f_1 f_2) &= f_1(a) d_a(f_2) + d_a(f_1) f_2(a). \end{aligned} \tag{3.4}$$

Now $d_a(f)(b)$ is well defined for $f \in k[V] = k[Y_1, \ldots, Y_r]/I$ and $b \in T_{V,a}$. It vanishes for $f$ a constant function, so we may as well restrict to $\mathfrak{m}_a/I$, the maximal ideal in $k[V]$ corresponding to $a$, which we shall denote $\mathfrak{m}_{V,a}$. Furthermore, by (3.4), $d_a(f_1 f_2)(b) = 0$ for $f_1, f_2 \in \mathfrak{m}_{V,a}$. So we obtain a well-defined bilinear evaluation map

$$\mathfrak{m}_{V,a}/\mathfrak{m}_{V,a}^2 \times T_{V,a} \to k. \tag{3.5}$$

**Proposition 3.4.1** *The pairing* (3.5) *induces an isomorphism*

$$T_{V,a} \cong (\mathfrak{m}_{V,a}/\mathfrak{m}_{V,a}^2)^*.$$

*Proof.* It suffices to show that the pairing (3.5) is perfect. Since every linear form in $T_{V,a}$ is the restriction to $T_{V,a}$ of a linear polynomial in $\mathfrak{m}_a$, it remains to show that if $f \in \mathfrak{m}_a$ satisfies $d_a(f)(b) = 0$ for all $b \in T_{V,a}$ then $f \in \mathfrak{m}_a^2 + I$. By definition of $T_{V,a}$, given such an $f$ there exists $\phi \in I$ such that $d_a(f) = d_a(\phi)$. Examining the Taylor expansion of $f - \phi$ about $a$, we see that $f - \phi \in \mathfrak{m}_a^2$ as desired. □

We shall need the following lemma.

**Lemma 3.4.2** *Let $V$ be a locally closed subset of $\mathbb{A}^r$ and let $\rho\colon \mathbb{A}^1 \to V$ be a rational map of the form*

$$\rho(t) = a + t^i b + t^{i+1} c(t)$$

*where $c(t)$ is defined at $t = 0$. Then $b \in T_{V,a}$.*

*Proof.* We may as well assume that $V$ is closed. If $f \in k[Y_1, \dots, Y_r]$ vanishes at $a$ then

$$(f \circ \rho)(t) = f(a + t^i b + t^{i+1} c(t)).$$

Consider the Taylor expansion of $f$ around $a$:

$$f(x) = d_a(f)(x - a) + \sum_{j \le k} \lambda_{j,k}(x_j - a_j)(x_k - a_k) + \cdots .$$

Then

$$(f \circ \rho)(t) = d_a(f)(b)t^i + d_a(f)(c(t))t^{i+1} + \sum_{j \le k} \lambda_{j,k}(t^i b_j + t^{i+1} c_j(t))$$
$$(t^i b_k + t^{i+1} c_k(t)) + \cdots .$$

If $f$ is in the ideal defining $V$ then we must have $(f \circ \rho)(t) = 0$. Comparing coefficients of $t^i$, this implies that $d_a(f)(b) = 0$. Thus $b \in T_{V,a}$. □

## 3.5 Presheaves and Sheaves

We want to keep track of regular functions on open subsets of a variety, and we do this by means of sheaves.

**Definition 3.5.1** Let $X$ be a topological space. A *presheaf* $\mathcal{F}$ of abelian groups, or rings, or modules, or more generally of objects in a category $\mathcal{C}$, assigns to each open subset $U \subseteq X$ an object $\mathcal{F}(U)$ in $\mathcal{C}$ and to each inclusion $U \subseteq U' \subseteq X$

a *restriction map* $\mathcal{F}(U') \to \mathcal{F}(U)$. This must send the inclusion $U \subseteq U$ to the identity map on $\mathcal{F}(U)$, and a composite inclusion $U \subseteq U' \subseteq U''$ to the composite $\mathcal{F}(U'') \to \mathcal{F}(U') \to \mathcal{F}(U)$. In other words, $\mathcal{F}$ is a contravariant functor on the category of open sets and inclusions. A *morphism* of presheaves is a natural transformation of functors.

For a presheaf to be a sheaf, the category $\mathcal{C}$ is required to have pullbacks. A presheaf $\mathcal{F}$ on $X$ is then defined to be a *sheaf* if the following *gluing condition* is satisfied. Given an open cover $\{U_\alpha\}$ of an open set $U$, $\mathcal{F}(U)$ is the pullback of the restriction maps $U_\alpha \to U_\alpha \cap U_\beta$. If $\mathcal{C}$ is a category of abelian groups, rings, modules, etc., we can express this as a condition on elements. Given elements $x_\alpha \in \mathcal{F}(U_\alpha)$ such that the restrictions of $x_\alpha$ and $x_\beta$ to $\mathcal{F}(U_\alpha \cap U_\beta)$ agree for all $\alpha$ and $\beta$, there exists a unique $x \in \mathcal{F}(U)$ that restricts to $x_\alpha$ on each $U_\alpha$.

**Example 3.5.2** Let $V$ be a quasi-projective variety. The *structure sheaf* $\mathcal{O}_V$ of $V$ assigns to each open subset $U \subseteq V$ the $k$-algebra of regular functions $U \to k$. If $U \subseteq U' \subseteq V$, the restriction map $\mathcal{O}_V(U') \to \mathcal{O}_V(U)$ is just given by restriction of functions. The gluing condition is easy to verify in this case.

**Definition 3.5.3** A *ringed space* $(X, \mathcal{O}_X)$ is a topological space $X$ together with a sheaf of commutative rings $\mathcal{O}_X$.

**Example 3.5.4** If $V$ is a quasi-projective variety and $\mathcal{O}_V$ is its structure sheaf then $(V, \mathcal{O}_V)$ is a ringed space.

**Definition 3.5.5** If $\mathcal{F}$ is a sheaf on a space $X$, we write $\Gamma(X, \mathcal{F})$ for $\mathcal{F}(X)$, and we refer to its elements as *global sections* of $\mathcal{F}$.

**Definition 3.5.6** If $U$ is an open subset of $X$ and $\mathcal{F}$ is a sheaf on $X$, we write $\mathcal{F}|_U$ for the restriction of $\mathcal{F}$ to $U$. This is a sheaf on $U$ in an obvious way.

The following lemma says that sheaves may be constructed locally. This is not true of presheaves.

**Lemma 3.5.7** *Suppose we are given an open cover $\{U_\alpha\}$ of a space $X$, sheaves $\mathcal{F}_\alpha$ on $U_\alpha$, and isomorphisms*

$$\phi_{\alpha,\beta}\colon \mathcal{F}_\alpha|_{U_\alpha \cap U_\beta} \xrightarrow{\cong} \mathcal{F}_\beta|_{U_\alpha \cap U_\beta},$$

*such that $\phi_{\alpha,\alpha}$ is the identity, and such that the diagram*

$$\begin{array}{ccc} \mathcal{F}_\alpha|_{U_\alpha \cap U_\beta \cap U_\gamma} & \xrightarrow{\phi_{\alpha,\gamma}} & \mathcal{F}_\gamma|_{U_\alpha \cap U_\beta \cap U_\gamma} \\ & {}_{\phi_{\alpha,\beta}}\searrow \qquad \nearrow_{\phi_{\beta,\gamma}} & \\ & \mathcal{F}_\beta|_{U_\alpha \cap U_\beta \cap U_\gamma} & \end{array}$$

*commutes for all $\alpha, \beta, \gamma$. Then there exists a sheaf $\mathcal{F}$ on $X$ and isomorphisms $\phi_\alpha : \mathcal{F}|_{U_\alpha} \to \mathcal{F}_\alpha$ such that the diagram*

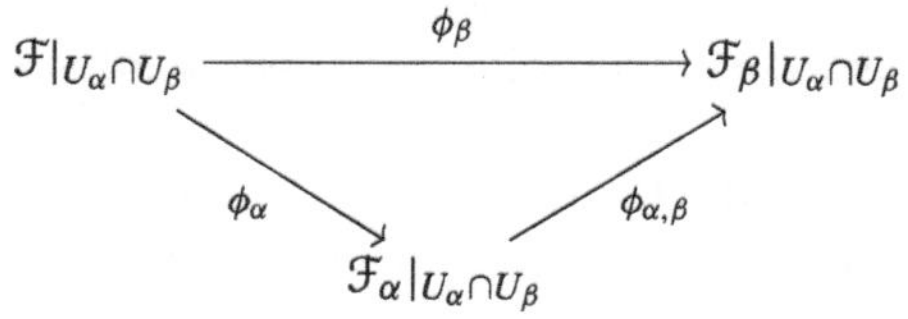


*commutes for all $\alpha, \beta$.*

*The sheaf $\mathcal{F}$ is unique up to natural isomorphism.*

*Proof.* By the gluing condition, an element of $\mathcal{F}(U)$ is uniquely determined by the choice of an element $x_\alpha \in \mathcal{F}_\alpha(U \cap U_\alpha)$ for each $\alpha$ such that $\phi_{\alpha,\beta}(x_\alpha|_{U \cap U_\alpha \cap U_\beta}) = x_\beta|_{U \cap U_\alpha \cap U_\beta}$ for all $\alpha, \beta$. One checks that this uniquely determines a sheaf $\mathcal{F}$ satisfying the given conditions. □

## 3.6 Stalks and Sheafification

**Definition 3.6.1** The *stalk* of a presheaf $\mathcal{F}$ at a point $x \in X$ is defined to be

$$\mathcal{F}_x = \varinjlim_{x \in U} \mathcal{F}(U).$$

**Example 3.6.2** In the case of the structure sheaf $\mathcal{O}_V$ on a quasi-projective variety $V$ the stalk $\mathcal{O}_{V,x}$ at $x \in V$ is the ring of germs of regular functions at $x$. It is a local ring, whose maximal ideal $\mathfrak{m}_x$ consists of the germs of functions vanishing at $x$. We write $k(x)$ for the field $\mathcal{O}_{V,x}/\mathfrak{m}_x$.

**Definition 3.6.3** The inclusion of the category of sheaves on $X$ into the category of presheaves on $X$ has a left adjoint called *sheafification*. If $\mathcal{F}$ is a presheaf and $U \subseteq X$ is open, then the sheafification $\mathcal{F}^+$ of $\mathcal{F}$ assigns to $U$ the functions $s$ from $U$ to the union of the stalks $\bigcup_{x \in U} \mathcal{F}_x$ such that

(1) each $s(x) \in \mathcal{F}_x$, and

(2) for each $x \in U$ there is an open neighbourhood $x \in U' \subseteq U$ and $f \in \mathcal{F}(U)$ whose image in each stalk $\mathcal{F}_y$ is equal to $s(y)$ for all $y \in U'$.

There is an obvious map $\mathcal{F} \to \mathcal{F}^+$ which is the unit of the adjunction. If $\mathcal{F}$ is already a sheaf, then $\mathcal{F} \to \mathcal{F}^+$ is an isomorphism.

The sheafification functor is particularly useful because many constructions on sheaves naturally produce presheaves, which then need to be sheafified.

For example, if $\mathcal{F}$ and $\mathcal{F}'$ are two sheaves of abelian groups on $X$, then the presheaf whose value on an open set $U$ is the tensor product $\mathcal{F}(U) \otimes \mathcal{F}'(U)$ is usually not a sheaf. We write $\mathcal{F} \otimes \mathcal{F}'$ for its sheafification.

As another example, given a morphism $\phi\colon \mathcal{F} \to \mathcal{F}'$ of sheaves of abelian groups on $X$, the kernel is already a sheaf. The presheaf cokernel, assigning to each open set $U \subseteq X$ the abelian group $\mathcal{F}'(U)/\phi(\mathcal{F}(U))$, is not necessarily a sheaf. We write $\mathcal{F}'/\phi(\mathcal{F})$ for the sheafification of this presheaf. With this definition of quotient sheaf, sheaves of abelian groups on $X$ form an abelian category.

**Remark 3.6.4** Closely related to the fact that quotients need sheafification is the following. If $0 \to \mathcal{F} \to \mathcal{F}' \to \mathcal{F}'' \to 0$ is a short exact sequence of sheaves on $X$ then

$$0 \to \Gamma(X, \mathcal{F}) \to \Gamma(X, \mathcal{F}') \to \Gamma(X, \mathcal{F}'')$$

is exact. In other words, the global section functor from sheaves to abelian groups is *left exact*, but not necessarily right exact. We shall examine its right derived functors in Section 3.10.

## 3.7 The Language of Schemes

We shall sometimes use the language of schemes rather than varieties where this is the appropriate generality. In this section, we gather the necessary definitions. For the purpose of this work, the reader who does not wish to think about schemes can usually just think of varieties, with the proviso that these must include generic points as described earlier in this chapter.

We saw in Section 3.2 that $\mathsf{Spec}\, R$, for $R$ a commutative ring, is a generalisation of affine variety which emphasises the role of generic points for closed irreducible subvarieties. The analogue of the sheaf of regular functions is the following.

**Definition 3.7.1** Let $X = \mathsf{Spec}\, R$. We define a sheaf of commutative rings $\mathcal{O}_X$ as follows. For $\mathfrak{p} \in X$, let $R_\mathfrak{p}$ be the localisation of $R$ at $\mathfrak{p}$. If $U \subseteq X$ is open then $\mathcal{O}_X(U)$ is the set of functions $s$ from $U$ to $\bigcup_{\mathfrak{p}\in U} R_\mathfrak{p}$, such that

(1) each $s(\mathfrak{p}) \in R_\mathfrak{p}$, and

(2) for each $\mathfrak{p} \in U$ there is an open neighbourhood $\mathfrak{p} \in U' \subseteq U$ and elements $x, y \in R$ such that for each $\mathfrak{q} \in U'$ we have $y \notin \mathfrak{q}$ and $s(\mathfrak{q}) = x/y \in R_\mathfrak{q}$.

We call $\mathcal{O}_X$ the *structure sheaf* on $X$. With the obvious ring operations and restriction maps, $(X, \mathcal{O}_X)$ is a ringed space, called the *spectrum* of $R$.

**Definition 3.7.2** An *affine scheme* is a ringed space $(X, \mathcal{O}_X)$ which is isomorphic to the spectrum of a commutative ring.

A *scheme* is a ringed space $(X, \mathcal{O}_X)$ which is locally affine. In other words, each $x \in X$ has an open neighbourhood $x \in U \subseteq X$ such that $(U, \mathcal{O}_X|_U)$ is an affine scheme.

A scheme $(X, \mathcal{O}_X)$ is *reduced* if each $\mathcal{O}_X(U)$ is a *reduced ring*; in other words, $\mathcal{O}_X(U)$ has no non-zero nilpotent elements.

A scheme is *Noetherian* if it has a finite cover by open affine subsets of the form $\mathsf{Spec}\, R$ with $R$ Noetherian.

**Example 3.7.3** An affine variety $V = \mathsf{Spec}\, k[Y_1, \ldots, Y_r]/I \subseteq \mathbb{A}^r$ together with the sheaf of regular functions $\mathcal{O}_V$ is an example of an affine scheme. It is reduced and Noetherian.

**Example 3.7.4** If $V = \mathsf{Proj}\, k[Y_1, \ldots, Y_r]/I \subseteq \mathbb{P}^{r-1}$ is a projective variety, with its sheaf of regular functions $\mathcal{O}_V$, then $(V, \mathcal{O}_V)$ is a reduced Noetherian scheme. This is because if $U_i$ is the $i$th affine patch of $\mathbb{P}^{r-1}$ then $\{(V \cap U_i, \mathcal{O}_V|_{V\cap U_i})\}$ is an open cover of $V$ by reduced Noetherian affine schemes.

**Definition 3.7.5** If $(X, \mathcal{O}_X)$ is a scheme, then the stalk $\mathcal{O}_{X,x}$ of $\mathcal{O}_X$ at $x \in X$ is a local ring. A ringed space with this property is called a *locally ringed space*.

If $(X, \mathcal{O}_X)$ is a locally ringed space and $x \in X$, we write $k(x)$ for the quotient of $\mathcal{O}_{X,x}$ by its maximal ideal $\mathfrak{m}_x$.

## 3.8 Sheaves of Modules

The material in this section is mostly taken from §2 of FAC [224]; see also §3.2–3.4 of EGA II [142] and §II.5 of Hartshorne [147].

If $(X, \mathcal{O}_X)$ is a ringed space, an *$\mathcal{O}_X$-module* $\mathcal{F}$ is a sheaf of abelian groups on $X$ together with an $\mathcal{O}_X(U)$-module structure on each $\mathcal{F}(U)$, compatible with the restriction maps on $\mathcal{O}_X$ and on $\mathcal{F}$.

For a morphism of $\mathcal{O}_X$-modules $\mathcal{F} \to \mathcal{F}'$, it is required that each $\mathcal{F}(U) \to \mathcal{F}'(U)$ is an $\mathcal{O}_X(U)$-module homomorphism, as well as requiring that $\mathcal{F} \to \mathcal{F}'$ is a morphism of sheaves of abelian groups. With these definitions, the $\mathcal{O}_X$-modules form an abelian category.

If $\mathcal{F}$ and $\mathcal{F}'$ are sheaves of $\mathcal{O}_X$-modules, we write $\mathcal{F} \otimes_{\mathcal{O}_X} \mathcal{F}'$ for the sheafification of the presheaf assigning to an open set $U \subseteq X$ the $\mathcal{O}_X(U)$-module $\mathcal{F}(U) \otimes_{\mathcal{O}_X(U)} \mathcal{F}'(U)$. The presheaf assigning to $U \subseteq X$ the $\mathcal{O}_X(U)$-module $\mathrm{Hom}_{\mathcal{O}_X(U)}(\mathcal{F}(U), \mathcal{F}'(U))$ is already a sheaf, denoted $\mathcal{H}om_{\mathcal{O}_X}(\mathcal{F}, \mathcal{F}')$, with global sections $\Gamma(X, \mathcal{H}om_{\mathcal{O}_X}(\mathcal{F}, \mathcal{F}')) = \mathrm{Hom}_{\mathcal{O}_X}(\mathcal{F}, \mathcal{F}')$. Thus $\mathcal{F} \otimes_{\mathcal{O}_X} \mathcal{F}'$ and $\mathcal{H}om_{\mathcal{O}_X}(\mathcal{F}, \mathcal{F}')$ are $\mathcal{O}_X$-modules, and we have a natural isomorphism

$$\mathcal{H}om_{\mathcal{O}_X}(\mathcal{F} \otimes_{\mathcal{O}_X} \mathcal{F}', \mathcal{F}'') \cong \mathcal{H}om_{\mathcal{O}_X}(\mathcal{F}, \mathcal{H}om_{\mathcal{O}_X}(\mathcal{F}', \mathcal{F}'')).$$

**Definition 3.8.1** Let $(X, \mathcal{O}_X)$ be a locally ringed space. If $\mathcal{F}$ is an $\mathcal{O}_X$-module, the *fibre* of $\mathcal{F}$ at a point $x \in X$ is defined to be the $k(x)$-vector space $\mathcal{F}_x \otimes_{\mathcal{O}_{X,x}} k(x)$.

For an affine scheme $(V, \mathcal{O}_V)$ with $V = \mathsf{Spec}\, R$, the global section functor for $\mathcal{O}_V$-modules has a left adjoint, constructed as follows.

**Definition 3.8.2** Let $V = \mathsf{Spec}\, R$ be an affine scheme. If $M$ is an $R$-module, we write $M^\sim$ for the $\mathcal{O}_V$-module defined as follows. If $U \subseteq V$ is open, then $M^\sim(U)$ is the set of functions $s$ from $U$ to the union of the localisations $\bigcup_{\mathfrak{p}\in U} M_\mathfrak{p}$, such that

(1) each $s(\mathfrak{p}) \in M_\mathfrak{p}$, and

(2) for each $\mathfrak{p} \in U$ there is an open neighbourhood $\mathfrak{p} \in U' \subseteq U$ and elements $m \in M$, $f \in R$ such that for each $\mathfrak{q} \in U'$ we have $f \notin \mathfrak{q}$ and $s(\mathfrak{q}) = m/f \in M_\mathfrak{q}$.

With the obvious restriction maps, this is an $\mathcal{O}_V$-module.

**Lemma 3.8.3** *Let $V = \mathsf{Spec}\, R$ be an affine scheme. If $M$ is an $R$-module and $\mathcal{F}$ is an $\mathcal{O}_V$-module, we have natural isomorphisms*

$$\mathrm{Hom}_{\mathcal{O}_V}(M^\sim, \mathcal{F}) \cong \mathrm{Hom}_R(M, \Gamma(V, \mathcal{F}))$$

*giving an adjunction between the functors $^\sim$ and $\Gamma(V, -)$.*

**Definition 3.8.4** Let $(X, \mathcal{O}_X)$ be a scheme. An $\mathcal{O}_X$-module $\mathcal{F}$ is *quasi-coherent* if $X$ can be covered by affine open subsets $U_\alpha = \mathsf{Spec}\, R_\alpha$ such that the restriction of $\mathcal{F}$ to each $U_\alpha$ is isomorphic to $M_\alpha^\sim$ for some $R_\alpha$-module $M_\alpha$. If the $M_\alpha$ can be taken to be finitely generated then $\mathcal{F}$ is *coherent.*

Quasi-coherence is equivalent to the statement that the restriction of $\mathcal{F}$ to every affine open subset $U = \mathsf{Spec}\, R \subseteq V$ is of the form $M^\sim$ for some $R$-module $M$. Given suitable Noetherian hypotheses, coherence is equivalent to the same condition with $M$ finitely generated. See Proposition II.5.4 of Hartshorne [147].

In particular, a quasi-coherent $\mathcal{O}_V$-module for $V = \mathsf{Spec}\, R$ affine is precisely one of the form $M^\sim$ for $M$ a $R$-module.

**Theorem 3.8.5** *Let $V = \mathsf{Spec}\, R$ be an affine scheme. The functors $^\sim$ and $\Gamma(V, -)$ give an equivalence of categories between $R$-modules and quasi-coherent $\mathcal{O}_V$-modules. This restricts to an equivalence between finitely generated $R$-modules and coherent $\mathcal{O}_V$-modules provided $R$ is Noetherian.*

*Proof.* See Corollary II.5.5 of Hartshorne [147]. □

## 3.9 Coherent Sheaves on Projective Varieties

There is a similar but more complicated equivalence of categories in the projective case, using graded modules.

**Definition 3.9.1** Let $R = k[Y_1, \dots, Y_r]/I$ with $\deg(Y_i) = 1$ and $I$ a homogeneous ideal, and let $V = \mathsf{Proj}\, R$. Then $V$ has an affine open cover $V \cap U_i = \mathsf{Spec}\, R[Y_i^{-1}]_0$, $i = 1, \dots, r$. If $M$ is a graded $R$-module then $M[Y_i^{-1}]_0$ is an $R[Y_i^{-1}]_0$-module. Let $\mathcal{F}_i$ be the $\mathcal{O}_{V\cap U_i}$-module $M[Y_i^{-1}]_0^\sim$. Then the hypotheses of Lemma 3.5.7 are satisfied, and the $\mathcal{F}_i$ glue together to give an $\mathcal{O}_V$-module $M^\sim$.

**Example 3.9.2** If $M$ is a graded $R$-module and $j \in \mathbb{Z}$, we define the $j$th *shift* of $M$, denoted $M(j)$, via $M(j)_i = M_{i+j}$. In particular, regarding $R$ as an $R$-module, this gives us an $R$-module $R(j)$. We write $\mathcal{O}_V(j)$ for the $\mathcal{O}_V$-module $R(j)^\sim$. The sheaf $\mathcal{O}_V(1)$ is called the *twisting sheaf* associated with the chosen embedding $V \hookrightarrow \mathbb{P}^{r-1}$. For any $\mathcal{O}_V$-module $\mathcal{F}$ we write $\mathcal{F}(j)$ for $\mathcal{O}_V(j) \otimes_{\mathcal{O}_V} \mathcal{F}$. It is easy enough to check that $\mathcal{O}_V(j) \otimes_{\mathcal{O}_V} \mathcal{O}_V(j') \cong \mathcal{O}_V(j+j')$, and so we have $\mathcal{F}(j)(j') = \mathcal{F}(j+j')$. We call $\mathcal{F}(j)$ the $j$th *twist* of $\mathcal{F}$. For example, $M(j)^\sim$ is naturally isomorphic to $M^\sim(j)$.

**Definition 3.9.3** In the other direction, we have a functor $\Gamma_*$ from $\mathcal{O}_V$-modules to graded $R$-modules given as follows. If $\mathcal{F}$ is an $\mathcal{O}_V$-module, we make a graded vector space $\Gamma_*(\mathcal{F})$ whose degree $j$ elements are given by the global sections of the $j$th twist,

$$\Gamma_j(\mathcal{F}) = \Gamma(V, \mathcal{F}(j)), \quad j \in \mathbb{Z}.$$

We give $\Gamma_*(\mathcal{F})$ a graded $R$-module structure as follows. If $r \in R_i$ then it determines a global section of $\mathcal{O}_V(i)$. If $x$ is a global section of $\mathcal{F}(j)$ then $rx \in \Gamma_{i+j}(\mathcal{F})$ defined by the map

$$\Gamma(V, \mathcal{O}_V(i)) \otimes_k \Gamma(V, \mathcal{F}(j)) \to \Gamma(V, \mathcal{F}(i+j))$$

induced by the natural isomorphism

$$\mathcal{O}_V(i) \otimes_{\mathcal{O}_V} \mathcal{F}(j) \xrightarrow{\cong} \mathcal{F}(i+j).$$

**Lemma 3.9.4** (i) *There are natural isomorphisms*

$$\mathrm{Hom}_{\mathcal{O}_V}(M^\sim, \mathcal{F}) \cong \mathrm{Hom}_R(M, \Gamma_*(\mathcal{F}))$$

*giving an adjunction between the functors* $^\sim$ *and* $\Gamma_*$.

(ii) *If $\mathcal{F}$ is a quasi-coherent $\mathcal{O}_V$-module then the counit of the adjunction is a natural isomorphism* $\Gamma_*(\mathcal{F})^\sim \to \mathcal{F}$.

*Proof.* In FAC [224, §59], see also EGA II [142, §3.3], natural maps $\alpha_M \colon M \to \Gamma_*(M^\sim)$ and $\beta_{\mathcal{F}} \colon \Gamma_*(\mathcal{F})^\sim \to \mathcal{F}$ are constructed, and in FAC Propositions 59.7, 59.8, EGA II (3.3.5.2) it is shown that the composites

$$M^\sim \xrightarrow{\alpha_M^\sim} \Gamma_*(M^\sim)^\sim \xrightarrow{\beta_{M^\sim}} M^\sim$$

$$\Gamma_*(\mathcal{F}) \xrightarrow{\alpha_{\Gamma_*(\mathcal{F})}} \Gamma_*(\Gamma_*(\mathcal{F})^\sim) \xrightarrow{\Gamma_*(\beta_{\mathcal{F}})} \Gamma_*(\mathcal{F})$$

are isomorphisms. By Theorem IV.2 of Mac Lane [190], this implies part (i) of the lemma. Part (ii) is proved in FAC §65, see also EGA II [142, (3.4.4)], Hartshorne [147, Prop. 5.15]. □

**Warning 3.9.5** Even if $M$ is finitely generated, $\Gamma_*(M^\sim)$ need not be, because there may be infinitely many negative degrees in which it is non-zero.

**Definition 3.9.6** If $M$ is a graded $R$-module and $d \in \mathbb{Z}$, we write $M_{\geq d}$ for the graded $R$-module $\bigoplus_{j \geq d} M_j$ obtained by replacing the terms in $M$ of degree strictly less than $d$ by zero.

**Lemma 3.9.7** *If $M$ is a finitely generated graded $R$-module then the unit of the adjunction $M \to \Gamma_*(M^\sim)$ is an isomorphism in large enough degrees, in the sense that there exists $d \in \mathbb{Z}$ such that $M_{\geq d} \to \Gamma_*(M^\sim)_{\geq d}$ is an isomorphism.*

*Proof.* An element of $\Gamma_j(M^\sim)$ may be described as follows. For each affine open set $U_i \subseteq \mathbb{P}^{r-1}$ defined by $Y_i$, we must name a fraction $m_i/Y_i^{n_i}$ with $\deg(m_i) = j + n_i$, $n_i \geq 0$, in such a way that for $i \neq j$ we have $Y_j^{n_j} m_i = Y_i^{n_i} m_j$. Two such elements are equivalent, $(m_1/Y_1^{n_1}, \dots, m_r/Y_r^{n_r}) \sim (m'_1/Y_1^{n'_1}, \dots, m'_r/Y_r^{n'_r})$, if each $Y_i^{n'_i} m_i - Y_i^{n_i} m'_i$ is annihilated by some power of $Y_i$. The map $\beta_M$ takes $m \in M_j$ to $(m/1, \dots, m/1)$.

The kernel of $\beta_M$ is the set of $m \in M$ annihilated by some power of each $Y_i$, and hence by every element of large enough degree. Since $M$ is finitely generated, this submodule is also finitely generated, and hence non-zero only in a finite number of degrees.

Similarly, for any $(m_1/Y_1^{n_1}, \dots, m_r/Y_r^{n_r})$ in $\Gamma_j(M)$, multiplication by $Y_i^{n_i}$ takes it to an element equivalent to $(m_i/1, \dots, m_i/1)$. So every element of the cokernel is annihilated by some power of each $Y_i$, and hence by every element of large enough degree. But the cokernel is not necessarily finitely generated, so a further argument is still necessary.

We note that since $M$ is finitely generated, it has a finite filtration with filtered quotients of the form $R/\mathfrak{p}(n)$, where $\mathfrak{p}$ is a homogeneous prime ideal and $n \in \mathbb{Z}$. A short exact sequence of modules $0 \to M \to M' \to M'' \to 0$ gives rise to a diagram

$$\begin{array}{ccccccc}
0 & \longrightarrow & M & \longrightarrow & M' & \longrightarrow & M'' & \longrightarrow & 0 \\
 & & \downarrow & & \downarrow & & \downarrow & & \\
0 & \longrightarrow & \Gamma_*(M^\sim) & \longrightarrow & \Gamma_*(M'^\sim) & \longrightarrow & \Gamma_*(M''^\sim) & &
\end{array}$$

and so it suffices to prove the theorem with $M = R/\mathfrak{p}(n)$, or equivalently with $M = R/\mathfrak{p}$. Replacing $R$ by $R/\mathfrak{p}$, we may assume that $\mathfrak{p} = 0$, $R$ is a graded integral domain and $M = R$, so that $M^\sim = \mathcal{O}_V$. In this case, $\alpha\colon R \to \Gamma_*(\mathcal{O}_V)$ is an embedding of $R$ as a subring of $\bigcap_{i=1}^r R[Y_i^{-1}]$, where the intersection takes place in the graded field of fractions of $R$.

Let $x = (x_1/Y_1^{n_1}, \dots, x_r/Y_r^{n_r}) \in \Gamma_j(\mathcal{O}_V)$ with $j \geq 0$. Then for some $n \geq 0$, every element of $R_{\geq n}$ sends $x$ into $R$, and since $j \geq 0$, it sends it into $R_{\geq n}$. So by induction on $q$, for every $q \geq 1$ we have $Y_1^n x^q \in R_{\geq n}$. Thus for all $q \geq 1$ we have $x^q \in Y_1^{-n} R$, which is a finitely generated $R$-module. So $x$ is in the integral closure of $R$ in its homogeneous field of fractions, which is a finitely generated $R$-module.

Thus $\Gamma_{\geq 0}(\mathcal{O}_V)$ is a finitely generated $R$-module. So the part of the cokernel in non-negative degrees is also finitely generated, and is annihilated by $R_{\geq n}$, and is hence zero in all large enough degrees. □

**Definition 3.9.8** We say that a graded $R$-module $M$ is *quasi-finitely generated* if for some $d \in \mathbb{Z}$, $M_{\geq d}$ is finitely generated.[1]

We say that $M$ has *finite length* if the total dimension of $M$ as a $k$-vector space is finite, and that $M$ has *quasi-finite length* if for some $d \in \mathbb{Z}$, $M_{\geq d}$ is the zero module.

We write $\mathsf{mod}(R)$, $\mathsf{qmod}(R)$, $\mathsf{fl}(R)$, $\mathsf{qfl}(R)$ for the categories of $R$-modules which are finitely generated, respectively quasi-finitely generated, finite length, quasi-finite length.

**Proposition 3.9.9** *If $\mathcal{F}$ is coherent then $\Gamma_*(\mathcal{F})$ is quasi-finitely generated. Furthermore, $\mathcal{F} \cong M^\sim$ for a suitable finitely generated $R$-module $M$.*

*Proof.* By a theorem of Serre (see Hartshorne [147, Theorem II.5.17]) $\mathcal{F}(n)$ is generated by a finite number of global sections for some large enough $n \in \mathbb{Z}$. Let $M$ be the submodule of $\Gamma_*(\mathcal{F})$ generated by these global sections. Then $M^\sim \to \Gamma_*(\mathcal{F})^\sim$ is an isomorphism. By Lemma 3.9.4 (ii), $\Gamma_*(\mathcal{F})^\sim \to \mathcal{F}$ is also an isomorphism, so $\mathcal{F} \cong M^\sim$, and $M$ is finitely generated. It now follows from Lemma 3.9.7 that there exists $d \in \mathbb{Z}$ such that $M_{\geq d} \to \Gamma_*(\mathcal{F})_{\geq d}$ is an isomorphism, and so $\Gamma_*(\mathcal{F})$ is quasi-finitely generated. □

**Definition 3.9.10** We write $\mathsf{mod}(R)/\mathsf{fl}(R)$ for the category whose objects are the same as those in $\mathsf{mod}(R)$, but whose morphisms are given by

$$\mathrm{Hom}_{\mathsf{mod}(R)/\mathsf{fl}(R)}(M, M') = \lim_{d\to\infty} \mathrm{Hom}_R(M_{\geq d}, M'_{\geq d}).$$

In other words, a morphism from $M$ to $M'$ in $\mathsf{mod}(R)/\mathsf{fl}(R)$ is given by a homomorphism $M_{\geq d} \to M'_{\geq d}$ for some $d \in \mathbb{Z}$, and two such morphisms are equal if they agree for some $d' \geq d$.

Similarly we form the category $\mathsf{qmod}(R)/\mathsf{qfl}(R)$ where the objects are the same as those in $\mathsf{qmod}(R)$ and the morphisms are defined in the same way as above.

Another way of saying this is that $\mathsf{fl}(R)$, respectively $\mathsf{qfl}(R)$, is a Serre subcategory of $\mathsf{mod}(R)$, respectively $\mathsf{qmod}(R)$, with Serre quotients $\mathsf{mod}(R)/\mathsf{fl}(R)$ and $\mathsf{qmod}(R)/\mathsf{qfl}(R)$.

**Lemma 3.9.11** *The inclusion $\mathsf{mod}(R) \to \mathsf{qmod}(R)$ induces an equivalence of categories*

$$\mathsf{mod}(R)/\mathsf{fl}(R) \xrightarrow{\cong} \mathsf{qmod}(R)/\mathsf{qfl}(R).$$

[1] In FAC [224, §56], this is referred to as "$M$ vérifie la condition (TF)."

*Proof.* This follows from the following easily verified facts.

(1) Every quasi-finitely generated module $M$ is isomorphic in $\mathsf{qmod}(R)/\mathsf{qfl}(R)$ to the finitely generated module $M_{\geq d}$ for $d$ large enough.

(2) A finitely generated module is isomorphic to the zero module in $\mathsf{qmod}(R)/\mathsf{qfl}(R)$ if and only if it has finite length. □

**Theorem 3.9.12** *The functors* $\sim$ *and* $\Gamma_*$ *induce an equivalence of categories between* $\mathsf{qmod}(R)/\mathsf{qfl}(R)$ *and the category of coherent sheaves on* $V = \mathsf{Proj}\, R$.

*Proof.* This follows from Lemma 3.9.4, Lemma 3.9.7 and Proposition 3.9.9. □

**Corollary 3.9.13** *The functors* $\sim$ *and* $\Gamma_{\geq 0}$ *induce an equivalence of categories between* $\mathsf{mod}(R)/\mathsf{fl}(R)$ *and coherent sheaves on* $V = \mathsf{Proj}\, R$.

*Proof.* This follows from Theorem 3.9.12 and Lemma 3.9.11. □

The case of greatest interest to us is the following.

**Corollary 3.9.14** *The functors* $\sim$ *and* $\Gamma_{\geq 0}$ *induce an equivalence of categories between* $\mathsf{mod}(k[Y_1, \dots, Y_r])/\mathsf{fl}(k[Y_1, \dots, Y_r])$ *and coherent sheaves on* $\mathbb{P}^{r-1}$. □

## 3.10 Cohomology of Sheaves

Let $\mathcal{O}$ be a sheaf of rings on a space $X$, and consider the category of $\mathcal{O}$-modules. We saw in Remark 3.6.4 that a short exact sequence of $\mathcal{O}$-modules $0 \to \mathcal{F} \to \mathcal{F}' \to \mathcal{F}'' \to 0$ gives rise to an exact sequence

$$0 \to \Gamma(X, \mathcal{F}) \to \Gamma(X, \mathcal{F}') \to \Gamma(X, \mathcal{F}'').$$

In other words, $\Gamma(X, -)$ is left exact, but not right exact. So it has right derived functors $H^i(X, -)$, which we now describe.

**Lemma 3.10.1** *The category of* $\mathcal{O}$*-modules has enough injectives.*

*Proof.* See Hartshorne [147, Proposition III.2.2]. □

**Example 3.10.2** If $V = \mathsf{Spec}\, R$ is an affine variety and $I$ is an injective $R$-module then $I^{\sim}$ is an injective $\mathcal{O}_V$-module.

It follows that every $\mathcal{O}$-module $\mathcal{F}$ has an *injective resolution*, namely a sequence

$$0 \to \mathcal{I}_0 \to \mathcal{I}_1 \to \mathcal{I}_2 \to \cdots$$

which is exact except in degree zero, where the kernel is isomorphic to $\mathcal{F}$. Applying the global section functor, we obtain a complex

$$0 \to \Gamma(X, \mathcal{I}_0) \to \Gamma(X, \mathcal{I}_1) \to \Gamma(X, \mathcal{I}_2) \to \cdots .$$

The cohomology of this complex in degree $i$ is denoted

$$H^i(X, \mathcal{F}).$$

A morphism of sheaves $\mathcal{F} \to \mathcal{F}'$ extends to a morphism of resolutions, unique up to homotopy,

$$\begin{array}{ccccccccc} 0 & \longrightarrow & \mathcal{I}_0 & \longrightarrow & \mathcal{I}_1 & \longrightarrow & \mathcal{I}_2 & \longrightarrow & \cdots \\ & & \downarrow & & \downarrow & & \downarrow & & \\ 0 & \longrightarrow & \mathcal{I}'_0 & \longrightarrow & \mathcal{I}'_1 & \longrightarrow & \mathcal{I}'_2 & \longrightarrow & \cdots \end{array}$$

and hence there is an induced map

$$H^i(X, \mathcal{F}) \to H^i(X, \mathcal{F}').$$

In particular, given two resolutions of the same sheaf, extending the identity map induces a natural isomorphism from $H^i(X, \mathcal{F})$ defined using one resolution to that defined using the other, so that this definition of cohomology of sheaves is independent of the choice of resolution up to natural isomorphism.

Since $\Gamma$ is left exact, we have a natural isomorphism

$$H^0(X, \mathcal{F}) \cong \Gamma(X, \mathcal{F}).$$

Given a short exact sequence of sheaves $0 \to \mathcal{F} \to \mathcal{F}' \to \mathcal{F}'' \to 0$, the horseshoe lemma constructs a short exact sequence of resolutions, and hence a long exact sequence in cohomology

$$\begin{aligned} 0 &\to \Gamma(X, \mathcal{F}) \to \Gamma(X, \mathcal{F}') \to \Gamma(X, \mathcal{F}'') \\ &\to H^1(X, \mathcal{F}) \to H^1(X, \mathcal{F}') \to H^1(X, \mathcal{F}'') \\ &\to H^2(X, \mathcal{F}) \to H^2(X, \mathcal{F}') \to H^2(X, \mathcal{F}'') \to \cdots . \end{aligned}$$

**Example 3.10.3** Let $V = \mathsf{Spec}\, R$ be an affine variety and let $M$ be an $R$-module. If

$$0 \to I_0 \to I_1 \to I_2 \to \cdots$$

is an injective resolution of $M$ as an $R$-module, then

$$0 \to I_0^\sim \to I_1^\sim \to I_2^\sim \to \cdots$$

is an injective resolution of $M^\sim$ because $^\sim$ is exact. Taking global sections gives us back the injective resolution of $M$. It follows that for $i > 0$ we have

$$H^i(V, M^\sim) = 0.$$

Thus quasi-coherent $\mathcal{O}_V$-modules with $V$ affine only have cohomology in degree zero.

This gives rise to a way of computing cohomology using a *Čech complex*. Let $\mathcal{O}$ be a sheaf of rings on a space $X$. Given a finite open cover $\mathfrak{U} = \{U_1, \ldots, U_r\}$ of a space $X$, and a sheaf $\mathcal{F}$ on $X$, we set

$$\check{C}^n(\mathfrak{U}, \mathcal{F}) = \bigoplus_{1 \le i_0 < \cdots < i_n \le r} \mathcal{F}(U_{i_0} \cap \cdots \cap U_{i_n}),$$

(so $\check{C}^n(\mathfrak{U}, \mathcal{F}) = 0$ for $n \geq r$).[2] We form a complex

$$0 \to \check{C}^0(\mathfrak{U}, \mathcal{F}) \xrightarrow{d} \check{C}^1(\mathfrak{U}, \mathcal{F}) \xrightarrow{d} \check{C}^2(\mathfrak{U}, \mathcal{F}) \xrightarrow{d} \cdots$$

via the alternating sum of the restriction maps

$$(dx)_{i_0,\ldots,i_{n+1}} = \sum_{j=0}^{n+1} (-1)^j x_{i_0,\ldots \overset{j}{\uparrow} \ldots, i_{n+1}} |_{U_{i_0} \cap \cdots \cap U_{i_{n+1}}}.$$

We write $\check{H}^i(\mathfrak{U}, \mathcal{F})$ for the cohomology of this Čech complex.

**Theorem 3.10.4** *Let $\mathfrak{U} = \{U_1, \ldots, U_r\}$ be an affine open cover of a quasi-projective variety $V$ and let $\mathcal{F}$ be a quasi-coherent $\mathcal{O}_V$-module. Then there are natural isomorphisms*

$$H^i(V, \mathcal{F}) \cong \check{H}^i(\mathfrak{U}, \mathcal{F}).$$

*In particular, $H^i(V, \mathcal{F}) = 0$ for $i \geq r$.*

*Proof.* See Theorem III.4.5 of Hartshorne [147]. □

**Corollary 3.10.5** *Let $V \subseteq \mathbb{P}^{r-1}$ be a projective variety and $\mathcal{F}$ a quasi-coherent $\mathcal{O}_V$-module. Then $H^i(V, \mathcal{F}) = 0$ for $i \geq r$.*

*Proof.* Let $U_i$ be the affine open subset of $V$ given by intersecting with the $i$th affine patch of $\mathbb{P}^{r-1}$, $1 \leq i \leq r$. Apply Theorem 3.10.4 to this open cover. □

2 These definitions extend in an obvious way to infinite totally ordered open covers, but we shall not need this.

# 4

# Jordan Type

In this chapter, following Friedlander, Pevtsova and Suslin [128] we introduce maximal Jordan type, generic Jordan type and modules of constant Jordan type, in the context of the representation theory of elementary abelian $p$-groups.

## 4.1 Nilvarieties

Suppose we are given a matrix $A \in \mathrm{Mat}_{n\times m}(k[V])$. Then given an integer $s$ with $0 \le s \le n$, for a point $v \in V$, the condition that

$$\mathrm{rank}\,(A(v)) \le s$$

is given by the vanishing of the set of $(s+1)\times(s+1)$ minors of $A$ (one for each subset of size $s+1$ of the rows and subset of size $s+1$ of the columns). Using the discussion in Section 3.2, we obtain the following.

**Theorem 4.1.1** *Given $A = A(\bar{Y}_1, \dots, \bar{Y}_r) \in \mathrm{Mat}_{n\times m}(k[V])$, the following are equivalent:*

(1) $\mathrm{rank}\,(A) = s$*, where $A$ is regarded as an element of* $\mathrm{Mat}_{n\times m}(k(V))$.
(2) $\mathrm{rank}\,(A(t_1, \dots, t_r)) = s$*, where $t_1, \dots, t_r$ are elements of some field extension $K$ of $k$ such that the ideal of polynomial relations satisfied by $t_1, \dots, t_r$ is the prime ideal $\mathfrak{p}$ defining $V$.*
(3) $\mathrm{rank}\,(A(v)) \le s$ *for all points $v \in V$, and there exists $v \in V$ with* $\mathrm{rank}\,(A(v)) = s$.
(4) $\mathrm{rank}\,(A(v)) = s$ *for all $v$ in a dense open subset of $V$.*

*Proof.* This follows from Lemma 3.2.1. □

We shall need the following consequence of Theorem 4.1.1 when we study generic kernels in Section 5.7 and generic images in Section 5.10.

**Proposition 4.1.2** *Let $A_1, A_2 \in \mathrm{Mat}_{n\times m}(k)$, regarded as maps from $k^m$ to $k^n$.*

(1) *Suppose that for all $\lambda, \mu \in k$, not both zero, $\lambda A_1 + \mu A_2$ is injective. Then*

$$\bigcap_{(\lambda,\mu)\neq(0,0)} \operatorname{Im}(\lambda A_1 + \mu A_2) = 0.$$

(2) *Suppose that for all $\lambda, \mu \in k$, not both zero, $\lambda A_1 + \mu A_2$ is surjective. Then*

$$\sum_{(\lambda,\mu)\neq(0,0)} \operatorname{Ker}(\lambda A_1 + \mu A_2) = k^m.$$

*Proof.* Since these statements are dual, it suffices to prove (1). Suppose that $0 \neq v \in k^n$ is in the image of $\lambda A_1 + \mu A_2$ for all $(\lambda, \mu) \neq (0, 0)$. This condition is equivalent to the statement that the augmented matrix $(\lambda A_1 + \mu A_2 | v)$ has rank $m$ for all $(\lambda, \mu) \neq (0, 0)$, and so by Theorem 4.1.1 it remains true over any extension field of $k$. In particular, it holds over the field of Laurent series $k[[t]][t^{-1}]$. So there exist $u_i \in k^m$ ($i \in \mathbb{Z}$) such that $(A_1 + tA_2) \sum_i u_i t^i = v$, with $u_i = 0$ for $i$ sufficiently large negative. Comparing coefficients of powers of $t$, we see that $u_i = 0$ for $i < 0$ and that $A_1 u_0 = v$, $A_2 u_i = A_1 u_{i+1}$ ($i \geq 0$). Since $k^m$ is finite dimensional, there exists a non-trivial linear relation between the $u_i$ with $i \geq 0$. Choose $s$ such that $u_s \in \langle u_0, \ldots, u_{s-1}\rangle$. Then

$$A_2 u_s \in \langle A_2 u_0, \ldots, A_2 u_{s-1}\rangle = \langle A_1 u_1, \ldots, A_1 u_s\rangle \subseteq \langle A_1 u_0, \ldots, A_1 u_{s-1}\rangle.$$

Then for all $(\lambda, \mu) \neq (0, 0)$ the map

$$\lambda A_1 + \mu A_2 \colon \langle u_0, \ldots, u_{s-1}\rangle \to \langle A_1 u_0, \ldots, A_1 u_{s-1}\rangle$$

is injective, and hence an isomorphism. This is impossible, because it implies that $A_1^{-1} A_2$ has no eigenvectors on $\langle u_0, \ldots, u_{s-1}\rangle$. □

Another consequence of Theorem 4.1.1 of more immediate use to us is the following.

**Corollary 4.1.3** *Let $A_1, \ldots, A_r \in \mathrm{Mat}_n(k)$. Then the following are equivalent:*

(1) *For all $\lambda_1, \ldots, \lambda_r \in k$ the matrix $\lambda_1 A_1 + \cdots + \lambda_r A_r$ is nilpotent.*
(2) *For all field extensions $K$ of $k$ and for all $t_1, \ldots, t_r \in K$, the matrix $t_1 A_1 + \cdots + t_r A_r$ is nilpotent.*
(3) *The matrix $t_1 A_1 + \cdots + t_r A_r \in \mathrm{Mat}_n(k(t_1, \ldots, t_r))$ is nilpotent, where $t_r, \ldots, t_r$ are independent transcendentals over $k$.*

*Proof.* This follows by applying Theorem 4.1.1 to the powers of $\lambda_1 A_1 + \cdots + \lambda_r A_r$ with $V = \mathbb{A}^r(k)$. □

**Definition 4.1.4** We say that $A_1, \ldots, A_r$ span a *nilvariety* of rank $r$ if the equivalent conditions of Corollary 4.1.3 are satisfied.

**Example 4.1.5** If $A_1, \ldots, A_r$ are commuting nilpotent matrices then they span a nilvariety. If, in addition, $k$ has characteristic $p$ and the $p$th powers of $A_1, \ldots, A_r$ are zero, then this nilvariety corresponds to the representation of the elementary abelian $p$-group $\langle g_1, \ldots, g_r \rangle \cong (\mathbb{Z}/p)^r$ given by $g_i \mapsto I + A_i$.

**Example 4.1.6** The matrices

$$A_1 = \begin{pmatrix} 0 & 1 & 0 \\ 0 & 0 & 1 \\ 0 & 0 & 0 \end{pmatrix} \quad \text{and} \quad A_2 = \begin{pmatrix} 0 & 0 & 0 \\ 1 & 0 & 0 \\ 0 & -1 & 0 \end{pmatrix}$$

span a nilvariety, but do not commute. We have

$$(Y_1 A_1 + Y_2 A_2)^3 = 0$$

for all values of $Y_1$ and $Y_2$. In fact, for all values of $Y_1$ and $Y_2$, not both zero, the matrix $Y_1 A_1 + Y_2 A_2$ has a single Jordan block with eigenvalue zero. We shall see this example again in Section 10.6.

**Example 4.1.7** Similar examples of larger dimension may also be constructed. The following examples of dimension five and seven with a single Jordan block are taken from a paper of Causa, Re and Teodorescu [85]. In each case we give the matrix $Y_1 A_1 + Y_2 A_2$:

$$\begin{pmatrix} 0 & Y_1 & 0 & 0 & 0 \\ 2Y_2 & 0 & Y_1 & 0 & 0 \\ 0 & -Y_2 & 0 & Y_1 & 0 \\ 0 & 0 & Y_2 & 0 & Y_1 \\ 0 & 0 & 0 & -2Y_2 & 0 \end{pmatrix}, \qquad \begin{pmatrix} 0 & 0 & -Y_2 & 0 & 0 \\ 0 & 0 & Y_1 & 0 & -Y_2 \\ 0 & 0 & 0 & 0 & Y_1 \\ Y_1 & Y_2 & 0 & 0 & 0 \\ Y_2 & 0 & 0 & -Y_1 & 0 \end{pmatrix},$$

$$\begin{pmatrix} 0 & Y_1 & 0 & 0 & 0 & 0 & 0 \\ -4Y_2 & 0 & Y_1 & 0 & 0 & 0 & 0 \\ 0 & 2Y_2 & 0 & Y_1 & 0 & 0 & 0 \\ 0 & 0 & -Y_2 & 0 & Y_1 & 0 & 0 \\ 0 & 0 & 0 & Y_2 & 0 & Y_1 & 0 \\ 0 & 0 & 0 & 0 & -2Y_2 & 0 & Y_1 \\ 0 & 0 & 0 & 0 & 0 & 4Y_2 & 0 \end{pmatrix}.$$

It is shown in [85] that in each odd dimension $n$ the variety of matrices looking like the first and third of these examples but with possibly different coefficients below the diagonal is a complete intersection of dimension $(n-3)/2$, and that there are no even-dimensional examples. See Theorems 10.10.1 and 10.10.2.

**Corollary 4.1.8** *For $A, B_1, \ldots, B_r \in \mathrm{Mat}_n(k)$, the following are equivalent.*

(1) $\mathrm{rank}\,(A) = \mathrm{rank}\,(A + t_1 B_1 + \cdots + t_r B_r)$, *where $t_1, \ldots, t_r$ are independent transcendentals over $k$.*

(2) $\operatorname{rank}(A) \geq \operatorname{rank}(A+\lambda_1 B_1+\cdots+\lambda_r B_r)$ *for every* $(\lambda_1, \ldots, \lambda_r) \in \mathbb{A}^r(K)$, *for every extension field* $K$ *of* $k$.
(3) $\operatorname{rank}(A) \geq \operatorname{rank}(A+\lambda_1 B_1+\cdots+\lambda_r B_r)$ *for every* $(\lambda_1, \ldots, \lambda_r) \in \mathbb{A}^r(k)$.
(4) $\operatorname{rank}(A) = \operatorname{rank}(A+\lambda_1 B_1+\cdots+\lambda_r B_r)$ *for* $(\lambda_1, \ldots, \lambda_r)$ *in a dense open subset of* $\mathbb{A}^r(k)$.

*Proof.* This also follows from Theorem 4.1.1; the open subset where the ranks are equal is non-empty because it contains the origin, and is therefore dense. □

**Definition 4.1.9** Let $A, B_1, \ldots, B_r \in \operatorname{Mat}_n(k)$. We say that $A$ is *rank maximal* with respect to $B_1, \ldots, B_r$ if the equivalent conditions of Corollary 4.1.8 are satisfied.

**Remark 4.1.10** It follows from Corollary 4.1.8 that if $A$ is rank maximal with respect to $B_1, \ldots, B_r$ and $\lambda_1, \ldots, \lambda_r$ are such that

$$\operatorname{rank}(A) = \operatorname{rank}(A + \lambda_1 B_1 + \cdots + \lambda_r B_r)$$

then $A + \lambda_1 B_1 + \cdots + \lambda_1 B_r$ is rank maximal with respect to $A, B_1, \ldots, B_r$.

## 4.2 Matrices and Tangent Spaces

In preparation for the proof of Gerstenhaber's Theorem 4.3.1, in this section we examine the tangent space of a conjugacy class in $\operatorname{Mat}_n(k)$ at a matrix $A$.

**Definition 4.2.1** Let $A \in \operatorname{Mat}_n(k)$. We write $\mathcal{C}c\ell(A)$ for the locally closed subvariety of $\operatorname{Mat}_n(k)$ consisting of matrices conjugate to $A$. In other words, it is the set of matrices of the form $BAB^{-1}$ with $B \in GL_n(k)$.

**Proposition 4.2.2** *The tangent space* $T_{A,\mathcal{C}c\ell(A)}$ *consists of the matrices of the form* $WA - AW$ *with* $W \in \operatorname{Mat}_n(k)$.

*Proof.* Consider the morphism of algebraic varieties $\mathbb{A}^1 \to \mathcal{C}c\ell(A)$ given by

$$t \mapsto (I + tW)A(I + tW)^{-1} = A + t(WA - AW) + t^2 C(t)$$

where $C(t) \in \operatorname{Mat}_n(k(t))$ is defined at $t = 0$. It follows from Lemma 3.4.2 that $WA - AW \in T_{\mathcal{C}c\ell(A),A}$.

The morphism $GL_n(k) \to \mathcal{C}c\ell(A)$ given by $B \mapsto BAB^{-1}$ induces a surjective map $T_{GL_n(k),I} \to T_{\mathcal{C}c\ell(A),A}$, and so all elements of $T_{\mathcal{C}c\ell(A),A}$ are of this form. □

**Proposition 4.2.3** *Let* $U(t) \in GL_n(k(t))$ *be a matrix of rational functions, and suppose that* $A \in \operatorname{Mat}_n(k)$ *satisfies*

$$U(t)AU(t)^{-1} = A + t^j B + t^{j+1} C(t)$$

*where* $B \in \mathrm{Mat}_n(k)$ *and* $C(t) \in \mathrm{Mat}_n(k(t))$ *such that* $C(0)$ *is defined. Then* $B \in T_{\mathcal{C}\ell(A),A}$, *and there is a matrix* $W$ *such that* $B = WA - AW$.

*Proof.* We begin by remarking that $U(0)$ does not have to be defined, and even if it is defined, it does not have to be invertible. However, $\det U(t) \in k(t)$ has only a finite number of zeros and poles, so there is a dense open subset of $\mathbb{A}^1(k)$ consisting of those $t$ such that $U(t)AU(t)^{-1}$ is defined. We can also see from the right-hand side of the equation that this definition extends to a regular map in an open neighbourhood of $t = 0$ that takes the value $A$ at $t = 0$. The function

$$f(t) = A + t^j B + t^{j+1} C(t)$$

is therefore a rational map from $\mathbb{A}^1$ to $\mathcal{C}\ell(A) \subseteq \mathbb{A}^{n^2}$ that is defined at $t = 0$. It follows by Lemma 3.4.2 that we have $B \in T_{\mathcal{C}\ell(A),A}$. The existence of $W$ then follows from Proposition 4.2.2. □

## 4.3 A Theorem of Gerstenhaber

**Theorem 4.3.1** (Gerstenhaber [131, Prop. 1.9]) *Let* $A, B \in \mathrm{Mat}_n(k)$. *Then the following are equivalent:*

(1) *If* $t$ *is transcendental over* $k$ *then* $A \sim A + tB$ *as elements of* $\mathrm{Mat}_n(k(t))$.
(2) *There exists a finite sequence of matrices* $W_1, W_2, \ldots, W_s$ *such that the following equations hold:*

$$W_1 A - AW_1 = B, \quad W_i A - AW_i = BW_{i-1} \textit{ for } i = 2, \ldots, s, \textit{ and } BW_s = 0.$$

(3) *There exists an infinite sequence of matrices* $W_1, W_2, \ldots$ *such that the following equations hold:*

$$W_1 A - AW_1 = B, \quad W_i A - AW_i = BW_{i-1} \textit{ for } i \geq 2.$$

*Proof.* Since there are missing details in [131], we have chosen to write out the proof in detail. If (1) holds, then there is a matrix $W(t) \in \mathrm{Mat}_n(k(t))$ such that

$$W(t)AW(t)^{-1} = A + tB.$$

Rewriting this as

$$W(t)A = (A + tB)W(t) \tag{4.1}$$

and comparing coefficients, we see that (3) is the statement that $W(t)$ may be chosen to be a power series in $t$ of the form

$$I + W_0 t + W_1 t^2 + \cdots \tag{4.2}$$

while (2) is the statement that $W(t)$ may be chosen to be a polynomial in $t$ of this form. It is clear that (2) implies (1), so we shall show that (1) implies (3), which in turn implies (2).

We begin by showing that (1) implies (3). Given $U(t) \in \mathrm{Mat}_n(k(t))$ conjugating $A$ to $A + tB$, we shall inductively construct a power series $W(t) \in \mathrm{Mat}_n(k[[t]])$ of the form (4.2) which also conjugates $A$ to $A + tB$.

Now examine the equation $U(t)AU(t)^{-1} = A+tB$. Applying Proposition 4.2.3 to this equation, we see that there exists a matrix $W_1$ such that $W_1A - AW_1 = B$.

Next, examine the expression

$$(I + tW_1)^{-1}(A + tB)(I + tW_1) = A + t^2BW_1 + \cdots$$

as elements of $\mathrm{Mat}_n(k[[t]])$. Since this matrix is also conjugate to $A$ in $\mathrm{Mat}_n(k(t))$, it follows that $BW_1 \in T_{\mathcal{C}\ell(A),A}$ and again applying Proposition 4.2.3, there exists a matrix $W_2$ such that $W_2A - AW_2 = B$.

Proceeding by induction, we obtain matrices $W_1, W_2, \ldots$ such that for all $i \geq 1$ we have

$$\begin{aligned}(I + tW_1 + \cdots + t^{i-1}W_{i-1})^{-1}(A + tB)(I + tW_1 + \cdots + t^{i-1}W_{i-1})\\ = A + t^iBW_{i-1} + \cdots .\end{aligned}$$

Therefore $BW_{i-1} \in T_A$, and this allows us to find a matrix $W_i$ such that $W_iA - AW_i = BW_{i-1}$.

To see that (3) implies (2), we note that given any sequence $W_1', W_2', \ldots$ satisfying (3) and a sequence $K_1, K_2, \ldots$ of matrices that commute with $A$, the sequence $W_1, W_2, \ldots$ defined by

$$W_i = W_i' + W_{i-1}'K_1 + \cdots + W_1'K_{i-1} + K_i$$

also satisfies (3).

For $s$ sufficiently large, the subspace of $\mathrm{Mat}_n(k)$ spanned by $I, W_1', W_2', \ldots$ contains $W_s'$, and so we may find scalar multiples of the identity $K_1, \ldots, K_s$ such that

$$W_s = W_s' + W_{s-1}'K_1 + \cdots + W_1'K_{s-1} + K_s = 0.$$

Then we can take $W_i = 0$ for $i \geq s$, and we see that (2) is satisfied. □

**Corollary 4.3.2** *Let $A, B \in \mathrm{Mat}_n(k)$, and let $C \in \mathrm{Mat}_n(k)$ be a matrix commuting with $A$ and $B$. If $A \sim A + tB$ in $\mathrm{Mat}_n(k(t))$ then also $A \sim A + tBC$.*

*Proof.* The matrices $C^iW_i$ satisfy:

$$(CW_1)A - A(CW_1) = BC, \quad (C^iW_i)A - A(C^iW_i) = (BC)(C^{i-1}W_{i-1})$$
$$\text{for } i = 2, \ldots, s, \quad \text{and } (BC)(C^sW_s) = 0.$$
□

The version in several variables works the same way, but with more notation.

**Theorem 4.3.3** *Let $A, B_1, \ldots, B_r \in \mathrm{Mat}_n(k)$. Then the following are equivalent:*

(1) *If $t_1, \ldots, t_r$ are independent transcendentals over $k$ then $A \sim A + t_1B_1 + \cdots + t_rB_r$ as elements of $\mathrm{Mat}_n(k(t_1, \ldots, t_r))$.*

(2) *There exist matrices* $W_{i_1,\ldots,i_r}$, *one for each sequence* $(i_1, \ldots, i_r)$ *of integers, such that*

$$W_{i_1,\ldots,i_r} A - A W_{i_1,\ldots,i_r} = B_1 W_{i_1-1,\ldots,i_r} + \cdots + B_r W_{i_1,\ldots,i_r-1},$$

*and furthermore,* $W_{i_1,\ldots,i_r} = 0$ *if and only of the indices are negative,* $W_{0,\ldots,0} = I$, *and:*

$$W_{i_1,\ldots,i_r} = 0 \text{ for all but finitely many values of } (i_1, \ldots, i_r).$$

(3) *There exist matrices* $W_{i_1,\ldots,i_r}$ *as in (2) but without the last finiteness condition.*

*Proof.* This proof follows the same lines as the proof of Theorem 4.4.10. This time, the conjugating matrix is given by

$$W(t_1, \ldots, t_r) = \sum_{(i_1,\ldots,i_r)} W_{i_1,\ldots,i_r} t_1^{i_1} \ldots t_r^{i_r}$$

and the conjugation equation is given by

$$W(t_1, \ldots, t_r)A = (A + t_1 B_1 + \cdots + t_r B_r) W(t_1, \ldots, t_r). \tag{4.3}$$

□

## 4.4 Dominance Order and Nilpotent Jordan Types

**Definition 4.4.1** The *dominance order* is the partial order on partitions of $n$ defined as follows: $[a_1] \ldots [a_t] \succeq [b_1] \ldots [b_t]$ if and only if for all $i$ with $1 \le i \le t$

$$\sum_{j=1}^{i} a_j \ge \sum_{j=1}^{i} b_j.$$

**Example 4.4.2** It is easy to check that $[3]\,[1] \succeq [2]\,[2]$. More generally, if $a_{i-1} > a_i$ and $a_j > a_{j+1}$ (with $i < j$) then

$$[a_1] \ldots [a_i + 1] \ldots [a_j - 1] \ldots [a_t] \succeq [a_1] \ldots [a_i] \ldots [a_j] \ldots [a_t].$$

It is not difficult to verify that $[a_1] \ldots [a_t] \succeq [b_1] \ldots [b_t]$ if and only if we can move from one to the other through a sequence of such steps (see for example James and Kerber [171, §1.4]).

**Lemma 4.4.3** *Conjugation of partitions reverses the dominance ordering:* $\mathbf{a} \succeq \mathbf{b}$ *if and only if* $\mathbf{b}' \succeq \mathbf{a}'$. *The latter condition may be written as follows:*

$$\forall\, s > 0 \quad \sum_j \min(a_j, s) \le \sum_j \min(b_j, s).$$

*Proof.* See James and Kerber [171, §1.4]. □

If $A$ is a nilpotent matrix then all its eigenvalues are zero, so its Jordan canonical form is a diagonal sum of matrices $J_a$, where $J_a$ is a Jordan block of length $a$ with eigenvalue zero, namely the $a \times a$ matrix

$$\begin{pmatrix} 0 & 0 & \cdots & 0 & 0 \\ 1 & 0 & & 0 & 0 \\ 0 & 1 & \cdots & 0 & 0 \\ \vdots & & \ddots & & \vdots \\ 0 & 0 & \cdots & 1 & 0 \end{pmatrix}.$$

We write $J_{[a_1]\dots[a_t]}$ for an $n \times n$ matrix which is a diagonal sum of Jordan blocks of lengths $a_1, \dots, a_t$ with $a_1 \geq \cdots \geq a_t$. So letting $GL(n,k)$ act on $\mathrm{Mat}_n(k)$ by conjugation, this puts the $GL(n,k)$-orbits of nilpotent matrices in one to one correspondence with partitions of $n$.

**Definition 4.4.4** We put a partial order on nilpotent $n \times n$ matrices, which we also call the *dominance order* as follows. If $A, B \in \mathrm{Mat}_n(k)$ are nilpotent, we write $A \succeq B$ if for all $s > 0$

$$\mathrm{rank}\,(A^s) \geq \mathrm{rank}\,(B^s).$$

This depends only on the Jordan canonical forms of $A$ and $B$, and so we can regard it as a partial order on nilpotent Jordan types.

**Lemma 4.4.5** *The dominance order on nilpotent matrices corresponds to the dominance order on partitions, in the sense that $J_{[a_1]\dots[a_t]} \succeq J_{[b_1]\dots[b_t]}$ if and only if $[a_1]\dots[a_t] \succeq [b_1]\dots[b_t]$.*

*Proof.* Let $A = J_{[a_1]\dots[a_t]}$, $B = J_{[b_1]\dots[b_t]}$. Then using Lemma 4.4.3 we have

$$\begin{aligned} A \succeq B &\iff \forall s > 0 \quad \mathrm{nullity}(A^s) \leq \mathrm{nullity}(B^s) \\ &\iff \forall s > 0 \quad \sum_j \min(a_j, s) \leq \sum_j \min(b_j, s) \\ &\iff [a_1]\dots[a_t] \succeq [b_1]\dots[b_t]. \end{aligned}$$

□

**Definition 4.4.6** We write $A \sim B$ to denote that $n \times n$ matrices $A$ and $B$ are conjugate; namely if there exists an invertible matrix $S$ such that $S^{-1}AS = B$.

**Lemma 4.4.7** *For nilpotent $n \times n$ matrices $A$ and $B$, we have $A \succeq B$ and $B \succeq A$ if and only if $A \sim B$.*

*Proof.* This follows from the fact that $A \sim B$ if and only if $A$ and $B$ have the same Jordan canonical form. □

The following proposition is due to Gerstenhaber [131, Prop. 1.6]. See also Gerstenhaber [133], Hesselink [153], O'Halloran [201].

**Proposition 4.4.8** *If $A$ and $B$ are nilpotent $n \times n$ matrices then the following are equivalent:*

(1) $A \succeq B$.
(2) *The $GL_n(k)$-orbit of $B$ is in the closure of the $GL_n(k)$-orbit of $A$.*

*Proof.* Since $\operatorname{rank}(A) \leq s$ is a closed condition (see §3.2), (2) imples (1). To prove the converse, it suffices by the observation of Example 4.4.2 to prove that if $a_1 \geq a_2 > 0$ then $J_{[a_1][a_2]}$ is in the closure of the orbit of $J_{[a_1+1][a_2-1]}$. To see this, write down a one-parameter family of matrices of the form

$$\begin{pmatrix} J_{a_1} & 0 \\ U & J_{a_2} \end{pmatrix}$$

where

$$U = \begin{pmatrix} 0 & \dots & 0 \\ \vdots & & \vdots \\ 0 & \dots & t \end{pmatrix}$$

is a matrix with $t$ in the bottom right-hand corner and all other entries equal to zero. When $t = 0$ this is $J_{[a_1][a_2]}$, while if $t \neq 0$, computing ranks of powers shows that it is conjugate to $J_{[a_1+1][a_2-1]}$. □

**Example 4.4.9** Consider the one-parameter family of matrices

$$\begin{pmatrix} 0 & 0 & 0 & 0 \\ 1 & 0 & 0 & 0 \\ 0 & 0 & 0 & 0 \\ 0 & t & 1 & 0 \end{pmatrix}.$$

When $t = 0$ this has Jordan type $[2]^2$, whereas when $t \neq 0$ it has Jordan type $[3]\,[1]$. This shows that $J_{[2]^2}$ is in the closure of the orbit of $J_{[3]\,[1]}$.

**Theorem 4.4.10** *Let $A, B \in \operatorname{Mat}_n(k)$ be matrices spanning a nilvariety. Then the following are equivalent:*

(1) *For all $\lambda \in k$, $A \succeq A + \lambda B$.*
(2) *For all extension fields $K$ of $k$ and for all $t \in K$, $A \succeq A + tB$.*
(3) *If $t$ is transcendental over $k$ then $A \sim A + tB$ as elements of $\operatorname{Mat}_n(k(t))$.*
(4) *There exists a finite sequence of matrices $W_1, W_2, \dots, W_s$ such that the following equations hold:*

$$W_1A - AW_1 = B, \quad W_iA - AW_i = BW_{i-1} \text{ for } i = 2, \dots, s, \text{ and } BW_s = 0.$$

(5) *There exists an infinite sequence of matrices $W_1, W_2, \dots$ such that the following equations hold:*

$$W_1A - AW_1 = B, \quad W_iA - AW_i = BW_{i-1} \text{ for } i \geq 2.$$

*Proof.* This follows from Theorem 4.3.1 and Corollary 4.1.8. □

**Definition 4.4.11** Let $A, B \in \mathrm{Mat}_n(k)$ span a nilvariety. We say that $A$ is *maximal* with respect to $B$ if the equivalent conditions of the theorem above are satisfied.

**Theorem 4.4.12** *Let $A, B_1, \ldots, B_r \in \mathrm{Mat}_n(k)$ be matrices spanning a nilvariety. Then the following are equivalent:*

(1) *For all $\lambda_1, \ldots, \lambda_r \in k$, $A \succeq A + \lambda_1 B_1 + \cdots + \lambda_r B_r$.*
(2) *For all extension fields $K$ of $k$ and for all $t_1, \ldots, t_r \in K$, $A \succeq A + t_1 B_1 + \cdots + t_r B_r$.*
(3) *If $t_1, \ldots, t_r$ are independent transcendentals over $k$ then $A \sim A + t_1 B_1 + \cdots + t_r B_r$ as elements of $\mathrm{Mat}_n(k(t_1, \ldots, t_r))$.*
(4) *There exist matrices $W_{i_1,\ldots,i_r}$, one for each sequence $(i_1, \ldots, i_r)$ of integers, such that*

$$W_{i_1,\ldots,i_r} A - A W_{i_1,\ldots,i_r} = B_1 W_{i_1-1,\ldots,i_r} + \cdots + B_r W_{i_1,\ldots,i_r-1},$$

*and furthermore, $W_{i_1,\ldots,i_r} = 0$ if and only of the indices are negative, $W_{0,\ldots,0} = I$, and:*

$$W_{i_1,\ldots,i_r} = 0 \textit{ for all but finitely many values of } (i_1, \ldots, i_r).$$

(5) *There exist matrices $W_{i_1,\ldots,i_r}$ as in (4) but without the last finiteness condition.*

*Proof.* This follows from Theorem 4.3.3 and Corollary 4.1.8. □

**Definition 4.4.13** Let $A, B_1, \ldots, B_r \in \mathrm{Mat}_n(k)$ span a nilvariety. We say that $A$ is *maximal* with respect to $B_1, \ldots, B_r$ if the equivalent conditions of the theorem above are satisfied.

**Remark 4.4.14** Just as in Remark 4.1.10, it follows from Theorem 4.4.12 that if $A$ is maximal with respect to $B_1, \ldots, B_r$ and $\lambda_1, \ldots, \lambda_r$ are such that

$$A \sim A + \lambda_1 B_1 + \cdots + \lambda_r B_r$$

then $A + \lambda_1 B_1 + \cdots + \lambda_1 B_r$ is maximal with respect to $A, B_1, \ldots, B_r$.

**Theorem 4.4.15** *Let $A, B_1, \ldots, B_r \in \mathrm{Mat}_n(k)$ span a nilvariety, and let $C_1, \ldots, C_r \in \mathrm{Mat}_n(k)$ be matrices commuting with each other and with $A, B_1, \ldots, B_r$. If $A$ is maximal with respect to $B_1, \ldots, B_r$ then it is also maximal with respect to $B_1, \ldots, B_r, B_1C_1, \ldots, B_rC_r$.*

*Proof.* Look at the equation

$$\begin{aligned}&\tilde{W}(t_1, \ldots, t_r, s_1, \ldots, s_r)A \\ &\quad = (A + t_1 B_1 + \cdots + t_r B_r + s_1 B_1 C_1 + \cdots + s_r B_r C_r)\tilde{W}(t_1, \ldots, t_r, s_1, \ldots, s_r).\end{aligned} \tag{4.4}$$

Setting

$$\tilde{W}(t_1, \ldots, t_r, s_1, \ldots, s_r) = \sum \tilde{W}_{i_1,\ldots,i_r,j_1,\ldots,j_r} t_1^{i_1} \ldots t_r^{i_r} s_1^{j_1} \ldots s_r^{j_r},$$

we see by comparing coefficients that we can satisfy this equation by setting

$$\tilde{W}_{i_1,\ldots,i_r,j_1,\ldots,j_r} = C_1^{j_1} \ldots C_r^{j_r} W_{i_1+j_1,\ldots,i_r+j_r}.$$

Namely, if $W(t_1, \ldots, t_r)$ satisfies Equation (4.3) then $\tilde{W}(t_1, \ldots, t_r, s_1, \ldots, s_r)$ satisfies Equation (4.4). □

**Remark 4.4.16** A different, inductive proof of the above theorem is given in [128].

The crucial theorem of Friedlander, Pevtsova and Suslin [128, Theorem 1.9], which appears below as Theorem 4.4.18, is a strengthening of Theorem 4.4.15 in the case where $C$ is nilpotent. We begin with the single variable case.

**Proposition 4.4.17** *Suppose that $A, B \in \mathrm{Mat}_n(k)$ span a nilvariety and $C \in \mathrm{Mat}_n(k)$ is a nilpotent matrix commuting with $A$ and $B$.*

(1) *If $A$ is rank maximal with respect to $B$ then* $\mathrm{rank}\,(A) = \mathrm{rank}\,(A + BC)$.
(2) *If $A$ is maximal with respect to $B$ then $A \sim A + BC$.*

*Proof.* (1) Let $W = \sum_{\mu \in k^\times} \mathsf{Ker}\,(A + \mu B)$, the span of the kernels of $A + \mu B$ as $B$ runs over the non-zero elements of $k$. On each $\mathsf{Ker}\,(A + \mu B)$, and hence on $W$, the matrices $A$ and $B$ have the same image. So for all $\lambda \in k$ we have $(A + \lambda B)(W) \subseteq A(W)$, with equality for all but finitely many values of $\lambda \in k$.

Since $A$ is rank maximal with respect to $B$, we have

$$\dim \mathsf{Ker}\,(A + \lambda B) = \dim \mathsf{Ker}\, A$$

for all but finitely many values of $\lambda \in k$.

Combining these, for all but finitely many values of $\lambda \in k^\times$ we have

$$\dim \mathsf{Ker}\, A|_W = \dim \mathsf{Ker}\,(A + \lambda B)|_W = \dim \mathsf{Ker}\,(A + \lambda B) = \dim \mathsf{Ker}\, A,$$

where the second equality follows from the definition of $W$. It follows that $\mathsf{Ker}\,(A) \subseteq W$.

Now $C$ commutes with $A$ and $B$, and hence $C(\mathsf{Ker}\,(A+\mu B)) \subseteq \mathsf{Ker}\,(A+\mu B)$. On $\mathsf{Ker}\,(A+\mu B)$, $A$ acts in the same way as $-\mu B$ and so $A + BC$ acts in the same way as $A(I - \mu^{-1}C)$. Since $C$ is nilpotent, $I - \mu^{-1}C$ is invertible and so $A$ and $A + BC$ have the same image. It follows that $A(W) = (A + BC)(W)$.

Next, we show that $\mathsf{Ker}\,(A + BC) \subseteq W$. To do this, we prove by downward induction on $i$ that if $v \in \mathsf{Ker}\,(A + BC)$ then $C^i v \in W$, this being trivially true for large $i$ because $C$ is nilpotent: if $C^{i+1}v \in W$ then $AC^i v = -BC^{i+1}v \in B(W) = A(W)$. So there exists $w \in W$ such that $AC^i v = Aw$. Then $C^i v = (C^i v - w) + w \in \mathsf{Ker}\,(A) + W = W$.

We now have $\mathsf{Ker}\,(A) \subseteq W$, $\mathsf{Ker}\,(A+BC) \subseteq W$, and $A(W) = (A+BC)(W)$. So $\mathsf{Ker}\,(A)$ and $\mathsf{Ker}\,(A+BC)$ have the same codimension in $W$ and hence also in $k^n$, and so $\operatorname{rank}(A) = \operatorname{rank}(A+BC)$.

(2) This follows by applying (1) to the powers of $A$ and of $A+BC$. □

**Theorem 4.4.18** (Friedlander, Pevtsova and Suslin [128]) *Suppose that $A, B_1, \dots, B_r$ span a nilvariety and $C_1, \dots, C_r \in \mathrm{Mat}_r(k)$ are nilpotent matrices commuting with each other and with $A, B_1, \dots, B_r$. If $A$ is maximal with respect to $B_1, \dots, B_r$ then*

$$A \sim A + B_1C_1 + \cdots + B_rC_r.$$

*Proof.* By Theorem 4.4.15, $A$ is maximal with respect to $B_1, \dots, B_r, B_1C_1, \dots, B_rC_r$. By Proposition 4.4.17 we have $A \sim A + B_1C_1$ and so by Remark 4.4.14, it follows that $A + B_1C_1$ is also maximal with respect to $B_1, \dots, B_r, B_1C_1, \dots, B_rC_r$. Continuing by induction, we obtain

$$A \sim A + B_1C_1 \sim A + B_1C_1 + B_2C_2 \sim \cdots \sim A + B_1C_1 + \cdots + B_rC_r. \quad \square$$

## 4.5 Generic and Maximal Jordan Type

We now return to the representation theory of an elementary abelian $p$-group

$$E = \langle g_1, \dots, g_r \rangle \cong (\mathbb{Z}/p)^r$$

or an exterior algebra $\Lambda = \Lambda(X_1, \dots, X_r)$. Our notation is the same as in Section 1.1. In particular, in the case of an elementary abelian group we set $X_i = g_i - 1 \in J(kE)$. In either case, if

$$\alpha = (\lambda_1, \dots, \lambda_r) \in \mathbb{A}^r(k),$$

we write $X_\alpha = \sum_i \lambda_i X_i$ for the corresponding element of $J = J(kE)$ or $J(\Lambda)$, $g_\alpha = 1 + X_\alpha$, and if $\alpha \neq 0$, we write $E_\alpha = \langle g_\alpha \rangle$ for the corresponding cyclic shifted subgroup. We also recall the convention that when we talk of a $kE$-module or a $\Lambda$-module, we mean a *finitely generated* module.

Let $M$ be a $kE$-module or a $\Lambda$-module. Then for each $x \in J$ we can look at the Jordan type of the nilpotent matrix $x_M$ representing the action of $x$ on $M$ with respect to some given basis. In general, if $x, x' \in J$ with $x - x' \in J^2$ then it can happen that $x_M$ and $x'_M$ have different Jordan types. However, we'll show that generically this does not happen.

**Example 4.5.1** Let $p = 2$ and $r = 3$, and let $M$ be the four-dimensional $kE$-module given by

$$g_1 \mapsto \begin{pmatrix} 1&0&0&0\\ 1&1&0&0\\ 0&0&1&0\\ 0&0&1&1 \end{pmatrix} \quad g_2 \mapsto \begin{pmatrix} 1&0&0&0\\ 0&1&0&0\\ 1&0&1&0\\ 0&1&0&1 \end{pmatrix} \quad g_3 \mapsto \begin{pmatrix} 1&0&0&0\\ 0&1&0&0\\ 0&0&1&0\\ 1&0&0&1 \end{pmatrix}.$$

Then $X_3$ has Jordan type $[2]\,[1]^2$ while $X_3 + X_1X_2$ has Jordan type $[1]^4$. Note that $X_3 \equiv X_3 + X_1X_2 \pmod{J^2(kE)}$.

**Definition 4.5.2** We say that $x \in J$ has *maximal Jordan type* on $M$ if the Jordan type of $x_M$ is maximal in the dominance order (Definition 4.4.4) among the matrices of elements of $J$.

**Theorem 4.5.3** (Friedlander, Pevtsova and Suslin) *Let $M$ be a $kE$-module or a $\Lambda$-module, on which $J$ does not act as zero.*

(1) *The elements of $J^2$ are not of maximal Jordan type on $M$.*
(2) *All elements of maximal Jordan type on $M$ have the same Jordan type, and this is the same as the Jordan type of the generic element of $J$ on $M$.*
(3) *The set of elements of $J$ of maximal Jordan type on $M$ is a union of cosets of $J^2$ in $J$, and is the inverse image in $J$ of a homogeneous dense open subset of $(J/J^2) \setminus \{0\}$.*

*Proof.* Choose $0 \neq \beta \in \mathbb{A}^r(k)$ such that the Jordan type of $(X_\beta)_M$ is maximal among all $(X_\alpha)_M$. Then $(X_\beta)_M$ is maximal with respect to $(X_1)_M, \dots, (X_r)_M$ (Definition 4.4.11). It follows from Corollary 4.1.8 that there is a dense open subset $U$ of $\mathbb{A}^r(k)$ such that the Jordan type of $(X_\alpha)_M$ for $\alpha \in U$ is the same as the Jordan type of $(X_\beta)_M$. Since $J$ does not act as zero on $M$, this set $U$ does not contain the origin. Furthermore, there is a unique Jordan type for which this happens, because any two dense open subsets of $\mathbb{A}^r(k)$ have non-empty intersection.

Any element $y \in J^2$ can be written in the form $y = X_1C_1 + \cdots + X_rC_r$ with $C_1, \dots, C_r \in J$. It follows from Theorem 4.4.15 that $(X_\beta)_M$ is maximal with respect to

$$(X_1)_M, \dots, (X_r)_M, (X_1C_1)_M, \dots, (X_rC_r)_M$$

so that for all $\alpha \in \mathbb{A}^r(k)$ we have $(X_\beta)_M \succeq (X_\alpha + y)_M$. Since every element of $J$ can be written as $X_\alpha + y$ for suitable $\alpha$ and $y$, it follows that $X_\beta$ has maximal Jordan type on $M$.

It follows from Theorem 4.4.18 that $(X_\beta)_M \sim (X_\beta + y)_M$. On the other hand, if $(X_\beta)_M \sim (X_\alpha + y)_M$ then $(X_\alpha + y)_M$ is maximal with respect to $(X_1)_M, \dots, (X_r)_M$. Applying the same argument, we find that $(X_\alpha + y)_M \sim (X_\alpha)_M$ and so $\alpha \in U$. In particular it follows that $\alpha \neq 0$. Thus we have shown that the elements of maximal Jordan type form the inverse image in $J$ of a dense open subset of $(J/J^2) \setminus \{0\}$ and all have the same Jordan type. Finally, it follows from Theorem 4.4.12 with $A = (X_\beta)_M$ that this Jordan type is the same as that of the generic element of $J$. □

**Definition 4.5.4** The *generic Jordan type* of a $kE$-module or $\Lambda$-module $M$ is defined to be the Jordan type of the generic element of $J$ on $M$. Theorem 4.5.3 shows that this is the same as the that of the elements of maximal Jordan type on $M$, and these elements are given by the inverse image in $J$ of a dense open subset

$$U_{\max}(M) \subseteq (J/J^2) \setminus \{0\}.$$

We refer to this set $U_{\max}(M)$ as the *maximal Jordan set* of $M$, and we regard it as a subset of $\mathbb{A}^r(k)$. It is the complement of a closed homogeneous subset of affine space, so it defines an open subvariety $\bar{U}_{\max}(M) \subseteq \mathbb{P}^{r-1}(k)$.

**Remark 4.5.5** We can similarly talk about the *generic rank* of a $kE$-module or $\Lambda$-module $M$, which is the rank of the generic element of $J$ on $M$. This is equal to the maximum rank of an element of $J$ on $M$, and also equal to the maximal rank of $X_\alpha$ on $M$ for $\alpha \in \mathbb{A}^r(k)$. We say that such an element of $J$ has *maximal rank* on $M$.

The following theorem shows that $U_{\max}(M)$ is well behaved under field extensions.

**Theorem 4.5.6** *Let $K$ be a field containing $k$. If $M$ is a $kE$-module or a $\Lambda$-module and*

$$\alpha = (\lambda_1, \dots, \lambda_r) \in \mathbb{A}^r(K)$$

*then the Jordan canonical form of $X_\alpha \in J(KE)$ or $J(K \otimes_k \Lambda)$ on $K \otimes_k M$ is equal to the generic Jordan type of $M$ if and only if there exists $f \in k[Y_1, \dots, Y_r]$ such that $f(\alpha) = 0$ but $f(\beta) = 0$ for all $\beta \in U_{\max}(M)$. Otherwise its Jordan canonical form is dominated by the generic Jordan type of $M$.*

*Proof.* This follows from Lemma 3.2.1 and the fact that Jordan canonical form of $X_\alpha$ is determined by the vanishing of polynomials given by minors of the action on $M$. □

**Corollary 4.5.7** *Let $M$ be a $kE$-module or a $\Lambda$-module and $K$ be an algebraically closed extension field of $k$. Then the generic Jordan types of $M$ and $K \otimes_k M$ are the same.*

*Let $I$ be the ideal of $k[Y_1, \dots, Y_r]$ of functions vanishing on $\mathbb{A}^r(k) \setminus U_{\max}(M)$. Then $K \otimes_k I$ is the ideal of $K[Y_1, \dots, Y_r]$ of functions vanishing on $\mathbb{A}^r(K) \setminus U_{\max}(K \otimes_k M)$.* □

**Proposition 4.5.8** *The generic Jordan type of $M \oplus N$ is the direct sum of the generic Jordan types of $M$ and $N$. Furthermore,*

$$U_{\max}(M \oplus N) = U_{\max}(M) \cap U_{\max}(N).$$

*Proof.* This is an easy consequence of the definition and Theorem 4.5.3. □

**Proposition 4.5.9** *If the generic Jordan type of $M$ is*

$$[p]^{n_p}\,[p-1]^{n_{p-1}}\dots[i]^{n_i}\dots[1]^{n_1}$$

*then that of $\Omega(M)$ is*

$$[p]^{m_p}\,[p-1]^{n_1}\dots[p-i]^{n_i}\dots[1]^{n_{p-1}}$$

*for some value of $m_p$. Furthermore, we have $U_{\max}(\Omega(M)) = U_{\max}(M)$.*

*Proof.* This follows easily from the fact that $\Omega$ and restriction to a cyclic shifted subgroup commute up to projective summands. □

**Corollary 4.5.10** *The generic Jordan type of $\Omega^2(M)$ is the same as that of $M$ apart from possibly the number of Jordan blocks of length $p$.* □

**Remark 4.5.11** We shall see in Section 8.5 that if $M$ has constant Jordan type (see Section 5.1) then the modules $\Omega^2(M)$ and $M$ are distinguished by the vector bundles on projective space associated to them.

**Theorem 4.5.12** *Let $n < p$. For a $kE$-module $M$, the following are equivalent:*

(1) *The generic Jordan type of $M$ has all Jordan blocks of length at most $n$.*
(2) *For all $\alpha \in \mathbb{A}^r(k)$, the action of $X_\alpha$ on $M$ has all Jordan blocks of length at most $n$.*
(3) *For all extension fields $K$ of $k$ and all $\alpha \in \mathbb{A}^r(K)$, the action of $X_\alpha$ on $K \otimes_k M$ has all Jordan blocks of length at most $n$.*
(4) $\mathsf{Rad}^n(M) = 0$.

*Proof.* This follows from Theorem 1.17.1 and Theorem 4.5.6. □

## 4.6 Tensor Products

The tensor product of modules for a cyclic group of order $p$ were discussed in Section 2.3. In particular, it was shown there that the isomorphism type of the tensor product does not depend on whether we use the usual group theoretic tensor product $M \otimes_k N$ or the restricted Lie tensor product $M \mathbin{\tilde{\otimes}_k} N$.

**Theorem 4.6.1** (Friedlander, Pevtsova and Suslin) *If $M$ and $N$ are two $kE$-modules then*

(1) *The generic Jordan type of $M \otimes_k N$ is the tensor product of the generic Jordan types of $M$ and $N$.*
(2) $U_{\max}(M \otimes_k N) \supseteq U_{\max}(M) \cap U_{\max}(N)$.

*The same statements hold for the restricted Lie tensor product $M \tilde{\otimes}_k N$.*

*Proof.* We begin by proving the theorem for $M \tilde{\otimes}_k N$. Let $\alpha \in U_{\max}(M) \cap U_{\max}(N)$. We have

$$(M \tilde{\otimes}_k N){\downarrow_{E_\alpha}} \cong M{\downarrow_{E_\alpha}} \tilde{\otimes}_k N{\downarrow_{E_\alpha}}$$

and so the isomorphism type of this restriction does not depend on which such $\alpha$ we use. It follows from Theorem 4.5.3 that this dense open subset is contained in $U_{\max}(M \tilde{\otimes}_k N)$, and that the generic Jordan type of $M \tilde{\otimes}_k N$ is the tensor product of those of $M$ and $N$.

To prove the theorem for $M \otimes_k N$, we argue as follows. Let $\alpha \in U_{\max}(M) \cap U_{\max}(N)$. Then $(X_\alpha)_M$ is maximal with respect to $(X_1)_M, \dots, (X_r)_M$, and so by Theorem 4.4.12 if $t_1, \dots, t_r$ are independent transcendentals we have

$$(X_\alpha)_M \sim (X_\alpha)_M + t_1(X_1)_M + \cdots + t_r(X_r)_M. \tag{4.5}$$

Similarly, if $t'_1, \dots, t'_r$ are further independent transcendentals we have

$$(X_\alpha)_N \sim (X_\alpha)_N + t'_1(X_1)_N + \cdots + t'_r(X_r)_N. \tag{4.6}$$

So we have

$$\begin{aligned}(X_\alpha)_{M\tilde{\otimes}_k N} &= (X_\alpha)_M \otimes I_N + I_M \otimes (X_\alpha)_N \\ &\sim (X_\alpha)_M \otimes I_N + I_M \otimes (X_\alpha)_N \\ &\quad + t_1(X_1)_M \otimes I_N + \cdots + t_r(X_r)_M \otimes I_N \\ &\quad + I_M \otimes t'_1(X_1)_N + \cdots + I_M \otimes t'_r(X_r)_N.\end{aligned} \tag{4.7}$$

Indeed, if a matrix $A$ effects the conjugation (4.5) and $B$ effects (4.6) then $A \otimes B$ effects the conjugation (4.7).

Now if $X_\alpha = \sum_i \lambda_i X_i$ then $(X_\alpha)_{M\otimes_k N}$ is equal to $(X_\alpha)_{M\tilde{\otimes}_k N} + \sum_i \lambda_i (X_i)_M \otimes (X_i)_N$ plus a sum of terms of the form $Y_M \otimes Z_N$ where $Y$ and $Z$ are in $J(kE)$. Such a term can be written as $(Y_M \otimes I_N)(I_M \otimes Z_N)$, and is therefore a sum of products of nilpotent matrices with matrices appearing in (4.7). All the matrices commute, so by Theorem 4.4.15 we have

$$(X_\alpha)_{M\tilde{\otimes}_k N} \sim (X_\alpha)_{M\otimes_k N}.$$

The version of the theorem for the group theoretic tensor product now follows. □

**Corollary 4.6.2** *If $\lambda$ is a partition of $d$ with $d < p$ and $S^\lambda$ is the Schur functor corresponding to $\lambda$ (see Section 1.16) then the generic Jordan type of $S^\lambda(M)$ is $S^\lambda$ of the generic Jordan type of $M$, and $U_{\max}(S^\lambda(M)) \supseteq U_{\max}(M)$.*

*Proof.* It follows from the discussion in Section 1.16 that if $d < p$ then $S^\lambda(M)$ is a direct summand of $M^{\otimes d}$. It follows from Theorem 4.6.1 that $M^{\otimes d}$ has generic Jordan type equal to the tensor $d$th power of the generic Jordan type of $M$, with $U_{\max}(M^{\otimes d}) \supseteq U_{\max}(M)$. The corollary now follows by applying Proposition 4.5.8. □

If $d \geq p$ the statement in the corollary is false. The following examples are the results of discussions with Kay Jin Lim.

Let $P$ be the regular representation of $kE$, of dimension $p^r$. Then looking at the permutation basis $e_1, \ldots, e_{p^r}$ for $kE$, we see that the exterior $p$th power $\Lambda^p(P)$ is a direct sum in which most of the summands are free. The non-free summands correspond to wedge products $e_{i_1} \wedge \cdots \wedge e_{i_p}$ where the corresponding group elements form a coset of a cyclic subgroup. These summands are isomorphic to the permutation modules on cyclic subgroups. In particular, if $r > 1$ then $\Lambda^p(P)$ is generically free. So its generic Jordan type is $[p]^{\frac{1}{p}\binom{p^r}{p}}$. On the other hand, $P$ has generic Jordan type $[p]^{p^{r-1}}$, and applying $\Lambda^p$ to this gives $[p]^{\frac{1}{p}\binom{p^r}{p}-p^{r-2}}[1]^{p^{r-1}}$.

Now look at the exact sequence

$$0 \to \Omega(k) \to P \to k \to 0.$$

Taking the symmetric $p$th power of the chain complex $P \to k$ (in degrees one and zero), we obtain a complex whose homology is $\Lambda^p(\Omega(k))$ in degree $p$. This gives us an exact sequence

$$0 \to \Lambda^p(\Omega(k)) \to \Lambda^p(P) \to \Lambda^{p-1}(P) \to \cdots \to P \to k \to 0.$$

Since $\Lambda^{p-1}(P), \ldots, P$ are projective, this gives us a short exact sequence

$$0 \to \Lambda^p(\Omega(k)) \to \Lambda^p(P) \to \Omega^{p-1}(k) \to 0.$$

The generic Jordan type of $\Omega^{p-1}(k)$ is of the form $[p]^?[1]$, while the generic Jordan type of $\Lambda^p(P)$ is of the form $[p]^?$. It follows that the generic Jordan type of $\Lambda^p(\Omega(k))$ is of the form $[p]^?[p-1]$ while $\Lambda^p$ of the generic Jordan type of $\Omega(k)$ is of the form $[p]^?[1]^{p^{r-1}-1}$. The set $U_{\max}(\Lambda^p(P))$ is the complement of the subvariety corresponding to the union of the $(p^r-1)/(p-1)$ cyclic subgroups of $E$.

For an explicit example, let $p = 3$ and $E = (\mathbb{Z}/3)^2$. Then the generic Jordan type of $\Omega(k)$ is $[3]^2[2]$, and the generic Jordan type of $\Lambda^3(\Omega(k))$ is $[3]^{18}[2]$, whereas $\Lambda^3$ of the generic Jordan type of $\Omega(k)$ is $[3]^{18}[1]^2$. The set $U_{\max}(\Lambda^3(\Omega(k)))$ is the complement of the subvariety corresponding to the union of the four cyclic subgroups of $E$.

# 5

# Modules of Constant Jordan Type

## 5.1 Introduction and Definitions

Modules of constant Jordan type were introduced by Carlson, Friedlander and Pevtsova [81]. The definition is as follows.

**Definition 5.1.1** We say that a finite-dimensional $kE$-module $M$ has *constant Jordan type* $[a_1]\dots[a_t]$ if the restriction of $M$ to every cyclic shifted subgroup has Jordan canonical form given by blocks of length $a_1, \dots, a_t$. The notation here is the same as in Section 4.4.

**Remark 5.1.2** This definition appears to depend on a choice of basis for the elementary abelian group $E$. In fact the notion is independent of choice of basis, as we shall see in Lemma 5.1.7.

**Definition 5.1.3** Similarly, we say that a nilvariety of rank $r$ spanned by matrices $A_1, \dots, A_r$ has *constant Jordan type* $[a_1]\dots[a_t]$ if for all $\lambda_1, \dots, \lambda_r \in k$, not all zero, the matrix

$$\lambda_1 A_1 + \cdots + \lambda_r A_r$$

has Jordan canonical form given by blocks of length $a_1, \dots, a_t$.

**Example 5.1.4** Suppose that $A_1, \dots, A_r$ span a linear variety of $m \times n$ matrices of constant rank $s$. Then the $(n+m) \times (n+m)$ matrices

$$B_i = \begin{pmatrix} 0 & 0 \\ A_i & 0 \end{pmatrix}, \qquad 1 \le i \le r$$

commute, and every linear combination squares to zero. The $kE$-module $M$ given by

$$X_i \mapsto B_i, \qquad g_i \mapsto I + B_i \qquad (1 \le i \le r)$$

satisfies $\mathsf{Rad}^2(M) = 0$, and has constant Jordan type $[2]^s[1]^{n+m-2s}$.

Examples 4.1.6 and 4.1.7 give two-dimensional spaces of matrices of constant rank. Here are some larger-dimensional examples. Consider the space spanned by the $5 \times 5$ matrices

$$Y_1A_1 + Y_2A_2 + Y_3A_3 = \begin{pmatrix} 0 & 0 & 0 & Y_1 & Y_2 \\ 0 & 0 & Y_1 & Y_2 & Y_3 \\ 0 & -Y_1 & 0 & Y_3 & 0 \\ -Y_1 & -Y_2 & -Y_3 & 0 & 0 \\ -Y_2 & -Y_3 & 0 & 0 & 0 \end{pmatrix}.$$

It is easy to check that every non-zero linear combination of $A_1$, $A_2$ and $A_3$ has rank four. This gives a $k(\mathbb{Z}/p)^3$-module of dimension 10 and constant Jordan type $[2]^4[1]^2$. The diagram for this module is

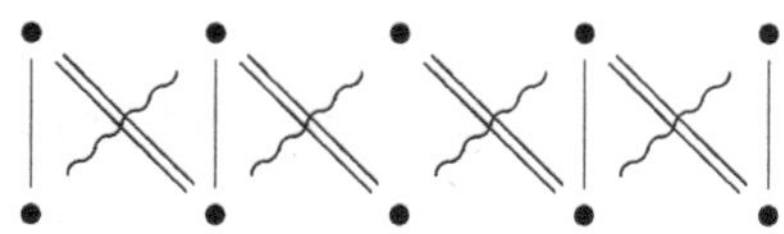

In this diagram, we have to keep in mind that the edges on the left-hand side have a minus sign associated with them.

A larger example (Westwick [248]) is given by the $10 \times 10$ matrices of constant rank 8

$$\begin{aligned} &Y_1A_1 + Y_2A_2 + Y_3A_3 + Y_4A_4 \\ &= \begin{pmatrix} 0 & 0 & 0 & 0 & 0 & 0 & 0 & Y_1 & Y_2 & 0 \\ 0 & 0 & 0 & 0 & 0 & 0 & Y_1 & Y_2 & 0 & Y_3 \\ 0 & 0 & 0 & 0 & 0 & -Y_1 & Y_2 & 0 & Y_3 & Y_4 \\ 0 & 0 & 0 & 0 & Y_1 & Y_2 & 0 & Y_3 & Y_4 & 0 \\ 0 & 0 & 0 & -Y_1 & 0 & 0 & Y_3 & -Y_4 & 0 & 0 \\ 0 & 0 & Y_1 & -Y_2 & 0 & 0 & Y_4 & 0 & 0 & 0 \\ 0 & -Y_1 & -Y_2 & 0 & -Y_2 & -Y_4 & 0 & 0 & 0 & 0 \\ -Y_1 & -Y_2 & 0 & -Y_3 & Y_4 & 0 & 0 & 0 & 0 & 0 \\ -Y_2 & 0 & -Y_3 & -Y_4 & 0 & 0 & 0 & 0 & 0 & 0 \\ 0 & -Y_3 & -Y_4 & 0 & 0 & 0 & 0 & 0 & 0 & 0 \end{pmatrix}, \end{aligned}$$

giving a $k(\mathbb{Z}/p)^4$-module of dimension 20 and constant Jordan type $[2]^8[1]^4$. The diagram for this module, with the same caveat about minus signs, and with the direction of an edge indicating which $X_i$ is involved, is

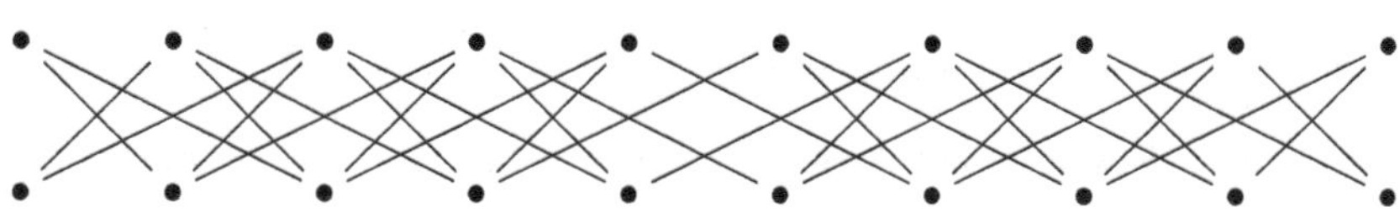

Further examples of modules of constant Jordan type will appear throughout this chapter as well as Chapter 9.

**Remark 5.1.5** Papers on modules of constant Jordan type include Baland [25, 26], Benson [41, 42, 44, 43], Benson and Pevtsova [53], Carlson and Friedlander

[80], Carlson, Friedlander and Pevtsova [81], Carlson, Friedlander and Suslin [82], Friedlander and Pevtsova [127].

Papers on linear varieties of matrices of constant rank and on nilvarieties of matrices of constant rank include Atkinson and Westwick [19], Beasley [37], Boralevi, Faenzi and Mezzetti [60], Fania and Mezzetti [117], Gerstenhaber [130, 131, 132, 133, 134], Ilic and Landsberg [166], Manivel and Mezzetti [191], Sylvester [231], Westwick [245, 246, 247, 248].

In Section 10.1 we use the theory of Chern numbers to prove theorems of Sylvester and Westwick about matrices of constant rank, and apply this to modules of constant Jordan type.

The following proposition shows that it is not always obvious whether or not a module has constant Jordan type.

**Proposition 5.1.6** *Let $E = (\mathbb{Z}/p)^2$ with $p \geq 3$, and let $M_n$ $(n \geq 2)$ be the module defined by the diagram*

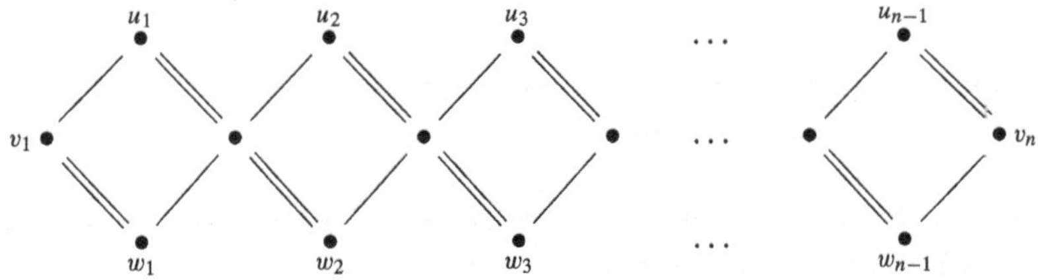


*More explicitly, $M_n$ has basis elements $u_1, \dots, u_{n-1}, v_1, \dots, v_n, w_1, \dots, w_{n-1}$ with*

$$X_1(u_i) = v_i \quad X_2(u_i) = v_{i+1} \quad X_1(v_{i+1}) = X_2(v_i) = w_i \quad (1 \leq i \leq n-1)$$

*and all other basis elements sent to zero by $X_1$ and $X_2$. Then $M_n$ has constant Jordan type if and only if $n$ is divisible by $p$. The Jordan type is $[3]^{n-2}[2]^2$.*

*Proof.* Let $0 \neq (\lambda_1, \lambda_2) \in \mathbb{A}^2(k)$. The rank of $\lambda_1 X_1 + \lambda_2 X_2$ is equal to $2n-2$. The rank of $(\lambda_1 X_1 + \lambda_2 X_2)^2$ is equal to $n-1$ if either $\lambda_1$ or $\lambda_2$ is zero. If they are both non-zero, the rank is either $n$ or $n-1$. The latter happens if and only if some linear combination of the $u_i$ is in the kernel. This happens if and only if $n$ is divisible by $p$, and then the kernel is the linear span of the element

$$\lambda_1^{n-2}u_1 - 2\lambda_1^{n-3}\lambda_2 u_2 + 3\lambda_1^{n-4}\lambda_2^2 u_3 - \cdots - (n-1)\lambda_2^{n-2}u_{n-1}. \qquad \square$$

**Lemma 5.1.7** *The following are equivalent.*

(1) *$M$ has constant Jordan type.*
(2) *The maximal Jordan set $U_{\max}(M)$ is equal to $\mathbb{A}^r(k) \setminus \{0\}$.*
(3) *The restriction of $M$ to every element of $J(kE) \setminus J^2(kE)$ has the same Jordan canonical form.*

*Proof.* This follows immediately from the definitions together with Theorem 4.5.3. □

**Theorem 5.1.8** *Let $M$ be a $kE$-module of constant Jordan type and let $K$ be an algebraically closed field containing $k$. Then $M \otimes_k K$ is a $KE$-module of the same constant Jordan type.*

*Proof.* This follows immediately from Corollary 4.5.7. □

**Theorem 5.1.9** *Let $M$ and $N$ be $kE$-modules.*

(1) *$M$ and $N$ both have constant Jordan type if and only if $M \oplus N$ has constant Jordan type.*
(2) *If $M$ and $N$ have constant Jordan type then so does $M \otimes_k N$.*
(3) *If $M$ has constant Jordan type and $\lambda$ is a partition of $d$ with $d < p$ then $S^\lambda(M)$ (where $S^\lambda$ is the Schur functor corresponding to $\lambda$ as in Section 1.16) has constant Jordan type.*

*The Jordan type of the sum, tensor product and $S^\lambda(M)$ is obtained by applying the same operation to the Jordan type of $M$ (and $N$).*

*Proof.* (1) It is obvious that if $M$ and $N$ have constant Jordan type then so does $M \oplus N$. The converse follows from Proposition 4.5.8.

(2) This follows from Theorem 4.6.1.

(3) This follows from Corollary 4.6.2. □

**Warning 5.1.10** It follows from part (3) of the theorem that if $M$ has constant Jordan type then the exterior powers $\Lambda^d(M)$ and the symmetric powers $S^d(M)$ have constant Jordan type when $d < p$. This does NOT remain true for $d \geq p$. The examples at the end of Section 4.6 show that if a module $M$ has constant Jordan type then $\Lambda^p(M)$ does not necessarily have constant Jordan type.

**Definition 5.1.11** (Carlson, Friedlander and Pevtsova [81].). The *stable constant Jordan type* of $M$ is obtained by omitting the Jordan blocks of length $p$. So $M$ has stable constant Jordan type $[a_1] \dots [a_t]$ if the restriction of $M$ to every cyclic shifted subgroup has Jordan canonical form given by blocks of length $a_1, \dots, a_t$ together with some number (possibly zero) of blocks of length $p$. In the case where $t = 0$, we write $\varnothing$ for the empty partition, to denote this stable constant Jordan type.

**Proposition 5.1.12** *If $M$ has constant Jordan type then so do $\Omega(M)$ and $\Omega^{-1}(M)$. If the stable Jordan type of $M$ is $[a_1] \dots [a_t]$ then the stable Jordan type of $\Omega(M)$ and $\Omega^{-1}(M)$ are both $[p - a_1] \dots [p - a_t]$.*

*Proof.* This follows from the fact that for $0 \neq \alpha \in \mathbb{A}^r(k)$ the module $\Omega(M){\downarrow}_{X_\alpha}$ is stably isomorphic to $\Omega(M{\downarrow}_{X_\alpha})$ and $\Omega^{-1}(M){\downarrow}_{X_\alpha}$ is stably isomorphic to $\Omega^{-1}(M{\downarrow}_{X_\alpha})$. □

**Lemma 5.1.13** *Given modules of constant Jordan type, the stable constant Jordan type of a direct sum or tensor product is the same operation on the stable constant Jordan types.*

*The same is true for symmetric and exterior dth powers and more generally for Schur functors $S^\lambda$ with $\lambda$ a partition of $d$, provided $d < p$.*

*Proof.* The first part follows from Theorem 5.1.9 together with the observation that the tensor product of a Jordan block of length $p$ with anything is a sum of Jordan blocks of length $p$. The second part then follows from Corollary 4.6.2. □

In the case where all Jordan blocks have length $p$, in other words the stable Jordan type is the empty partition $\varnothing$, we have the following.

**Proposition 5.1.14** *If $M$ has stable constant Jordan type $\varnothing$ then $M$ is projective.*

*Proof.* This is a restatement of Dade's Lemma 1.9.5. □

**Corollary 5.1.15** *If $M$ has stable constant Jordan type $\varnothing$ then the dimension of $M$ is divisible by $p^r$.*

**Remark 5.1.16** We shall see in Section 10.6 that for $p = 2$, a nilvariety of rank $r$ of stable constant Jordan type $\varnothing$ has to have dimension divisible by $2^r$. Example 4.1.6 shows that the same is not true if $p$ is odd.

At the other extreme, when there are no Jordan blocks of length $p$, we have the following.

**Proposition 5.1.17** *If $M$ has constant Jordan type with all Jordan blocks of length at most $n$, with $n < p$, then $\mathsf{Rad}^n(M) = 0$.*

*Proof.* This follows from Theorem 4.5.12. □

**Theorem 5.1.18** *Suppose that $r \geq 2$. If $M$ has constant Jordan type with all blocks of the same length then either $M$ is projective or $E$ acts trivially on $M$.*

*Proof.* If all the blocks have length $p$ then we can apply Dade's Lemma 1.9.5 to deduce that $M$ is projective. So we need to show that if all blocks are of equal length less than $p$ then $E$ acts trivially on $M$.

Let $n$ be the lengths of the Jordan blocks, with $n < p$. Then by Theorem 4.5.12 we have $\mathsf{Rad}^n(M) = 0$. If $n \geq 2$, it follows that all non-zero linear combinations of $X_1$ and $X_2$ induce an isomorphism from $M/\mathsf{Rad}(M)$ to $\mathsf{Rad}(M)/\mathsf{Rad}^2(M)$. Using $X_1$ to identify these two spaces, it follows that $X_2$ has no eigenvalues on this space. So $M/\mathsf{Rad}(M) = 0$, and hence $M = 0$.

So we are left with the case $n = 1$, in which case $E$ acts trivially on $M$. □

## 5.2 Homogeneous Modules

Our first examples of modules of constant Jordan type are the modules $\Omega^n(k)$, $n \in \mathbb{Z}$. The reason why they are of constant Jordan type is that they are homogeneous. The Jordan type is $[1]$ if $n$ is even and $[p-1]$ if $n$ is odd. Note that if $p = 2$ then $[1] = [p-1]$ and there is no dependence on the parity of $n$.

**Definition 5.2.1** A $kE$-module $M$ is said to be *homogenous* if given any maximal shifted subgroup $E'$ with choice of basis, the corresponding automorphism $\rho\colon kE \to kE$ (see Definition 1.3.1) satisfies $\rho^*(M) \cong M$. Recall that the group of all such automorphisms is isomorphic to $GL(r, k)$.

We shall see in Chapter 8 that there are many modules of constant Jordan type that are not homogeneous. An example is also given later in this section.

**Proposition 5.2.2** *The class of homogeneous $kE$-modules is closed under direct sums and summands.*

*Proof.* Closure under direct sums is clear, so we prove closure under direct summands. If $M$ is homogeneous and $M = M_1 \oplus \cdots \oplus M_t$ with the $M_i$ indecomposable, then for $\rho \in GL(r, k)$, $\rho^*(M) \cong M$ and so by the Krull–Schmidt theorem, $\rho^*(M_1)$ is isomorphic to some $M_i$. Let $H$ be the subgroup of $GL(r, k)$ consisting of those $\rho$ such that $\rho^*(M_1) \cong M_1$. Then for $\rho$, $\rho'$ in different left cosets of $H$ in $GL(r, k)$, the isomorphism type of $\rho^*(M_1)$ and $(\rho')^*(M_1)$ are not the same. Therefore $H$ is a closed subgroup of finite index in $GL(r, k)$. Since $GL(r, k)$ is connected, this implies that $H = GL(r, k)$ and the proposition is proved. □

**Proposition 5.2.3** *If $M$ is a homogeneous $kE$-module and $j > 0$ then $\mathsf{Rad}^j(M)$, $M/\mathsf{Rad}^j(M)$, $\mathsf{Soc}^j(M)$, $M/\mathsf{Soc}^j(M)$, $\Omega(M)$ and $\Omega^{-1}(M)$ are all homogeneous.* □

**Proposition 5.2.4** *If $M$ is homogeneous then $M$ has constant Jordan type.*

*Proof.* The automorphisms $\rho$ in the definition of homogeneity are transitive on cyclic shifted subgroups. □

**Corollary 5.2.5** *The modules $\Omega^n(k)$ have constant Jordan type for $n \in \mathbb{Z}$.* □

It is easy to construct a large supply of homogeneous modules. For example we can start with any homogeneous module, such as $k$, and allow ourselves to apply the operations $\Omega^n$ for $n \in \mathbb{Z}$, $\mathsf{Rad}^i$ and $\mathsf{Soc}^i$, and more generally any quotient of one submodule of the form $\mathsf{Rad}^i$ or $\mathsf{Soc}^i$ by another contained in it.

**Example 5.2.6** The modules $kE/J^2(kE)$ and $\mathsf{Soc}^2(kE)$ are both homogeneous modules of constant Jordan type $[2]\,[1]^{r-1}$.

More generally, we have the following lemma, which will be used in the proof of Proposition 5.3.11.

**Lemma 5.2.7** *For $j \le p$, the modules $kE/J^j(kE)$ and $\mathsf{Soc}^j(kE)$ are homogeneous modules of constant Jordan type*

$$[j]\,[j-1]^{r-1}[j-2]^{\binom{r}{2}}[j-3]^{\binom{r+1}{3}}\dots[1]^{\binom{r+j-3}{j-1}}.$$

*Proof.* This is easy to see using the explicit basis for $kE/J^j(kE)$ consisting of monomials in $X_1, \dots, X_r$ of degree at most $j-1$ and looking at the action of $X_1$. Using the obvious Jordan decomposition, each Jordan block has a unique monomial not involving $X_1$. So the number of Jordan blocks of length $j-i$ is equal to the number of monomials in $X_2, \dots, X_r$ of degree $i$, namely $\binom{r-2+i}{i}$. The module $\mathsf{Soc}^j(kE)$ is dual, so it has the same Jordan type. □

**Example 5.2.8** If $p = 3$ and $E = (\mathbb{Z}/3)^2$ then $\Omega(k)$ was described in Section 1.11. This has constant Jordan type $[3]^2[2]$. Taking the quotient by the socle gives us a module with the following diagram:

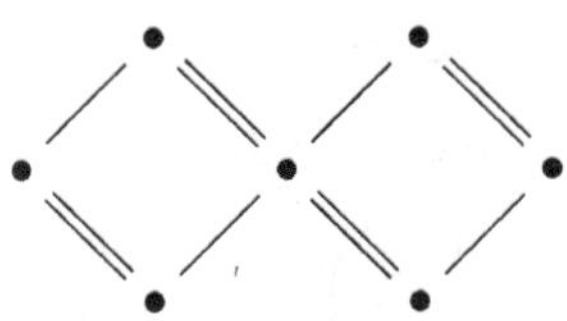

This is a module of constant Jordan type $[3]\,[2]^2$.

As another example, $\mathsf{Soc}^3(\Omega^2(k))$ has the following diagram:

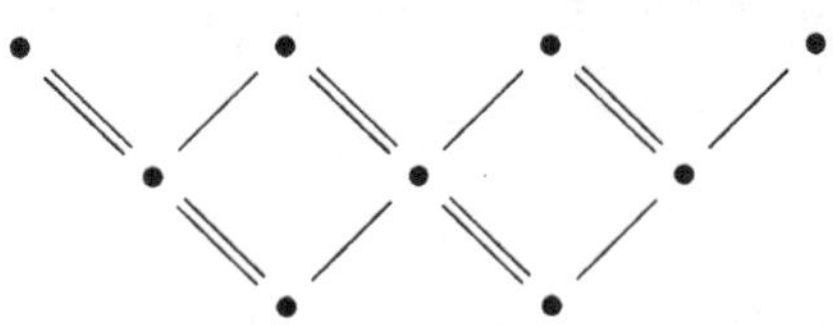

This is a homogeneous module of constant Jordan type $[3]^2[2]\,[1]$.

Be warned, however, that it is not easy to tell whether a module is homogeneous, or even whether it has constant Jordan type, just by looking at the diagram. See for example Proposition 5.1.6.

**Example 5.2.9** Let $p \geq 3$ and let $M$ be the module with diagram

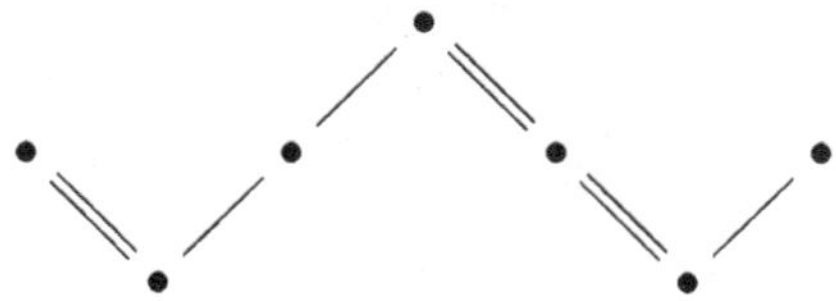

Then $M$ has constant Jordan type $[3][2][1]^2$. However, $\mathsf{Rad}(M)$ does not have constant Jordan type. It follows from Proposition 5.2.3 that $M$ is not homogeneous. It also follows from this example and its dual that if $M$ has constant Jordan type and $j > 0$ then $\mathsf{Rad}^j(M)$, $M/\mathsf{Rad}^j(M)$, $\mathsf{Soc}^j(M)$ and $M/\mathsf{Soc}^j(M)$ do not necessarily have constant Jordan type.

**Remark 5.2.10** Another operation preserving homogeneity is as follows. Given a non-projective indecomposable module $M$, there is an almost split sequence

$$0 \to \Omega^2 M \to M' \to M \to 0,$$

unique up to isomorphism, and if $M$ is homogeneous then so is $M'$. See for example Section 4.12 of [39]. Carlson, Friedlander and Pevtsova [81] observe furthermore that for $r \geq 2$, the almost split sequence splits on every cyclic shifted subgroup. So if $M$ has constant Jordan type then so does $M'$, and the Jordan type of $M'$ is just a direct sum of two copies of that of $M$.

**Example 5.2.11** We end this section with an example of a module of constant Jordan type with a large group of automorphisms that is nonetheless not homogeneous. Let $p = 2$ and $r = 4$, so that

$$E = (\mathbb{Z}/2)^4 = \langle g_1, \ldots, g_4 \rangle.$$

Let $M$ be the six-dimensional module given by

$$g_1 \mapsto \begin{pmatrix} 1 & 0 & 0 & 0 & 0 & 0 \\ 1 & 1 & 0 & 0 & 0 & 0 \\ 0 & 0 & 1 & 0 & 0 & 0 \\ 0 & 0 & 0 & 1 & 0 & 0 \\ 0 & 0 & 0 & 0 & 1 & 0 \\ 0 & 0 & 0 & 0 & 1 & 1 \end{pmatrix} \qquad g_2 \mapsto \begin{pmatrix} 1 & 0 & 0 & 0 & 0 & 0 \\ 0 & 1 & 0 & 0 & 0 & 0 \\ 1 & 0 & 1 & 0 & 0 & 0 \\ 0 & 0 & 0 & 1 & 0 & 0 \\ 0 & 0 & 0 & 0 & 1 & 0 \\ 0 & 0 & 0 & 1 & 0 & 1 \end{pmatrix}$$

$$g_3 \mapsto \begin{pmatrix} 1 & 0 & 0 & 0 & 0 & 0 \\ 0 & 1 & 0 & 0 & 0 & 0 \\ 0 & 0 & 1 & 0 & 0 & 0 \\ 1 & 0 & 0 & 1 & 0 & 0 \\ 0 & 0 & 0 & 0 & 1 & 0 \\ 0 & 0 & 1 & 0 & 0 & 1 \end{pmatrix} \qquad g_4 \mapsto \begin{pmatrix} 1 & 0 & 0 & 0 & 0 & 0 \\ 0 & 1 & 0 & 0 & 0 & 0 \\ 0 & 0 & 1 & 0 & 0 & 0 \\ 0 & 0 & 0 & 1 & 0 & 0 \\ 1 & 0 & 0 & 0 & 1 & 0 \\ 0 & 1 & 0 & 0 & 0 & 1 \end{pmatrix}.$$

It is easy to check that $M$ has constant Jordan type $[2]^2[1]^2$. The module is easier to visualise using the diagram

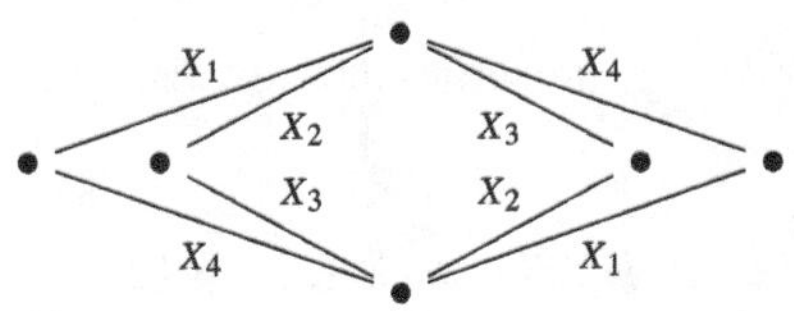


This module is not homogeneous, because the elements $\rho \in GL(4, k)$ satisfying $\rho^*(M) \cong M$ have to fix a nondegenerate symplectic form in four variables corresponding to the matrix with ones down the trailing diagonal and zeros elsewhere. To see where this symplectic form comes from, note that $M/\mathsf{Soc}(M) \cong kE/J^2(kE)$. Consider the diagram

$$\begin{array}{ccccccccc}
 & & 0 & & & & k & & \\
 & & \downarrow & & & & \downarrow & & \\
0 & \longrightarrow & \Omega(M)/\mathsf{Soc}^2\Omega(M) & \longrightarrow & kE/J^3(kE) & \longrightarrow & M & \longrightarrow & 0 \\
 & & \downarrow & & \| & & \downarrow & & \\
0 & \longrightarrow & J^2(kE)/J^3(kE) & \longrightarrow & kE/J^3(kE) & \longrightarrow & M/\mathsf{Soc}(M) & \longrightarrow & 0 \\
 & & \downarrow & & & & \downarrow & & \\
 & & k & & & & 0 & &
\end{array}$$

in which the switchback map is an isomorphism. We have

$$J^2(kE)/J^3(kE) \cong \Lambda^2(J(kE)/J^2(kE)) = \Lambda^2(\mathbb{A}^r(k)),$$

and so the map $J^2(kE)/J^3(kE) \to k$ gives us the symplectic form $\Lambda^2(\mathbb{A}^r(k)) \to k$ that has to be preserved by $\rho$.

Note that the module described in this way for $p$ odd does not have constant Jordan type, because for example $(X_1 + X_4)^2$ does not act as zero whereas $X_1^2$ does.

This example will reappear in Section 9.6 in the context of the null correlation bundle on projective 3-space. and again in Section 10.5 concerning restrictions on possible Jordan types.

This example satisfies a weaker condition than homogeneity, which still has some of the consequences that are true for homogeneous modules.

**Definition 5.2.12** A $kE$-module $M$ is said to be *n-homogeneous* $(1 \leq n \leq r)$ if given an elementary abelian group $E'$ of rank $n$ and embeddings as shifted subgroups $\phi_1, \phi_2 \colon E' \to kE$ we have $\phi_1^*(M) \cong \phi_2^*(M)$.

For $n = r$, this definition is equivalent to the definition of homogeneity, while for $n = 1$ it is equivalent to contant Jordan type.

The proofs of the following properties are essentially the same as for homogeneous modules.

**Proposition 5.2.13** *The class of $n$-homogeneous $kE$-modules is closed under direct sums and summands.* □

**Proposition 5.2.14** *If $M$ is an $n$-homogeneous $kE$-module then $\Omega(M)$ and $\Omega^{-1}(M)$ are $n$-homogeneous.* □

**Proposition 5.2.15** *If $M$ is $n$-homogeneous with $n \geq 1$ then $M$ has constant Jordan type.* □

**Example 5.2.16** Example 5.2.11 is 1-homogeneous, but not 2-homogeneous.

**Example 5.2.17** Let $n \leq r$, and let $M$ be the $kE$-module with generators $m_1, \ldots, m_n$ satisfying the following relations:

(i) For all $i, j, k$, $X_i X_j m_k = 0$; i.e., $J^2(kE).M = 0$.
(ii) $X_1 m_1 + \cdots + X_n m_n = 0$.

Then the restriction of $M$ to a shifted subgroup $E' \leq kE$ is isomorphic to the corresponding module for that subgroup plus some copies of $kE'/J^2(kE')$ and of the trivial module, if $X_1, \ldots, X_n \in E'$, and to a direct sum of $n$ copies of $kE'/J^2(kE')$ plus some copies of the trivial module otherwise.

It follows that $M$ is $(n-1)$-homogeneous but not $n$-homogeneous.

## 5.3 An Exact Category

Carlson and Friedlander [80] put a structure of exact category on the modules of constant Jordan type. In this section we describe this structure.

**Lemma 5.3.1** *Let $A$ be a $k$-algebra. A short exact sequence of finite-dimensional $A$-modules*

$$0 \to M_1 \to M_2 \to M_3 \to 0$$

*splits if and only if $M_2 \cong M_1 \oplus M_3$.*

*Proof.* One direction is obvious. For the other direction, suppose that $M_2 \cong M_1 \oplus M_3$. Consider the sequence

$$0 \to \mathrm{Hom}_A(M_3, M_1) \to \mathrm{Hom}_A(M_3, M_2) \to \mathrm{Hom}_A(M_3, M_3).$$

If $M_2 \cong M_1 \oplus M_3$ then the dimension of the middle term is the sum of the dimensions of the outside terms, so the right-hand map is surjective. An inverse image in $\mathrm{Hom}_A(M_3, M_2)$ of the identity map in $\mathrm{Hom}_A(M_3, M_3)$ is a splitting for the sequence. □

**Lemma 5.3.2** *Let*

$$0 \to M_1 \to M_2 \to M_3 \to 0$$

*be a short exact sequence of* $kE$*-modules of modules of constant Jordan type. Then the following are equivalent:*

(1) *The sequence splits on restriction to* $X_\alpha$ *for every closed point* $0 \neq \alpha \in \mathbb{A}^r$.
(2) *There exists a point* $0 \neq \alpha$ *of* $\mathbb{A}^r(K)$ *for some extension field* $K$ *(i.e., a generic point for some irreducible subvariety other than* $\{0\}$*, cf. §3.2), such that the sequence splits on extending coefficients to* $K$ *and restricting to* $X_\alpha$.

*Proof.* By Lemma 5.3.1, the sequence splits for $0 \neq \alpha \in \mathbb{A}^r$ if and only if the Jordan type of $X_\alpha$ on $M_2$ is the sum of the Jordan types on $M_1$ and $M_3$. Since these are modules of constant Jordan type, this condition is independent of $\alpha$. □

**Definition 5.3.3** A short exact sequence

$$0 \to M_1 \to M_2 \to M_3 \to 0$$

of $kE$-modules is *locally split* if for every $0 \neq \alpha \in \mathbb{A}^r$ the restriction of the sequence to $X_\alpha$ splits.

**Remark 5.3.4** Not every short exact sequence of modules of constant Jordan type is locally split. For example, if $0 \leq i < j < \ell \leq p$ then the sequence

$$0 \to J^j(kE)/J^\ell(kE) \to J^i(kE)/J^\ell(kE) \to J^i(kE)/J^j(kE)$$

is not locally split because the Jordan type of the middle is not the sum of the Jordan types of the outside terms.

On the other hand, there exist non-split short exact sequences of modules of constant Jordan type that are locally split. For example, the almost split sequences of Remark 5.2.10 have this property.

**Example 5.3.5** Another class of examples comes from Tate cohomology. For any $n, j \in \mathbb{Z}$ we have

$$\hat{H}^n(E, k) \cong \mathrm{Ext}^1_{kE}(\Omega^{j+n-1}(k), \Omega^j(k)).$$

If $n < 0$ then the corresponding sequence

$$0 \to \Omega^j(k) \to M \to \Omega^{j+n-1}(k) \to 0$$

is a locally split sequence of modules of constant Jordan type. This is because elements of negative degree restrict to zero on every proper shifted subgroup. In case $n = -1$ these are almost split sequences, but for $n < -1$ the module $M$ is usually not homogeneous. We shall examine these examples in more detail in Section 8.12.

**Definition 5.3.6** (Quillen [208]) An *exact category* is an additive category with a class of *short exact sequences* $M_1 \to M_2 \to M_3$ satisfying the following properties:

(1) Any sequence isomorphic to a short exact sequence is a short exact sequence.
(2) If $M_1 \to M_2 \to M_3$ is a short exact sequence then $M_1 \to M_2$ is the kernel of $M_2 \to M_3$ and $M_2 \to M_3$ is the cokernel of $M_1 \to M_2$.
(3) For every pair of objects $M_1$ and $M_3$, the split exact sequence $M_1 \to M_1 \oplus M_3 \to M_3$ is a short exact sequence.
We say that a map $M_1 \to M_2$ appearing as the first map in a short exact sequence is an *admissible monomorphism*, and that a map $M_2 \to M_3$ appearing as the second map in a short exact sequence is an *admissible epimorphism*.
(4) The pushout of any map and an admissible monomorphism exists and is an admissible monomorphism.
(5) The pullback of any map and an admissible epimorphism exists and is an admissible epimorphism.
(6) The composite of two admissible monomorphisms is an admissible monomorphism.
(7) The composite of two admissible epimorphisms is an admissible epimorphism.
(8) If $M \to M_1$ is a morphism with a cokernel, and $M_1 \to M_2$ and the composite $M \to M_1 \to M_2$ are both admissible monomorphisms, then so is $M \to M_1$.
(9) If $M_3 \to M$ is a morphism with a kernel, and $M_2 \to M_3$ and the composite $M_2 \to M_3 \to M$ are both admissible epimorphisms then so is $M_3 \to M$.

**Remark 5.3.7** Keller [175, Appendix A] has shown that these axioms are somewhat redundant. It suffices to assume that (1), (2), (4), (5) and (7) hold, and that the identity map $0 \to 0$ is an admissible epimorphism; (3), (6), (8) and (9) follow from these statements. A survey of the theory of exact categories can be found in Bühler [68].

**Proposition 5.3.8** *The category of finite-dimensional $kE$-modules and locally split exact sequences is an exact category. Similarly, the category of modules of constant Jordan type and locally split exact sequences is an exact category.*

*Proof.* The requisite properties hold for split exact sequences, and therefore for locally split exact sequences. □

**Definition 5.3.9** We write $\mathsf{cJt}(kE)$ for the exact category of modules of constant Jordan type and locally split exact sequences.

**Definition 5.3.10** We write $K_0(\mathsf{cJt}(kE))$ for the Grothendieck group whose generators are the symbols $[M]$ corresponding to the isomorphism classes of modules $M$ of constant Jordan type, and with relations $[M_1] - [M_2] + [M_3]$ where $0 \to M_1 \to M_2 \to M_3 \to 0$ is a locally split exact sequence.

There is a homomorphism of abelian groups

$$\mathsf{JType}\colon K_0(\mathsf{cJt}(kE)) \to \mathbb{Z}^p$$

defined as follows. If $M$ is a module of constant Jordan type $[p]^{a_p} \dots [1]^{a_1}$ then

$$\mathsf{JType}[M] = (a_p, \dots, a_1).$$

We can make $K_0(\mathsf{cJt}(kE))$ and $\mathbb{Z}^p$ into commutative rings via tensor product; the tensor structure on $\mathbb{Z}^p$ being given by the tensor product of modules for a cyclic group of order $p$. Recall that this is described in Section 2.3 and does not depend on which Hopf algebra structure is used, by Theorem 2.3.2. With this definition, $\mathsf{JType}$ is a ring homomorphism.

The following is Proposition 4.1 of [80].

**Proposition 5.3.11** *The homomorphism* $\mathsf{JType}$ *is surjective.*

*Proof.* This is easy to see using the modules $kE/J^j(kE)$ with $1 \le j \le p$ and the description of their Jordan type given in Lemma 5.2.7. The only property we need of this is that the Jordan type of $kE/J^j(kE)$ contains a single block of length $j$ together with some smaller ones, and this property is easy to verify. □

## 5.4 Endotrivial Modules

In this section, we examine modules of stable constant Jordan type $[1]$ and $[p-1]$, and find that these are exactly the endotrivial modules.

**Definition 5.4.1** If $G$ is a finite group, we say that a $kG$-module $M$ is *endotrivial* if $\mathrm{End}_k(M)$ is stably isomorphic to the trivial module. In other words,

$$M \otimes_k M^* \cong \mathrm{End}_k(M) \cong k \oplus (\text{projective}).$$

This definition comes from Dade [98], where he classified the endotrivial modules over an abelian $p$-group.

**Theorem 5.4.2** (Dade, Theorem 10.1 of [98]) *The endotrivial modules for an abelian $p$-group are precisely the modules which are stably isomorphic to $\Omega^n(k)$ for $n \in \mathbb{Z}$.*

**Proposition 5.4.3** (Carlson, Friedlander and Pevtsova [81]) *A $kE$-module $M$ has stable constant Jordan type $[1]$ if and only if $M$ is stably isomorphic to $\Omega^n(k)$ for some integer $n$, and $n$ is even in case $p$ is odd.*

*Proof.* One direction of this is given in Corollary 5.2.5. So assume that $M$ has stable constant Jordan type $[1]$. By Theorem 5.1.9, $M \otimes_k M^*$ has constant Jordan type. By Lemma 5.1.13, $\mathrm{End}_k(M) \cong M \otimes_k M^*$ again has stable constant Jordan type $[1]$. There are maps

$$k \to \mathrm{End}_k(M) \to k$$

given by the inclusion of the scalar matrices and the matrix trace. The dimension of $M$ is congruent to 1 modulo $p$, so the composite of these two maps is the identity. Thus

$$M \otimes_k M^* \cong k \oplus N$$

where $N$ is a module of stable constant Jordan type $\varnothing$. By Proposition 5.1.14, $N$ is a projective module, so $M$ is endotrivial. By Theorem 5.4.2, it follows that $M$ is stably isomorphic to $\Omega^n(k)$ for some $n$. If $p$ is odd and $n$ is odd, the stable constant Jordan type of $\Omega^n(k)$ is $[p-1]$ rather than $[1]$. So if $p$ is odd then $n$ is even. □

**Proposition 5.4.4** *If $p$ is odd then a $kE$-module $M$ has stable constant Jordan type $[p-1]$ if and only if $M \cong \Omega^n(k)$ for some odd integer $n$.*

*Proof.* The proof of this is essentially the same, after noting that for a cyclic group of order $p$ we have

$$J_{p-1} \otimes_k J_{p-1} \cong (p-2)J_p \oplus J_1.$$

This is because $J_{p-1} \cong \Omega(k)$ and $\Omega(k) \otimes_k \Omega(k)$ is stably isomorphic to $\Omega^2(k) \cong k \cong J_1$. Alternatively, this formula is a special case of the formulas given at the end of Section 2.3. □

Putting these together, we obtain the following theorem.

**Theorem 5.4.5** *Let $M$ be a $kE$-module. The following are equivalent:*

(1) *$M$ has stable constant Jordan type $[1]$ or $[p-1]$.*
(2) *$M$ is endotrivial.*
(3) *$M \cong \Omega^n(k)$ for some $n \in \mathbb{Z}$.* □

## 5.5 Wild Representation Type

In this section, we show that the modules of constant Jordan type usually have wild representation type. This has been independently proved by Bondarenko and Lytvynchuk [58]. See Section 1.2 for a discussion of representation type. We do this by taking a pair of endomorphisms $\phi$ and $\psi$ of a finite-dimensional vector space $V$ and constructing a module of constant Jordan type $M$ in such a way that $V$, $\phi$ and $\psi$ can be recovered functorially from $M$. We begin with an observation about common eigenvectors.

**Lemma 5.5.1** *The problem of classifying configurations of a finite-dimensional vector space $V$ with two endomorphisms $\phi$ and $\psi$ with no common eigenvector has wild representation type.*

*Proof.* Given a finite-dimensional vector space with endomorphisms $\phi$ and $\psi$, we construct another vector space $W = V \oplus V$ with endomorphisms $\Phi$ and $\Psi$ with no common eigenvector as follows. In terms of matrices, we take

$$\Phi = \begin{pmatrix} 1 & 0 \\ 0 & 0 \end{pmatrix} \qquad \Psi = \begin{pmatrix} \phi & 1 \\ 1 & \psi \end{pmatrix}.$$

The two copies of $V$ can be recovered as $W = \mathsf{Im}\,\Phi \oplus \mathsf{Ker}\,\Phi$, and so the off-diagonal components of $\Psi$ allow us to identify these two copies, and the diagonal components allow us to recover the endomorphisms $\phi$ and $\psi$. □

Next we discuss the quiver

$$Q = \quad \bullet \underset{\delta}{\overset{\beta}{\rightrightarrows}} \bullet \underset{\gamma}{\overset{\alpha}{\rightrightarrows}} \bullet$$

with relation $\alpha\beta = \gamma\delta$. Representations of $Q$ satisfying this relation are equivalent to representations of the quotient $kQ/I$ of the quiver algebra $kQ$ by the ideal $I = (\alpha\beta - \gamma\delta)$.

If $V$ is a finite-dimensional vector space with endomorphisms $\phi$ and $\psi$, we have a representation of $kQ/I$ given by

$$V \underset{\left(\begin{smallmatrix}0\\1\\0\end{smallmatrix}\right)}{\overset{\left(\begin{smallmatrix}1\\0\\0\end{smallmatrix}\right)}{\rightrightarrows}} V \oplus V \oplus V \underset{\left(\begin{smallmatrix}\phi&1&0\\\psi&0&1\end{smallmatrix}\right)}{\overset{\left(\begin{smallmatrix}1&0&0\\0&1&0\end{smallmatrix}\right)}{\rightrightarrows}} V \oplus V. \tag{5.1}$$

The summands in the middle term can be identified as $\mathsf{Im}\,(\beta) \oplus \mathsf{Im}\,(\delta) \oplus \ker(\alpha)$, and the summands in the right-hand term can be identified as $\mathsf{Im}\,(\alpha\beta) \oplus \mathsf{Im}\,(\alpha\delta)$. The maps $\alpha$, $\beta$, $\gamma$ and $\delta$ induce isomorphisms between the five copies of $V$ in these decompositions. It follows that the vector space $V$ as well as the maps $\phi$ and $\psi$ can be recovered from the representation of $kQ/I$.

**Remark 5.5.2** It is easy to check that this construction has the following property. For all $\lambda, \mu \in k$, not both zero, $\lambda\beta + \mu\delta$ is injective and $\lambda\gamma + \mu\alpha$ is surjective. Furthermore, $(\lambda\gamma + \mu\alpha)(\lambda\beta + \mu\delta)$ acts as

$$\begin{pmatrix} 2\lambda\mu + \lambda^2\phi \\ \mu^2 + \lambda^2\psi \end{pmatrix} : V \to V \oplus V.$$

As long as $\phi$ and $\psi$ have no common eigenvectors, this map is injective.

Next, let $E = \langle g_1, g_2 \rangle \cong \mathbb{Z}/p \times \mathbb{Z}/p$ with $p$ odd. Given a representation of $kQ/I$, we can make a $kE$-module with the same underlying vector space, and where $X_1 = g_1 - 1$ acts as $\beta + \gamma$ and $X_2 = g_2 - 1$ acts as $\delta + \alpha$. The relation $X_1X_2 = X_2X_1$ follows from the relation $\alpha\beta = \gamma\delta$, and $X_1^3$, $X_2^3$ act as zero.

More explicitly, the representations (5.1) have underlying vector space $V^{\oplus 6}$ with the following matrices for $X_1$ and $X_2$:

$$X_1 \mapsto \left(\begin{array}{c|ccc|cc} 0 & 0 & 0 & 0 & 0 & 0 \\ \hline 1 & 0 & 0 & 0 & 0 & 0 \\ 0 & 0 & 0 & 0 & 0 & 0 \\ 0 & 0 & 0 & 0 & 0 & 0 \\ \hline 0 & \phi & 1 & 0 & 0 & 0 \\ 0 & \psi & 0 & 1 & 0 & 0 \end{array}\right) \qquad X_2 \mapsto \left(\begin{array}{c|ccc|cc} 0 & 0 & 0 & 0 & 0 & 0 \\ \hline 0 & 0 & 0 & 0 & 0 & 0 \\ 1 & 0 & 0 & 0 & 0 & 0 \\ 0 & 0 & 0 & 0 & 0 & 0 \\ \hline 0 & 1 & 0 & 0 & 0 & 0 \\ 0 & 0 & 1 & 0 & 0 & 0 \end{array}\right). \tag{5.2}$$

Here is a diagram for such a module:

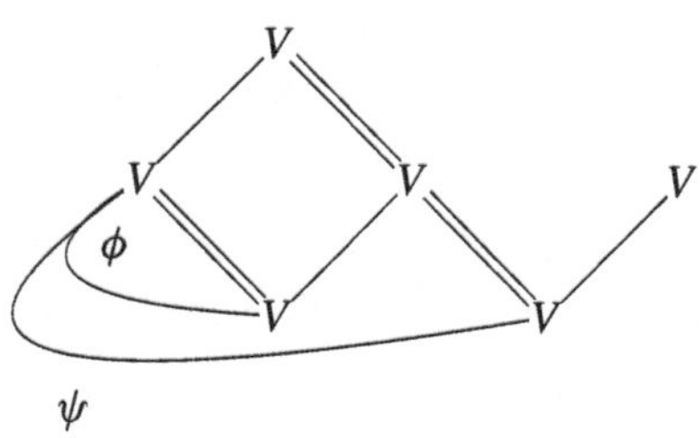


The representation of $kQ/I$ can be recovered from the socle layers of the $kE$-module.

**Theorem 5.5.3** *Let $p$ be an odd prime, $k$ an algebraically closed field of characteristic $p$ and $E = \mathbb{Z}/p \times \mathbb{Z}/p$. As long as $\phi$ and $\psi$ have no common eigenvectors, the matrices* (5.2) *define a $kE$-module of constant Jordan type* $[3]^n[2]^n[1]^n$ *where $n$ is the dimension of $V$.*

*Proof.* This is easy to check using Remark 5.5.2. ∎

**Corollary 5.5.4** *Let $p$ be an odd prime, $k$ an algebraically closed field of characteristic $p$ and $E = \mathbb{Z}/p \times \mathbb{Z}/p$. Then $kE$-modules of constant Jordan type $[3]^n[2]^n[1]^n$ are of wild representation type.*

*Proof.* This follows from Theorem 5.5.3 and Lemma 5.5.1. □

Next we give a different construction which works in rank at least three for any prime. Again we begin with a vector space $V$ with endomorphisms $\phi$ and $\psi$. Let

$$\alpha_1, \alpha_2, \alpha_3 \colon V^{\oplus 4} \to V^{\oplus 2}$$

be given by the following matrices:

$$\alpha_1 = \begin{pmatrix} 1 & 0 & 0 & 0 \\ 0 & 1 & 0 & 0 \end{pmatrix} \qquad \alpha_2 = \begin{pmatrix} 0 & 1 & 0 & 0 \\ 0 & 0 & 1 & 0 \end{pmatrix} \qquad \alpha_3 = \begin{pmatrix} \phi & 0 & 1 & \psi \\ 0 & 0 & 0 & 1 \end{pmatrix}.$$

The summands of $V^{\oplus 2}$ can be identified as

$$\alpha_1 \mathsf{Ker}\,(\alpha_2) \oplus \alpha_2 \mathsf{Ker}\,(\alpha_1).$$

The summands of $V^{\oplus 4}$ can be identified as

$$(\alpha_3^{-1}\alpha_1 \mathsf{Ker}\,\alpha_2 \cap \mathsf{Ker}\,\alpha_2) \oplus (\alpha_1^{-1}\alpha_2 \mathsf{Ker}\,\alpha_1 \cap \alpha_2^{-1}\alpha_1 \mathsf{Ker}\,\alpha_2 \cap \alpha_3^{-1}\alpha_1 \mathsf{Ker}\,\alpha_2)$$
$$\oplus (\mathsf{Ker}\,\alpha_1 \cap \alpha_3^{-1}\alpha_1 \mathsf{Ker}\,\alpha_2) \oplus (\mathsf{Ker}\,\alpha_1 \cap \mathsf{Ker}\,\alpha_2).$$

The maps $\alpha_1, \alpha_2, \alpha_3$ induce isomorphisms between the six copies of $V$ in these decompositions. It follows that the vector space $V$ and the maps $\phi$ and $\psi$ can be recovered from the maps $\alpha_1, \alpha_2, \alpha_3 \colon V^{\oplus 4} \to V^{\oplus 2}$.

If $E = \langle g_1, g_2, g_3 \rangle \cong (\mathbb{Z}/p)^3$ then we form a $kE$-module $M$ with underlying vector space $V^{\oplus 6}$ and where the elements $X_i = g_i - 1$ acts via $\alpha_i$. It is easy to check that every non-zero linear combination of $\alpha_1, \alpha_2, \alpha_3$ is surjective, so that $M$ has constant Jordan type $[2]^{2n}[1]^{2n}$. The matrices for this representation are

$$X_1 \mapsto \left(\begin{array}{cccc|cc} 0&0&0&0&0&0 \\ 0&0&0&0&0&0 \\ 0&0&0&0&0&0 \\ 0&0&0&0&0&0 \\ \hline 1&0&0&0&0&0 \\ 0&1&0&0&0&0 \end{array}\right) \quad X_2 \mapsto \left(\begin{array}{cccc|cc} 0&0&0&0&0&0 \\ 0&0&0&0&0&0 \\ 0&0&0&0&0&0 \\ 0&0&0&0&0&0 \\ \hline 0&1&0&0&0&0 \\ 0&0&1&0&0&0 \end{array}\right)$$

$$X_3 \mapsto \left(\begin{array}{cccc|cc} 0&0&0&0&0&0 \\ 0&0&0&0&0&0 \\ 0&0&0&0&0&0 \\ 0&0&0&0&0&0 \\ \hline \phi&0&1&\psi&0&0 \\ 0&0&0&1&0&0 \end{array}\right).$$

This example generalises easily to groups of higher rank. If $E$ has rank $r$ then we use maps $\alpha_1, \dots, \alpha_r \colon V^{\oplus(r+1)} \to V^{\oplus 2}$ given by

$$\alpha_1 = \begin{pmatrix} 1 & 0 & \dots & 0 \\ 0 & 1 & \dots & 0 \end{pmatrix}, \qquad \alpha_2 = \begin{pmatrix} 0 & 1 & 0 & \dots & 0 \\ 0 & 0 & 1 & \dots & 0 \end{pmatrix}, \quad \dots$$

$$\alpha_r = \begin{pmatrix} \phi & 0 & \dots & 1 & \psi \\ 0 & 0 & \dots & 0 & 1 \end{pmatrix}.$$

To conclude, we have proved the following theorem.

**Theorem 5.5.5** *If $r \geq 3$ or if $r \geq 2$ and $p \geq 3$ then the $kE$-modules of constant Jordan type have wild representation type.* □

**Remark 5.5.6** This theorem is best possible, because modules for $\mathbb{Z}/2 \times \mathbb{Z}/2$ in characteristic two have tame representation type and modules for $\mathbb{Z}/p$ in characteristic $p$ have finite representation type.

## 5.6 The Constant Image Property

Modules with the constant image property were introduced by Carlson and Friedlander [80] and further investigated by Carlson, Friedlander and Suslin [82] and by Worch [250]. We continue with the convention that $k$ is algebraically closed.

**Definition 5.6.1** A finitely generated $kE$-module has the *constant image property* if for every non-zero $\alpha \in \mathbb{A}^r(k)$ the image of $X_\alpha$ on $M$ is equal to $\mathsf{Rad}(M)$.

**Remark 5.6.2** The name comes from the fact that if the image of $X_\alpha$ on $M$ is independent of $\alpha$ for $0 \neq \alpha \in \mathbb{A}^r(k)$ then the image has to equal $\mathsf{Rad}(M)$.

**Lemma 5.6.3** *A $kE$-module $M$ has the constant image property if and only if for all $X \in J(kE) \setminus J^2(kE)$ we have $X.M = \mathsf{Rad}(M)$.*

*Proof.* One direction is clear. For the other direction, suppose that $M$ has the constant image property. Write $X = X_\alpha + Y$ with $\alpha \neq 0$ and $Y \in J^2(kE)$. Then $Y.M \subseteq \mathsf{Rad}^2(M)$ and

$$\mathsf{Rad}(M) = X_\alpha.M \subseteq X.M + Y.M,$$

and so $X.M = \mathsf{Rad}(M)$. □

**Lemma 5.6.4** *A $kE$-module $M$ has the constant image property if and only if the quotient module $M/\mathsf{Rad}^2(M)$ has the constant image property.*

*Proof.* This follows immediately from Lemma 5.6.3, because $X.M = \mathsf{Rad}(M)$ if and only if

$$X.(M/\mathsf{Rad}^2(M)) = \mathsf{Rad}(M)/\mathsf{Rad}^2(M) = \mathsf{Rad}(M/\mathsf{Rad}^2(M)). \qquad \square$$

**Lemma 5.6.5** *If $M$ is a $kE$-module with the constant image property and $K$ is an extension field of $k$ then $K \otimes_k M$ is a $KE$-module with the constant image property.*

*Proof.* This follows immediately from Theorem 4.1.1. $\square$

**Example 5.6.6** The $(2n-1)$-dimensional module for $(\mathbb{Z}/p)^2$ with diagram

has the constant image property. A generalisation of this example is the $W$-modules of Section 5.11.

**Proposition 5.6.7** *If $j \geq (r-1)(p-1)$ then $\mathsf{Rad}^j(kE)$ has the constant image property.*

*Proof.* Since the automorphism group of $kE$ is transitive on the $X_\alpha$ for $\alpha \neq 0$, we may assume that $X_\alpha = X_1$. So we must show that every element of $\mathsf{Rad}^{j+1}(kE)$ is in the image of $X_1$.

The submodule $\mathsf{Rad}^{j+1}(kE)$ has a basis consisting of the monomials $X_1^{a_1} \dots X_r^{a_r}$ with $0 \leq a_i < p$ and $\sum_i a_i \geq j+1 > (r-1)(p-1)$. This forces all $a_1$ to be greater than zero, and so this basis element is in the image of $X_1$. $\square$

**Remark 5.6.8** If $r = 2$, the modules of Proposition 5.6.7 are examples of the $W$-modules that we shall be discussing in Section 5.11.

**Lemma 5.6.9** *The class of modules with the constant image property is closed under finite sums, quotients (and, in particular, summands) and radicals.*

*Proof.* This follows immediately from the definition. $\square$

**Lemma 5.6.10** *If $M$ has constant image property then for every non-zero $\alpha \in \mathbb{A}^r(k)$ and for every $1 \leq j \leq p$, the image of $X_\alpha^j$ is equal to $\mathsf{Rad}^j(M)$. In particular we have $\mathsf{Rad}^p(M) = 0$.*

*Proof.* Since $kE$ is commutative we have

$$\begin{aligned} X_\alpha.\mathsf{Rad}^j(M) = X_\alpha.J(kE)^j.M = J(kE)^j.X_\alpha.M = J(kE)^j. \\ \mathsf{Rad}(M) = \mathsf{Rad}^{j+1}(M). \end{aligned}$$

The lemma now follows by induction. □

The following is Proposition 5.8 of [80].

**Theorem 5.6.11** *If $M$ is a $kE$-module with the constant image property then $M$ has constant Jordan type.*

*Proof.* This follows from Lemma 5.6.10, since the Jordan canonical form of a nilpotent matrix is determined by the dimensions of the images of powers of the matrix. □

**Theorem 5.6.12** *If $E$ has rank at least three then the collection of indecomposable modules with the constant image property has wild representation type.*

*Proof.* The modules constructed in Section 5.5 for rank at least three have the constant image property. □

The following theorem shows that the radical layers of a module with the constant image property are decreasing in dimension by at least $r - 1$ at each stage.

**Theorem 5.6.13** *If $M$ has the constant image property then*

$$\dim_k M/\mathsf{Rad}(M) \geq \dim_k \mathsf{Rad}(M)/\mathsf{Rad}^2(M) + r - 1.$$

*Proof.* Let $m = \dim_k M/\mathsf{Rad}(M)$ and $n = \dim_k \mathsf{Rad}(M)/\mathsf{Rad}^2(M)$. Consider the matrix $A$ for the action of

$$X_\alpha = \sum_{i=1}^{r} \lambda_i X_i \colon M/\mathsf{Rad}(M) \to \mathsf{Rad}(M)/\mathsf{Rad}^2(M).$$

This is an $n \times m$ matrix of homogeneous linear forms in the $\lambda_i$. By the definition of the constant image property, the matrix $A$ has full row rank provided the $\lambda_i$ are not all zero.

Let $z_1, \dots, z_n$ be variables, and consider the row vector $(z_1, \dots, z_n)$ times the matrix $A$. Then for all values of $z_1, \dots, z_n$, not all zero, and all values of $\lambda_1, \dots, \lambda_r$, not all zero, the product $(z_1, \dots, z_n)A$ is non-zero. So the $m$ equations saying that the products of $(z_1, \dots, z_n)$ with the columns of $A$ are zero have no common solutions. These $m$ equations define a subvariety of codimension at most $m$ in the projective variety $\mathbb{P}^{n-1} \times \mathbb{P}^{r-1}$. Since this subvariety is empty, it follows that

$$m \geq (n - 1) + (r - 1) + 1 = n + r - 1.$$ □

**Corollary 5.6.14** *If $M$ has the constant image property then for all $j \geq 1$ we have*

$$\dim_k \mathsf{Rad}^j(M)/\mathsf{Rad}^{j+1}(M) \leq \dim_k M/\mathsf{Rad}(M) - j(r-1).$$

*Proof.* This follows from Lemma 5.6.9 and Theorem 5.6.13. □

**Corollary 5.6.15** *If $M$ has the constant image property and radical length $d$ then*

$$\dim_k M \geq \binom{d}{2}(r-1) + d.$$

*Proof.* If $M$ has radical length $d$ then $\dim_k \mathsf{Rad}^{d-1}(M) \geq 1$ and so for $0 \leq j \leq d-1$ we have

$$\dim_k \mathsf{Rad}^j(M)/\mathsf{Rad}^{j+1}(M) \geq (d-1-j)(r-1) + 1.$$

Now sum over $j$. □

**Example 5.6.16** The following family of examples shows that Theorem 5.6.13 is sharp. Let $n \geq 1$ and consider the $n \times (n+r-1)$ matrix

$$U_{n,r} = \begin{pmatrix} \lambda_1 & \lambda_2 & & \cdots & \lambda_r & 0 & \cdots & 0 \\ 0 & \lambda_1 & \lambda_2 & & \cdots & \lambda_r & & 0 \\ \vdots & & \ddots & \ddots & & & \ddots & \vdots \\ 0 & 0 & \cdots & \lambda_1 & \lambda_2 & & \cdots & \lambda_r \end{pmatrix}.$$

Let $M$ be the representation of $(\mathbb{Z}/p)^r$ of dimension $2n+r-1$ for which $\lambda_1 X_1 + \cdots + \lambda_r X_r$ acts as

$$\begin{pmatrix} \mathbf{0} & \mathbf{0} \\ U_{n,r} & \mathbf{0} \end{pmatrix}.$$

Then $M$ has the constant image property, with

$$\dim M/\mathsf{Rad}(M) = n + r - 1, \qquad \dim \mathsf{Rad}(M) = n.$$

In particular, $M$ has constant Jordan type $[2]^n[1]^{r-1}$. Here is a diagram for the module $M$ in the case $r = 3$, $n = 4$:

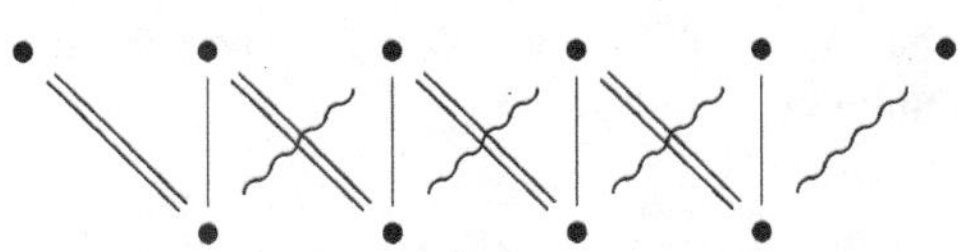

Finally, we have the following property of modules with the constant image property.

**Theorem 5.6.17** *Suppose that $E$ has rank at least two, and let $M$ be a $kE$-module with the constant image property. Then*

$$\sum_{0\neq\alpha\in\mathbb{A}^r(k)} \mathsf{Ker}\,(X_\alpha, M) = M.$$

*Proof.* Consider the matrices $A_1, A_2$ representing the maps induced by $X_1, X_2$ from $M/\mathsf{Rad}(M)$ to $\mathsf{Rad}(M)/\mathsf{Rad}^2(M)$. These matrices satisfy the conditions of Proposition 4.1.2 (1), and so $M/\mathsf{Rad}(M)$ is equal to the sum of the kernels of $\lambda A_1 + \mu A_2$ for $(\lambda, \mu) \neq (0, 0)$.

If $\bar{m} \in M/\mathsf{Rad}(M)$ is in the kernel of $X_\alpha$ then $X_\alpha m \in \mathsf{Rad}^2(M)$, and so there exists $m' \in \mathsf{Rad}(M)$ such that $X_\alpha m = X_\alpha m'$. Thus $X_\alpha(m - m') = 0$. It follows that every element of the kernel of $X_\alpha \colon M/\mathsf{Rad}(M) \to \mathsf{Rad}(M)/\mathsf{Rad}^2(M)$ lifts to an element of the kernel of $X_\alpha$ on $M$.

It follows that $\mathsf{Rad}(M)+\sum_{0\neq\alpha\in\mathbb{A}^r(k)} \mathsf{Ker}\,(X_\alpha) = M$, and hence by Nakayama's lemma we have $\sum_{0\neq\alpha\in\mathbb{A}^r(k)} \mathsf{Ker}\,(X_\alpha) = M$. □

## 5.7 The Generic Kernel

For rank two elementary abelian groups there is a notion closely related to the constant image property, namely the generic kernel of a module. Throughout this section, we set $E = \langle g_1, g_2\rangle \cong \mathbb{Z}/p \times \mathbb{Z}/p$, $X_1 = g_1 - 1$ and $X_2 = g_2 - 1$ in $kE$. The results in this section are due to Carlson, Friedlander and Suslin [82]. We shall use the methods developed here in Section 5.13 to prove a special case of a conjecture of Rickard.

**Definition 5.7.1** Let $M$ be a $kE$-module. If

$$0 \neq \alpha = (\lambda_1, \lambda_2) \in \mathbb{A}^2(k),$$

then the kernel of $X_\alpha = \lambda_1 X_1 + \lambda_2 X_2$ on $M$ only depends on the image $\bar{\alpha}$ of $\alpha$ in $\mathbb{P}^1(k)$. We define the *generic kernel* $\mathfrak{K}(M)$ to be the intersection over all dense open subsets $S$ of $\mathbb{P}^1(k)$ (i.e., subsets whose complement is finite) of the sum of $\mathsf{Ker}\,(X_\alpha, M)$ for $\bar{\alpha} \in S$.

**Remark 5.7.2** Since $M$ is finite dimensional, there exists a dense open subset $S \subseteq \mathbb{P}^1(k)$ such that $M = \sum_{\bar{\alpha}\in S} \mathsf{Ker}\,(X_\alpha, M)$. The subset $S$ may be taken to exclude any given finite set of points.

**Lemma 5.7.3** *Let $M$ be a $kE$-module. Then we have $\mathfrak{K}(\mathfrak{K}(M)) = \mathfrak{K}(M)$.*

*Proof.* This is clear from the definition. □

**Theorem 5.7.4** *Let $M$ be a $kE$-module. Then*

(1) *$\mathfrak{K}(M)$ has the constant image property.*
(2) *If $N$ is a submodule of $M$ with the constant image property then $N \subseteq \mathfrak{K}(M)$.*

*Proof.* (1) By the lemma, without loss of generality we can assume that $M = \mathfrak{K}(M)$. We need to show that for all $0 \neq \alpha, \beta \in \mathbb{A}^2(k)$ we have $X_\alpha M = X_\beta M$. Write

$$M = \sum_{\gamma \in S} \mathsf{Ker}\,(X_\gamma, M)$$

where $S$ is a dense open subset chosen to satisfy $\alpha, \beta \notin S$. For all $\gamma \in S$ there exist $\lambda_1, \lambda_2 \in k$, neither equal to zero, such that $\gamma = \lambda_1\alpha + \lambda_2\beta$. So

$$X_\alpha \mathsf{Ker}\,(X_\gamma, M) = X_\beta \mathsf{Ker}\,(X_\gamma, M).$$

It follows that $X_\alpha M = X_\beta M$.

(2) This follows from Theorem 5.6.17. □

**Corollary 5.7.5** *Let $M$ be a $kE$-module. Then $\mathfrak{K}(M)$ has constant Jordan type.*

*Proof.* This follows from Theorems 5.6.11 and 5.7.4 (1). □

**Theorem 5.7.6** *Let $M$ be a $kE$-module. Then $\mathsf{Ker}\,(X_\alpha, M) \subseteq \mathfrak{K}(M)$ if and only if $X_\alpha$ has maximal rank on $M$.*

*Proof.* First note that the set of $\bar{\alpha} \in \mathbb{P}^1(k)$ such that $X_\alpha$ has maximal rank on $M$ is a dense open subset of $\mathbb{P}^1(k)$. The subset $S$ may therefore be chosen so that for all $\bar{\alpha} \in S$ $X_\alpha$ has maximal rank on $M$. Now let $\alpha \in \mathbb{P}^1(k)$ and $\gamma \in S$. Then using the maximality of the rank of $X_\gamma$ and Theorem 5.7.4, we have

$$\begin{aligned}\dim \mathsf{Ker}\,(X_\alpha, M) \geq \dim \mathsf{Ker}\,(X_\gamma, M) &= \dim \mathsf{Ker}\,(X_\gamma, \mathfrak{K}(M)) \\ &= \dim \mathsf{Ker}\,(X_\alpha, \mathfrak{K}(M)).\end{aligned}$$

It follows that

$$\begin{aligned}\mathsf{Ker}\,(X_\alpha, M) \subseteq \mathfrak{K}(M) &\Leftrightarrow \dim \mathsf{Ker}\,(X_\alpha, M) = \dim \mathsf{Ker}\,(X_\alpha, \mathfrak{K}(M)) \\ &\Leftrightarrow \dim \mathsf{Ker}\,(X_\alpha, M) = \dim \mathsf{Ker}\,(X_\gamma, M).\end{aligned}$$ □

**Corollary 5.7.7** *If $M$ has constant Jordan type then for all $0 \neq \alpha \in \mathbb{A}^2(k)$ we have $\mathsf{Ker}\,(X_\alpha, M) \subseteq \mathfrak{K}(M)$.* □

**Proposition 5.7.8** *Let $M$ be a $kE$-module of constant Jordan type. If $0 \neq \alpha \in \mathbb{A}^2(k)$ and $m \in M$ such that $X_\alpha m$ is in the generic kernel $\mathfrak{K}(M)$ then for all $0 \neq \beta \in \mathbb{A}^2(k)$ we have $X_\beta m \in \mathfrak{K}(M)$. Thus we have*

$$J(kE).m \subseteq \mathfrak{K}(M).$$

*Proof.* If $X_\alpha m \in \mathfrak{K}(M)$ then $X_\alpha X_\beta m = X_\beta X_\alpha m \in \mathsf{Rad}\,\mathfrak{K}(M)$. So by Theorem 5.7.4 and Lemma 5.6.3 there exists $m' \in \mathfrak{K}(M)$ such that $X_\alpha m' = X_\alpha X_\beta m$. Thus $X_\alpha(X_\beta m - m') = 0$. By Corollary 5.7.7, this implies that $X_\beta m - m' \in \mathfrak{K}(M)$. Since also $m' \in \mathfrak{K}(M)$, this implies that $X_\beta m \in \mathfrak{K}(M)$. □

## 5.8 The Subquotient $\mathsf{Rad}^{-1}\mathfrak{K}(M)/\mathsf{Rad}^2\mathfrak{K}(M)$

In this section we show that for some purposes we can reduce to the case of a module of constant Jordan type of radical length three. We will use this construction in Section 5.13 to prove a special case of a conjecture of Rickard.

For the purposes of this section, we maintain the notation of the last section: $E = \langle g_1, g_2\rangle \cong \mathbb{Z}/p \times \mathbb{Z}/p$, $k$ is an algebraically closed field of characteristic $p$, and $X_1 = g_1 - 1$ and $X_2 = g_2 - 1$ in $kE$. If $M$ is a $kE$-module, we write $\mathsf{Rad}^{-1}\mathfrak{K}(M)$ for the submodule $\{m \in M \mid J(kE).m \in \mathfrak{K}(M)\}$. We are concerned with the subquotient $\mathsf{Rad}^{-1}\mathfrak{K}(M)/\mathsf{Rad}^2\mathfrak{K}(M)$.

**Example 5.8.1** If $M = kE$ then $\mathsf{Rad}^{-1}\mathfrak{K}(M)/\mathsf{Rad}^2\mathfrak{K}(M)$ is isomorphic to the module $M_p$ discussed in Proposition 5.1.6.

**Lemma 5.8.2** *Let $0 \neq \alpha \in \mathbb{A}^2(k)$ and let $M$ be a $kE$-module of constant Jordan type. Then $X_\alpha$ has the same number of Jordan blocks of length one on $\mathsf{Rad}^{-1}\mathfrak{K}(M)/\mathsf{Rad}^2\mathfrak{K}(M)$ as on $M$.*

*Proof.* The number of Jordan blocks of $X_\alpha$ on $M$ is the dimension of

$$\frac{\mathsf{Ker}\,(X_\alpha, M)}{\mathsf{Ker}\,(X_\alpha, M) \cap X_\alpha M} \cong \frac{\mathsf{Ker}\,(X_\alpha, M) + X_\alpha M}{X_\alpha M}.$$

By Corollary 5.7.7 we have

$$\mathsf{Ker}\,(X_\alpha, M) \subseteq \mathfrak{K}(M) \subseteq \mathsf{Rad}^{-1}\mathfrak{K}(M)$$

and by Proposition 5.7.8 we have

$$\mathsf{Ker}\,(X_\alpha, M) \cap X_\alpha M = \mathsf{Ker}\,(X_\alpha, \mathsf{Rad}^{-1}\mathfrak{K}(M)) \cap X_\alpha \mathsf{Rad}^{-1}\mathfrak{K}(M).$$

So

$$\begin{aligned}\frac{\mathsf{Ker}\,(X_\alpha, M)}{\mathsf{Ker}\,(X_\alpha, M) \cap X_\alpha M} &= \frac{\mathsf{Ker}\,(X_\alpha, \mathsf{Rad}^{-1}\mathfrak{K}(M))}{\mathsf{Ker}\,(X_\alpha, \mathsf{Rad}^{-1}\mathfrak{K}(M)) \cap X_\alpha \mathsf{Rad}^{-1}\mathfrak{K}(M)}\\ &\cong \frac{\mathsf{Ker}\,(X_\alpha, \mathsf{Rad}^{-1}\mathfrak{K}(M)) + X_\alpha \mathsf{Rad}^{-1}\mathfrak{K}(M)}{X_\alpha \mathsf{Rad}^{-1}\mathfrak{K}(M)}.\end{aligned}$$

Now if $m \in \mathsf{Rad}^{-1}\mathfrak{K}(M)$ and $X_\alpha m \in \mathsf{Rad}^2\mathfrak{K}(M)$ then by Theorem 5.7.4 and Lemma 5.6.10 there exists $m' \in \mathfrak{K}(M)$ such that $X_\alpha m = X_\alpha^2 m'$, and so $X_\alpha(m -$

$X_\alpha m') = 0$ and $X_\alpha m' \in X_\alpha \mathfrak{K}(M) \subseteq X_\alpha \mathsf{Rad}^{-1}\mathfrak{K}(M)$. It follows that the map

$$\frac{\mathsf{Ker}\,(X_\alpha, \mathsf{Rad}^{-1}\mathfrak{K}(M)) + X_\alpha \mathsf{Rad}^{-1}\mathfrak{K}(M)}{X_\alpha \mathsf{Rad}^{-1}\mathfrak{K}(M)} \to \frac{\mathsf{Ker}\left(X_\alpha, \dfrac{\mathsf{Rad}^{-1}\mathfrak{K}(M)}{\mathsf{Rad}^2\mathfrak{K}(M)}\right) + X_\alpha \dfrac{\mathsf{Rad}^{-1}\mathfrak{K}(M)}{\mathsf{Rad}^2\mathfrak{K}(M)}}{X_\alpha \dfrac{\mathsf{Rad}^{-1}\mathfrak{K}(M)}{\mathsf{Rad}^2\mathfrak{K}(M)}}$$

is an isomorphism. The dimension of this is the number of Jordan blocks of length one of $X_\alpha$ on $\mathsf{Rad}^{-1}\mathfrak{K}(M)/\mathsf{Rad}^2\mathfrak{K}(M)$. □

**Theorem 5.8.3** (Baland [26]) *Let $M$ be a $kE$-module of constant Jordan type. Then the subquotient $\mathsf{Rad}^{-1}\mathfrak{K}(M)/\mathsf{Rad}^2\mathfrak{K}(M)$ also has constant Jordan type, with the same number of Jordan blocks of length one, and the same total number of Jordan blocks.*

*Proof.* Let $0 \neq \alpha \in \mathbb{A}^2(k)$. Lemma 5.8.2 shows that the number of Jordan blocks of length one of $X_\alpha$ on $M$ is the same as on $\mathsf{Rad}^{-1}\mathfrak{K}(M)/\mathsf{Rad}^2\mathfrak{K}(M)$. Corollary 5.7.7, Theorem 5.7.4 and Lemma 5.6.10 show that the total number of Jordan blocks of $X_\alpha$ on $M$ is the same as on $\mathsf{Rad}^{-1}\mathfrak{K}(M)/\mathsf{Rad}^2\mathfrak{K}(M)$.

Since $\mathsf{Rad}^{-1}\mathfrak{K}(M)/\mathsf{Rad}^2\mathfrak{K}(M)$ has radical length at most three, the Jordan canonical form of $X_\alpha$ on $M$ has the form $[3]^a[2]^b[1]^c$. The quantity $3a + 2b + c$ is the dimension, and is therefore independent of $\alpha$. The quantities $a + b + c$ and $c$ are also independent of $\alpha$. Since the matrix

$$\begin{pmatrix} 3 & 2 & 1 \\ 1 & 1 & 1 \\ 0 & 0 & 1 \end{pmatrix} \in GL(3, \mathbb{Z})$$

has determinant one, it follows that $a$, $b$ and $c$ are independent of $\alpha$. □

**Example 5.8.4** Carlson, Friedlander and Suslin [82, Question 8.11] ask the following question. If a $kE$-module ($E$ is $\mathbb{Z}/p \times \mathbb{Z}/p$ as in the rest of this section) $M$ has constant Jordan type, is it true that $\mathsf{Rad}^{-i}\mathfrak{K}(M)$ has constant Jordan type for all $i \geq 0$? Baland [26] found the following example to show that this is not the case. Let $p \geq 5$ and let $M$ be the 26-dimensional module given by the following diagram.

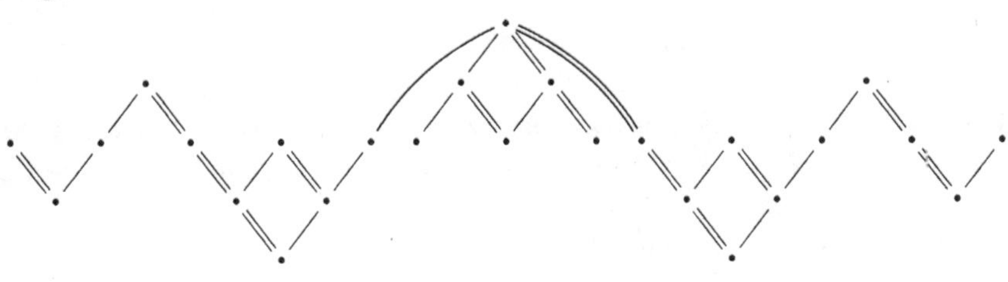

Note that in this diagram, the top vertex is taken by $X_1$ to the sum of the two vertices at the end of single lines from it, and by $X_2$ to the sum of the two vertices at the end of double lines from it.

It is straightforward to check that $M$ has constant Jordan type $[4]^2[3]\,[2]^5[1]^5$, by computing the ranks of powers of $\lambda_1 X_1 + \lambda_2 X_2$ for $\lambda_1, \lambda_2 \in k$ not both zero.

On the other hand, the codimension one submodule $\mathsf{Rad}^{-1}\mathfrak{K}(M)$ decomposes as a direct sum of three modules, two of which do not have constant Jordan type, and therefore $\mathsf{Rad}^{-1}\mathfrak{K}(M)$ does not have constant Jordan type.

## 5.9 The Constant Kernel Property

For the sake of completeness, in this and the next section we formulate the dual notions to the constant image property and the generic kernel. The proofs are mostly dual to those of Sections 5.6 and 5.7.

**Definition 5.9.1** A finite-dimensional $kE$-module $M$ has the *constant kernel property* if for every non-zero $\alpha \in \mathbb{A}^r(k)$ the kernel of $X_\alpha$ on $M$ is equal to $\mathsf{Soc}(M)$.

For the purposes of the following lemma, we use the Lie theoretic dual rather than the group theoretic one, so that duality commutes with restriction to cyclic shifted subgroups.

**Lemma 5.9.2** *A $kE$-module $M$ has the constant image property if and only if the dual module $M^{\tilde{*}}$ has the constant kernel property.*

*Proof.* The kernel of $X_\alpha$ on $M^{\tilde{*}}$ is the annihilator in $M^{\tilde{*}}$ of the image of $X_\alpha$ on $M$, and $\mathsf{Soc}(M^{\tilde{*}})$ is the annihilator of $\mathsf{Rad}(M)$. □

**Lemma 5.9.3** *A $kE$-module $M$ has the constant kernel property if and only if for all $X \in J(kE) \setminus J^2(kE)$ the kernel of $X$ on $M$ is equal to $\mathsf{Soc}(M)$.*

*Proof.* This is dual to Lemma 5.6.3. □

**Lemma 5.9.4** *A $kE$-module $M$ has the constant kernel property if and only if $\mathsf{Soc}^2(M)$ has the constant kernel property.*

*Proof.* This is dual to Lemma 5.6.4. □

**Lemma 5.9.5** *If $M$ is a $kE$-module with the constant kernel property and $K$ is an extension field of $k$ then $K \otimes_k M$ is a $KE$-module with the constant kernel property.*

*Proof.* This is dual to Lemma 5.6.5. □

**Example 5.9.6** The $(2n-1)$-dimensional module for $(\mathbb{Z}/p)^2$ with diagram

has the constant kernel property.

**Proposition 5.9.7** *If $j \le p$ then $kE/\mathsf{Rad}^j(kE)$ has the constant kernel property.*

*Proof.* Noting that $\mathsf{Rad}^j(kE) = \mathsf{Soc}^{r(p-1)+1-j}(kE)$, this is dual to Proposition 5.6.7. □

**Lemma 5.9.8** *The class of modules with the constant kernel property is closed under finite sums, submodules and socle quotients.* □

**Remark 5.9.9** Every $k(\mathbb{Z}/p)^2$-module $M$ with the constant kernel property and with $\mathsf{Rad}^d(M) = 0$ may be embedded in $W^*_{n,d}$ for some $n \ge d$.

**Lemma 5.9.10** *If $M$ has the constant kernel property then for every non-zero $\alpha \in \mathbb{A}^r(k)$ and for every $1 \le j \le p$, the kernel of $X^j_\alpha$ on $M$ is equal to $\mathsf{Soc}^j(M)$.*

*Proof.* This is dual to Lemma 5.6.10. □

**Theorem 5.9.11** *If $M$ is a $kE$-module with the constant kernel property then $M$ has constant Jordan type.*

*Proof.* This is dual to Theorem 5.6.11. □

**Theorem 5.9.12** *If $E$ has rank at least three then the collection of indecomposable modules with the constant kernel property has wild representation type.*

*Proof.* This is dual to Theorem 5.6.12. □

**Theorem 5.9.13** *If $M$ has the constant kernel property then*

$$\dim_k \mathsf{Soc}(M) \ge \dim_k \mathsf{Soc}^2(M)/\mathsf{Soc}(M) + r - 1.$$

*Proof.* This is dual to Theorem 5.6.13. □

**Corollary 5.9.14** *If $M$ has the constant kernel property then for all $j \ge 1$ we have*

$$\dim_k \mathsf{Soc}^{j+1}(M)/\mathsf{Soc}^j(M) \le \dim_k \mathsf{Soc}(M) - j(r-1).$$ □

**Theorem 5.9.15** *If $M$ has constant Jordan type then $M/\mathfrak{K}(M)$ has the constant kernel property.*

*Proof.* This follows from Proposition 5.7.8. □

**Theorem 5.9.16** *Suppose that $E$ has rank at least two, and let $M$ be a $kE$-module with the constant kernel property. Then*

$$\bigcap_{0\neq\alpha\in\mathbb{A}^r(k)} \mathsf{Im}\,(X_\alpha, M) = 0.$$

*Proof.* This is dual to Theorem 5.6.17. □

## 5.10 The Generic Image

For this section we assume that $E = \langle g_1, g_2\rangle$, $X_1 = g_1 - 1$, $X_2 = g_2 - 1$ in $kE$, and we develop the concept of generic image of a module.

**Definition 5.10.1** Let $M$ be a $kE$-module. If

$$0 \neq \alpha = (\lambda_1, \lambda_2) \in \mathbb{A}^2(k),$$

then the image of $X_\alpha = \lambda_1 X_1 + \lambda_2 X_2$ on $M$ only depends on the image $\bar{\alpha}$ of $\alpha$ in $\mathbb{P}^1(k)$. We define the generic image $\mathfrak{I}(M)$ to be the sum over all dense open subsets $S$ of $\mathbb{P}^1(k)$ of the intersection of $\mathsf{Im}\,(X_\alpha, M)$ for $\bar{\alpha} \in S$.

**Remark 5.10.2** Since $M$ is finite dimensional, there exists a dense open subset $S \subseteq \mathbb{P}^1(k)$ such that $M = \bigcap_{\bar{\alpha}\in S} \mathsf{Im}\,(X_\alpha, M)$. The subset $S$ may be taken to exclude any given finite set of points.

Again, for the purposes of the following lemma, we use the Lie theoretic dual rather than the group theoretic one, so that duality commutes with restriction to cyclic shifted subgroups.

**Lemma 5.10.3** *Let $M$ be a $kE$-module. Then $\mathfrak{I}(M^{\tilde{*}})$ is the annihilator in $M^{\tilde{*}}$ of $\mathfrak{K}(M)$ and $\mathfrak{K}(M^{\tilde{*}})$ is the annihilator of $\mathfrak{I}(M)$. Thus we have*

$$M^{\tilde{*}}/\mathfrak{I}(M^{\tilde{*}}) \cong \mathfrak{K}(M)^{\tilde{*}}, \qquad (M/\mathfrak{I}(M))^{\tilde{*}} \cong \mathfrak{K}(M^{\tilde{*}}),$$
$$\mathfrak{I}(M^{\tilde{*}}) \cong (M/\mathfrak{K}(M))^{\tilde{*}}, \qquad \mathfrak{I}(M)^{\tilde{*}} \cong M^{\tilde{*}}/\mathfrak{K}(M^{\tilde{*}}).$$

□

**Lemma 5.10.4** *Let $M$ be a $kE$-module. Then we have $\mathfrak{I}(M/\mathfrak{I}(M)) = 0$.* □

**Theorem 5.10.5** *Let $M$ be a $kE$-module. Then*

(1) *$M/\mathfrak{I}(M)$ has the constant kernel property.*

(2) *If $N$ is a submodule of $M$ such that $M/N$ has the constant kernel property then $N \supseteq \mathfrak{I}(M)$.*

*Proof.* This is dual to Theorem 5.7.4. □

**Corollary 5.10.6** *Let $M$ be a $kE$-module. Then $M/\mathfrak{I}(M)$ has constant Jordan type.* □

**Theorem 5.10.7** *Let $M$ be a $kE$-module. Then $\mathfrak{I}(M) \subseteq \mathsf{Im}\,(X_\alpha, M)$ if and only if $X_\alpha$ has maximal rank on $M$.*

*Proof.* This is dual to Theorem 5.7.6. □

**Corollary 5.10.8** *If $M$ has constant Jordan type then for all $0 \neq \alpha \in \mathbb{A}^2(k)$ we have $\mathfrak{I}(M) \subseteq \mathsf{Im}\,(X_\alpha)$.* □

**Proposition 5.10.9** *Let $M$ be a $kE$-module of constant Jordan type. If $m \in \mathfrak{I}(M)$ and $0 \neq \alpha, \beta \in \mathbb{A}^2(k)$ then there exists $m' \in \mathfrak{I}(M)$ such that $X_\beta m' = X_\alpha m$.*

*Proof.* This is almost dual to Proposition 5.7.8, but requires separate proof. By Corollary 5.10.8 there exists $m'' \in M$ such that $m = X_\beta m''$. We set $m' = X_\alpha m''$. By Theorem 5.10.5 the image of $m''$ in $M/\mathfrak{I}(M)$ lies in the socle, and so $X_\alpha m'' \in \mathfrak{I}(M)$ and $X_\beta m' = X_\beta X_\alpha m'' = X_\alpha X_\beta m'' = X_\alpha m$. □

**Theorem 5.10.10** *If $M$ has constant Jordan type then $\mathfrak{I}(M)$ has the constant image property.*

*Proof.* This is dual to Theorem 5.9.15. □

**Corollary 5.10.11** *If $M$ has constant Jordan type then $\mathfrak{I}(M) \subseteq \mathfrak{K}(M)$.*

*Proof.* This follows from Theorems 5.10.10 and 5.7.4 (2). □

**Lemma 5.10.12** *Suppose that $M$ has constant Jordan type. Then the following are equal:*

(1) *The number of Jordan blocks of each $X_\alpha$ on $M$.*
(2) *The number of Jordan blocks of each $X_\alpha$ on $\mathfrak{K}(M)$.*
(3) *The number of Jordan blocks of each $X_\alpha$ on $\mathfrak{K}(M)/\mathsf{Rad}^2\mathfrak{K}(M)$.*
(4) *The dimension of $\mathfrak{K}(M)/\mathsf{Rad}\,\mathfrak{K}(M)$.*
(5) *The number of Jordan blocks of each $X_\alpha$ on $M/\mathfrak{I}(M)$.*
(6) *The number of Jordan blocks of each $X_\alpha$ on $\mathsf{Soc}^2(M/\mathfrak{I}(M))$.*
(7) *The dimension of $\mathsf{Soc}(M/\mathfrak{I}(M))$.*

*Proof.* This follows from Corollaries 5.7.7 and 5.10.8 and Lemmas 5.6.3 and 5.9.3. □

## 5.11 $W$-Modules

If $n \geq d \geq 1$ and $d \leq p$, we define a module $W_{n,d}$ for $(\mathbb{Z}/p)^2$ with generators $v_1, \dots, v_n$ and relations

$$X_1 v_1 = 0, \qquad X_2 v_n = 0, \qquad X_2 v_i = X_1 v_{i+1} \qquad (1 \leq i < n),$$
$$X_1^d v_i = 0, \qquad X_2^d v_i = 0 \qquad (1 \leq i \leq n).$$

Then $W_{n,d}$ has the constant image property.

**Example 5.11.1** Here is the diagram for $W_{6,4}$ $(p \geq 5)$:

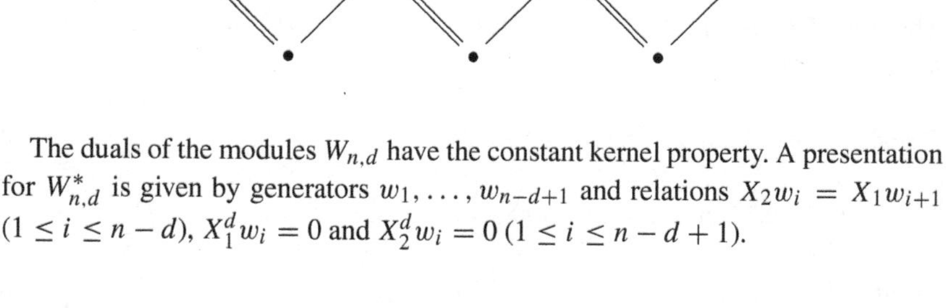

The duals of the modules $W_{n,d}$ have the constant kernel property. A presentation for $W_{n,d}^*$ is given by generators $w_1, \dots, w_{n-d+1}$ and relations $X_2 w_i = X_1 w_{i+1}$ $(1 \leq i \leq n-d)$, $X_1^d w_i = 0$ and $X_2^d w_i = 0$ $(1 \leq i \leq n-d+1)$.

**Lemma 5.11.2** *The module $W_{n,d}$ is of constant Jordan type*

$$[d]^{n-d+1}, [d-1], \dots, [2], [1].$$ □

**Remark 5.11.3** Carlson, Friedlander and Suslin [82] have proved that every $k(\mathbb{Z}/p)^2$-module $M$ with the constant image property and with $\mathsf{Rad}^d(M) = 0$ is a quotient of the module $W_{n,d}$ of Example 5.6.6 for some $n \geq d$.

It seems difficult to formulate an analogue of this statement for higher rank. The left-hand diagram below gives a one-parameter family of examples of modules for $(\mathbb{Z}/p)^3$ with the constant image property:

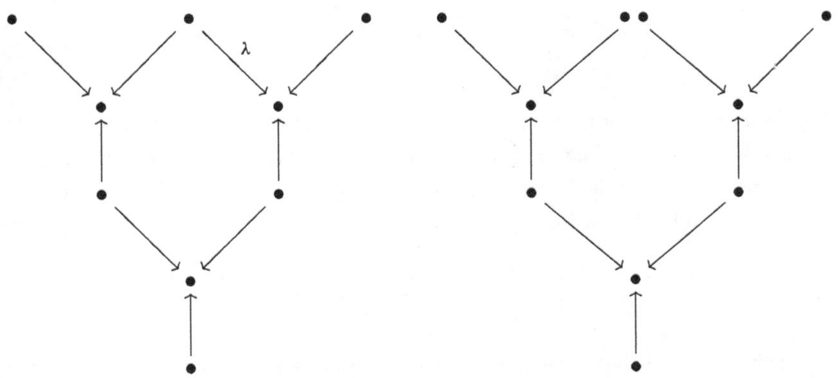


($0 \neq \lambda \in k$). Here, the actions of the elements $X_1$, $X_2$, $X_3$ are indicated by up, down-right and down-left arrows respectively. Note that the modules in this family are not quotients of the module on the right. So it is unlikely that a discrete set of modules analogous to the $W_{n,d}$ exists for higher rank, in such a way that they surject onto every module with the constant image property. This is probably related to the fact that for rank at least three the modules with the constant image property are of wild representation type, see Theorem 5.6.12.

## 5.12 Constant Jordan type with One Non-Projective Block

In Section 5.4, we proved that the only modules of stable constant Jordan type [1] and $[p-1]$ are the modules stably isomorphic to $\Omega^n(k)$.

Conjecture 9.5 of Carlson, Friedlander and Pevtsova [81] states that there cannot exist a module of stable constant Jordan type [2] for an elementary abelian $p$-group of rank at least two with $p > 3$. Such a module is characterised by the condition that its restriction to every cyclic shifted subgroup consists of a single Jordan block of length two together with some number of Jordan blocks of length $p$. In this section, we prove the following strengthened form of their conjecture, which is the main theorem of [42]. In Section 5.13 we prove a stronger statement (Corollary 5.14.2) by more complicated methods.

**Theorem 5.12.1** *Let $E = (\mathbb{Z}/p)^r$ with $r \geq 2$, and let $k$ be an algebraically closed field of characteristic $p$. If $2 \leq a \leq p-2$ then there does not exist a finite-dimensional $kE$-module of stable constant Jordan type $[a]$.*

*Proof.* Let $M$ be a $kE$-module of stable constant Jordan type $[a]$ with $2 \leq a \leq p-2$, and let $\dim_k M = np + a$. By Theorem 5.1.9, $\Lambda^{a+1}(M)$ is of stable constant Jordan type $\varnothing$, so by Proposition 5.1.14 it is projective. It follows that $\dim_k \Lambda^{a+1} M$ is divisible by $p^2$. This dimension is

$$\binom{np+a}{a+1} = \frac{(np+a)(np+a-1)\dots(np)}{(a+1)!}.$$

The only term in the numerator or the denominator of this fraction that is divisible by $p$ is the last term on the top, namely $np$. So we deduce that $n$ is a multiple of $p$.

On the other hand, by Theorem 2.8.1 and Proposition 5.1.14, $S^{p-a+1}M$ is also projective, and its dimension is

$$\binom{np+p}{p-a+1} = \frac{(np+p)(np+p-1)\dots(np+a)}{(p-a+1)!}.$$

The only term in the numerator or the denominator of this fraction that is divisible by $p$ is the first term on the top, namely $np+p$. Since $n$ is a multiple of $p$, this is not divisible by $p^2$. This contradicts the fact that this module is projective, and hence no such $M$ can exist. □

**Corollary 5.12.2** *Let $E = (\mathbb{Z}/p)^r$ and let $M$ be a finite-dimensional $kE$-module with stable generic Jordan type $[a]$, with $2 \le a \le p-2$. Then the generic Jordan set $U_{\max}(M)$ is the complement of a closed set of codimension one in affine space $\mathbb{A}^r(k)$.*

*Proof.* Let $V \subseteq \mathbb{A}^r(k)$ be the complement of $U_{\max}(M)$. If $V$ has codimension at least two then there is a linear subspace $W$ of $\mathbb{A}^r(k)$ of dimension two such that $V \cap W = \{0\}$. Choosing a basis for $W$ gives a rank two shifted subgroup $E'$, and then $kE'$ is a subalgebra of $kE$. The restriction of $M$ to $kE'$ has constant Jordan type $[a]$, contradicting Theorem 5.12.1. □

## 5.13 Rickard's Conjecture

The following conjecture was made by Rickard at MSRI in 2008, based on data from a large number of modules of constant Jordan type. The modules he was examining were all produced by the methods of Section 5.2, so they are all homogeneous, but the methods we know that produce inhomogeneous modules of constant Jordan type still seem to satisfy this conjecture. Throughout this section $E$ is an elementary abelian $p$-group of rank at least two and $k$ is an algebraically closed field of characteristic $p$.

**Conjecture 5.13.1** (Rickard's Conjecture) Suppose that $M$ is a $kE$-module of constant Jordan type $[a_1]\dots[a_t]$. If there is a value of $j$ with $1 \le j \le p-1$ such that none of the $a_i$ is equal to $j$ then the total number of $a_i$ that are greater than $j$ is a multiple of $p$.

In other words, if there is a missing size of Jordan block, then the total number of Jordan blocks of larger size is divisible by $p$.

This conjecture also rules out stable constant Jordan type $[a]$ for $2 \le a \le p-2$ (see Theorem 5.12.1) as follows. If there were such a module, then the conjecture with $j = a+1$ would imply that the number of Jordan blocks of length $p$ is divisible by $p$. On the other hand, the conjecture with $j = a-1$ would imply that this number is congruent to $-1$ modulo $p$. These incompatible congruences would show that such a module cannot exist.

There is a strengthened version of Rickard's conjecture which states that the number of $a_i$ that are greater than $j$ is a multiple not just of $p$ but of $p^{r-1}$. Rickard's conjecture reduces to the case $r = 2$, whereas the strengthened version does not. The following is easy to verify.

**Theorem 5.13.2** *If $M$ is a module of constant Jordan type that satisfies Rickard's conjecture then so is $\Omega(M)$. Similarly, if $M$ satisfies the strengthened version then so does $\Omega(M)$.* □

Following [44], we shall use the generic kernel to prove a special case of Rickard's conjecture, namely the case $j = 1$. Motivation for the proof of this theorem may be found in the proof of Dade's lemma in [98], and in the example of Proposition 5.1.6. Consequences of this theorem, and a related theorem in characteristic zero, are investigated in section 5.14.

**Theorem 5.13.3** *Let $M$ be a $kE$-module of constant Jordan type $[a_1]\dots[a_t]$, and suppose that none of the $a_i$ is equal to 1. Then $t$ is divisible by $p$.*

*Proof.* Let $\mathfrak{K}(M)$ be the generic kernel of $M$, as described in Section 5.7. By Lemma 5.10.12, it suffices to show that for each summand $N$ of $\mathfrak{K}(M)/\mathsf{Rad}^2\mathfrak{K}(M)$, $N/\mathsf{Rad}N$ has dimension divisible by $p$.

Let $0 \ne \alpha \in \mathbb{A}^2(k)$. Look at the maps of vector spaces induced by $X_\alpha$

$$\mathsf{Rad}^{-1}\mathfrak{K}(M)/\mathfrak{K}(M) \xrightarrow{X_\alpha} \mathfrak{K}(M)/\mathsf{Rad}\mathfrak{K}(M) \xrightarrow{X_\alpha} \mathsf{Rad}\mathfrak{K}(M)/\mathsf{Rad}^2\mathfrak{K}(M).$$

It follows from Lemma 5.6.3, Theorem 5.6.11 and Corollary 5.7.7 that the first of these maps is injective and the second is surjective.

It follows from Theorem 5.8.3 that $\mathsf{Rad}^{-1}\mathfrak{K}(M)/\mathsf{Rad}^2\mathfrak{K}(M)$ has constant Jordan type, $X_\alpha$ has no Jordan blocks of length one on it, and the total number of Jordan blocks on it is equal to $t$.

In order to prove the theorem, we may therefore replace $M$ by $\mathsf{Rad}^{-1}\mathfrak{K}(M)/\mathsf{Rad}^2\mathfrak{K}(M)$. Thus we may assume that $\mathsf{Rad}^2\mathfrak{K}(M) = 0$ and $\mathsf{Rad}M \subseteq \mathfrak{K}(M)$, which implies that $\mathsf{Rad}^3 M = 0$ so that every Jordan block of $X_\alpha$ on $M$ has length two or three.

The indecomposable modules for $kE/J^2(kE)$ are well understood in terms of the classification of modules for the Kronecker quiver. Since every $X_\alpha$ induces a surjective map from $\mathfrak{K}(M)/\mathsf{Rad}\mathfrak{K}(M)$ to $\mathsf{Rad}\mathfrak{K}(M)$, it follows that $\mathfrak{K}(M)$ is a direct sum of $W$-modules for $kE/J^2$ of the following form. Such a module has a basis $v_0, \dots, v_n, w_1, \dots, w_n$ with

$$X_1 v_0 = 0, \quad X_1 v_i = w_i, \quad X_1 w_i = 0,$$
$$X_2 v_{i-1} = w_i, \quad X_2 v_n = 0, \quad X_2 w_i = 0$$
$$(1 \le i \le n).$$

A diagram of such a module is as follows.

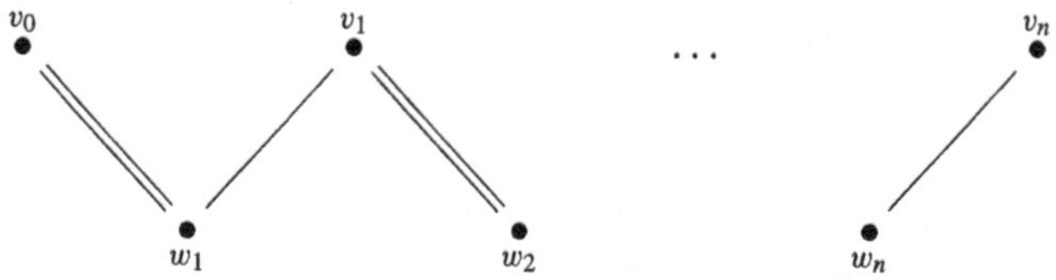

(5.3)

Let $N$ be an indecomposable summand of $\mathfrak{R}(M)$. Thus $N$ is a module of the form (5.3) with $n \ge 0$. For notation, we write $\bar{v}_0, \dots, \bar{v}_n$ for the images of $v_0, \dots, v_n$ in $\mathfrak{R}(M)/\mathsf{Rad}\mathfrak{R}(M)$. Let $0 \ne \alpha \in \mathbb{A}^2(k)$. Since $X_\alpha$ has no Jordan blocks of length one on $M$, the kernel of $X_\alpha$ on $N$ is in the image of $X_\alpha$ on $M$. Indeed, this remains true after field extension, see Theorem 5.1.8. Apply this to $X_\alpha = X_1 + tX_2$ after extending the coefficients to the field of Laurent power series $k[[t]][t^{-1}]$. The kernel of $X_\alpha$ on $N$ is spanned by the image of $X_\alpha$ on $N$ together with the vector

$$v_0 - tv_1 + t^2 v_2 - \cdots \pm t^n v_n.$$

It follows that there is a vector $\sum_i (-1)^i t^i \bar{u}_i \in M/\mathfrak{R}(M)$ whose image under $X_1 + tX_2$ is

$$\bar{v}_0 - t\bar{v}_1 + t^2 \bar{v}_2 - \cdots \pm t^n \bar{v}_n \in \mathfrak{R}(M)/\mathsf{Rad}\mathfrak{R}(M).$$

Since $X_1$ and $X_2$ are injective as maps from $M/\mathfrak{R}(M)$ to $\mathfrak{R}(M)/\mathsf{Rad}\mathfrak{R}(M)$, comparing coefficients of powers of $t$ shows that $\bar{u}_i$ is zero unless $i \ge 0$. For $i \ge n$ we have $X_2 \bar{u}_i = X_1 \bar{u}_{i+1}$.

The argument at this stage follows the lines of the proof of Proposition 4.1.2. If $\bar{u}_n \ne 0$ then let $m > n$ be as small as possible so that there is a linear relation of the form $\sum_{i=n}^m \lambda_i \bar{u}_i = 0$. Such a linear relation exists because $M$ is finite dimensional, and minimality of $m$ implies that $\bar{u}_n, \dots, \bar{u}_{m-1}$ are linearly independent and $\bar{u}_m$ is a linear combination of them. So

$$X_2 \bar{u}_m \in \langle X_2 \bar{u}_n, \dots, X_2 \bar{u}_{m-1} \rangle = \langle X_1 \bar{u}_{n+1}, \dots, X_1 \bar{u}_m \rangle \subseteq \langle X_1 \bar{u}_n, \dots, X_1 \bar{u}_{m-1} \rangle$$

as subspaces of $\mathfrak{R}(M)/\mathsf{Rad}\mathfrak{R}(M)$. It follows that for all $\lambda, \mu$ not both zero,

$$\lambda X_1 + \mu X_2 \colon \langle \bar{u}_n, \dots, \bar{u}_{m-1} \rangle \to \langle X_1 \bar{u}_n, \dots, X_1 \bar{u}_{m-1} \rangle$$

is injective, and hence an isomorphism. This is impossible because using $X_1$ to identify these two spaces, $X_2$ has to have a non-zero eigenvector for some eigenvalue. This contradiction shows that $\bar{u}_n = 0$.

Again using the fact that $X_1$ is injective and $X_2\bar{u}_i = X_1\bar{u}_{i+1}$ for $i \geq n$, it follows that $\bar{u}_i = 0$ for $i \geq n$. Thus we have

$$(X_1 + tX_2)(\bar{u}_0 - t\bar{u}_1 + t^2\bar{u}_2 - \cdots \mp t^{n-1}\bar{u}_{n-1}) = \bar{v}_0 - t\bar{v}_1 + t^2\bar{v}_2 - \cdots \pm t^n\bar{v}_n.$$

In particular it follows that $n > 0$. Comparing coefficients we get

$$X_1\bar{u}_0 = \bar{v}_0, \quad X_1\bar{u}_1 - X_2\bar{u}_0 = \bar{v}_1, \quad X_1\bar{u}_2 - X_2\bar{u}_1 = \bar{v}_2, \quad \cdots, \quad -X_2\bar{u}_{n-1} = \bar{v}_n.$$

Next, we lift $\bar{u}_0, \ldots, \bar{u}_{n-1}$ to elements $u_0, \ldots, u_{n-1}$ in $M$. Then the equation above holds without bars, modulo $\mathsf{Rad}\mathfrak{K}(M)$. Since $X_1$ is surjective as a map from $\mathfrak{K}(M)/\mathsf{Rad}\mathfrak{K}(M)$ to $\mathsf{Rad}\mathfrak{K}(M)$, the lifting can be done in such a way that

$$X_1u_0 = v_0, \quad X_1u_1 - X_2u_0 = v_1, \quad X_1u_2 - X_2u_1 = v_2, \quad \cdots, \quad -X_2u_{n-1} = v_n + w$$

with $w \in \mathsf{Rad}\mathfrak{K}(M)$. This shows that $X_1u_1 - 2v_1 = X_2u_0 - v_1$ is annihilated by $X_1$; for $2 \leq i \leq n-1$ we have

$$X_2(X_1u_{i-1} - iv_{i-1}) = X_1^2u_i - (i+1)w_i = X_1(X_1u_i - (i+1)v_i),$$

and

$$X_2(X_1u_{n-1} - nv_{n-1}) = -(n+1)w_n.$$

We therefore have the following relationships between elements of $\mathfrak{K}(M)$:

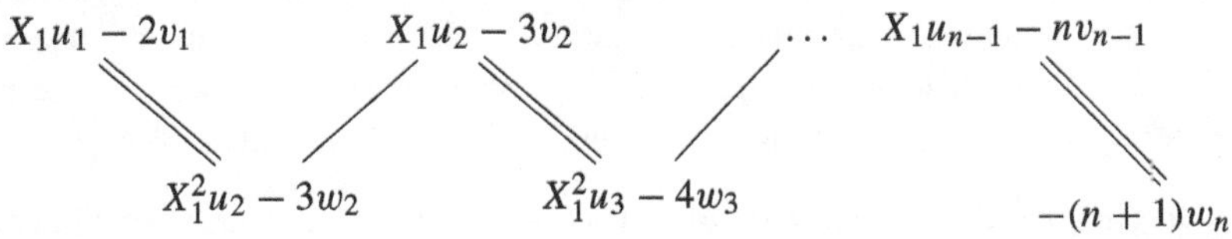


Here, we do not claim that the depicted elements are linearly independent, or even non-zero. However, it is easily checked that there cannot be such a chain of elements in $\mathfrak{K}M$ unless the coefficient of $w_n$ on the lower rightmost vector is zero. This is because

$$w_n \notin (X_2X_1^{-1})^{n-1}\{0\}_{\mathfrak{K}M},$$

as can be seen by projecting onto the summand $N$ of $\mathfrak{K}M$ (the notation here is that if $U$ is a subspace of $\mathfrak{K}M$ we denote by $X_1^{-1}U$ its inverse image under $X_1$ and by $X_2U$ its image under $X_2$).

It follows that $-(n+1)$ is zero in $k$, which implies that $n \equiv -1 \pmod{p}$. It follows that the dimension of $N/\mathsf{Rad}N$ is divisible by $p$. Since this is true for every indecomposable summand $N$ of $\mathfrak{K}M$, it follows that the dimension of $\mathfrak{K}M/\mathsf{Rad}\mathfrak{K}M$ is divisible by $p$. □

## 5.14 Consequences and Variations

In this section we deduce some consequences of the special case of Rickard's conjecture proved in Theorem 5.13.3, and then we investigate an analogue in characteristic zero. The first consequence is another special case of Rickard's conjecture given in the following corollary.

**Corollary 5.14.1** *Let $M$ be a $kE$-module of constant Jordan type $[a_1]\dots[a_t]$, and suppose that none of the $a_i$ is equal to $p-1$. Then the number of $a_i$ that are equal to $p$ is divisible by $p$.*

*Proof.* Let $M$ be a module of constant Jordan type $[a_1]\dots[a_t]$ with none of the $a_i$ equal to $p-1$. Then $\Omega(M)$ is a module of constant Jordan type where the number of blocks of length $p-j$ ($j \geq 1$) is equal to the number of blocks of $M$ of length $j$. The number of Jordan blocks of length $p$ in $\Omega(M)$ is congruent mod $p$ to minus the total number of Jordan blocks of $M$. Therefore, Corollary 5.14.1 for $M$ follows by applying Theorem 5.13.3 to $\Omega(M)$. □

**Corollary 5.14.2** *Let $M$ be a $kE$-module of stable constant Jordan type $[a_1]\dots[a_t]$, and suppose that none of the $a_i$ is equal to $1$ or $p-1$. Then $t$ is divisible by $p$.*

*Proof.* If $M$ has constant Jordan type with no blocks of length 1 or $p-1$ then we may apply Theorem 5.13.3 to deduce that the total number of Jordan blocks is divisible by $p$, and we may also apply Corollary 5.14.1 to deduce that the number of Jordan blocks of length $p$ is also divisible by $p$. Subtracting, we see that the number of Jordan blocks of length between 2 and $p-2$ is divisible by $p$. □

Corollary 5.14.2 gives a new proof of Theorem 5.12.1. Moreover, combining our main theorem with the methods of Section 5.12 gives new information.

**Theorem 5.14.3** *Suppose that $M$ has constant Jordan type with all stable Jordan blocks of distinct odd length or all of distinct even length. Then the stable Jordan type is either $[1]$ or $[p-1]$ or has at least four Jordan blocks.*

*Proof.* Suppose that all the stable Jordan blocks of $M$ are distinct and odd. By Theorem 5.1.9, $\Lambda^2(M)$ has constant Jordan type.

If $M$ has one stable Jordan block then by Theorem 5.13.3 the length of the Jordan block is equal to one.

If $M$ has two stable Jordan blocks then again applying Theorem 5.13.3, the restriction to a cyclic group is $J_a \oplus J_1$ plus projectives, with $a > 1$. So $\Lambda^2(M)$ has stable constant Jordan type $\Lambda^2(J_a \oplus J_1) = \Lambda^2(J_a) \oplus J_a$. By Proposition 2.8.6, this has at most $(p-1)/4$ non-projective summands, none of which is isomorphic to $J_1$, contradicting Theorem 5.13.3.

If $M$ has three stable Jordan blocks then the restriction to a cyclic group is $J_a \oplus J_b \oplus J_1$ plus projectives, with $a < b$. We have

$$\Lambda^2(J_a \oplus J_b \oplus J_1) \cong \Lambda^2(J_a) \oplus \Lambda^2(J_b) \oplus J_a \otimes J_b \oplus J_a \oplus J_b.$$

It follows from Proposition 2.8.6 that $\Lambda^2(J_a)$ and $\Lambda^2(J_b)$ each has at most $(p-3)/4$ non-projective indecomposable summands, and $J_a \otimes J_b$ has at most $(p-3)/2$. So $\Lambda^2(M)$ has at most $p-1$ non-projective summands, and that none of them is isomorphic to $J_1$. Again this contradicts Theorem 5.13.3.

It follows that if $M$ has more than one stable Jordan block then it has at least four.

Finally, if all the stable Jordan blocks are distinct and even, then $\Omega(M)$ has constant Jordan type with all stable Jordan blocks distinct and odd. So we apply the previous arguments to $\Omega(M)$. □

**Theorem 5.14.4** *If $p/2 < n < p-1$ there is no module of stable constant Jordan type $[n][1]$.*

*Proof.* If $M$ has stable constant Jordan type $[n][1]$ then because of the restriction on the value of $n$, there are no Jordan blocks of length $p-1$. So by Corollary 5.14.1, the number of Jordan blocks of length $p$ is divisible by $p$. By Proposition 2.8.6, $S^2(J_n)$ has $n-(p-1)/2$ copies of $J_p$. So $S^2(M)$ has no Jordan blocks of length $p-1$ and the number of Jordan blocks of length $p$ is congruent to $n-(p-1)/2$ modulo $p$. This contradicts Corollary 5.14.1 for $S^2(M)$. □

The same technique does not seem to dispose of stable constant Jordan type $[n][1]$ for $4 \leq n < p/2$ even, so for example $[4][1]$ for $p = 11$ is the smallest case that remains undecided.

Some further cases can be ruled out by the use of other Schur functors. For example, by applying the Schur functor $S^{[2][1]}$, we can rule out the case of stable constant Jordan type $[6][1]$ for $p = 13$ using the method of proof of Theorem 5.14.4. More generally the Schur functor $S^{[2][1]}$ rules out modules of stable constant Jordan type $[2n][1]$ when $p = 6n-5$ or $p = 6n-7$. This is because using the technique explained in Section 2.9 we see that for $p = 6n-5$ we have

$$S^{[2][1]}(J_{2n}) = 2J_p \oplus J_{p-3} \oplus \cdots$$

while for $p = 6n-7$ we have

$$S^{[2][1]}(J_{2n}) = 4J_p \oplus J_{p-3} \oplus \cdots$$

where the dots indicate smaller Jordan blocks.

**Theorem 5.14.5** *Let $M$ be a $kE$-module of constant Jordan type $[a_1]\dots[a_t]$ with none of the $a_i$ equal to one. Then at most $2t/p$ of the $a_i$ are equal to 2.*

*Proof.* Looking back at the proof of Theorem 5.13.3, we see that $\mathfrak{K}(M)/J^2\mathfrak{K}(M)$ is a direct sum of $W$-modules of the form (5.3) with $n+1$ divisible by $p$. There is at most one $a_i$ equal to 2 and $n$ of the $a_i$ greater than 2 for each such summand. □

As an application of Theorem 5.14.5, we completely classify the Jordan types that can occur for $(\mathbb{Z}/3)^2$. The corresponding problem for elementary abelian 2-groups is studied in Chapter 10 using the theory of vector bundles.

**Theorem 5.14.6** *Let $E = (\mathbb{Z}/3)^2$ and $k$ be an algebraically closed field of characteristic 3. Then there is a $kE$-module of constant Jordan type $[3]^a[2]^b[1]^c$ if and only if the following conditions hold:*

(1) *If $a = 0$ then $c \geq 1$.*
(2) *If $b = 0$ then $a$ is divisible by 3.*
(3) *If $c = 0$ then $a + b$ is divisible by 3 and $b \leq 2a$.*

*Proof.* Condition (1) holds by Theorem 5.1.18, condition (2) holds by Corollary 5.14.1, and condition (3) holds by Theorems 5.13.3 and 5.14.5.

To prove that these conditions are sufficient, we display the following modules of constant Jordan type:

(a) $kE$ of type $[3]^3$,
(b) $k$ of type $[1]$,
(c) $\mathsf{Rad}^2(kE)$ of type $[3]\,[2]\,[1]$,
(d) $\mathsf{Rad}(kE)$ of type $[3]^2[2]$,
(e) $\mathsf{Rad}(kE)/\mathsf{Soc}(kE)$ of type $[3]\,[2]^2$,
(f) $W$-modules $W_{n,2}$ (see Section 5.11) of type $[2]^{n-1}[1]$ $(n \geq 2)$.

If $a = 0$ then we use direct sums of modules of type (f). If $b = 0$ then we use direct sums of modules of type (a) and (b). If $c = 0$ then we use direct sums of modules of type (a), (d) and (e). Finally, if $a, b, c$ are all non-zero then we use direct sums of modules of all these types; first we use modules of type (a) to reduce to $a \leq 2$, then we use modules of type (b) and (f) to reduce to the cases (c), (d) and (e). □

Now we turn to a characteristic zero analogue of Rickard's conjecture, and again we prove a special case of it.

**Conjecture 5.14.7** Let $k$ be an algebraically closed field of characteristic zero. Let $A$ and $B$ be commuting nilpotent $d \times d$ matrices over $k$, and assume that all non-zero linear combinations of $A$ and $B$ have the same Jordan canonical form. If there is a Jordan block of length $j$ then there are Jordan blocks of all lengths up to and including $j$.

**Theorem 5.14.8** *Let $k$ be an algebraically closed field of characteristic zero. Let $A$ and $B$ be commuting nilpotent $d \times d$ matrices over $k$ with $d \geq 1$, and assume that all non-zero linear combinations of $A$ and $B$ have the same Jordan canonical form. Then there are Jordan blocks of length one.*

*Proof.* The concept of generic kernel works in this situation just as well as for representations of $\mathbb{Z}/p \times \mathbb{Z}/p$ in characteristic $p$. For the proof of the theorem, assume that there are no Jordan blocks of length one, and argue as in the proof of Theorem 5.13.3 up to the last paragraph, where $-(n+1)w_n$ is shown to be equal to zero in $k$. Since $w_n$ is non-zero, and $k$ has characteristic zero, it follows that $n = -1$, which contradicts the fact that $n \geq 0$. □

## 5.15 Further Conjectures

Let $E$ be an elementary abelian $p$-group of rank $r \geq 2$ and let $k$ be an algebraically closed field of characteristic $p$. In Section 5.13 we discussed Rickard's conjecture on the possible constant Jordan types of $kE$-modules. In this section we discuss some further conjectures.

If $p = 3$ then $k$ and $\Omega(k)$ have stable constant Jordan type [1] and [2] respectively, and therefore sums of these realise all possible stable constant Jordan types. So we shall assume for the sake of this section that $p \geq 5$.

For $(\mathbb{Z}/p)^2$, the only known restrictions on stable constant Jordan type are given in Section 5.13, although restrictions for larger rank will be given in the next chapter. It is not known, for example, whether there can exist a module of stable constant Jordan type [4] [1] if $p \geq 11$. This is the simplest type where the answer is not known.

The following conjecture is recorded in [81].

**Conjecture 5.15.1** (Suslin's Conjecture) Suppose that $M$ is a $kE$-module of constant Jordan type. If there is a Jordan block of length $j$ with $2 \leq j \leq p-1$ then there is either a Jordan block of length $j+1$ or one of length $j-1$.

This conjecture does rule out a module of stable constant Jordan type $[a]$ for $2 \leq a \leq p-2$, as in Theorem 5.12.1, as there is no Jordan block of length $a-1$ or $a+1$ in this case. It also rules out $[a]^n$ for all $n > 0$, for example.

Rickard's conjecture does not rule out $[a]^p$ for $2 \leq a \leq p-2$, whereas Suslin's conjecture does. On the other hand, Rickard's conjecture rules out stable constant Jordan type [3] [2] for $p \geq 5$ whereas Suslin's conjecture does not. So neither of these conjectures implies the other.

At the moment, we have no plausible conjecture which precisely characterises which set of stable constant Jordan types we expect to occur, even in the case $r = 2$.

We mention one more conjecture which is stated in [81]. This should be compared with Example 5.2.6.

**Conjecture 5.15.2** If a module has stable constant Jordan type $[2]\,[1]^j$ then $j \geq r-1$.

In Section 10.4 we prove a weak version of this conjecture, namely that $j \geq r - 2$ as long as $p$ is large enough, see Corollary 10.4.2. The proof involves vector bundles on projective space and their Chern numbers, which is our next topic.

# 6

# Vector Bundles on Projective Space

This chapter is by way of introduction to algebraic vector bundles on projective space for modular representation theorists. Background material can be found in the books of Hartshorne [147], Okonek, Schneider and Spindler [202] and Fulton [129]. The literature concerning vector bundles on projective space is extensive; see the references for a selection of relevant papers.

## 6.1 Definitions and First Properties

**Definition 6.1.1** Let $\mathcal{O}$ be a sheaf of rings on a space $X$. An $\mathcal{O}$-module is said to be *locally free* if $X$ has an open cover $\{U_\alpha\}$ such that $\mathcal{F}|_{U_\alpha}$ is a free $\mathcal{O}|_{U_\alpha}$-module for all $\alpha$. If $X$ is connected then the rank of $\mathcal{F}|_{U_\alpha}$ as a free $\mathcal{O}|_{U_\alpha}$-module does not depend on $\alpha$, and is called the *rank* of $\mathcal{F}$.

**Definition 6.1.2** If $V$ is a quasi-projective variety then a *vector bundle* is defined to be a locally free $\mathcal{O}_V$-module which has finite rank on each connected component. A homomorphism of vector bundles $\mathcal{F}_1 \to \mathcal{F}_2$ is just a homomorphism of sheaves from $\mathcal{F}_1$ to $\mathcal{F}_2$.

**Definition 6.1.3** The *dual* of a vector bundle $\mathcal{F}$ is the vector bundle $\mathcal{F}^\vee = \mathcal{H}om_{\mathcal{O}_V}(\mathcal{F}, \mathcal{O}_V)$. There are natural isomorphisms

$$\mathcal{F} \to \mathcal{F}^{\vee\vee}$$
$$\mathcal{F}_1^\vee \otimes_{\mathcal{O}_V} \mathcal{F}_2 \to \mathcal{H}om_{\mathcal{O}_V}(\mathcal{F}_1, \mathcal{F}_2).$$

In particular we have

$$\mathrm{Hom}_{\mathcal{O}_V}(\mathcal{F}_1, \mathcal{F}_2) \cong \Gamma(V, \mathcal{F}_1^\vee \otimes_{\mathcal{O}_V} \mathcal{F}_2).$$

**Definition 6.1.4** A *line bundle* on $V$ is a vector bundle of rank one. For example, the twists $\mathcal{O}_v(i)$ of the structure sheaf are line bundles on $V$. Under tensor product $- \otimes_{\mathcal{O}_V} -$, isomorphism classes of vector bundles on $V$ form an abelian group called the *Picard group* $\mathrm{Pic}(V)$.

The following definitions are analogous to the corresponding definition for modules.

**Definition 6.1.5** A non-zero vector bundle $\mathcal{F}$ on $X$ is said to be *decomposable* if it can be written as $\mathcal{F}_1 \oplus \mathcal{F}_2$ with $\mathcal{F}_1$ and $\mathcal{F}_2$ non-zero vector bundles. Otherwise it is said to be *indecomposable*.

**Theorem 6.1.6** (Atiyah [14]) *The Krull–Schmidt theorem holds for vector bundles over a connected complete algebraic variety (for example projective space).* □

**Definition 6.1.7** A non-zero vector bundle $\mathcal{F}$ on $X$ is said to be *irreducible* if given a short exact sequence of vector bundles

$$0 \to \mathcal{F}' \to \mathcal{F} \to \mathcal{F}'' \to 0,$$

either $\mathcal{F}' = 0$ or $\mathcal{F}'' = 0$.

However, the following definition is not the analogue of the corresponding module theoretic notion of simplicity; rather, it is the analogue of the conclusion of Schur's lemma.

**Definition 6.1.8** A non-zero vector bundle $\mathcal{F}$ on $X$ is said to be *simple* if every endomorphism of $\mathcal{F}$ is a scalar multiple of the identity. In other words, the natural map

$$k \to \mathrm{End}_X(\mathcal{F}) \cong H^0(X, \mathcal{F}^\vee \otimes \mathcal{F})$$

is an isomorphism. This is equivalent to the statement that

$$\dim_k H^0(X, \mathcal{F}^\vee \otimes \mathcal{F}) = 1.$$

**Lemma 6.1.9** *For $j \in \mathbb{Z}$, a vector bundle $\mathcal{F}$ is simple if and only if $\mathcal{F}(j)$ is simple.*

*Proof.* We have

$$\mathcal{F}(j)^\vee \otimes \mathcal{F}(j) \cong (\mathcal{F}^\vee \otimes \mathcal{O}(-j)) \otimes (\mathcal{F} \otimes \mathcal{O}(j)) \cong \mathcal{F}^\vee \otimes \mathcal{F}.$$ □

**Remark 6.1.10** The notion of algebraic vector bundle behaves quite differently from the notion of topological vector bundle. For example, every short exact sequence of topological vector bundles on a paracompact space splits, whereas the same is not at all true of algebraic vector bundles on a variety. Every complex algebraic vector bundle induces a topological one, but there are topological vector bundles which do not come from algebraic ones, and homomorphisms of topological ones not coming from algebraic homomorphisms.

A theorem of Serre [225] states that if $k = \mathbb{C}$, the field of complex numbers, then algebraic vector bundles over a complex projective variety are essentially the same as holomorphic vector bundles, regarding the variety as a complex manifold.

This theorem usually goes under the name GAGA, after the title of the paper. An extended discussion is contained in the book of Neeman [198].

## 6.2 Tests for Vector Bundles

The following test for when a coherent $\mathcal{O}_V$-module is a vector bundle comes from Exercise 5.8 in Section II.5 of Hartshorne [147] and Proposition 4.11 of Friedlander–Pevtsova [127]. Since we shall use it extensively, we provide a proof.

**Theorem 6.2.1** *Let $X$ be a reduced connected Noetherian scheme.*

(a) *If $\mathcal{F}$ is a coherent sheaf of $\mathcal{O}_X$-modules, then the following are equivalent.*
   (1) *$\mathcal{F}$ is a vector bundle.*
   (1) *The dimension of the fibre $\dim_{k(x)} \mathcal{F}_x \otimes_{\mathcal{O}_x} k(x)$ is constant for $x \in X$.*

(b) *If $f : \mathcal{F} \to \mathcal{F}'$ is a map of vector bundles on $X$ then the following are equivalent.*
   (1) *The cokernel of $f$ is a vector bundle of rank equal to $\operatorname{rank} \mathcal{F}' - n$.*
   (2) *The induced map of fibres*

$$\bar{f} : \mathcal{F}_x \otimes_{\mathcal{O}_x} k(x) \to \mathcal{F}'_x \otimes_{\mathcal{O}_x} k(x)$$

   *has constant rank $n$ for $x \in X$.*

*If these hold then the image of $f$ is a vector bundle of rank $n$.*

(c) *If $f : \mathcal{F} \to \mathcal{F}'$ is a map of vector bundles on $X$ then the following are equivalent.*
   (1) *$f$ is a locally split injection.*
   (2) *The induced map of fibres*

$$\bar{f} : \mathcal{F}_x \otimes_{\mathcal{O}_x} k(x) \to \mathcal{F}'_x \otimes_{\mathcal{O}_x} k(x)$$

   *is injective for all $x \in X$.*

*Proof.* Consider the function $\phi : X \to \mathbb{Z}$ given by $\phi(x) = \dim_{k(x)} \mathcal{F}_x \otimes_{\mathcal{O}_x} k(x)$. We shall show that for any coherent sheaf $\mathcal{F}$, the function $\phi$ is upper semicontinuous, i.e., that for any $n \in \mathbb{Z}$ the set $X_{\geq n} = \{x \in X \mid \phi(x) \geq n\}$ is closed, or equivalently, the complementary set $X_{<n} = \{x \in X \mid \phi(x) < n\}$ is open.

Since a set is open if and only if its intersection with each set from an open cover is open, we may assume that $X$ is an affine variety, $X = \operatorname{Spec} R$ with $R$ Noetherian and $\mathcal{F} = M^{\sim}$ for some finitely generated $R$-module $M$, see Theorem 3.8.5. If $x \in X_{<n}$ corresponds to a prime ideal $\mathfrak{p}$ of $R$, then $k(x)$ is the field of fractions $k(\mathfrak{p})$ of $R/\mathfrak{p}$ and $\mathcal{F}_x \otimes_{\mathcal{O}_x} k(x) = M_\mathfrak{p}/\mathfrak{p}M_\mathfrak{p}$. Let $m = \dim_{k(\mathfrak{p})}(M_\mathfrak{p}/\mathfrak{p}M_\mathfrak{p})$, so $m < n$ and let $\tilde{a}_1, \ldots, \tilde{a}_m \in M_\mathfrak{p}$ be elements such that $\bar{a}_1, \ldots, \bar{a}_m$ form a $k(\mathfrak{p})$-basis for $M_\mathfrak{p}/\mathfrak{p}M_\mathfrak{p}$. By Nakayama's lemma, $\tilde{a}_1, \ldots, \tilde{a}_m$ generate $M_\mathfrak{p}$. Clearing denominators, we may assume that $\tilde{a}_1, \dots, \tilde{a}_m$ are the images of $a_1, \dots, a_m \in M$ under the map $M \to M_{\mathfrak{p}}$.

Let $b_1, \dots, b_t$ be a set of generators for $M$. The image of each $b_i$ in $M_{\mathfrak{p}}$ is a linear combination of the $\tilde{a}_j$ with coefficients in $R_{\mathfrak{p}}$. There are only a finite number of denominators in these equations, and so there exists a single element $r \in R$ with $r \notin \mathfrak{p}$ such that each $rb_i$ is a linear combination of the $a_j$ with coefficients in $R$. Consider the open set $U$ of $X$ consisting of those primes $\mathfrak{q}$ not containing $r$. Then for $\mathfrak{q} \in U$ the images of $b_i$ are linear combinations of the $a_j$ with coefficients in $R_{\mathfrak{q}}$. So $M_{\mathfrak{q}}$ is generated by at most $m$ elements. It follows that $U \subseteq X_{<n}$. So we have shown that each point $\mathfrak{p} \in X_{<n}$ has an open neighbourhood $U \subseteq X_{<n}$, and it follows that $X_{<n}$ is open.

(a) (1) $\Rightarrow$ (2): If $\mathcal{F}$ is a vector bundle then for each $x \in X$ there is an open neighbourhood of $x$ on which $\mathcal{F}$ is free, and therefore on which $\phi$ is constant. It follows that for each $n \in \mathbb{Z}$ the set $X_{\geq n}$ is open. We already showed that it is closed, so since $X$ is connected it follows that $\phi$ is constant on $X$.

(2) $\Rightarrow$ (1): Assume that $\phi$ is constant on $X$, say with value $n$. We must prove that each point $x \in X$ has an open neighbourhood on which $\mathcal{F}$ is free of finite rank. So again we may assume that $X = \operatorname{Spec} R$ is affine, $\mathcal{F} = M^{\sim}$ for some finitely generated $R$-module $M$, and $x$ corresponds to some prime ideal $\mathfrak{p}$ of $R$. As before, choose $a_1, \dots, a_n \in M$ such that $\bar{a}_1, \dots, \bar{a}_n$ generate $M_{\mathfrak{p}}/\mathfrak{p}M_{\mathfrak{p}}$. Let $b_1, \dots, b_t$ be a set of generators for $M$. Then there exists $r \in R$ with $r \notin \mathfrak{p}$ such that the image of each $rb_i$ in $M_{\mathfrak{p}}$ is a linear combination of the images of the $a_j$. Writing $R_r$ for $R$ with $r$ inverted, the images of $b_1, \dots, b_t$ in $M_r$ are linear combinations of the images of $a_1, \dots, a_n$. It follows that $M_r$ is generated by the images of $a_1, \dots, a_n$. This gives us a surjective module homomorphism from a free module of rank $n$ over $R_r$ to $M_r$. Write $K$ for the kernel, so that we have a short exact sequence

$$0 \to K \to R_r^n \to M_r \to 0.$$

Localisation is exact, so for each prime ideal $\mathfrak{q}$ not containing $r$ we have an exact sequence

$$0 \to K_{\mathfrak{q}} \to R_{\mathfrak{q}}^n \to M_{\mathfrak{q}} \to 0.$$

Since $\phi$ is constant, $M_{\mathfrak{q}}/\mathfrak{q}M_{\mathfrak{q}}$ has dimension $n$ over $R_{\mathfrak{q}}/\mathfrak{q}R_{\mathfrak{q}}$ and so $K_{\mathfrak{q}} \subseteq \mathfrak{q}R_{\mathfrak{q}}^n$. This implies that the coordinates of an element of $K$ are in $\mathfrak{q}$. Since $X$ is reduced, $R_r$ has no nilpotent elements, so the intersection of the prime ideals of $R_r$ is zero. Thus $K = 0$ and $\mathcal{F}$ is free on the open subset of $X$ defined by $r$.

(b) Consider the exact sequence

$$\mathcal{F} \to \mathcal{F}' \to \operatorname{Coker} f \to 0. \tag{6.1}$$

Exactness of stalks and right exactness of tensor product shows that

$$\mathcal{F}_x \otimes_{\mathcal{O}_x} k(x) \to \mathcal{F}'_x \otimes_{\mathcal{O}_x} k(x) \to (\operatorname{Coker} f)_x \otimes_{\mathcal{O}_x} k(x) \to 0$$

is also exact. By part (a), $\mathsf{Coker}\, f$ is a vector bundle of rank equal to rank $\mathcal{F}' - n$ if and only if $(\mathsf{Coker}\, f)_x \otimes_{\mathcal{O}_x} k(x)$ has constant dimension equal to rank $\mathcal{F}' - n$ for $x \in X$, namely if and only if $\mathcal{F}_x \otimes_{\mathcal{O}_x} k(x) \to \mathcal{F}' \otimes_{\mathcal{O}_x} k(x)$ has constant rank $n$.

If $\mathsf{Coker}\, f$ is a vector bundle then the sequence (6.1) is locally split and so the image of $f$ is also a vector bundle.

(c) It is clear that (1) implies (2). To see that (2) implies (1), let $f\colon \mathcal{F} \to \mathcal{F}'$ be such that the induced map of fibres is injective for all $x \in X$. Then by part (b), the cokernel is a vector bundle, the sequence $\mathcal{F} \xrightarrow{f} \mathcal{F}' \to \mathsf{Coker}\, f \to 0$ is locally split, and so the image of $f$ is a subbundle of $\mathcal{F}'$. Thus

$$0 \to \mathsf{Ker}\, f \to \mathcal{F} \to \mathsf{Im}\, f \to 0$$

is locally split, and $(\mathsf{Ker}\, f)_x \otimes_{\mathcal{O}_x} k(x) = 0$ for all $x \in X$. So $(\mathsf{Ker}\, f)_x = 0$ for all $x \in X$ and hence $\mathsf{Ker}\, f = 0$. □

**Lemma 6.2.2** *Let*

$$0 \to \mathcal{F} \xrightarrow{\alpha} \mathcal{F}' \xrightarrow{\beta} \mathcal{F}'' \to 0$$

*be a short exact sequence of sheaves. If $\mathcal{F}''$ is a vector bundle then it follows that $\mathcal{F}$ is a vector bundle if and only if $\mathcal{F}'$ is a vector bundle.*

*Proof.* Since $\mathcal{F}''$ is locally free, the sequence is locally split. So $\mathcal{F}'$ is locally a direct sum of $\mathcal{F}$ and $\mathcal{F}''$. Since locally projective is the same as locally free, $\mathcal{F}$ is a vector bundle if and only if $\mathcal{F}'$ is a vector bundle. □

**Remark 6.2.3** In the lemma, it is not true that if $\mathcal{F}$ and $\mathcal{F}'$ are vector bundles then so is $\mathcal{F}''$. For example if $x$ is a non-zero function of degree one on projective space then we have a short exact sequence

$$0 \to \mathcal{O}(-1) \xrightarrow{x} \mathcal{O} \to \mathcal{O}/x \to 0$$

and $\mathcal{O}/x$ is not a vector bundle.

**Definition 6.2.4** We shall say that the image of $\alpha$ is a *subbundle* of $\mathcal{F}'$ if $\mathcal{F}$ and $\mathcal{F}'$ are bundles and the sequence is locally split. Under this condition, $\mathcal{F}''$ is also a vector bundle, called the *quotient bundle*.

Under the assumptions of Theorem 6.2.1, this condition may be checked using part (b) of that theorem.

We shall also need the following extended version of Lemma 6.2.2 in the proof of Theorem 8.4.11.

**Proposition 6.2.5** *Let $\mathcal{F}$ be a coherent sheaf of $\mathcal{O}_X$-modules with a filtration*

$$0 = \mathcal{F}_0 \subseteq \mathcal{F}_1 \subseteq \cdots \subseteq \mathcal{F}_n = \mathcal{F}.$$

*If the quotients $\mathcal{F}_i/\mathcal{F}_{i-1}$ are vector bundles for $1 \le i \le n$ then $\mathcal{F}_i/\mathcal{F}_j$ is a vector bundle for $0 \le j < i \le n$. In particular, $\mathcal{F}$ is a vector bundle.*

*Proof.* The proof is by induction on $n$. If $n < 2$ there is nothing to prove. If $n = 2$ it says that for the short exact sequence

$$0 \to \mathcal{F}_1 \to \mathcal{F} \to \mathcal{F}/\mathcal{F}_1 \to 0,$$

if both end terms are locally free then so is the middle. This follows from Lemma 6.2.2.

For $n > 2$ the inductive hypothesis shows that $\mathcal{F}_i/\mathcal{F}_j$ is locally free for $0 \le j < i \le n-1$. Now examine the sequence

$$0 \to \mathcal{F}_{n-1}/\mathcal{F}_j \to \mathcal{F}/\mathcal{F}_j \to \mathcal{F}/\mathcal{F}_{n-1} \to 0$$

and apply the case $n = 2$, to deduce the remaining cases. □

## 6.3 Vector Bundles on Projective Space

Let $k$ be an algebraically closed field, and let $\mathbb{P}^{r-1} = \mathsf{Proj}\, k[Y_1, \dots, Y_r]$, projective space of dimension $r-1$ over $k$. We use $r-1$ to denote the dimension because in later chapters $r$ will be the rank of a finite elementary abelian $p$-group $E$, and $k$ will be an algebraically closed field of characteristic $p$.

**Theorem 6.3.1** *The only line bundles on $\mathbb{P}^{r-1}$ are the twists of the structure sheaf $\mathcal{O}(i)$. Thus* $\mathrm{Pic}(\mathbb{P}^{r-1}) \cong \mathbb{Z}$.

Vector bundles on $\mathbb{P}^1$ were classified by Grothendieck [140].

**Theorem 6.3.2** *Every vector bundle on $\mathbb{P}^1$ is a direct sum of line bundles*

$$\mathcal{O}(i_1) \oplus \cdots \oplus \mathcal{O}(i_d).$$

*The integers $i_1, \dots, i_d$ are uniquely determined up to reordering.*

If $r \ge 3$, it is not difficult to produce indecomposable vector bundles on $\mathbb{P}^{r-1}$ of every rank at least $r-2$; see, for example, the examples of Tango described in Section 6.8. But it turns out to be much harder to produce any of smaller dimension. So there is a great deal of interest in vector bundles of rank $d$ satisfying $2 \le d \le r-3$. Here is a complete list of the known vector bundles with such ranks.

(1) The Horrocks–Mumford bundle of rank 2 on $\mathbb{P}^4$ with 15, 000 symmetries [161].
(2) Horrocks' Parent bundle of rank 3 on $\mathbb{P}^5$ [160].
(3) The Tango bundle of rank 2 on $\mathbb{P}^5$ in characteristic two (Tango [235]; see also [116, 170]).

(4) Some further rank 2 vector bundles on $\mathbb{P}^4$ and rank 3 bundles on $\mathbb{P}^5$ constructed by Kumar [182], Kumar, Peterson and Rao [183], in positive characteristic only.
(5) Bundles obtained from the above by twisting and by pulling back through self-maps of projective space.

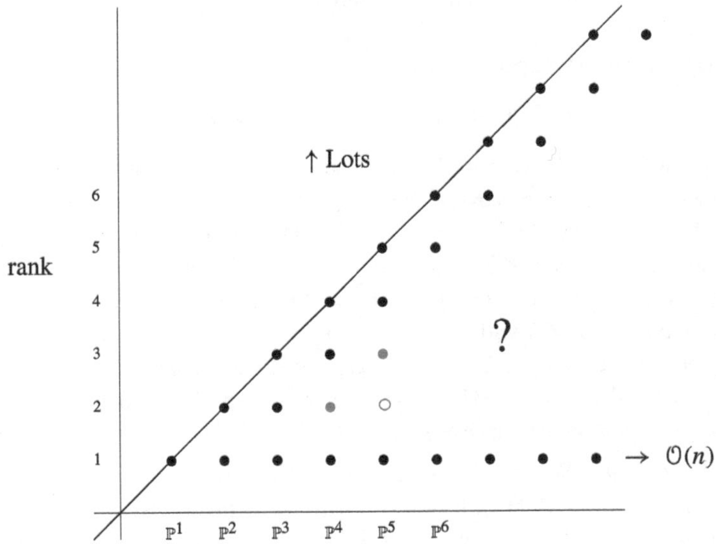


**Remark 6.3.3** There are no known restrictions on the possible ranks of indecomposable bundles on $\mathbb{P}^{r-1}$ for $r \geq 3$. For example, it is not known whether there exists an indecomposable vector bundle of rank 2 on $\mathbb{P}^5$ in characteristic not equal to 2, or on $\mathbb{P}^{r-1}$ ($r \geq 7$) in any characteristic.

**Remark 6.3.4** Sato [222] and Tyurin [238] (reprinted in [239]) proved that a vector bundle on $\mathbb{P}^{r-1}$ which extends to a vector bundle on $\mathbb{P}^{r+n-1}$ for all $n \geq 0$ has to be a direct sum of line bundles. A short proof of this "Babylonian tower theorem" using ideas from the proof of Kempf's splitting theorem [177] may be found in Coandă and Trautmann [93]. A closely related theorem on stably extendable bundles may be found in Coandă [92].

We return to the subject of Babylonian towers in the context of modules of constant Jordan type in Section 10.11.

In the next few sections, we discuss some of the vector bundles appearing in this section. The proofs of simplicity (and hence indecomposability) depend on cohomological computations from Sections 6.9 and 6.10, and so this will not happen until Section 6.11. The Horrocks–Mumford, Horrocks Parent, and Tango bundles will appear in Chapter 9.

## 6.4 The Tangent Bundle and the Euler Sequence

The first example of a vector bundle on $\mathbb{P}^{r-1}$ which is not a sum of line bundles is the *tangent bundle* or *sheaf of derivations* $\mathcal{T}$.

**Definition 6.4.1** The tangent bundle on $\mathbb{P}^{r-1}$ is defined as the cokernel of the map $\mathcal{O} \to \mathcal{O}(1)^{\oplus r}$ whose $i$th coordinate is given by multiplication by $Y_i$.

Thus we have a short exact sequence

$$0 \to \mathcal{O} \to \mathcal{O}(1)^{\oplus r} \to \mathcal{T} \to 0. \tag{6.2}$$

On the affine patch given by $Y_i \neq 0$, the $r-1$ local coordinate functions are $Y_j/Y_i$ with $j \neq i$ and the map is split by $(f_1, \ldots, f_r) \mapsto f_i/Y_i$. Thus the sequence is locally split, and so $\mathcal{T}$ is a vector bundle (cf. Definition 6.2.4).

We identify the fibre of $\mathcal{T}$ at a point $a \in \mathbb{P}^{r-1}$ with the tangent space $T_{\mathbb{P}^{r-1},a}$ as follows. If $a$ is in the affine patch $Y_i \neq 0$ then an element $g \in \mathfrak{m}_{\mathbb{P}^{r-1},a}$ is a polynomial in the $Y_j/Y_i$ with $j \neq i$, and so $g$ is a rational function of $Y_1, \ldots, Y_r$ of degree zero. Given $f_1, \ldots, f_r$ of degree one, we evaluate $\sum_i \frac{\partial g}{\partial Y_i}$ at $a$. This expression has degree zero, so its value at $a$ is well defined. It vanishes on products, so it only depends on the image of $g$ in $\mathfrak{m}_{\mathbb{P}^{r-1},a}/\mathfrak{m}^2_{\mathbb{P}^{r-1},a}$. It also only depends on the value of the $f_i$ at $a$. We have $\sum_i Y_i \frac{\partial g}{\partial Y_i} = 0$ so it is well defined on the fibre at $a$ of the cokernel of $\mathcal{O} \to \mathcal{O}(1)^{\oplus r}$. It is easy to check that this is then a perfect pairing, so it identifies the fibre of $\mathcal{T}$ at $a$ with the vector space dual of $\mathfrak{m}_{\mathbb{P}^{r-1},a}/\mathfrak{m}^2_{\mathbb{P}^{r-1},a}$, which is the tangent space $T_{\mathbb{P}^{r-1},a}$.

**Definition 6.4.2** The *Euler sequence* is the short exact sequence obtained by tensoring the sequence (6.2) with $\mathcal{O}(-1)$:

$$0 \to \mathcal{O}(-1) \to \mathcal{O}^{\oplus r} \to \mathcal{T}(-1) \to 0. \tag{6.3}$$

**Definition 6.4.3** The *cotangent bundle* or *sheaf of differentials* is the dual of the tangent bundle

$$\boldsymbol{\Omega} = \boldsymbol{\Omega}_{\mathbb{P}^{r-1}} = \mathcal{T}^\vee.$$

It is the kernel of the map $\mathcal{O}(-1)^{\oplus r} \to \mathcal{O}$ whose $i$th coordinate is given by multiplication by $Y_i$. The dual of the sequence (6.2) is the short exact sequence

$$0 \to \boldsymbol{\Omega} \to \mathcal{O}(-1)^{\oplus r} \to \mathcal{O} \to 0. \tag{6.4}$$

## 6.5 Homogeneity and Uniformity

The tangent and cotangent bundles are examples of homogeneous bundles on $\mathbb{P}^{r-1}$. These are the bundles that are invariant under the action of the general linear group on projective space, as we now explain.

The general linear group $GL(r, k)$ acts on the vector space spanned by $Y_1, \ldots, Y_r$ and hence on the graded polynomial ring $k[Y_1, \ldots, Y_r]$ and the projective space $\mathbb{P}^{r-1}$. The scalar matrices $k^\times$ act trivially on $\mathbb{P}^{r-1}$, so we get an action of $PGL(r, k) = GL(r, k)/k^\times$. Since $k$ is algebraically closed, every element has an $r$th root. So $GL(r, k)$ is generated by the special linear group $SL(r, k)$ and $k^\times$, and we have

$$PGL(r, k) \cong PSL(r, k) = SL(r, k)/(k^\times \cap SL(r, k)).$$

**Definition 6.5.1** A vector bundle $\mathcal{F}$ on $\mathbb{P}^{r-1}$ is said to be *homogeneous* if for every $\rho \in GL(r, k)$ (or equivalently for every $\rho \in SL(r, k)$) we have $\rho^*(\mathcal{F}) \cong \mathcal{F}$.

**Lemma 6.5.2** *We have* $H^2(SL(r, k), k^\times) = 0$.

*Proof.* See Steinberg [229]. □

**Proposition 6.5.3** *If $\mathcal{F}$ is a simple homogeneous vector bundle then the isomorphisms $i_\rho \colon \rho^*(\mathcal{F}) \cong \mathcal{F}$ can be chosen so that if $\rho, \rho' \in SL(r, k)$ then the following diagram commutes:*

$$\begin{array}{ccc} \mathcal{F} & \xrightarrow{\;i_\rho\;} & \rho^*(\mathcal{F}) \\ {\scriptstyle i_{\rho\rho'}}\downarrow & & \downarrow{\scriptstyle i_{\rho'}} \\ (\rho\rho')^*(\mathcal{F}) & = & (\rho')^*\rho^*(\mathcal{F}) \end{array}$$

*Proof.* Since $\mathcal{F}$ is simple, if we're given a system of isomorphisms $i_\rho$ then there are scalars $\lambda_{\rho,\rho'}$ such that $i_{\rho\rho'} = \lambda_{\rho,\rho'} i_{\rho'} i_\rho$. These form a 2-cocycle on $SL(r, k)$ for the bar resolution with coefficients in $k^\times$. By Lemma 6.5.2, this is a coboundary, so there are scalars $\mu_\rho$ such that $\lambda_{\rho,\rho'} = \mu_\rho \mu_{\rho'} \mu_{\rho\rho'}^{-1}$. Then replacing $i_\rho$ by $\mu_\rho i_\rho$ makes the diagram commute. □

Given a vector bundle $\mathcal{F}$ on $\mathbb{P}^{r-1}$, for each line $L \subseteq \mathbb{P}^{r-1}$ (i.e., linearly embedded copy of $\mathbb{P}^1$) we may restrict $\mathcal{F}$ to $L$ and decompose as a sum of line bundles

$$\mathcal{F}|L \cong \bigoplus_{i=1}^{s} \mathcal{O}_L(a_i).$$

The unordered set of $a_i$ with multiplicities is called the *splitting type* of $\mathcal{F}$ on $L$.

**Definition 6.5.4** A vector bundle $\mathcal{F}$ on $\mathbb{P}^{r-1}$ is said to be *uniform* if the splitting type is the same on all lines.

**Proposition 6.5.5** *If $\mathcal{F}$ is a homogeneous bundle on $\mathbb{P}^{r-1}$ then $\mathcal{F}$ is uniform.*

*Proof.* The group $GL(r, k)$ is transitive on lines in $\mathbb{P}^{r-1}$. □

**Remark 6.5.6** Homogeneous vector bundles on $\mathbb{P}^2$ are studied in the paper of Gorodentsev and Rudakov [137], in terms of solutions of the Markov equation

$$x^2 + y^2 + z^2 = 3xyz,$$

see also Rudakov [221].

More generally, homogeneous vector bundles on $\mathbb{P}^{r-1}$ are discussed in detail by Bondal and Kapranov [56], Hille [154] and Ottaviani and Rubei [205], where they are related to Higgs bundles and to representations of certain quivers with relations.

## 6.6 Monads and Subquotients

In the literature, vector bundles are often described using monads, which are nothing more than a convenient method of describing subquotients of a bundle.

**Definition 6.6.1** A *monad* over $\mathbb{P}^{r-1}$ (or over any other variety) is a complex

$$0 \to \mathcal{F}_1 \xrightarrow{\alpha} \mathcal{F}_2 \xrightarrow{\beta} \mathcal{F}_3 \to 0$$

of vector bundles such that $\alpha$ is injective, $\beta$ is surjective, $\beta\alpha = 0$, and the image of $\alpha$ is a subbundle of $\mathcal{F}_2$.

A monad gives rise to a commutative diagram with exact rows and columns as follows:

$$\begin{array}{ccccccccc}
 & & & & 0 & & 0 & & \\
 & & & & \downarrow & & \downarrow & & \\
0 & \longrightarrow & \mathcal{F}_1 & \longrightarrow & \mathsf{Ker}\,\beta & \longrightarrow & \mathcal{F} & \longrightarrow & 0 \\
 & & \| & & \downarrow & & \downarrow & & \\
0 & \longrightarrow & \mathcal{F}_1 & \xrightarrow{\alpha} & \mathcal{F}_2 & \longrightarrow & \mathsf{Coker}\,\alpha & \longrightarrow & 0 \\
 & & & & \downarrow & & \downarrow & & \\
 & & & & \mathcal{F}_3 & = & \mathcal{F}_3 & & \\
 & & & & \downarrow & & \downarrow & & \\
 & & & & 0 & & 0 & &
\end{array}$$

**Lemma 6.6.2** *The subquotient* $\mathcal{F} = \mathsf{Ker}\,\beta/\mathsf{Im}\,\alpha$ *is a vector bundle.*

*Proof.* Since by assumption $\mathsf{Im}\,\alpha$ is a subbundle of $\mathcal{F}_2$, it follows that $\mathsf{Coker}\,\alpha$ is a vector bundle. Applying Lemma 6.2.2 to the right-hand column of the display, we see that $\mathcal{F}$ is a vector bundle. □

**Definition 6.6.3** The vector bundle $\mathcal{F} = \operatorname{Ker}\beta/\operatorname{Im}\alpha$ is called the *cohomology* of the monad.

We shall see examples of monads in the next section, as well as in Sections 9.8 and 9.9.

## 6.7 The Null Correlation Bundle

Our next example is the *null correlation bundle* $\mathcal{F}_N$. This is an example of an indecomposable bundle of rank $r-2$ on $\mathbb{P}^{r-1}$ for $r$ even.

We form a monad

$$0 \to \mathcal{O}(-1) \to \mathcal{O}^{\oplus r} \to \mathcal{O}(1) \to 0 \tag{6.5}$$

where the first map is the same as in the Euler sequence, while the second map is given by multiplication by $(Y_r, -Y_{r-1}, \ldots, Y_2, -Y_1)$. It is easy to see by direct computation that the composite is zero:

$$(Y_r, -Y_{r-1}, \ldots, Y_2, -Y_1)(Y_1, Y_2, \ldots, Y_{r-1}, Y_r)^t = 0.$$

So we do indeed have a monad.

The second map above may be interpreted as being obtained from the first by means of the symplectic form on $\mathbb{A}^r$ defined by

$$\langle X_i, X_j\rangle = \begin{cases} (-1)^i & \text{if } i+j=r+1 \\ 0 & \text{otherwise.} \end{cases}$$

Any choice of non-degenerate symplectic form gives a monad.

The null correlation bundle $\mathcal{F}_N$ is defined to be the cohomology of this monad:

$$\begin{array}{ccccccccc}
 & & & & 0 & & 0 & & \\
 & & & & \downarrow & & \downarrow & & \\
0 & \longrightarrow & \mathcal{O}(-1) & \longrightarrow & \boldsymbol{\Omega}(1) & \longrightarrow & \mathcal{F}_N & \longrightarrow & 0 \\
 & & \| & & \downarrow & & \downarrow & & \\
0 & \longrightarrow & \mathcal{O}(-1) & \longrightarrow & \mathcal{O}^{\oplus r} & \longrightarrow & \mathcal{T}(-1) & \longrightarrow & 0. \\
 & & & & \downarrow & & \downarrow & & \\
 & & & & \mathcal{O}(1) & = & \mathcal{O}(1) & & \\
 & & & & \downarrow & & \downarrow & & \\
 & & & & 0 & & 0 & &
\end{array}$$

It is clear from the construction that $\mathcal{F}_N$ is a self-dual bundle:

$$\mathcal{F}_N^\vee \cong \mathcal{F}_N.$$

## 6.8 The Examples of Tango

In the last section we saw examples of indecomposable bundles of rank $r-2$ on $\mathbb{P}^{r-1}$ when $r$ is even. Tango [234] constructed indecomposable bundles of rank $r-2$ on $\mathbb{P}^{r-1}$ for all values of $r$. We present here a simplified version of his construction. Let $V$ be a vector space of dimension $r$, and consider the map $V \times V \to \Lambda^2 V$ given by $(x, y) \mapsto x \wedge y$. The image is the closed homogeneous subvariety corresponding to the cone on the Grassmannian of 2-dimensional subspaces of $V$, and hence has dimension $2r-3$ (one more than the dimension of the Grassmannian). So there exists a linear subspace $W$ of $\Lambda^2 V$ of codimension $2r-3$ whose intersection with the image is equal to the origin. In other words, $W$ contains no non-zero elements of the form $x \wedge y$.

Now look at the beginning of the Koszul complex on $\mathbb{P}^{r-1}$, suitably twisted, and let $\mathcal{E}$ be the cokernel:

$$0 \to \mathcal{O}(-2) \to \mathcal{O}(-1)^{\oplus r} \to \mathcal{O}^{\oplus\binom{r}{2}} \to \mathcal{E} \to 0.$$

Thus

$$\mathcal{E} \cong (\Lambda^2 \mathcal{T})(-2) \cong \mathbf{\Omega}^{r-2}(r-2)$$

of rank

$$\binom{r}{2} - (r-1) = \binom{r-1}{2}.$$

Then thinking of $\mathcal{O}^{\oplus\binom{r}{2}}$ as the bundle $\Lambda^2(V) \otimes_k \mathcal{O}$, the subspace $W$ of $\Lambda^2(V)$ defines a trivial subbundle $W \otimes_k \mathcal{O}$ which injects into $\mathcal{E}$. Letting $\mathcal{F}_W$ be the quotient of $\mathcal{E}$ by the image of this subbundle, we have a diagram with exact rows and columns

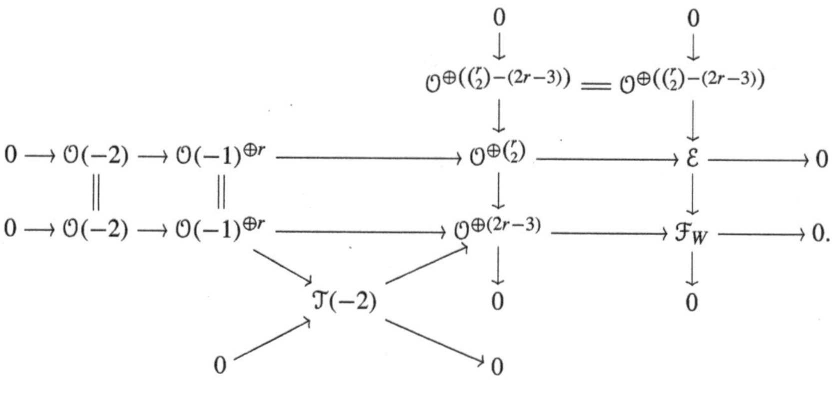


(6.6)

This bundle $\mathcal{F}_W$ is the example of Tango.

**Remark 6.8.1** The construction depends on the choice of a subspace $W$ of $\Lambda^2 V$ not containing any non-zero wedges, and so we obtain a large family of examples this way. The symmetry of the bundle is the same as the symmetry of the subspace $W$ in $\Lambda^2 V$.

Cascini [84] has shown that if we take $V$ to be the $(r-1)$st symmetric power of a vector space $U$ of dimension two then for $p$ large, decomposing $\Lambda^2 V$ as

$$\Lambda^2 V = S^{2(r-2)}U \oplus S^{2(r-4)}U \oplus S^{2(r-6)}U \oplus \cdots \oplus S^2 U$$

we may take

$$W = S^{2(r-4)}U \oplus S^{2(r-6)}U \oplus \cdots \oplus S^2 U$$

of codimension $2r-3$. Thus there are examples of Tango's construction with $SL(2, k)$ acting as symmetries if $p$ is large compared with $r$.

## 6.9 Cohomology of Projective Space

We shall need to make use of the cohomology of $\mathbb{P}^{r-1}$ with coefficients in twists of the structure sheaf in order to understand vector bundles. The following is Theorem III.5.1 in Hartshorne [147].

**Theorem 6.9.1** *Let $R = k[Y_1, \ldots, Y_r]$ and $\mathbb{P}^{r-1} = \mathsf{Proj}\, R$.*

(i) *The natural map $R \to \bigoplus_{j\in\mathbb{Z}} H^0(\mathbb{P}^{r-1}, \mathcal{O}(j))$ is an isomorphism of graded $R$-modules.*
(ii) *$H^i(\mathbb{P}^{r-1}, \mathcal{O}(j)) = 0$ for $0 < i < r-1$ and all $j \in \mathbb{Z}$.*
(iii) *$H^{r-1}(\mathbb{P}^{r-1}, \mathcal{O}(-r-j))$ is isomorphic to the $k$-vector space dual of $H^0(\mathbb{P}^{r-1}, \mathcal{O}(j))$.*

**Corollary 6.9.2** *The dimension of $H^i(\mathbb{P}^{r-1}, \mathcal{O}(j))$ is equal to $\binom{r+j-1}{j}$ if $i = 0$ and $j \geq 0$, $\binom{-j-1}{-r-j}$ if $i = r-1$ and $j \leq -r$, and zero otherwise.*

We illustrate this with a diagram for $r = 4$:

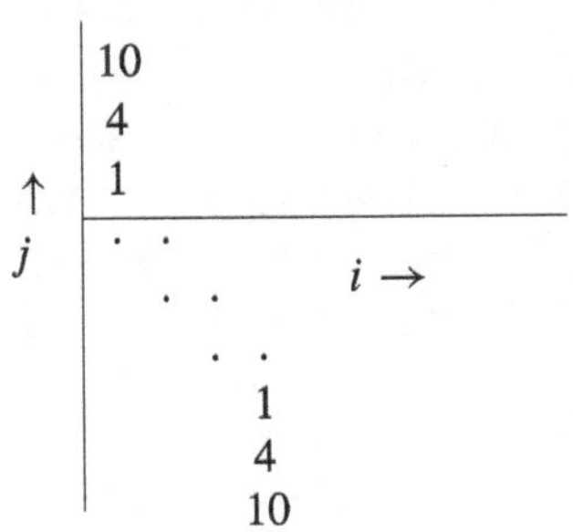

## 6.10 Differential Forms and Bott's Theorem

The exterior $m$th power of the cotangent bundle $\mathbf{\Omega}$ is the *bundle of differential $m$-forms*

$$\mathbf{\Omega}^m = \Lambda^m \mathbf{\Omega}.$$

The exterior powers of the sequence (6.4) give sequences

$$0 \to \mathbf{\Omega}^m \to \mathcal{O}(-m)^{\oplus\binom{r}{m}} \to \mathbf{\Omega}^{m-1} \to 0. \tag{6.7}$$

In particular for $m = r - 1$, the top exterior power is the *canonical bundle* and we obtain

$$\omega = \mathbf{\Omega}^{r-1} \cong \mathcal{O}(-r), \qquad \mathbf{\Omega}^{r-2} \cong \mathcal{T}(-r). \tag{6.8}$$

The differential forms are the kernels in the (exact) *Koszul complex*:

$$\begin{array}{ccccccccccccccc}
0 \to & \mathcal{O}(-r) & \to & \mathcal{O}(-r+1)^{\oplus r} & & \longrightarrow & & \mathcal{O}(-r+2)^{\oplus\binom{r}{r-2}} \cdots\cdots \mathcal{O}(-3)^{\oplus\binom{r}{3}} & & \longrightarrow & & \mathcal{O}(-2)^{\oplus\binom{r}{2}} & \longrightarrow & \mathcal{O}(-1)^{\oplus r} & \to \mathcal{O} \to 0 \\
& \| & & & \searrow & & \nearrow & & \searrow & & \nearrow \quad \searrow & & \nearrow & & \\
& \omega & & & & \mathbf{\Omega}^{r-2} & & & & \mathbf{\Omega}^{2} & & \mathbf{\Omega} & & &
\end{array}$$

The $r$-fold shift of this complex is isomorphic to its dual:

$$\begin{array}{ccccccccccc}
0 \to \mathcal{O} \to & \mathcal{O}(1)^{\oplus r} & \longrightarrow & \mathcal{O}(2)^{\oplus\binom{r}{2}} & \longrightarrow & \mathcal{O}(3)^{\oplus\binom{r}{3}} \cdots\cdots \mathcal{O}(r-2)^{\oplus\binom{r}{r-2}} & & \longrightarrow & & \mathcal{O}(r-1)^{\oplus r} & \to \mathcal{O}(r) \to 0 \\
& \searrow & \nearrow \quad \searrow & & \nearrow & & \searrow & & \nearrow & & \\
& & \mathcal{T} & \Lambda^2\mathcal{T} & & & & \Lambda^{r-2}\mathcal{T} & & &
\end{array}$$

and we have

$$(\mathbf{\Omega}^m)^\vee \cong \mathbf{\Omega}^{r-1-m}(r) \cong \Lambda^m \mathcal{T}. \tag{6.9}$$

**Theorem 6.10.1** (Bott's formula)

$$\dim_k H^i(\mathbb{P}^{r-1}, \mathbf{\Omega}^m(j))$$
$$= \begin{cases} \binom{j+r-m-1}{j}\binom{j-1}{m} & \textit{for } i = 0, 0 \le m < r, j > m \\ 1 & \textit{for } j = 0, 0 \le m = i < r \\ \binom{m-j}{-j}\binom{-j-1}{r-m-1} & \textit{for } i = r-1, 0 \le m < r, j \le m - r \\ 0 & \textit{otherwise.} \end{cases}$$

*Proof.* Bott [64] proved a more general statement in characteristic 0 (see also [202]), but this theorem also holds in any characteristic. We prove this by induction using the long exact cohomology sequences of the short exact sequences of sheaves (6.4) and (6.7). The cases $m = 0$ and $m = r - 1$ come from Corollary 6.9.2 and the isomorphism $\mathbf{\Omega}^{r-1} \cong \mathcal{O}(-r)$. By downwards induction on $m$ we have $H^i(\mathbb{P}^{r-1}, \mathbf{\Omega}^m(j)) = 0$ for $i > 0$ and $j > 0$. It follows that we have short exact sequences

$$0 \to H^0(\mathbb{P}^{r-1}, \mathbf{\Omega}^m(j)) \to H^0(\mathbb{P}^{r-1}, \mathcal{O}(j-m)^{\oplus\binom{r}{m}}) \to H^0(\mathbb{P}^{r-1}, \mathbf{\Omega}^{m-1}(j)) \to 0$$

for $j > 0$, which inductively gives the dimension of $H^0(\mathbb{P}^{r-1}, \mathbf{\Omega}^m(j))$ for $j > 0$.

Dually, by (upwards) induction on $m$ we have $H^i(\mathbb{P}^{r-1}, \Omega^m(j)) = 0$ for $i < r-1$ and $j < 0$. So we have short exact sequences

$$0 \to H^{r-1}(\mathbb{P}^{r-1}, \Omega^m(j)) \to H^{r-1}(\mathbb{P}^{r-1}, \mathcal{O}(j-m)^{\oplus\binom{r}{m}}) \to H^{r-1}(\mathbb{P}^{r-1}, \Omega^{m-1}(j)) \to 0$$

for $j < 0$, which inductively gives the dimension of $H^{r-1}(\mathbb{P}^{r-1}, \Omega^m(j))$ for $j < 0$.

Finally, in the remaining case $j = 0$, it again follows by induction on $m$ that the connecting homomorphisms give isomorphisms

$$H^i(\mathbb{P}^{r-1}, \Omega^m) \to H^{i+1}(\mathbb{P}^{r-1}, \Omega^{m+1})$$

so that these groups are isomorphic to $k$ when $i = m$ and zero otherwise. □

We illustrate Bott's formula with a diagram for $r = 4$:

$m = 0$

| $j \uparrow$ / $i \to$ | 0 | 1 | 2 | 3 |
|---|---|---|---|---|
| 5 | 56 | | | |
| 4 | 35 | | | |
| 3 | 20 | | | |
| 2 | 10 | | | |
| 1 | 4 | | | |
| 0 | 1 | | | |
| −1 | | | | |
| −2 | | | | |
| −3 | | | | |
| −4 | | | | 1 |
| −5 | | | | 4 |

$m = 1$

| $j \uparrow$ / $i \to$ | 0 | 1 | 2 | 3 |
|---|---|---|---|---|
| 5 | 84 | | | |
| 4 | 45 | | | |
| 3 | 20 | | | |
| 2 | 6 | | | |
| 1 | | | | |
| 0 | | 1 | | |
| −1 | | | | |
| −2 | | | | |
| −3 | | | | 4 |
| −4 | | | | 15 |
| −5 | | | | 36 |

$m = 2$

| $j \uparrow$ / $i \to$ | 0 | 1 | 2 | 3 |
|---|---|---|---|---|
| 5 | 36 | | | |
| 4 | 15 | | | |
| 3 | 4 | | | |
| 2 | | | | |
| 1 | | | | |
| 0 | | | 1 | |
| −1 | | | | |
| −2 | | | | 6 |
| −3 | | | | 20 |
| −4 | | | | 45 |
| −5 | | | | 84 |

$m = 3$

| $j \uparrow$ / $i \to$ | 0 | 1 | 2 | 3 |
|---|---|---|---|---|
| 5 | 4 | | | |
| 4 | 1 | | | |
| 3 | | | | |
| 2 | | | | |
| 1 | | | | |
| 0 | | | | 1 |
| −1 | | | | 4 |
| −2 | | | | 10 |
| −3 | | | | 20 |
| −4 | | | | 35 |
| −5 | | | | 56 |

## 6.11 Simplicity

Next, we use Bott's formula to prove the simplicity of the bundles $\Omega^m(j)$ (these include the tangent and cotangent bundles), the null correlation bundle and the examples of Tango.

**Theorem 6.11.1** $H^0(\mathbb{P}^{r-1}, \Omega \otimes \mathcal{T}) \cong k$.

*Proof.* Tensor $\Omega$ with the sequence (6.2) to get a sequence

$$0 \to \Omega \to \Omega(1)^{\oplus r} \to \Omega \otimes \mathcal{T} \to 0.$$

Taking cohomology, we obtain an exact sequence

$$\cdots \to H^0(\mathbb{P}^{r-1}, \Omega(1)^{\oplus r}) \to H^0(\mathbb{P}^{r-1}, \Omega \otimes \mathcal{T}) \to H^1(\mathbb{P}^{r-1}, \Omega) \to H^1(\mathbb{P}^{r-1}, \Omega(1)^{\oplus r}) \to \cdots .$$

By Theorem 6.10.1, we have

$$H^0(\mathbb{P}^{r-1}, \boldsymbol{\Omega}(1)) = H^1(\mathbb{P}^{r-1}, \boldsymbol{\Omega}(1)) = 0, \qquad H^1(\mathbb{P}^{r-1}, \boldsymbol{\Omega}) \cong k. \qquad \square$$

**Corollary 6.11.2** *The tangent bundle $\mathcal{T}$ and the cotangent bundle $\boldsymbol{\Omega}$ of $\mathbb{P}^{r-1}$ are simple.*

*Proof.* Using the theorem, we have

$$\mathrm{End}_{\mathbb{P}^{r-1}}(\mathcal{T}) \cong H^0(\mathbb{P}^{r-1}, \mathcal{T}^\vee \otimes \mathcal{T}) = H^0(\mathbb{P}^{r-1}, \boldsymbol{\Omega} \otimes \mathcal{T}) \cong k.$$

Similarly $\mathrm{End}_{\mathbb{P}^{r-1}}(\boldsymbol{\Omega}) \cong H^0(\mathbb{P}^{r-1}, \boldsymbol{\Omega} \otimes \mathcal{T}) \cong k$. □

More generally, we have the following.

**Proposition 6.11.3** *The bundles $\boldsymbol{\Omega}^m(j)$ are simple for $m \geq 0$ and $j \in \mathbb{Z}$.*

*Proof.* By Lemma 6.1.9, we may assume that $j = 0$. The case $m = 0$ is clear, and the case $m = 1$ is dealt with in Corollary 6.11.2. The general case is proved by induction on $m$ using the exact sequence (6.7). Namely, we tensor $(\boldsymbol{\Omega}^{m-1})^\vee$ with the sequence (6.7), take the long exact sequence in cohomology and use Bott's formula (Theorem 6.10.1) to deduce that

$$H^0(\mathbb{P}^{r-1}, (\boldsymbol{\Omega}^{m-1})^\vee \otimes \boldsymbol{\Omega}^{m-1}) \cong H^1(\mathbb{P}^{r-1}, (\boldsymbol{\Omega}^{m-1})^\vee \otimes \boldsymbol{\Omega}^m).$$

Then tensor the dual of the sequence (6.7), take the long exact sequence in cohomology and use Bott's formula to deduce that

$$H^0(\mathbb{P}^{r-1}, (\boldsymbol{\Omega}^m)^\vee \otimes \boldsymbol{\Omega}^m) \cong H^1(\mathbb{P}^{r-1}, (\boldsymbol{\Omega}^{m-1})^\vee \otimes \boldsymbol{\Omega}^m).$$

Thus by the inductive hypothesis

$$H^0(\mathbb{P}^{r-1}, (\boldsymbol{\Omega}^m)^\vee \otimes \boldsymbol{\Omega}^m) \cong H^0(\mathbb{P}^{r-1}, (\boldsymbol{\Omega}^{m-1})^\vee \otimes \boldsymbol{\Omega}^{m-1}) \cong k. \qquad \square$$

**Theorem 6.11.4** *The null correlation bundle is simple.*

*Proof.* Tensor $\mathcal{F}_N$ with the right hand column of the above diagram to get an exact sequence

$$0 \to \mathcal{F}_N \otimes \mathcal{F}_N \to \mathcal{F}_N \otimes \mathcal{T}(-1) \to \mathcal{F}_N(1) \to 0$$

which shows that

$$\dim_k H^0(\mathbb{P}^{r-1}, \mathcal{F}_N \otimes \mathcal{F}_N) \leq \dim_k H^0(\mathbb{P}^{r-1}, \mathcal{F}_N \otimes \mathcal{T}(-1)). \tag{6.10}$$

On the other hand, tensoring the top row with $\mathcal{T}(-1)$ gives an exact sequence

$$0 \to \mathcal{T}(-2) \to \boldsymbol{\Omega} \otimes \mathcal{T} \to \mathcal{F}_N \otimes \mathcal{T}(-1) \to 0.$$

The long exact sequence in cohomology for this gives

$$\cdots \to H^0(\mathbb{P}^{r-1}, \boldsymbol{\Omega} \otimes \mathcal{T}) \to H^0(\mathbb{P}^{r-1}, \mathcal{F}_N \otimes \mathcal{T}(-1)) \to H^1(\mathbb{P}^{r-1}, \mathcal{T}(-2)) \to \cdots.$$

Now by (6.8) we have $\mathcal{T}(-2) \cong \mathbf{\Omega}^{r-2}(r-2)$, so that by Bott's formula (Theorem 6.10.1) we have

$$H^1(\mathbb{P}^{r-1}, \mathcal{T}(-2)) = 0.$$

Thus

$$\dim_k H^0(\mathbb{P}^{r-1}, \mathcal{F}_N \otimes \mathcal{T}(-1)) \leq \dim_k H^0(\mathbb{P}^{r-1}, \mathbf{\Omega} \otimes \mathcal{T}). \tag{6.11}$$

Combining Theorem 6.11.1 with (6.10) and (6.11) and using the self-duality of $\mathcal{F}_N$, we get

$$\mathrm{End}_{\mathbb{P}^{r-1}}(\mathcal{F}_N) = H^0(\mathbb{P}^{r-1}, \mathcal{F}_N^\vee \otimes \mathcal{F}_N) \cong H^0(\mathbb{P}^{r-1}, \mathcal{F}_N \otimes \mathcal{F}_N) \cong k. \qquad \square$$

Next, we examine the examples of Tango.

**Lemma 6.11.5** *Suppose that there is a short exact sequence of vector bundles*

$$0 \to \mathcal{O}^{\oplus m} \to \mathcal{A} \to \mathcal{B} \to 0$$

*with $m \geq 0$. If $\mathcal{A}$ is simple and $H^1(\mathbb{P}^{r-1}, \mathcal{A}^\vee) = 0$ then $\mathcal{B}$ is simple.*

*Proof.* Tensor the dual of the given short exact sequence with $\mathcal{B}$ to get

$$0 \to \mathcal{B}^\vee \otimes \mathcal{B} \to \mathcal{A}^\vee \otimes B \to \mathcal{B}^{\oplus m} \to 0.$$

Thus

$$\dim_k H^0(\mathbb{P}^{r-1}, \mathcal{B}^\vee \otimes \mathcal{B}) \leq \dim_k H^0(\mathbb{P}^{r-1}, \mathcal{A}^\vee \otimes \mathcal{B}).$$

Similarly, we tensor $\mathcal{A}^\vee$ with the given short exact sequence to get

$$0 \to (\mathcal{A}^\vee)^{\oplus m} \to \mathcal{A}^\vee \otimes \mathcal{A} \to \mathcal{A}^\vee \otimes \mathcal{B} \to 0.$$

Since $H^1(\mathbb{P}^{r-1}, \mathcal{A}^\vee) = 0$, we obtain

$$\dim_k H^0(\mathbb{P}^{r-1}, \mathcal{A}^\vee \otimes \mathcal{B}) \leq \dim_k H^0(\mathbb{P}^{r-1}, \mathcal{A}^\vee \otimes \mathcal{A}) = 1.$$

Thus

$$\dim_k H^0(\mathbb{P}^{r-1}, \mathcal{B}^\vee \otimes \mathcal{B}) = 1. \qquad \square$$

**Proposition 6.11.6** *Tango's vector bundle $\mathcal{F}_W$ is simple.*

*Proof.* Apply Lemma 6.11.5 to the last column of diagram (6.8), and use the fact that

$$H^1(\mathbb{P}^{r-1}, \mathcal{E}^\vee) = H^1(\mathbb{P}^{r-1}, \mathbf{\Omega}^2(2)) = 0$$

by Bott's formula (Theorem 6.10.1). $\square$

**Remark 6.11.7** The simplicity of the bundle (Proposition 6.11.6) does not depend on the dimension of $W$; using a smaller-dimensional subspaces produces simple bundles of larger rank.

## 6.12 Hilbert's Syzygy Theorem

We shall make repeated use of Hilbert's Syzygy Theorem, so it seems wise to give a brief exposition here for the sake of the uninitiated. We begin with the Koszul complex.

Let $R$ be a commutative $k$-algebra and $x_1, \ldots, x_n$ be a sequence of elements generating an ideal $I$ of $R$. We say that $x_1, \ldots, x_n$ is a *regular sequence* if each $x_i$ is not a zero divisor in $R/(x_1, \ldots, x_{i-1})$. The *Koszul complex* over $R$ for the sequence $x_1, \ldots, x_n$ is the tensor product over $R$ of the complexes

$$0 \to R \xrightarrow{x_i} R \to 0.$$

For example, if $n = 2$, this is the complex

$$0 \to R \xrightarrow{(x_2, -x_1)} R^{\oplus 2} \xrightarrow{\binom{x_1}{x_2}} R \to 0.$$

It follows from the fact that $x_1, \ldots, x_n$ is a regular sequence that the homology of the Koszul complex is concentrated in degree zero, and equal to $R/I$. Thus the following is an exact sequence:

$$0 \to R \to R^{\oplus n} \to R^{\oplus \binom{n}{n-2}} \to \cdots \to R^{\oplus \binom{n}{2}} \to R^{\oplus n} \to R \to R/I \to 0.$$

Now let $R = k[Y_1, \ldots, Y_r]$ and let

$$S = R \otimes_k R = k[Y_1, \ldots, Y_r, Y_1', \ldots, Y_r'].$$

Then $S$-modules are the same as $(R, R)$-bimodules. Let $I$ be the ideal in $S$ generated by the regular sequence $Y_1 - Y_1', \ldots, Y_r - Y_r'$. Then $S/I \cong R$ as an $(R, R)$-bimodule. So the Koszul complex for this regular sequence is a resolution of $R$ as an $(R, R)$-bimodule, and we have an exact sequence

$$0 \to S \to S^{\oplus r} \to \cdots \to S^{\oplus \binom{r}{2}} \to S^{\oplus r} \to S \to R \to 0. \tag{6.12}$$

Such a complex is called a *Hochschild resolution* of $R$.

Now if $M$ is a left $R$-module, we may tensor $M$ over $R$ with this Hochschild resolution to obtain a complex whose terms are

$$S^{\oplus \binom{r}{j}} \otimes_R M \cong R^{\oplus \binom{r}{j}} \otimes_k M.$$

So we obtain a free resolution of $M$ as a left $R$-module. This proves the following theorem.

**Theorem 6.12.1** (Hilbert's Syzygy Theorem) *Every left module over* $R = k[Y_1, \ldots, Y_r]$ *has a free resolution of length at most* $r$ *over* $R$. □

**Corollary 6.12.2** *Every finitely generated* $R$*-module has a resolution of length at most* $r$ *over* $R$ *by finitely generated free modules.*

*Proof.* By Hilbert's basis theorem, $R$ is a Noetherian ring. So if $M$ is a finitely generated $R$-module then $M$ has a free resolution by finitely generated $R$-modules. By the extended Schanuel lemma it follows from the Hilbert Syzygy Theorem that the $r$th kernel in such a resolution is a finitely generated projective module.

By a theorem of Quillen and Suslin [209, 230] projective $R$-modules are free. An easier theorem of Serre [226] states that finitely generated projective modules are stably finitely generated free, which is enough here. In all our applications, the modules are graded with respect to the standard grading on the polynomial ring, in which case we can use the much more obvious statement that a direct summand of a free graded module is again free. □

**Example 6.12.3** The Koszul complex for the regular sequence $Y_1, \dots, Y_r$ in $R$ gives such a resolution for $k$, considered as an $R$-module with all $Y_i$ acting as zero. So we have an exact sequence:

$$0 \to R \to \cdots \to R^{\oplus\binom{r}{2}} \to R \to k \to 0.$$

The analogue for graded modules is as follows. We say that a graded $R$-module is free if it is isomorphic to a direct sum of shifts of $R$.

**Theorem 6.12.4** *Every left graded module over $R = k[Y_1, \dots, Y_r]$ has a resolution of length at most $r$ by free graded modules.*

*Every finitely generated graded $R$-module has a resolution of length at most $r$ over $R$ by finitely generated free graded modules.*

*Proof.* The Hochschild resolution (6.12) for $R = k[Y_1, \dots, Y_r]$ is a resolution by graded free bimodules, provided we shift the $i$th term $S^{\oplus\binom{r}{i}}$ so that the generators are in degree $i$. If $M$ is a graded module then tensoring with the Hochschild resolution gives a resolution by free graded $R$-modules.

Now when we apply Schanuel's lemma in the proof of Corollary 6.12.2, the $r$th kernel is a finitely generated projective graded module. For graded $R$-modules, it is very easy to show that projective graded modules are free, without recourse to the theorem of Quillen and Suslin, so this part of the proof is easier. □

**Corollary 6.12.5** *Let $\mathcal{O}$ be the structure sheaf on $\mathbb{P}^{r-1}$. Every coherent sheaf of $\mathcal{O}$-modules has a resolution of length at most $r$ in which each term is a finite direct sum of line bundles $\mathcal{O}(j)$.*

*Given a morphism $\mathcal{F} \to \mathcal{F}'$ of coherent sheaves, the resolutions may be chosen so that it lifts to a morphism of resolutions, unique up to chain homotopy.*

*Proof.* Let $\mathcal{F}$ be a coherent sheaf of $\mathcal{O}$-modules. Then by Theorem 3.9.12, there is a finitely generated graded $R$-module $M$ such that $\mathcal{F} \cong M^{\sim}$. By Theorem 6.12.4, $M$ has a resolution of the form

$$0 \to F_r \to \cdots \to F_1 \to F_0 \to M \to 0$$

where the $F_i$ are free graded modules. This gives an exact sequence of sheaves

$$0 \to F_r^{\sim} \to \cdots \to F_1^{\sim} \to F_0^{\sim} \to \mathcal{F} \to 0$$

where each $F_i^{\sim}$ is a direct sum of line bundles $\mathcal{O}(j)$.

By the definition of morphisms in the quotient by a Serre subcategory, given a morphism $\mathcal{F} \to \mathcal{F}'$, we may choose $R$-modules $M$ and $M'$ with $M^{\sim} \cong \mathcal{F}$ and $M'^{\sim} \cong \mathcal{F}'$, and a morphism of $R$-modules $M \to M'$ inducing the given morphism of sheaves. The last statement of the corollary now follows from the usual comparison theorem for free resolutions. □

# 7

# Chern Classes

In this chapter we give a purely algebraic development of the theory of Chern classes for coherent sheaves on projective space. Our treatment is inspired by the book of Hirzebruch [156].

## 7.1 Chern Classes of Graded Modules

Recall from Theorem 3.9.12 that if $\mathcal{O}$ is the structure sheaf on $\mathbb{P}^{r-1}$ then there is an equivalence of categories between coherent sheaves of $\mathcal{O}$-modules and finitely generated graded modules over $R = k[Y_1, \ldots, Y_r]$ modulo finite length modules. So finding invariants of coherent sheaves of $\mathcal{O}$-modules is equivalent to finding invariants of finitely generated graded $R$-modules that respect the equivalence relation.

We define the *Poincaré series* (also known as the *Hilbert series*) of a finitely generated graded $R$-module $M = \bigoplus_{j \in \mathbb{Z}} M_j$ to be

$$p_M(t) = \sum_{j \in \mathbb{Z}} t^j \dim_k M_j.$$

Note that since $M$ is finitely generated, $M_j = 0$ for $j$ sufficiently large negative, and so $p_M(t)$ is a Laurent power series in $t$.

**Lemma 7.1.1** (Hilbert, Serre) *The Poincaré series of a finitely generated graded $R$-module $M$ takes the form*

$$p_M(t) = \frac{f(t)}{(1-t)^r}$$

*where $f(t)$ is a Laurent polynomial.*

*If $M$ is a finite length module then $p_M(t)$ is itself a Laurent polynomial, so that the Laurent polynomial $f(t) = (1-t)^r p_M(t)$ is divisible by $(1-t)^r$.*

*Proof.* See for example Theorem 11.1 of Atiyah and Macdonald [16]. □

**Definition 7.1.2** The number $f(1)$, namely the sum of the coefficients of the numerator of the Poincaré series, is called the *rank* of $M$. It is not hard to see that this is equal to the dimension of the (ungraded) vector space

$$k(Y_1, \dots, Y_r) \otimes_{k[Y_1,\dots,Y_r]} M$$

over the field $k(Y_1, \dots, Y_r)$.

**Lemma 7.1.3** *If* $0 \to M_1 \to M_2 \to M_3 \to 0$ *is a short exact sequence of finitely generated graded* $R$*-modules then*

$$p_{M_2}(t) = p_{M_1}(t) + p_{M_3}(t).$$

**Definition 7.1.4** The *Chow ring* of $R = k[Y_1, \dots, Y_r]$ is defined to be the truncated polynomial ring $A^*(R) = \mathbb{Z}[h]/(h^r)$.

**Remark 7.1.5** The Chow ring [90] $A^*(X)$ of an algebraic variety $X$ is the analogue of the cohomology ring of a topological space. The formal definition involves algebraic cycles and rational equivalence. The multiplication is given by intersection of algebraic cycles in general position. More precise details may be found, for example, in Appendix A of Hartshorne [147], or in Fulton [129]. We do not need this general definition; our purpose is to give a formal algebraic definition of the Chern polynomial of a vector bundle on projective space, as an element of the truncated polynomial ring $A^*(\mathbb{P}^{r-1}) = A^*(R) = \mathbb{Z}[h]/(h^r)$.

**Definition 7.1.6** If $M$ is a finitely generated graded $R$-module with Poincaré series

$$p_M(t) = \frac{\sum_j a_j t^j}{(1-t)^r}$$

where the numerator is a Laurent polynomial, then the *Chern polynomial* of $M$ is defined to be

$$c(M) = \prod_j (1 + jh)^{a_j} \in A^*(R) = \mathbb{Z}[h]/(h^r).$$

If we want to emphasise the role of the variable $h$, we write $c(M, h)$.

The *Chern numbers* are the coefficients $c_i(M) \in \mathbb{Z}$ of powers of $h$ in this polynomial:

$$c(M) = 1 + c_1(M)h + \cdots + c_{r-1}(M)h^{r-1};$$

we also use the convention $c_0(M) = 1$.

The *Chern character* of $M$ is defined to be

$$\mathsf{Ch}(M) = \sum_j a_j e^{jh} \in A^*_{\mathbb{Q}}(R) = \mathbb{Q} \otimes_{\mathbb{Z}} A^*(R) = \mathbb{Q}[h]/(h^r).$$

From the definition, it follows that the Chern character is determined by the rank and the Chern numbers:

$$\begin{aligned}\mathsf{Ch}(M) &= \text{rank}\,(M) + c_1 h + \tfrac{1}{2}(c_1^2 - 2c_2)h^2 + \tfrac{1}{6}(c_1^3 - 3c_1c_2 + 3c_3)h^3 \\ &\quad + \tfrac{1}{24}(c_1^4 - 4c_1^2c_2 + 4c_1c_3 + 2c_2^2 - 4c_4) + \cdots .\end{aligned}$$

**Proposition 7.1.7** *If* $0 \to M_1 \to M_2 \to M_3 \to 0$ *is a short exact sequence of finitely generated graded* $R$*-modules then*

(i) $c(M_2) = c(M_1)c(M_3)$ *as elements of* $\mathbb{Z}[h]/(h^r)$*; in other words,*

$$c_j(M_2) = \sum_{i=0}^{j} c_i(M_1)c_{j-i}(M_3).$$

(ii) $\mathsf{Ch}(M_2) = \mathsf{Ch}(M_1) + \mathsf{Ch}(M_3)$ *as elements of* $\mathbb{Q}[h]/(h^r)$.

*Proof.* Both statements follow from Lemma 7.1.3. □

## 7.2 Chern Classes of Coherent Sheaves on $\mathbb{P}^{r-1}$

Recall from Section 3.9 that there is an equivalence of categories between coherent sheaves on $\mathbb{P}^{r-1}$ and the quotient category of finitely generated graded modules over $k[Y_1, \ldots, Y_r]$ modulo finite length modules. So let us regard finitely generated $k[Y_1, \ldots, Y_r]$-modules $M$ and $M'$ as equivalent if for $d$ large enough $M_{\geq d}$ and $M'_{\geq d}$ are isomorphic. Our goal is to show that equivalent modules have the same Chern polynomial and Chern character. This will enable us to define Chern polynomial and Chern character for coherent sheaves on $\mathbb{P}^{r-1}$. In the case of the Chern character this is easy, so we treat this first.

**Lemma 7.2.1** *If* $M$ *and* $M'$ *are equivalent finitely generated graded* $R$*-modules for the equivalence relation of Theorem 3.9.12 then there are finite length modules* $M_1$ *and* $M_2$*, a graded* $R$*-module* $M''$ *and short exact sequences*

$$0 \to M'' \to M \to M_1 \to 0$$
$$0 \to M'' \to M' \to M_2 \to 0.$$

*Proof.* Recall that $M$ and $M'$ are equivalent if and only if there exists $d \in \mathbb{Z}$ such that $M_{\geq d} \cong M'_{\geq d}$. So we set $M'' = M_{\geq d}$. □

**Theorem 7.2.2** *If* $M$ *and* $M'$ *are equivalent finitely generated graded* $R$*-modules then*

$$\mathsf{Ch}(M) = \mathsf{Ch}(M') \in \mathbb{Q}[h]/(h^r).$$

*Proof.* By Proposition 7.1.7 and Lemma 7.2.1, it suffices to show that if $M$ is a finite length module then $\mathsf{Ch}(M) = 0$. Again applying Proposition 7.1.7, it suffices to show this for $M$ equal to a copy of $k$ in degree $n$. In this case $p_M(t) = t^n$, $f(t) = t^n(1-t)^r$ and

$$\mathsf{Ch}(M) = \sum_j (-1)^j \binom{r}{j} e^{(j+n)h} = e^{nh}(1-e^h)^r.$$

Since $1 - e^h$ is divisible by $h$, $\mathsf{Ch}(M)$ is divisible by $h^r$ and is hence zero in $\mathbb{Q}[h]/(h^r)$. □

To prove that equivalent modules have the same Chern polynomial, we use an interesting congruence for polynomials, namely

$$\prod_{j=0}^{r} (1+jh)^{(-1)^j\binom{r}{j}} \equiv 1 \pmod{h^r}.$$

We prove a slight generalisation of this.

**Proposition 7.2.3** *For $r \geq 1$ we have*

$$\prod_{j=0}^{r} (1+jh)^{(-1)^j\binom{r}{j}} \equiv 1 - (r-1)!\,h^r \pmod{h^{r+1}}.$$

*Proof.* Let $f_r(h) = \prod_{j=0}^{r} (1+jh)^{(-1)^j\binom{r}{j}}$. Then using the identities $\binom{r}{j} = \binom{r-1}{j} + \binom{r-1}{j-1}$ and $\sum_{j=0}^{r-1}(-1)^j\binom{r-1}{j} = 0$, it is easy to check that $f_r(h)$ satisfies the following recurrence relation:

$$f_r(h) = \frac{f_{r-1}(h)}{f_{r-1}\left(\dfrac{h}{1+h}\right)}.$$

We work by induction on $r$. If $r = 1$ the formula states that $1/(1+h) \equiv 1 - h \pmod{h^2}$, which is clearly true. So assume that $r \geq 2$ and the statement is true for $r-1$. Then for some value of $a$ we have

$$f_{r-1}(h) = 1 - (r-2)!\,h^{r-1} + ah^r + \cdots$$

and so

$$\begin{aligned} f_{r-1}\left(\frac{h}{1+h}\right) &= f_{r-1}(h - h^2 + h^3 - \cdots) \\ &= 1 - (r-2)!\,(h - h^2 + h^3 - \ldots)^{r-1} \\ &\quad + a(h - h^2 + h^3 - \cdots)^r + \cdots \\ &= 1 - (r-2)!\,h^{r-1} + (a + (r-1)!)h^r + \cdots. \end{aligned}$$

Dividing, we get

$$f_r(h) = 1 - (r-1)!\,h^r + \cdots$$

as required. □

**Corollary 7.2.4** *For $r \geq 1$, we have*

(i) *for any value of $a$,*

$$\prod_{j=0}^{r} (a + jh)^{(-1)^j \binom{r}{j}} \equiv 1 - (r-1)!a^{-r}h^r \pmod{h^{r+1}},$$

(ii) *for any value of $n$,*

$$\prod_{j=0}^{r} (1 + (j+n)h)^{(-1)^j \binom{r}{j}} \equiv 1 - (r-1)!\,h^r \pmod{h^{r+1}}.$$

*Proof.* (i) Replace $h$ by $a^{-1}h$ in Proposition 7.2.3 and notice that

$$\prod_{j=0}^{r} a^{(-1)^j \binom{r}{j}} = 1$$

so that we can multiply every term in the product by $a$ without changing the answer.

(ii) This follows by putting $a = 1 + nh$ in part (i) and noticing that $(1+nh)^{-1}$ is congruent to 1 modulo $h$. □

**Theorem 7.2.5** *If $M$ and $M'$ are equivalent finitely generated graded $R$-modules then*

$$c(M) = c(M').$$

*Proof.* As in the proof of Theorem 7.2.2, it suffices to prove that if $M$ is a copy of $k$ in degree $n$. In this case, we have

$$c(M) = \prod_{j=0}^{r} (1 + (j+n)h)^{(-1)^j \binom{r}{j}} \equiv 1 \pmod{h^r}$$

by Corollary 7.2.4 (ii). □

**Definition 7.2.6** Let $\mathcal{O}$ be the structure sheaf of $\mathbb{P}^{r-1}$. If $\mathcal{F}$ is a coherent sheaf of $\mathcal{O}$-modules, we define

$$c(\mathcal{F}) = c(M), \qquad \mathsf{Ch}(\mathcal{F}) = \mathsf{Ch}(M),$$

where $M$ is any finitely generated graded $R$-module satisfying $M^{\sim} \cong \mathcal{F}$, see Theorem 3.9.12. By Theorems 7.2.2 and 7.2.5, this is well defined. We thus talk about the *Chern polynomial*, *Chern numbers* and *Chern character* of a coherent sheaf.

## 7.3 Some Computations

In this section, we compute Chern polynomial and Chern character of some vector bundles on $\mathbb{P}^{r-1}$. Our starting point is the following example.

**Example 7.3.1** It follows from the definitions that we have $c(\mathcal{O}(a)) = 1 + ah \in \mathbb{Z}[h]/(h^r)$ and $\mathsf{Ch}(\mathcal{O}(a)) = e^{ah} \in \mathbb{Q}[h]/(h^r)$.

The following is a direct consequence of Proposition 7.1.7.

**Proposition 7.3.2** *If* $0 \to \mathcal{F}_1 \to \mathcal{F}_2 \to \mathcal{F}_3 \to 0$ *is a short exact sequence of coherent sheaves on* $\mathbb{P}^{r-1}$ *then*

(i) $c(\mathcal{F}_2) = c(\mathcal{F}_1)c(\mathcal{F}_3)$ *as elements of* $\mathbb{Z}[h]/(h^r)$.
(ii) $\mathsf{Ch}(\mathcal{F}_2) = \mathsf{Ch}(\mathcal{F}_1) + \mathsf{Ch}(\mathcal{F}_3)$ *as elements of* $\mathbb{Q}[h]/(h^r)$. □

**Example 7.3.3** Let $\mathcal{T}$ and $\boldsymbol{\Omega}$ be the tangent and cotangent bundes. Applying Proposition 7.3.2 to the short exact sequences (6.2) and (6.4), we obtain $c(\mathcal{T}) = (1+h)^r$, $c(\boldsymbol{\Omega}) = (1-h)^r$ in $\mathbb{Z}[h]/(h^r)$, and $\mathsf{Ch}(\mathcal{T}) = re^h - 1$, $\mathsf{Ch}(\boldsymbol{\Omega}) = re^{-h} - 1$ in $\mathbb{Q}[h]/(h^r)$.

**Example 7.3.4** Tango's vector bundle $\mathcal{F}_W$ has rank $r-2$ and Chern polynomial

$$(1-2h)/(1-h)^r.$$

**Lemma 7.3.5** *Given a monad on* $\mathbb{P}^{r-1}$ *as in Definition 6.6.1, we have*

(1) $\operatorname{rank}\mathcal{F} = \operatorname{rank}\mathcal{F}_2 - \operatorname{rank}\mathcal{F}_1 - \operatorname{rank}\mathcal{F}_3$.
(2) $c(\mathcal{F}) = c(\mathcal{F}_2)c(\mathcal{F}_1)^{-1}c(\mathcal{F}_3)^{-1}$ *as elements of* $\mathbb{Z}[h]/(h^r)$.
(3) $\mathsf{Ch}(\mathcal{F}) = \mathsf{Ch}(\mathcal{F}_2) - \mathsf{Ch}(\mathcal{F}_1) - \mathsf{Ch}(\mathcal{F}_3)$ *as elements of* $\mathbb{Q}[h]/(h^r)$.

*Proof.* This follows immediately from Proposition 7.3.2. □

For example, the null correlation bundle of Section 6.7 is defined in terms of a monad, so we may now compute its Chern polynomial.

**Proposition 7.3.6** *The Chern polynomial of the null correlation bundle is equal to*

$$(1-h^2)^{-1} = 1 + h^2 + \cdots + h^{r-2}.$$

*Proof.* It follows from Lemma 7.3.5 that

$$c(\mathcal{F}_N) = c(\mathcal{O})c(\mathcal{O}(-1))^{-1}c(\mathcal{O}(1))^{-1} = (1-h)^{-1}(1+h)^{-1}. \qquad \square$$

**Theorem 7.3.7** *Let $\mathcal{O}$ be the structure sheaf of $\mathbb{P}^{r-1}$. There exists a unique function assigning to each coherent sheaf $\mathcal{F}$ of $\mathcal{O}$-modules a polynomial $c(\mathcal{F}) \in \mathbb{Z}[h]/(h^r)$ satisfying the following properties.*

(1) *If $0 \to \mathcal{F}_1 \to \mathcal{F}_2 \to \mathcal{F}_3 \to 0$ is a short exact sequence of coherent sheaves then*
$$c(\mathcal{F}_2) = c(\mathcal{F}_1)c(\mathcal{F}_3).$$

(2) $c(\mathcal{O}(a)) = 1 + ah$.

*Proof.* The existence of such a function is proved in Definition 7.2.6 via Theorem 7.2.5. For the uniqueness, let $c$ be such a function. By (1), $c$ is multiplicative over direct sums. Let $\mathcal{F} = M^{\sim}$. By Corollary 6.12.5 to Hilbert's Syzygy Theorem 6.12.1 there is an exact sequence of sheaves
$$0 \to F_r^{\sim} \to \cdots \to F_0^{\sim} \to \mathcal{F} \to 0 \tag{7.1}$$
where each $F_i^{\sim}$ is a direct sum of line bundles. We then have
$$c(\mathcal{F}) = \prod_{i=0}^{r} c(F_i^{\sim})^{(-1)^i}.$$
Since each $F_i^{\sim}$ is a direct sum of twists of $\mathcal{O}$, this determines $c(\mathcal{F})$ using (2) and multiplicativity over direct sums. □

**Remark 7.3.8** If $\mathcal{F}$ is a vector bundle then $c_i(\mathcal{F}) = 0$ for $i$ larger than the rank of $\mathcal{F}$. It does not seem easy to see this from the algebraic point of view we have developed here.

In applying Proposition 7.3.2 (i) and Remark 7.3.8, it is useful to have the following elementary lemma.

**Lemma 7.3.9** *A polynomial in $\mathbb{Z}[h]/(h^r)$ factorises as a product of polynomials such that the sum of the degrees is less than $r$ if and only if its unique lift to a polynomial of degree less than $r$ factorises non-trivially in $\mathbb{Z}[h]$.*

*Proof.* Such a factorisation $f(h) = f_1(h)f_2(h)$ in $\mathbb{Z}[h]/(h^r)$ lifts to a congruence modulo $h^r$ between polynomials of degree less than $r$ in $\mathbb{Z}[h]$. This congruence is therefore an equality. □

## 7.4 Restriction of Vector Bundles

**Definition 7.4.1** Let $f : X \to Y$ be a morphism of schemes. If $\mathcal{G}$ is a sheaf on $X$ then $f_*\mathcal{G}$ is the sheaf on $Y$ given by
$$f_*\mathcal{G}(U) = \mathcal{G}(f^{-1}(U))$$

for $U \subseteq Y$ open. If $\mathcal{G}$ is an $\mathcal{O}_X$-module then $f_*\mathcal{G}$ is an $\mathcal{O}_Y$-module via the natural map $\mathcal{O}_Y \to f_*\mathcal{O}_X$ of sheaves of rings on $Y$. We call $f_*\mathcal{G}$ the *direct image* of $\mathcal{G}$ under the morphism $f$.

The direct image functor is left exact, but usually not right exact. So it has right derived functors $R^i f_*\mathcal{G}$.

**Definition 7.4.2** If $\mathcal{F}$ is a sheaf on $Y$ then $f^{-1}\mathcal{F}$ is the sheaf on $X$ given by

$$f^{-1}\mathcal{F}(U) = \varinjlim_{V \supseteq f(U)} \mathcal{F}(V).$$

Here, the colimit is taken over the directed system of open sets $V$ containing $f(U)$.
If $\mathcal{F}$ is a sheaf of $\mathcal{O}_Y$-modules then $f^*\mathcal{F}$ is the sheaf of $\mathcal{O}_X$-modules given by

$$f^*\mathcal{F} = f^{-1}\mathcal{F} \otimes_{f^{-1}\mathcal{O}_Y} \mathcal{O}_X.$$

We call $f^*\mathcal{F}$ the *inverse image* of $\mathcal{F}$ along the morphism $f$. If $f \colon X \to Y$ is the inclusion of a subscheme then $f^*\mathcal{F}$ is the *restriction* of $\mathcal{F}$ to $X$.

The following lemma is straightforward, and may be found in any of the standard textbooks.

**Lemma 7.4.3** *The functors $f_*$ and $f^*$ are adjoint:*

$$\mathrm{Hom}_{\mathcal{O}_X}(f^*\mathcal{F}, \mathcal{G}) \cong \mathrm{Hom}_{\mathcal{O}_Y}(\mathcal{F}, f_*\mathcal{G}).$$

While $f^{-1}$ is an exact functor from sheaves on $Y$ to sheaves on $X$, tensor product is only right exact. So the functor $f^*$ is right exact and has left derived functors

$$L_i f^*\mathcal{F} = \mathcal{T}or_i^{f^{-1}\mathcal{O}_Y}(f^{-1}\mathcal{F}, \mathcal{O}_X).$$

On the other hand, since $f^*(\mathcal{O}_Y^{\oplus n}) \cong \mathcal{O}_X^{\oplus n}$, $f^*$ takes vector bundles of rank $n$ to vector bundles of rank $n$. Furthermore, since exactness is a local property, $f^*$ is exact on vector bundles.

Let $r' < r$, and let $i \colon \mathbb{P}^{r'-1} \hookrightarrow \mathbb{P}^{r-1}$ be the inclusion induced by the map of graded rings

$$\pi \colon k[Y_1, \dots, Y_r] \to k[Y_1, \dots, Y_{r'}]$$

with $\pi(Y_i) = Y_i$ $(1 \le i \le r')$, $\pi(Y_i) = 0$ $(r' < i \le r)$. Then

$$i^*(\mathcal{O}_{\mathbb{P}^{r-1}}(a)) \cong \mathcal{O}_{\mathbb{P}^{r'-1}}(a).$$

If $\mathcal{F}$ is a vector bundle over $\mathbb{P}^{r-1}$ then by Corollary 6.12.5 it has a finite resolution by direct sums of twists of $\mathcal{O}_{\mathbb{P}^{r-1}}$. By the exactness of restriction on vector bundles, $i^*(\mathcal{F})$ has a resolution by direct sums of the same twists of $\mathcal{O}_{\mathbb{P}^{r'-1}}$. It follows using the proof of Theorem 7.3.7 that $c(i^*(\mathcal{F})) \in \mathbb{Z}[h]/(h^{r'})$ is the image of $c(\mathcal{F}) \in$

$\mathbb{Z}[h]/(h^r)$ under the quotient map taking $h$ to $h$. Similarly, $\mathsf{Ch}(i^*(\mathcal{F}))$ is the image of $\mathsf{Ch}(\mathcal{F})$ under the corresponding map $\mathbb{Q}[h]/(h^r) \to \mathbb{Q}[h]/(h^{r'})$.

The way to express this is to write

$$i^* \colon A^*(\mathbb{P}^{r-1}) \to A^*(\mathbb{P}^{r'-1})$$

for the map of Chow rings $\mathbb{Z}[h]/(h^r) \to \mathbb{Z}[h]/(h^{r'})$ taking $h$ to $h$. In terms of this we have

$$c(i^*(\mathcal{F})) = i^*(c(\mathcal{F})).$$

We also write

$$i^* \colon A^*_{\mathbb{Q}}(\mathbb{P}^{r-1}) \to A^*_{\mathbb{Q}}(\mathbb{P}^{r'-1})$$

for the map of rational Chow rings $\mathbb{Q}[h]/(h^r) \to \mathbb{Q}[h]/(h^{r'})$ taking $h$ to $h$. In terms of this we have

$$\mathsf{Ch}(i^*(\mathcal{F})) = i^*(\mathsf{Ch}(\mathcal{F})).$$

For more general coherent sheaves on $\mathbb{P}^{r-1}$ this formula does not hold, because restriction is not exact. However, what this means is that we should instead use the *total left derived functor*

$$\mathsf{L}i^* \colon \mathsf{D}^b(\mathsf{Coh}(\mathbb{P}^{r-1})) \to \mathsf{D}^b(\mathsf{Coh}(\mathbb{P}^{r'-1})).$$

In this case the total left derived functor may be applied by replacing a sheaf (or a bounded complex) with a finite resolution by direct sums of line bundles before restricting.

We summarise in the following theorem.

**Theorem 7.4.4** *Suppose that $r' < r$ and let $i \colon \mathbb{P}^{r'-1} \to \mathbb{P}^{r-1}$ be the inclusion. Let*

$$i^* \colon \mathbb{Z}[h]/(h^r) \to \mathbb{Z}[h]/(h^{r'})$$

*be the ring homomorphism taking $h$ to $h$.*

(i) *If $\mathcal{F}$ is a vector bundle on $\mathbb{P}^{r-1}$ then $c(i^*(\mathcal{F})) = i^*(c(\mathcal{F}))$ and $\mathsf{Ch}(i^*(\mathcal{F})) = i^*(\mathsf{Ch}(\mathcal{F}))$.*
(ii) *If $\mathcal{F}$ is a coherent sheaf on $\mathbb{P}^{r-1}$ then $c(\mathsf{L}i^*(\mathcal{F})) = i^*(c(\mathcal{F}))$ and $\mathsf{Ch}(\mathsf{L}i^*(\mathcal{F})) = i^*(\mathsf{Ch}(\mathcal{F}))$.* □

## 7.5 Chern Numbers of Twists and Duals

**Lemma 7.5.1** *The Chern polynomial of a twisted vector bundle $\mathcal{F}(i) = \mathcal{F} \otimes_{\mathcal{O}} \mathcal{O}(i)$ is given by*

$$c(\mathcal{F}(i), h) = (1+ih)^{\mathrm{rank}\,(\mathcal{F})} c\left(\mathcal{F}, \frac{h}{1+ih}\right) = \sum_{n=0}^{\mathrm{rank}\,(\mathcal{F})} c_n(\mathcal{F}) h^n (1+ih)^{\mathrm{rank}\,(\mathcal{F})-n}. \tag{7.2}$$

*Thus the Chern numbers of $\mathcal{F}(i)$ are given by*

$$c_m(\mathcal{F}(i)) = \sum_{j=0}^{m} i^j \binom{\operatorname{rank}(\mathcal{F}) - m + j}{j} c_{m-j}(\mathcal{F}).$$

*The Chern character of $\mathcal{F}(i)$ is given by*

$$\mathsf{Ch}(\mathcal{F}(i)) = e^{ih}\mathsf{Ch}(\mathcal{F}).$$

*Proof.* Let $\mathcal{F} = M^{\sim}$, where $M$ is a finitely generated graded $R$-module with Poincaré series $\sum_j a_j t^j/(1-t)^r$. Then using the definition of the Chern polynomial we have

$$\begin{aligned} c(\mathcal{F}(i), h) &= \prod_j (1 + (j+i)h)^{a_j} \\ &= (1+ih)^{\sum_j a_j} \prod_j \left(\frac{1+(j+i)h}{1+ih}\right)^{a_j} \\ &= (1+ih)^{\sum_j a_j} \prod_j \left(1 + \frac{jh}{1+ih}\right)^{a_j} \\ &= (1+ih)^{\operatorname{rank}(\mathcal{F})} c\left(\mathcal{F}, \frac{h}{1+ih}\right). \end{aligned}$$

This gives the formula for the Chern numbers.

For the Chern character we have

$$\mathsf{Ch}(\mathcal{F}(i)) = \sum_j a_j e^{(j+i)h} = e^{ih} \sum_j a_j e^{jh} = e^{ih}\mathsf{Ch}(\mathcal{F}). \qquad \square$$

More explicitly, the lemma says that if $\mathcal{F}$ has rank $s$ then

$$\begin{aligned} c_1(\mathcal{F}(i)) &= c_1(\mathcal{F}) + is \\ c_2(\mathcal{F}(i)) &= c_2(\mathcal{F}) + i(s-1)c_1(\mathcal{F}) + i^2\binom{s}{2} \\ c_3(\mathcal{F}(i)) &= c_3(\mathcal{F}) + i(s-2)c_2(\mathcal{F}) + i^2\binom{s-1}{2}c_1(\mathcal{F}) + i^3\binom{s}{3} \end{aligned}$$

and so on.

**Lemma 7.5.2** *The Chern polynomial and Chern character of the dual bundle $\mathcal{F}^\vee$ are given by*

$$c(\mathcal{F}^\vee, h) = c(\mathcal{F}, -h), \qquad \mathsf{Ch}(\mathcal{F}^\vee, h) = \mathsf{Ch}(\mathcal{F}, -h).$$

*Proof.* By Corollary 6.12.5 there is a resolution of $\mathcal{F}$ by sums of line bundles

$$0 \to \bigoplus_{j=1}^{m_r} \mathcal{O}(a_{r,j}) \to \cdots \to \bigoplus_{j=1}^{m_1} \mathcal{O}(a_{1,j}) \to \bigoplus_{j=1}^{m_0} \mathcal{O}(a_{0,j}) \to \mathcal{F} \to 0.$$

Dualising, and using local freeness, we get an exact sequence

$$0 \to \mathcal{F}^\vee \to \bigoplus_{j=1}^{m_0} \mathcal{O}(-a_{0,j}) \to \bigoplus_{j=1}^{m_1} \mathcal{O}(-a_{1,j}) \to \cdots \to \bigoplus_{j=1}^{m_r} \mathcal{O}(-a_{r,j}) \to 0.$$

Thus

$$c(\mathcal{F}, h) = \prod_{i=0}^{r} \prod_{j=1}^{m_i} (1 + a_{i,j}h)^{(-1)^i} \quad \text{while} \quad c(\mathcal{F}^\vee, h) = \prod_{i=0}^{r} \prod_{j=1}^{m_i} (1 - a_{i,j}h)^{(-1)^i}.$$

Similarly,

$$\mathsf{Ch}(\mathcal{F}, h) = \sum_{i=0}^{r} (-1)^i \sum_{j=1}^{m_i} e^{a_{i,j}h} \quad \text{while} \quad \mathsf{Ch}(\mathcal{F}^\vee, h) = \sum_{i=0}^{r} (-1)^i \sum_{j=1}^{m_i} e^{-a_{i,j}h}.$$

□

## 7.6 Chern Roots

Throughout this section, let $\mathcal{O}$ be the structure sheaf of $\mathbb{P}^{r-1}$, and let $\mathcal{F}$ be a coherent sheaf of $\mathcal{O}$-modules with Chern polynomial $c(\mathcal{F}) = c(\mathcal{F}, h) \in \mathbb{Z}[h]/(h^r)$. Then $c(\mathcal{F})$ lifts uniquely to a polynomial in $\mathbb{Z}[h]$ of degree at most $r - 1$. By abuse of notation we denote this element of $\mathbb{Z}[h]$ and its image in $\mathbb{C} \otimes_{\mathbb{Z}} \mathbb{Z}[h] = \mathbb{C}[h]$ also by $c(\mathcal{F})$ or $c(\mathcal{F}, h)$. In $\mathbb{C}[h]$ it factors as a product of linear factors

$$c(\mathcal{F}) = \prod_j (1 + \alpha_j h).$$

The complex numbers $\alpha_j$ obtained this way are called the *Chern roots* of $\mathcal{F}$. It is clear that the $\alpha_j$ are algebraic integers.

**Example 7.6.1** If $\mathcal{F}$ is a direct sum of line bundles,

$$\mathcal{F} = \bigoplus_j \mathcal{O}(a_j)$$

then

$$c(\mathcal{F}) = \prod_j (1 + a_j h)$$

and so the Chern roots are the $a_j$.

**Remarks 7.6.2** (i) We do not regard the *number* of Chern roots as well defined, since we can always pad them out with a list of zeros. If $\mathcal{F}$ is a vector bundle then it is sometimes convenient to regard the number of Chern roots as being equal to rank $(\mathcal{F})$, as this makes many formulas easier. But this is not possible for coherent sheaves, as the degree of the polynomial $c(\mathcal{F})$ may be larger than the rank of $\mathcal{F}$.

(ii) In the case where $\mathcal{F}$ is a vector bundle, there is a way of regarding $\mathcal{F}$ as formally having a filtration by line bundles with first Chern classes $\alpha_j$. The precise way to formulate this is called the *splitting principle*, see for example Remark 3.2.3 of Fulton [129]. We do not need to use this point of view in our development of the subject.

The following way to rewrite Lemma 7.5.1 is the content of Exercise 3.2.2 of [129], where it is regarded as a consequence of the splitting principle.

**Lemma 7.6.3** *The Chern roots of a twist of a coherent sheaf are given by*

$$c(\mathcal{F}(i)) = (1 + ih)^m \prod_j (1 + (\alpha_j + i)h), \tag{7.3}$$

*where m is equal to* rank $(\mathcal{F})$ *minus the number of Chern roots used in the product.*

*Proof.* By Lemma 7.5.1 we have

$$\begin{aligned} c(\mathcal{F}(i), h) &= (1 + ih)^{\operatorname{rank}(\mathcal{F})} c\left(\mathcal{F}, \frac{h}{1 + ih}\right) \\ &= (1 + ih)^{\operatorname{rank}(\mathcal{F})} \prod_j \left(1 + \alpha_j \frac{h}{1 + ih}\right) \\ &= (1 + ih)^m \prod_j (1 + ih) \left(1 + \alpha_j \frac{h}{1 + ih}\right) \\ &= (1 + ih)^m \prod_j (1 + (\alpha_j + i)h). \end{aligned}$$

□

## 7.7 Power Sums

We continue with the notation of the last section. In this section we define invariants $s_i(\mathcal{F})$ $(1 \le i \le r - 1)$ which will turn out to be the sum of the $i$ th powers of the Chern roots.

**Definition 7.7.1** Let $\mathcal{F}$ be a coherent sheaf on $\mathbb{P}^{r-1}$. We define $s(\mathcal{F}, h) \in \mathbb{Z}[h]/(h^r)$ by the formula

$$-s(\mathcal{F}, -h) = \frac{hc'(\mathcal{F}, h)}{c(\mathcal{F}, h)},$$

where $c'(\mathcal{F}, h)$ denotes the (formal) derivative of $c(\mathcal{F}, h)$ with respect to $h$. The numbers $s_i(\mathcal{F})$ $(1 \le i \le r - 1)$ are defined by

$$s(\mathcal{F}, h) = \sum_{i \ge i} s_i(\mathcal{F}) h^i = s_1(\mathcal{F})h + s_2(\mathcal{F})h^2 + \cdots .$$

**Remark 7.7.2** The definition (7.7.1) for $s(\mathcal{F}, h)$ can be rewritten as

$$jc_j = \sum_{i=1}^{j-1} (-1)^{i-1} c_{j-i} s_i.$$

This equation is called *Newton's identity.*

Thus each $s_i = s_i(\mathcal{F})$ is some polynomial in the Chern numbers with integer coefficients. For example,

$$\begin{aligned} s_1 &= c_1 \\ s_2 &= c_1^2 - 2c_2 \\ s_3 &= c_1^3 - 3c_1c_2 + 3c_3 \\ s_4 &= c_1^4 - 4c_1^2c_2 + 4c_1c_3 + 2c_2^2 - 4c_4. \end{aligned}$$

Inverting this process allows us to express the Chern numbers as polynomials in the $s_i$ with rational coefficients:

$$\begin{aligned} c_1 &= s_1 \\ c_2 &= \tfrac{1}{2}(s_1^2 - s_2) \\ c_3 &= \tfrac{1}{6}(s_1^3 - 3s_1s_2 + 2s_3) \\ c_4 &= \tfrac{1}{24}(s_1^4 - 6s_1^2s_2 + 3s_2^2 + 8s_1s_3 - 6s_4). \end{aligned}$$

The coefficients here may be read off from the conjugacy class sizes and signs of permutations in the corresponding symmetric group. The same expressions without the signs give the complete symmetric functions of the Chern roots instead of the elementary symmetric functions.

**Lemma 7.7.3** *If* $0 \to \mathcal{F}_1 \to \mathcal{F}_2 \to \mathcal{F}_3 \to 0$ *is a short exact sequence of vector bundles then for* $1 \le i \le r-1$ *we have*

$$s_i(\mathcal{F}_2) = s_i(\mathcal{F}_1) + s_i(\mathcal{F}_3).$$

*Proof.* We have

$$c(\mathcal{F}_2, h) = c(\mathcal{F}_1, h)c(\mathcal{F}_3, h).$$

Differentiating with respect to $h$, we get

$$c'(\mathcal{F}_2, h) = c'(\mathcal{F}_1, h)c(\mathcal{F}_3, h) + c(\mathcal{F}_1, h)c'(\mathcal{F}_3, h).$$

Multiply by $h$ and divide by $c(\mathcal{F}_2, h)$ to get

$$\frac{hc'(\mathcal{F}_2, h)}{c(\mathcal{F}_2, h)} = \frac{hc'(\mathcal{F}_1, h)}{c(\mathcal{F}_1, h)} + \frac{hc'(\mathcal{F}_3, h)}{c(\mathcal{F}_3, h)}.$$

Thus by Definition 7.7.1 we have

$$-s(\mathcal{F}_2, -h) = -s(\mathcal{F}_1, -h) - s(\mathcal{F}_3, -h)$$

and so

$$s(\mathcal{F}_2, h) = s(\mathcal{F}_1, h) + s(\mathcal{F}_3, h).$$

Now compare coefficients of $h^j$. □

**Proposition 7.7.4** *Let $\mathcal{F} = M^\sim$ with $M$ a finitely generated graded $R$-module with Poincaré series*

$$\frac{\sum_j a_j t^j}{(1-t)^r}$$

*where the numerator is a Laurent polynomial. Then for $1 \le i \le r-1$ we have*

$$s_i(\mathcal{F}) = \sum_j a_j j^i.$$

*Proof.* By definition we have

$$\begin{aligned} c(\mathcal{F}, h) &= \prod_j (1 + jh)^{a_j} \\ \log c(\mathcal{F}, h) &= \sum_j a_j \log(1 + jh). \end{aligned}$$

Differentiating both sides with respect to $h$, we obtain

$$\begin{aligned} \frac{c'(\mathcal{F}, h)}{c(\mathcal{F}, h)} &= \sum_j a_j \frac{j}{1+jh} \\ -s(\mathcal{F}, -h) = \frac{hc'(\mathcal{F}, h)}{c(\mathcal{F}, h)} &= \sum_j a_j \frac{jh}{1+jh} \\ s(\mathcal{F}, h) &= \sum_j a_j \frac{jh}{1-jh} \\ &= \sum_j a_j (jh + j^2h^2 + \cdots) \\ s_i(\mathcal{F}) &= \sum_j a_j j^i. \end{aligned}$$

□

**Proposition 7.7.5** *For $1 \le i \le r-1$ the $s_i(\mathcal{F})$ are given in terms of the Chern roots as the power sums*

$$s_i(\mathcal{F}) = \sum_j \alpha_j^i.$$

*Proof.* Working in $\mathbb{C}[h]$ we have

$$c(\mathcal{F}, h) = \prod_j (1 + \alpha_j h)$$

$$\log c(\mathcal{F}, h) = \sum_j \log(1 + \alpha_j h).$$

Differentiating both sides with respect to $h$, we obtain

$$\begin{aligned}
\frac{c'(\mathcal{F}, h)}{c(\mathcal{F}, h)} &= \sum_j \frac{\alpha_j}{1 + \alpha_j h} \\
-s(\mathcal{F}, -h) &= \frac{hc'(\mathcal{F}, h)}{c(\mathcal{F}, h)} = \sum_j \frac{\alpha_j h}{1 + \alpha_j h} \\
s(\mathcal{F}, h) &= \sum_j \frac{\alpha_j h}{1 - \alpha_j h} \\
&= \sum_j (\alpha_j h + \alpha_j^2 h^2 + \cdots) \\
s_i(\mathcal{F}) &= \sum_j \alpha_j^i.
\end{aligned}$$

□

**Theorem 7.7.6** *Let $p(x) \in \mathbb{C}[x]$ be a polynomial of degree $< r$ with $p(0) = 0$. Then*

(i) $\sum_j p(\alpha_j) = \sum_j a_j p(j)$.
(ii) *If $p(x)$ satisfies $n \in \mathbb{Z} \Rightarrow p(n) \in \mathbb{Z}$, then we have $\sum_j p(\alpha_j) \in \mathbb{Z}$.*

*Proof.* (i) Both sides are linear in $p(x)$, so it suffices to prove the equation for $p(x) = x^i$ $(i > 0)$. In this case it follows from Propositions 7.7.4 and 7.7.5.

(ii) This follows from (i), since the coefficients $a_j$ are integers. □

**Corollary 7.7.7** *Let $\mathcal{F}$ be a coherent sheaf on $\mathbb{P}^{r-1}$ with Chern roots $\alpha_j$. Then the Chern character is given by*

$$\mathsf{Ch}(\mathcal{F}) = \text{rank}\,(\mathcal{F}) + \sum_j (e^{\alpha_j h} - 1).$$

*Proof.* By definition

$$\begin{aligned}
\mathsf{Ch}(\mathcal{F}) &= \sum_j a_j e^{jh} \\
&= \text{rank}\,(\mathcal{F}) + \sum_j a_j (e^{jh} - 1) \\
&= \text{rank}\,(\mathcal{F}) + \sum_j \sum_{n=1}^{r-1} \frac{a_j j^n h^n}{n!}.
\end{aligned}$$

Applying Theorem 7.7.6 (i) to the polynomials $x^n/n!$ $(1 \leq n \leq r-1)$, we deduce that

$$\begin{aligned}\mathsf{Ch}(\mathcal{F}) &= \operatorname{rank}(\mathcal{F}) + \sum_j \sum_{n=1}^{r-1} \frac{\alpha_j^n h^n}{n!} \\ &= \operatorname{rank}(\mathcal{F}) + \sum_j (e^{\alpha_j h} - 1).\end{aligned}$$ □

**Remark 7.7.8** We remarked before (Remark 7.6.2 (i)) that the number of Chern roots is not regarded as well defined. If $\mathcal{F}$ is a vector bundle and we insist that there are exactly rank $(\mathcal{F})$ Chern roots then the formula in the corollary reads

$$\mathsf{Ch}(\mathcal{F}) = \sum_j e^{\alpha_j h}.$$

**Example 7.7.9** Let $m$ be a positive integer. Then the polynomial

$$p(x) = \binom{x}{m} = \frac{x(x-1)\dots(x-m+1)}{m!}$$

satisfies $n \in \mathbb{Z} \Rightarrow p(n) \in \mathbb{Z}$ and $p(0) = 0$. More generally, so does the polynomial $\binom{x+s}{m}$ for $s \in \mathbb{Z}$. So these are candidates for use in Theorem 7.7.6 (ii).

**Corollary 7.7.10** *Let $\mathcal{F}$ be a coherent sheaf on $\mathbb{P}^{r-1}$ with Chern roots $\alpha_j$. If $s$ and $m$ are integers with $1 \leq m \leq r-1$ then*

$$\sum_j \binom{\alpha_j + s}{m} \in \mathbb{Z}.$$

**Example 7.7.11** For coherent sheaves on $\mathbb{P}^1$ and $\mathbb{P}^2$, this gives no condition at all on the Chern numbers. For any coherent sheaf on $\mathbb{P}^3$ the conditions are equivalent to

$$c_1c_2 + c_3 \equiv 0 \pmod 2.$$

Horrocks [159] has shown that given any $c_1$ and $c_2$ with $c_1c_2$ even, there exists a vector bundle of rank two on $\mathbb{P}^3$ with these Chern numbers.

For a bundle of rank two on $\mathbb{P}^4$ the conditions are equivalent to

$$c_2(c_2 + 1 - 3c_1 - 2c_1^2) \equiv 0 \pmod{12}.$$

These congruences are usually regarded as consequences of the Hirzebruch–Riemann–Roch theorem, which we discuss in the next section. They will be used in Sections 10.5–10.9 to obtain restrictions on modules of constant Jordan type.

The following proposition tells us that Corollary 7.7.10 contains all the information to be obtained from the theorem.

**Proposition 7.7.12** *Let $p(x) \in \mathbb{C}[x]$. Then the following are equivalent:*

(i) $n \in \mathbb{Z} \Rightarrow p(n) \in \mathbb{Z}$.
(ii) $p(x)$ *is in the* $\mathbb{Z}$*-span of the polynomials*

$$\binom{x}{m} = \frac{x(x-1)\dots(x-m+1)}{m!}$$

$(0 \le m \le \deg(p))$.

**Remark 7.7.13** Here, $\binom{x}{0}$ is taken to be 1, and if we exclude this from (ii) and take $1 \le m \le \deg(p)$ instead, then we obtain the polynomials satisfying condition (i) with $p(0) = 0$.

*Proof.* Clearly (ii) $\Rightarrow$ (i). To prove that (i) $\Rightarrow$ (ii), we work by induction on $\deg(p)$ using the calculus of finite differences as follows. If $p(x) \in \mathbb{C}[x]$, we define

$$\Delta p(x) = p(x+1) - p(x).$$

Then $\deg(\Delta p) = \deg(p) - 1$. We also have

$$\Delta\binom{x}{m+1} = \binom{x+1}{m+1} - \binom{x}{m+1} = \binom{x}{m}.$$

By the inductive hypothesis we have

$$\Delta p(x) = \sum_{i=0}^{\deg(p)-1} b_m \binom{x}{m}$$

with $b_m \in \mathbb{Z}$. We also have

$$\Delta \sum_{m=0}^{\deg(p)-1} b_m \binom{x}{m+1} = \sum_{m=0}^{\deg(p)-1} b_m \binom{x}{m}$$

and so

$$\Delta\left(p(x) - \sum_{m=0}^{\deg(p)-1} b_m \binom{x}{m+1}\right) = 0.$$

It follows that

$$p(x) - \sum_{m=0}^{\deg(p)-1} b_m \binom{x}{m+1}$$

is a constant, and this constant must be an integer. So $p(x)$ is of the required form. □

## 7.8 The Hirzebruch–Riemann–Roch Theorem

Hirzebruch [156] developed a generalised version of the Riemann–Roch theorem that applies to vector bundles on a non-singular algebraic variety. We shall provide a quick proof of the Hirzebruch–Riemann–Roch theorem for vector bundles on $\mathbb{P}^{r-1}$. As remarked in the previous section, the congruences on Chern numbers coming from this theorem are already contained in Corollary 7.7.10.

**Definition 7.8.1** The *Euler characteristic* of a vector bundle $\mathcal{F}$ on $\mathbb{P}^{r-1}$ is defined to be

$$\chi(\mathcal{F}) = \sum_{i \geq 0} (-1)^i \dim_k H^i(\mathbb{P}^{r-1}, \mathcal{F}).$$

Since $H^i(\mathbb{P}^{r-1}, \mathcal{F}) = 0$ for $i \geq r$, this is a finite sum.

**Lemma 7.8.2** *If* $0 \to \mathcal{F}_1 \to \mathcal{F}_2 \to \mathcal{F}_3 \to 0$ *is a short exact sequence of vector bundles on* $\mathbb{P}^{r-1}$ *then*

$$\chi(\mathcal{F}_2) = \chi(\mathcal{F}_1) + \chi(\mathcal{F}_3).$$

*Proof.* This follows from the long exact sequence in cohomology of this short exact sequence of vector bundles. □

**Theorem 7.8.3** (Schwarzenberger, Appendix I, §22.4 of Hirzebruch [156]; see also §I.6.1 of Okonek, Schneider and Spindler [202]) *Let* $\mathcal{F}$ *be a coherent sheaf on* $\mathbb{P}^{r-1}$. *The Euler characteristic of* $\mathcal{F}$ *is given in terms of the Chern roots by the formula*

$$\chi(\mathcal{F}) = \operatorname{rank}(\mathcal{F}) + \sum_j \left( \binom{\alpha_j + r - 1}{r - 1} - 1 \right).$$

**Remark 7.8.4** Provided the number of Chern roots is taken to be equal to the rank of $\mathcal{F}$ (see Remark 7.6.2 (i)), this simplifies to

$$\chi(\mathcal{F}) = \sum_j \binom{\alpha_j + r - 1}{r - 1}.$$

*Proof.* First we treat the case of a line bundle $\mathcal{O}(a)$. In this case the statement is that

$$\chi(\mathcal{O}(a)) = \binom{a + r - 1}{r - 1}.$$

This is true by Corollary 6.9.2.

By Lemmas 7.7.3 and 7.8.2 both sides are additive over short exact sequences. So it remains true for direct sums of line bundles.

By Corollary 6.12.5 to Hilbert's Syzygy Theorem 6.12.1, $\mathcal{F}$ has a resolution

$$0 \to F_n^{\sim} \to \cdots \to F_0^{\sim} \to \mathcal{F} \to 0$$

where each $F_i^{\sim}$ is a direct sum of line bundles and $n \leq r$. We proceed by induction on $n$. If $n = 0$ then $\mathcal{F}$ is a direct sum of line bundles, and this case is already treated. If $n > 0$, let $\mathcal{F}'$ be the kernel of $F_0^{\sim} \to \mathcal{F}$, so that we have a short exact sequence

$$0 \to \mathcal{F}' \to F_0^{\sim} \to \mathcal{F} \to 0.$$

Then $\mathcal{F}'$ has such a resolution of length $n - 1$, so by the inductive hypothesis the theorem is true for $\mathcal{F}'$. It is also true for $F_0^{\sim}$, since this is a direct sum of line bundles. So by additivity of both sides over short exact sequences, it is true for $\mathcal{F}$. □

**Remark 7.8.5** Schwartzenberger's theorem can be seen as telling us how to recover the Poincaré series of a graded $k[Y_1, \dots, Y_r]$-module $M$ from the Chern roots of the coherent sheaf $M^{\sim}$. Of course, this can only be achieved up to the Poincaré series of a finite length module, namely up to the addition of a polynomial.

The way the theorem tells us this information is as follows. As long as $n$ is large enough, we have $H^i(\mathbb{P}^{r-1}, M^{\sim}(n)) = 0$ for $i > 0$ and so $\chi(M^{\sim}(n)) = \dim H^0(\mathbb{P}^{r-1}, M^{\sim}(n))$. Furthermore, for $n$ large enough we have $H^0(\mathbb{P}^{r-1}, M^{\sim}(n)) \cong M_n$. It follows that provided we can take the number of Chern roots to be the rank, then the Poincaré series $p_M(t)$ differs from the power series

$$\sum_{n \geq 0} \sum_j \binom{\alpha_j + n + r - 1}{r - 1} t^n$$

by a polynomial.

**Example 7.8.6** An instanton bundle (see Section 9.8) $\mathcal{F}$ of quantum number $k$ on $\mathbb{P}^{r-1}$ ($r$ even) has rank $r - 2$ and Chern polynomial $1/(1 - t^2)^k$. We can behave as though the Chern roots are $+1$ and $-1$, each with multiplicity $-k$, and $0$ with multiplicity $r - 2 + 2k$, even though this doesn't make literal sense. We then have

$$\chi(\mathcal{F}(n)) = -k\binom{n + r}{r - 1} + (r - 2 + 2k)\binom{n + r - 1}{r - 1} - k\binom{n + r - 2}{r - 1}.$$

The Poincaré series of a corresponding graded module therefore differs by a polynomial from

$$\frac{-kt^{-1} + (r - 2 + 2k) - kt}{(1 - t)^r}.$$

**Theorem 7.8.7** (Hirzebruch–Riemann–Roch for $\mathbb{P}^{r-1}$) *Let $\mathcal{F}$ be a coherent sheaf on $\mathbb{P}^{r-1}$. Then $\chi(\mathcal{F})$ is equal to the coefficient of $h^{r-1}$ in the power series*

$$\left(\frac{h}{1 - e^{-h}}\right)^r \mathsf{Ch}(\mathcal{F}).$$

*Proof.* By Cauchy's integral formula, the coefficient of $h^{r-1}$ in this power series is

$$\frac{1}{2\pi i}\oint\left(\frac{h}{1-e^{-h}}\right)^r \mathsf{Ch}(\mathcal{F})\,\frac{dh}{h^r}$$

where the integral is over a small anticlockwise circle around the origin in the complex plane. Using Corollary 7.7.7, we get

$$\frac{1}{2\pi i}\oint\frac{\operatorname{rank}(\mathcal{F})+\sum_j(-1+e^{\alpha_j h})}{(1-e^{-h})^r}\,dh.$$

Now substitute $z=1-e^{-h}$. We have

$$dz=e^{-h}\,dh,\qquad dh=dz/(1-z),\qquad e^{\alpha_j h}=1/(1-z)^{\alpha_j}.$$

First we compute

$$\frac{1}{2\pi i}\oint\frac{dh}{(1-e^{-h})^r}=\frac{1}{2\pi i}\oint\frac{dz}{z^r(1-z)}=\frac{1}{2\pi i}\oint(z^{-r}+z^{-r+1}+\ldots)\,dz=1.$$

Therefore the constant term of the Chern polynomial comes out of the integral, and we get

$$\begin{aligned}
&\operatorname{rank}(\mathcal{F})+\sum_j\left(-1+\frac{1}{2\pi i}\oint\frac{dz}{z^r(1-z)^{\alpha_j+1}}\right)\\
&\quad=\operatorname{rank}(\mathcal{F})+\sum_j\left(-1+\frac{1}{2\pi i}\oint z^{-r}(1+(\alpha_j+1)z\right.\\
&\qquad\left.+\binom{\alpha_j+2}{2}z^2+\cdots)\,dz\right)\\
&\quad=\operatorname{rank}(\mathcal{F})+\sum_j\left(-1+\binom{\alpha_j+r-1}{r-1}\right).
\end{aligned}$$

By Theorem 7.8.3 this is equal to $\chi(\mathcal{F})$. □

**Remark 7.8.8** It is worth making some remarks about this substitution $z=1-e^{-h}$ in this proof. Since $z=h+O(h^2)$, if $h$ traverses a small anticlockwise circle around the origin, $z$ traverses a deformation of this circle that is small compared with the radius. So for small enough radius, the winding number of this contour around the origin is still equal to one. Since the integrand has an isolated pole at $z=0$, by Cauchy's theorem this contour gives the same integral as the anticlockwise circle in the $z$ plane.

**Remark 7.8.9** The power series

$$\left(\frac{h}{1-e^{-h}}\right)^r$$

in the Hirzebruch–Riemann–Roch theorem for $\mathbb{P}^{r-1}$ is playing the role of the *Todd class* of the tangent bundle. We discuss the tangent bundle in Section 6.4. Note that *whatever* the coherent sheaf $\mathcal{F}$ is, the Todd class of the tangent bundle is what appears in the formula. But the Todd class is indeed defined for any vector bundle $\mathcal{F}$, and is given in terms of the Chern roots or Chern numbers by the formula

$$\begin{aligned}\mathsf{Td}(\mathcal{F}) &= \prod_j \frac{\alpha_j h}{1-e^{-\alpha_j h}}\\ &= 1 + \tfrac{1}{2}c_1 h + \tfrac{1}{12}(c_2 + c_1^2)h^2 + \tfrac{1}{24}c_1c_2h^3\\ &\quad + \tfrac{1}{720}(-c_4 + c_1c_3 + 3c_2^2 + 4c_1^2c_2 - c_1^4)h^4\\ &\quad + \tfrac{1}{1440}c_1(-c_4 + c_1c_3 + 3c_2^2 - c_1^2c_2)h^5 + \cdots .\end{aligned}$$

Note that we have

$$\frac{x}{1-e^{-x}} = \sum_{j=0}^{\infty} \frac{B_j x^j}{j!}$$

where the $B_j$ are the Bernoulli numbers:

$$B_0 = 1, \quad B_1 = \tfrac{1}{2}, \quad B_2 = \tfrac{1}{6}, \quad B_4 = -\tfrac{1}{30}, \quad B_6 = \tfrac{1}{42}, \quad \text{etc.,}$$

and $B_j$ is zero for $j$ odd, $j > 1$.[1]

The general statement of the Hirzebruch–Riemann–Roch theorem is that if $X$ is a non-singular projective variety (or a compact complex manifold) and $\mathcal{F}$ is a holomorphic vector bundle over $X$ then $\chi(\mathcal{F})$ is the evaluation of $\mathsf{Td}(\mathcal{T})\mathsf{Ch}(\mathcal{F})$ at the fundamental class of $X$.

Grothendieck produced a relative version of this theorem, known as the Grothendieck–Riemann–Roch theorem. This was written up and published by Borel and Serre [62]. The proof of Grothendieck–Riemann–Roch in its full generality starts by proving Hirzebruch–Riemann–Roch for projective space, see Proposition 10 of [62].

## 7.9 Chern Numbers and the Frobenius Map

Let $k$ be an algebraically closed field of characteristic $p > 0$ and let $R = k[Y_1, \dots, Y_r]$. The Frobenius map $F \colon \mathbb{P}^{r-1} \to \mathbb{P}^{r-1}$ is the map of projective varieties induced by the map of $k$-algebras $R \to R$ sending $Y_i$ to $Y_i^p$. In this section we examine the inverse image of a coherent sheaf along $F$ and its effect on the Chern polynomial.

**Definition 7.9.1** We write $\Phi R$ for the subring $k[Y_1^p, \dots, Y_r^p]$ of $R$. If $M$ be a finitely generated graded $R$-module, we write $\Phi M$ for the graded vector space given by

[1] Beware that Hirzebruch [156] uses a somewhat unconventional indexing for the Bernoulli numbers.

$$(\Phi M)_j = \begin{cases} M_{j/p} & p|j \\ 0 & \text{otherwise.} \end{cases}$$

If $m \in M_i$, we write $\Phi(m)$ for the corresponding element of $(\Phi M)_{ip}$. We make $\Phi M$ into a $\Phi R$-module via

$$Y_i^p \Phi(m) = \Phi(Y_i m).$$

Finally, we write $FM$ for the graded $R$-module

$$FM = R \otimes_{\Phi R} \Phi M.$$

If $\mathcal{F}$ is a coherent sheaf of $\mathcal{O}$-modules, where $\mathcal{O}$ is the structure sheaf of $\mathbb{P}^{r-1}$, then according to Theorem 3.9.12 there is a finitely generated graded $R$-module $M$ with $\mathcal{F} = M^{\sim}$. The inverse image $F^*(\mathcal{F})$ of $\mathcal{F}$ along the Frobenius map $F\colon \mathbb{P}^{r-1} \to \mathbb{P}^{r-1}$ is then defined by

$$F^*(\mathcal{F}) = (FM)^{\sim}.$$

**Example 7.9.2** If $\mathcal{F} = \mathcal{O}(a)$ then $F^*\mathcal{F} = \mathcal{O}(pa)$.

**Lemma 7.9.3** *$R$ is a free $\Phi R$-module on the basis consisting of monomials $Y_1^{i_1} \dots Y_r^{i_r}$ with $0 \le i_j \le p-1$ for $1 \le j \le r$.*

*Proof.* This is clear by examining the basis of monomials in $R$. □

**Proposition 7.9.4** *If*

$$0 \to M_1 \to M_2 \to M_3 \to 0$$

*is a short exact sequence of $R$-modules then so is*

$$0 \to FM_1 \to FM_2 \to FM_3 \to 0.$$

*If*

$$0 \to \mathcal{F}_1 \to \mathcal{F}_2 \to \mathcal{F}_3 \to 0$$

*is a short exact sequence of sheaves of $\mathcal{O}$-modules, then so is*

$$0 \to F^*(\mathcal{F}_1) \to F^*(\mathcal{F}_2) \to F^*(\mathcal{F}_3) \to 0.$$

*Proof.* This follows from Lemma 7.9.3, since tensoring with a free module is exact. □

**Remark 7.9.5** Recall from Corollary 6.12.5 that every coherent sheaf $\mathcal{F}$ of $\mathcal{O}$-modules has a resolution of length at most $r$ by sums of line bundles. It follows from the Proposition that replacing each line bundle $\mathcal{O}(a)$ by $\mathcal{O}(pa)$ and replacing each $Y_i$ by $Y_i^p$ in the matrices defining the maps, we obtain a resolution of $F^*(\mathcal{F})$.

**Theorem 7.9.6** *Let $\mathcal{F}$ be a coherent sheaf of $\mathcal{O}$-modules, and let $F$ be the Frobenius map.*

(i) *The Chern polynomial of $F^*(\mathcal{F})$ is given by*

$$c(F^*(\mathcal{F}), h) = c(\mathcal{F}, ph).$$

(ii) *The Chern numbers of $F^*(\mathcal{F})$ are given by*

$$c_i(F^*(\mathcal{F})) = p^i c_i(\mathcal{F}).$$

(iii) *If the Chern roots of $\mathcal{F}$ are $\alpha_j$ then the Chern roots of $F^*(\mathcal{F})$ are $p\alpha_j$.*

(iv) *The Chern character of $F^*(\mathcal{F})$ is given by*

$$\mathsf{Ch}(F^*(\mathcal{F}), h) = \mathsf{Ch}(\mathcal{F}, ph).$$

*Proof.* (i) Let $\mathcal{F} = M^\sim$, where $M$ is a finitely generated graded $R$-module. Write the Poincaré series for $M$ in the form

$$p_M(t) = \frac{\sum_j a_j t^j}{(1-t)^r}$$

where $\sum_j a_j t^j$ is a Laurent polynomial in $t$. Then

$$p_{\Phi M}(t) = \frac{\sum_j a_j t^{pj}}{(1-t^p)^r}.$$

By Lemma 7.9.3, we have

$$p_{FM}(t) = (1 + t + \cdots + t^{p-1})^r p_{\Phi M}(t) = \frac{\sum_j a_j t^{pj}}{(1-t)^r}.$$

Thus

$$c(F^*(\mathcal{F}), h) = \prod_j (1 + pjh)^{a_j} = c(\mathcal{F}, ph).$$

(ii) and (iii) follow immediately from (i), and then (iv) follows using Corollary 7.7.7. □

# 8

# Modules of Constant Jordan Type and Vector Bundles

## 8.1 Introduction

The construction of algebraic vector bundles (i.e., locally free sheaves) on projective space from modules of constant Jordan type over an elementary abelian $p$-group was introduced by Friedlander and Pevtsova [127]. In this chapter we give a refined version of their construction and investigate its properties.

Let $E$ be an elementary abelian $p$-group of rank $r$ and let $k$ be an algebraically closed field of characteristic $p$. In this chapter we introduce functors $\mathcal{F}_i$ from finitely generated $kE$-modules of constant Jordan type to vector bundles over projective space $\mathbb{P}^{r-1}$. We prove that given any vector bundle $\mathcal{F}$ of rank $s$ on $\mathbb{P}^{r-1}$, there is a $kE$-module $M$ of stable constant Jordan type $[1]^s$ such that $\mathcal{F}_1(M) \cong \mathcal{F}$ if $p = 2$, and such that $\mathcal{F}_1(M) \cong F^*(\mathcal{F})$ if $p$ is odd. Here, $F \colon \mathbb{P}^{r-1} \to \mathbb{P}^{r-1}$ is the Frobenius map. We prove that the theorem cannot be improved if $p$ is odd, because if $M$ is any module of stable constant Jordan type $[1]^s$ then the Chern numbers $c_1, \dots, c_{p-2}$ of $\mathcal{F}_1(M)$ are divisible by $p$. We also prove various formulas such as $\mathcal{F}_{p-i}(\Omega M) \cong \mathcal{F}_i(M)(-p+i)$, and in Chapter 9 we shall give a number of explicit examples of vector bundles obtained from modules of constant Jordan type. In Chapter 11, we shall return to this topic, and give another interpretation of the main construction of this chapter in terms of the Orlov correspondence relating the stable module category $\mathsf{stmod}(kE)$ to a certain category of maximal Cohen–Macaulay modules over a certain hypersurface singularity.

In more detail, we shall define functors $\mathcal{F}_i$ $(1 \le i \le p)$ from finitely generated $kE$-modules of constant Jordan type to vector bundles on projective space $\mathbb{P}^{r-1}$, capturing the sum of the socles of the length $i$ Jordan blocks. We prove the following realisation theorem in Section 8.9.

**Theorem 8.1.1** (Benson and Pevtsova [53]) *Given any vector bundle $\mathcal{F}$ of rank $s$ on $\mathbb{P}^{r-1}$, there exists a finitely generated $kE$-module $M$ of stable constant Jordan type $[1]^s$ such that*

(i) *if $p = 2$, then $\mathcal{F}_1(M) \cong \mathcal{F}$.*
(ii) *if $p$ is odd, then $\mathcal{F}_1(M) \cong F^*(\mathcal{F})$, the inverse image of $\mathcal{F}$ along the Frobenius morphism $F \colon \mathbb{P}^{r-1} \to \mathbb{P}^{r-1}$.*

For $p = 2$ the theorem may be thought of as a version of the Bernstein–Gelfand–Gelfand correspondence [55], since the group algebra of an elementary abelian 2-group in characteristic two is isomorphic to an exterior algebra. See Section 8.13 for a clarification of this. But for $p$ odd it says something new and interesting. In particular, it is striking that part (ii) of Theorem 8.1.1 cannot be strengthened to say that $\mathcal{F}_1(M) \cong \mathcal{F}$. We shall prove in Section 10.2 that for a module $M$ of stable constant Jordan type $[1]^s$, the Chern numbers $c_m(\mathcal{F}_1(M))$ are divisible by $p$ for $1 \leq m \leq p-2$.

## 8.2 The Operator $\theta$

Recall that $k$ is an algebraically closed field of characteristic $p$, $E = \langle g_1, \ldots, g_r\rangle$ is an elementary abelian $p$-group of rank $r$ and $X_i = g_i - 1 \in J(kE)$ for $1 \leq i \leq r$. If

$$0 \neq \alpha = (\lambda_1, \ldots, \lambda_r) \in \mathbb{A}^r,$$

we define

$$X_\alpha = \lambda_1 X_1 + \cdots + \lambda_r X_r \in kE,$$

an element of $J(kE)$ satisfying $X_\alpha^p = 0$.

We write $k[Y_1, \ldots, Y_r]$ for the coordinate ring $k[\mathbb{A}^r]$, where the $Y_i$ are the linear functions defined by $Y_i(X_j) = \delta_{ij}$ (Kronecker delta). We write $\mathbb{P}^{r-1}$ for the corresponding projective space. Let $\mathcal{O}$ be the structure sheaf on $\mathbb{P}^{r-1}$. If $\mathcal{F}$ is a sheaf of $\mathcal{O}$-modules and $j \in \mathbb{Z}$, we write $\mathcal{F}(j)$ for the $j$th twist $\mathcal{F} \otimes_{\mathcal{O}} \mathcal{O}(j)$. If $M$ is a finitely generated $kE$-module, we write $\widetilde{M}$ for the trivial vector bundle $M \otimes_k \mathcal{O}$, so that $\widetilde{M}(j) = M \otimes_k \mathcal{O}(j)$.[1] Friedlander and Pevtsova [127] define maps of vector bundles $\theta_M \colon \widetilde{M}(j) \to \widetilde{M}(j+1)$, for all $j \in \mathbb{Z}$, by the formula

$$\theta_M(m \otimes f) = \sum_{i=1}^{r} X_i(m) \otimes Y_i f.$$

If the module $M$ is clear from the context, we write $\theta$ for $\theta_M$.

**Remark 8.2.1** If we think of $M$ as given by a degree $n$ matrix representation with coefficients in $k$, then the space of homomorphisms from $\widetilde{M}$ to $\widetilde{M}(1)$ is equal to $\mathrm{Mat}_n(\mathrm{Hom}(\mathcal{O}, \mathcal{O}(1)))$. Since $\mathrm{Hom}(\mathcal{O}, \mathcal{O}(1))$ is the vector space spanned by $Y_1, \ldots, Y_r$, it follows that $\theta_M$ can be thought of as a matrix whose entries are linear forms in $Y_1, \ldots, Y_r$. Namely, if $\phi_M \colon kE \to \mathrm{Mat}_n(k)$ gives the matrix representation for $M$ then the matrix representing $\theta_M$ is

$$\sum_i Y_i \phi_M(X_i) \in \mathrm{Mat}_n(k[Y_1, \ldots, Y_r]).$$

[1] The use of a wide tilde here is to distinguish it from the tilde construction of a sheaf from a graded module as given in Section 3.8.

**Example 8.2.2** Let $E = \langle g_1, g_2\rangle \cong (\mathbb{Z}/p)^2$, $kE = k[X_1, X_2]/(X_1^p, X_2^p)$ and let $M$ be the three-dimensional $kE$-module given by

$$g_1 \mapsto \begin{pmatrix} 1 & 0 & 0 \\ 1 & 1 & 0 \\ 0 & 0 & 1 \end{pmatrix} \qquad g_2 \mapsto \begin{pmatrix} 1 & 0 & 0 \\ 0 & 1 & 0 \\ 1 & 0 & 1 \end{pmatrix}.$$

Then

$$X_1 \mapsto \begin{pmatrix} 0 & 0 & 0 \\ 1 & 0 & 0 \\ 0 & 0 & 0 \end{pmatrix} \qquad X_2 \mapsto \begin{pmatrix} 0 & 0 & 0 \\ 0 & 0 & 0 \\ 1 & 0 & 0 \end{pmatrix},$$

and the map $\theta_M$ is represented by the matrix

$$\begin{pmatrix} 0 & 0 & 0 \\ Y_1 & 0 & 0 \\ Y_2 & 0 & 0 \end{pmatrix} \in \mathrm{Mat}_3(k[Y_1, Y_2]).$$

It can be seen that in some sense $\theta_M$ represents the action of the generic cyclic shifted subgroup of $E$.

**Remark 8.2.3** More generally, a linear variety of matrices of constant rank, such as for example a nilvariety of constant Jordan type (see Definition 5.1.3) gives rise to an operator $\theta$ via exactly the same formula. So Example 4.1.6 gives rise to the following matrix:

$$\theta = \begin{pmatrix} 0 & Y_1 & 0 \\ Y_2 & 0 & Y_1 \\ 0 & -Y_2 & 0 \end{pmatrix}.$$

The reason this matrix does not represent the operator $\theta$ for a representation of an elementary abelian $p$-group is that its specialisations, for different values of $Y_1$ and $Y_2$, do not commute.

## 8.3 The Action of $\theta$ on Fibres

Let $M$ be a finitely generated $kE$-module and let

$$0 \neq \alpha = (\lambda_1, \ldots, \lambda_r) \in \mathbb{A}^r$$

with image

$$\bar{\alpha} = (\lambda_1 : \cdots : \lambda_r) \in \mathbb{P}^{r-1}.$$

To find the fibre $\mathcal{F} \otimes_{\mathcal{O}} k(\bar{\alpha})$ of a sheaf $\mathcal{F}$ at $\bar{\alpha}$, we must choose $\ell$ with $\lambda_\ell \neq 0$ and look at the affine patch $U_\ell \subseteq \mathbb{P}^{r-1}$ defined by the non-vanishing of the $\ell$th coordinate. Then the fibre of $\mathcal{F}$ at $\bar{\alpha}$ is

$$\mathcal{F}(U_\ell) \otimes_{\mathcal{O}(U_\ell)} k$$

where

$$\mathcal{O}(U_\ell) = k[U_\ell] = k[Y_1 Y_\ell^{-1}, \dots \overset{\ell}{\uparrow} \dots, Y_r Y_\ell^{-1}]$$

and $k$ is regarded as a $k[U_\ell]$-module via evaluation at $\bar{\alpha}$.

Applying this to $\mathcal{O}$, the fibre of $\mathcal{O}$ at $\bar{\alpha}$ is

$$k[U_\ell] \otimes_{k[U_\ell]} k = k.$$

On the other hand, for $\mathcal{O}(j)$ the sections are the degree $j$ functions

$$\mathcal{O}(j)(U_\ell) = k[Y_1 Y_\ell^{-1}, \dots \overset{\ell}{\uparrow} \dots, Y_r Y_\ell^{-1}] Y_\ell^j$$

and the fibre of $\mathcal{O}(j)$ at $\bar{\alpha}$ is

$$k[Y_1 Y_\ell^{-1}, \dots \overset{\ell}{\uparrow} \dots, Y_r Y_\ell^{-1}] Y_\ell^j \otimes_{k[U_\ell]} k.$$

If we normalise by choosing a particular value for $\lambda_\ell$, for example $\lambda_\ell = 1$, we obtain an identification of the left side of this tensor product with $k[U_\ell]$ and hence an identification of the fibre with $k$. Changing the normalisation to some other value of $\lambda_\ell$ changes this identification by the $j$th power of this factor.

To summarise, identifying the fibre of $\mathcal{O}(j)$ at $\bar{\alpha}$ with $k$, we need to choose a point $\alpha \in \mathbb{A}^r$ lying above $\bar{\alpha}$.

Since $\widetilde{M}(j) = M \otimes_k \mathcal{O}(j)$, a choice of $\alpha \in \mathbb{A}^r$ lying above $\bar{\alpha} \in \mathbb{P}^{r-1}$ lets us identify the fibre of $\widetilde{M}(j)$ at $\bar{\alpha}$ with $M$. In terms of this identification the action of $\theta$ on the fibre at $\bar{\alpha}$ sends

$$m \otimes 1 \otimes 1 \in M \otimes_k k[U_\ell] \otimes_{k[U_\ell]} k \cong M$$

to

$$\begin{aligned} \sum_i X_i m \otimes Y_i \otimes 1 &= \sum_i X_i m \otimes 1 \otimes \lambda_i \\ &= \sum_i \lambda_i X_i m \otimes 1 \otimes 1 \\ &= X_\alpha m \otimes 1 \otimes 1. \end{aligned}$$

We summarise in a theorem.

**Theorem 8.3.1** *Let $M$ be a $kE$-module and let $\theta\colon \widetilde{M}(j) \to \widetilde{M}(j+1)$ be the operator defined in Section 8.2. If $\bar{\alpha} = (\lambda_1 : \cdots : \lambda_r) \in \mathbb{P}^{r-1}$ then a choice of $\alpha = (\lambda_1, \dots, \lambda_r) \in \mathbb{A}^r$ lying over $\bar{\alpha}$ gives an identification of the fibre of $\widetilde{M}(j)$ at $\bar{\alpha}$ with $M$. After making this identification, the action of $\theta$ on the fibre at $\bar{\alpha}$ is given by the action of $X_\alpha$ on $M$.* □

**Remark 8.3.2** The same technique can be used to find the action on the fibre at the generic point for an irreducible subvariety of $\mathbb{P}^{r-1}$. Let $\bar{\alpha}$ be the generic point for a homogeneous prime ideal $\mathfrak{p} \subseteq k[Y_1, \dots, Y_r]$. As in Section 3.2, we get a closed point $\bar{\beta} \in \mathbb{P}^{r-1}_{k(\bar{\alpha})}$ whose image is $\bar{\alpha}$. The fibre of $\widetilde{M}(j)$ at $\bar{\alpha}$ is the same as the fibre of the bundle $\widetilde{M \otimes_k k(\bar{\alpha})}(j)$ at $\bar{\beta}$, namely $M \otimes_k k(\bar{\alpha})$. Lift $\bar{\beta}$ to an element $\beta = (\lambda_1, \ldots, \lambda_r) \in \mathbb{A}^r_{k(\bar{\alpha})}$. Then the action of $\theta$ on the fibre of $\widetilde{M}(j)$ at $\bar{\alpha}$ is given by the action of $X_\beta = \lambda_1 X_1 + \cdots + \lambda_r X_r$ on $M \otimes_k k(\bar{\alpha})$. By abuse of notation we shall write $X_\alpha$ for this $X_\beta$.

## 8.4 The Functors $\mathcal{F}_i$ and $\mathcal{F}_{i,j}$

We define functors $\mathcal{F}_{i,j}$ for $0 \le j < i \le p$, from finitely generated $kE$-modules (or nilvarieties) $M$ to coherent sheaves on $\mathbb{P}^{r-1}$, by taking the following subquotients of $\widetilde{M}$:

$$\mathcal{F}_{i,j}(M) = \frac{\operatorname{Ker}\theta_M^{j+1} \cap \operatorname{Im}\theta_M^{i-j-1}}{(\operatorname{Ker}\theta_M^{j+1} \cap \operatorname{Im}\theta_M^{i-j}) + (\operatorname{Ker}\theta_M^{j} \cap \operatorname{Im}\theta_M^{i-j-1})}.$$

We then define

$$\mathcal{F}_i(M) = \mathcal{F}_{i,0}(M) = \frac{\operatorname{Ker}\theta_M \cap \operatorname{Im}\theta_M^{i-1}}{\operatorname{Ker}\theta_M \cap \operatorname{Im}\theta_M^{i}}.$$

For a point $0 \ne \alpha \in \mathbb{A}^r$ with residue field $k(\bar{\alpha})$ and the corresponding operator

$$X_\alpha \colon M \otimes_k k(\bar{\alpha}) \to M \otimes_k k(\bar{\alpha}),$$

we also define

$$\mathcal{F}_{i,\alpha}(M) = \frac{\operatorname{Ker} X_\alpha \cap \operatorname{Im} X_\alpha^{i-1}}{\operatorname{Ker} X_\alpha \cap \operatorname{Im} X_\alpha^{i}}.$$

Note that $\mathcal{F}_{i,\alpha}(M)$ only depends on the image $\bar{\alpha} \in \mathbb{P}^{r-1}$ of $\alpha \in \mathbb{A}^r$.

**Example 8.4.1** Suppose that $M$ has the constant image property. Then $\operatorname{Im}\theta^i = \widetilde{M_i}$ where $M_i = \operatorname{Rad}^i M$. Thus

$$\mathcal{F}_i(M) \cong \frac{(\operatorname{Ker}\theta \cap \operatorname{Im}\theta^{i-1}) + \operatorname{Im}\theta^i}{\operatorname{Im}\theta^i} = \frac{(\operatorname{Ker}\theta \cap \widetilde{M_{i-1}}) + \widetilde{M_i}}{\widetilde{M_i}}$$

can be thought of as the kernel of the map induced by $\theta$ from $\widetilde{M_{i-1}/M_i}$ to $\widetilde{M_i/M_{i+1}}$. So we have a short exact sequence

$$0 \to \mathcal{F}_i(M) \to \mathcal{O}^{\oplus a_{i-1}} \xrightarrow{\theta} \mathcal{O}(1)^{\oplus a_i} \to 0$$

where $a_i = \dim(\operatorname{Rad}^i M/\operatorname{Rad}^{i+1} M)$. Thus the Chern polynomials are given by

$$c(\mathcal{F}_i(M)) = \frac{1}{(1+h)^{a_i}}.$$

**Example 8.4.2** Suppose that $M$ has the constant kernel property. Then $\operatorname{Ker}\theta = \widetilde{\operatorname{Soc} M}$ is a direct sum of copies of $\mathcal{O}$, and $\operatorname{Ker}\theta \cap \operatorname{Im}\theta^i = \theta^i(\widetilde{M_{i+1}/M_i})$ is a direct sum of copies of $\mathcal{O}(-i)$, where $M_i = \operatorname{Soc}^i M$. So there is a short exact sequence

$$0 \to \mathcal{O}(-i)^{\oplus b_{i+1}} \xrightarrow{\theta} \mathcal{O}(-i+1)^{\oplus b_i} \to \mathcal{F}_i(M) \to 0$$

where $b_i = \dim(\mathsf{Soc}^i M/\mathsf{Soc}^{i-1} M)$. Thus the Chern polynomials are given by

$$c(\mathcal{F}_i(M)) = \frac{(1-(i-1)h)^{b_i}}{(1-ih)^{b_{i+1}}}.$$

The relationship between these two examples will be explained by Theorem 8.5.5.

In the next Proposition we show that functors $\mathcal{F}_i$ take modules of constant Jordan type to algebraic vector bundles and that they commute with specialisation.

**Proposition 8.4.3**

(1) *Let $M$ be a $kE$-module of constant Jordan type $[p]^{a_p}\dots[1]^{a_1}$. Then the sheaf $\mathcal{F}_i(M)$ is a vector bundle of rank $a_i$.*

(2) *Let $f\colon M \to N$ be a map of modules of constant Jordan type. For any point $\bar{\alpha} = (\lambda_1 : \cdots : \lambda_r) \in \mathbb{P}^{r-1}$ with residue field $k(\bar{\alpha})$ we have a commutative diagram*

$$\begin{array}{ccc} \mathcal{F}_i(M)\otimes_{\mathcal{O}} k(\bar{\alpha}) & \xrightarrow{\mathcal{F}_i(f)} & \mathcal{F}_i(N)\otimes_{\mathcal{O}} k(\bar{\alpha}) \\ \downarrow{\scriptstyle\cong} & & \downarrow{\scriptstyle\cong} \\ \mathcal{F}_{i,\alpha}(M) & \longrightarrow & \mathcal{F}_{i,\alpha}(N). \end{array}$$

*Proof.* (1) Since the module $M$ is fixed throughout the proof, we shall use $\theta$ to denote $\theta_M$.

Note that $\mathsf{Ker}\,\theta \cap \mathsf{Im}\,\theta^i = \mathsf{Ker}\,\{\theta\colon \mathsf{Im}\,\theta^i \to \mathsf{Im}\,\theta^{i+1}\}$. Hence, we have a short exact sequence

$$0 \longrightarrow \mathsf{Ker}\,\theta \cap \mathsf{Im}\,\theta^i \longrightarrow \mathsf{Im}\,\theta^i \xrightarrow{\ \theta\ } \mathsf{Im}\,\theta^{i+1} \longrightarrow 0\,. \tag{8.1}$$

Since $M$ has constant Jordan type, $\mathsf{Im}\,\theta^i$ is a vector bundle by Theorem 6.2.1. Therefore, specialisation of the sequence (8.1) at any point $\bar{\alpha} = (\lambda_1 : \cdots : \lambda_r)$ of $\mathbb{P}^{r-1}$ yields a short exact sequence of vector spaces

$$0 \to (\mathsf{Ker}\,\theta \cap \mathsf{Im}\,\theta^i)\otimes_{\mathcal{O}} k(\bar{\alpha}) \to \mathsf{Im}\,\theta^i \otimes_{\mathcal{O}} k(\bar{\alpha}) \to \mathsf{Im}\,\theta^{i+1}\otimes_{\mathcal{O}} k(\bar{\alpha}) \to 0. \tag{8.2}$$

By Theorem 6.2.1, $\mathsf{Im}\,\theta^i \otimes_{\mathcal{O}} k(\bar{\alpha}) \cong \mathsf{Im}\,\{X_\alpha^i : M \to M\}$. In particular, the dimension of fibres of $\mathsf{Im}\,\theta^i$ is constant and equals $\sum_{j=i+1}^{p} a_j(j-i)$. We can rewrite the sequence (8.2) as

$$0 \longrightarrow (\mathsf{Ker}\,\theta \cap \mathsf{Im}\,\theta^i)\otimes_{\mathcal{O}} k(\bar{\alpha}) \longrightarrow \mathsf{Im}\,X_\alpha^i \xrightarrow{\ X_\alpha\ } \mathsf{Im}\,X_\alpha^{i+1} \longrightarrow 0\,.$$

Hence the fibre of $\mathsf{Ker}\,\theta \cap \mathsf{Im}\,\theta^i$ at a point $\bar{\alpha}$ equals $\mathsf{Ker}\,X_\alpha \cap \mathsf{Im}\,X_\alpha^i$. In particular, $\mathsf{Ker}\,\theta \cap \mathsf{Im}\,\theta^i$ has fibres of constant dimension, equal to

$$\sum_{j=i+1}^{p} a_j(j-i) - \sum_{j=i+2}^{p} a_j(j-i-1) = \sum_{j=i+1}^{p} a_j.$$

Applying Theorem 6.2.1 again, we conclude that $\mathsf{Ker}\,\theta \cap \mathsf{Im}\,\theta^i$ is a vector bundle of rank $\sum_{j=i+1}^{p} a_j$.

Consider the short exact sequence that defines $\mathcal{F}_i(M)$:

$$0 \longrightarrow \mathsf{Ker}\,\theta \cap \mathsf{Im}\,\theta^i \longrightarrow \mathsf{Ker}\,\theta \cap \mathsf{Im}\,\theta^{i-1} \longrightarrow \mathcal{F}_i(M) \longrightarrow 0.$$

Specialising at $\bar{\alpha}$, we get

$$\begin{array}{ccccccc}
(\mathsf{Ker}\,\theta \cap \mathsf{Im}\,\theta^i) \otimes_{\mathcal{O}} k(\bar{\alpha}) & \longrightarrow & (\mathsf{Ker}\,\theta \cap \mathsf{Im}\,\theta^{i-1}) \otimes_{\mathcal{O}} k(\bar{\alpha}) & \longrightarrow & \mathcal{F}_i(M) \otimes_{\mathcal{O}} k(\bar{\alpha}) & \longrightarrow & 0 \\
\| & & \| & & \| & & \\
\mathsf{Ker}\,X_\alpha \cap \mathsf{Im}\,X_\alpha^i & \longrightarrow & \mathsf{Ker}\,X_\alpha \cap \mathsf{Im}\,X_\alpha^{i-1} & \longrightarrow & \mathcal{F}_i(M) \otimes_{\mathcal{O}} k(\bar{\alpha}) & \longrightarrow & 0
\end{array}$$

The first arrow of the bottom row is clearly an injection. Hence

$$\dim(\mathcal{F}_i(M) \otimes_{\mathcal{O}} k(\bar{\alpha})) = \sum_{j=i}^{p} a_j - \sum_{j=i+1}^{p} a_j = a_i$$

for any point $\bar{\alpha} \in \mathbb{P}^{r-1}$. Applying Theorem 6.2.1 again, we conclude that $\mathcal{F}_i(M)$ is locally free (of rank $a_i$).

Statement (2) follows immediately by applying the last diagram to both $M$ and $N$. □

The following proposition examines the exactness properties of the functors $\mathcal{F}_i$. Recall from Section 5.3 that $\mathsf{cJt}(kE)$ is the exact category whose objects are the modules of constant Jordan type and whose exact sequences are the locally split exact sequences of such modules. We write $\mathsf{Vec}(\mathbb{P}^{r-1})$ for the category of vector bundles over $\mathbb{P}^{r-1}$ and morphisms of vector bundles.

**Proposition 8.4.4** *The functors $\mathcal{F}_i \colon \mathsf{cJt}(kE) \to \mathsf{Vec}(\mathbb{P}^{r-1})$ are exact for $1 \leq i \leq p$.*

*Proof.* Let $1 \leq i \leq p$ and let

$$0 \to M_1 \to M_2 \to M_3 \to 0$$

be a locally split exact sequence of modules of constant Jordan type. Then the Jordan type of the middle term is the sum of the Jordan types of the end terms. Hence

$$\operatorname{rank} \mathcal{F}_i(M_2) = \operatorname{rank} \mathcal{F}_i(M_1) + \operatorname{rank} \mathcal{F}_i(M_3). \tag{8.3}$$

Consider the map $\mathcal{F}_i(M_2) \to \mathcal{F}_i(M_3)$. By Proposition 8.4.3 if $\bar{\alpha} \in \mathbb{P}^{r-1}$ then

$$\mathcal{F}_i(M_2) \otimes_{\mathcal{O}} k(\bar{\alpha}) \to \mathcal{F}_i(M_3) \otimes_{\mathcal{O}} k(\bar{\alpha})$$

is surjective for all $\bar{\alpha} \in \mathbb{P}^{r-1}$. So by Theorem 6.2.1, $\mathcal{F}_i(M_2) \to \mathcal{F}_i(M_3)$ is surjective. The same argument applied to $\mathcal{F}_i(M_1) \to \mathcal{F}_i(M_2)$ shows that it is injective. Finally, equation (8.3) implies exactness in the middle. □

**Remark 8.4.5** In Example 8.5.6 and Section 8.12, we shall see that the functors $\mathcal{F}_i$ do not in general take locally split exact sequences of modules of constant Jordan type to split exact sequences of vector bundles.

**Lemma 8.4.6** *Let $M$ be a $kE$-module. The trivial bundle $\widetilde{M}$ has a filtration in which the filtered quotients are isomorphic to $\mathcal{F}_{i,j}(M)$ for $0 \le j < i \le p$.*

*Proof.* We consider two filtrations on $\widetilde{M}$, by kernels and image:

$$0 \subseteq \mathsf{Ker}\,\theta_M \subseteq \ldots \subseteq \mathsf{Ker}\,\theta_M^{p-1} \subseteq \widetilde{M}$$
$$0 = \mathsf{Im}\,\theta_M^p \subseteq \mathsf{Im}\,\theta_M^{p-1} \subseteq \ldots \subseteq \mathsf{Im}\,\theta_M \subseteq \mathsf{Im}\,\theta_M^0 = \widetilde{M}.$$

To simplify notation, we set $\mathcal{K}_j = \mathsf{Ker}\,\theta_M^i$ and $\mathcal{I}_i = \mathsf{Im}\,\theta_M^{p-i}$. Using the standard refinement procedure, we refine the kernel filtration by the image filtration:

$$\mathcal{K}_j \subseteq (\mathcal{K}_{j+1} \cap \mathcal{I}_1) + \mathcal{K}_j \subseteq \ldots \subseteq (\mathcal{K}_{j+1} \cap \mathcal{I}_\ell) + \mathcal{K}_j \subseteq (\mathcal{K}_{j+1} \cap \mathcal{I}_{\ell+1}) + \mathcal{K}_j \subseteq \ldots \subseteq \mathcal{K}_{j+1}.$$

For any three sheaves $A, B, C$ with $B \subseteq A$, the second isomorphism theorem and the modular law imply that

$$\frac{A+C}{B+C} = \frac{A+(B+C)}{B+C} \cong \frac{A}{A \cap (B+C)} = \frac{A}{B+(A\cap C)}.$$

Hence, we can identify the subquotients of the refined kernel filtration above as

$$\frac{(\mathcal{K}_{j+1} \cap \mathcal{I}_{\ell+1}) + \mathcal{K}_j}{(\mathcal{K}_{j+1} \cap \mathcal{I}_\ell) + \mathcal{K}_j} \cong \frac{\mathcal{K}_{j+1} \cap \mathcal{I}_{\ell+1}}{(\mathcal{K}_{j+1} \cap \mathcal{I}_\ell) + (\mathcal{K}_j \cap \mathcal{I}_{\ell+1})}.$$

Setting $i = p - \ell + j$, we get that the latter quotient is precisely $\mathcal{F}_{i,j}(M)$ (note that when $j > \ell$, the corresponding subquotient is trivial). □

Essentially the same proof gives the following.

**Lemma 8.4.7** *Let $M$ be a $kE$-module.*

(1) *The subsheaf $\mathsf{Ker}\,(\theta_M^i) \subseteq \widetilde{M}$ has a filtration whose filtered quotients are isomorphic to $\mathcal{F}_{j,\ell}(M)$ with $\ell \le i$.*

(2) *The subsheaf $\mathsf{Im}\,(\theta_M^i) \subseteq \widetilde{M}$ has a filtration whose filtered quotients are isomorphic to $\mathcal{F}_{j,\ell}(M)$ with $j - \ell > i$.* □

**Lemma 8.4.8** *For $0 \le j < i$, we have $\mathcal{F}_{i,j}(M) \cong \mathcal{F}_i(M)(j)$.*

*Proof.* For $0 < j < i$, the map $\theta_M \colon \widetilde{M} \to \widetilde{M}(1)$ induces a natural isomorphism $\mathcal{F}_{i,j}(M) \to \mathcal{F}_{i,j-1}(M)(1)$. Since $\mathcal{F}_{i,0} = \mathcal{F}_i$, the result follows by induction on $j$. □

**Theorem 8.4.9** *Let $M$ be a $kE$-module. The trivial sheaf $\widetilde{M}$ has a filtration in which the filtered quotients are isomorphic to $\mathcal{F}_i(M)(j)$ for $0 \le j < i \le p$.*

*Proof.* This follows from Lemmas 8.4.6 and 8.4.8. □

**Lemma 8.4.10** *Suppose that $M$ has constant Jordan type $[p]^{a_p} \dots [1]^{a_1}$. Then the sheaf $\mathcal{F}_{i,j}(M)$ is a vector bundle of rank $a_i$.*

*Proof.* This follows from Proposition 8.4.3 and Lemma 8.4.8. □

I'd like to thank Julia Pevtsova for help with the proof of the following theorem.

**Theorem 8.4.11** *Let $M$ be a $kE$-module. Then the following are equivalent.*

(1) *$M$ has constant Jordan type.*
(2) *$\mathcal{F}_i(M)$ is a vector bundle for each $i$ with $1 \le i \le p$.*

*Proof.* If $M$ has constant Jordan type then by Proposition 8.4.3 (1) $\mathcal{F}_i(M)$ is a vector bundle for $1 \le i \le p$. Conversely, if $\mathcal{F}_i(M)$ is a vector bundle for $1 \le i \le p$ then also $\mathcal{F}_{j,\ell}(M)$ is a vector bundle for $0 \le \ell < j \le p$ by Lemma 8.4.8. By Lemma 8.4.7 (2), $\mathsf{Coker}\,(\theta^i_M)$ has a filtration in which the filtered subquotients are isomorphic to $\mathcal{F}_{j,\ell}(M)$ with $j - \ell \le i$. So by Proposition 6.2.5 $\mathsf{Coker}\,(\theta^i_M)$ is a vector bundle for $1 \le i \le p$. Since tensor product is right exact, for $0 \ne \alpha \in \mathbb{A}^r(k)$ we have

$$\mathsf{Coker}\,(\theta^i_M \otimes_{\mathcal{O}} k(\bar{\alpha})) \cong \mathsf{Coker}\,(\theta^i_M) \otimes_{\mathcal{O}} k(\bar{\alpha}) \cong \mathsf{Coker}\,(X^i_\alpha, M)$$

and so $X^i_\alpha$ has constant rank on $M$ for $1 \le i \le p$. Thus $M$ has constant Jordan type. □

**Remark 8.4.12** It follows from the proof of Proposition 8.4.3 that the subquotient functors $\mathcal{F}_{i,j}$ are linked as follows:

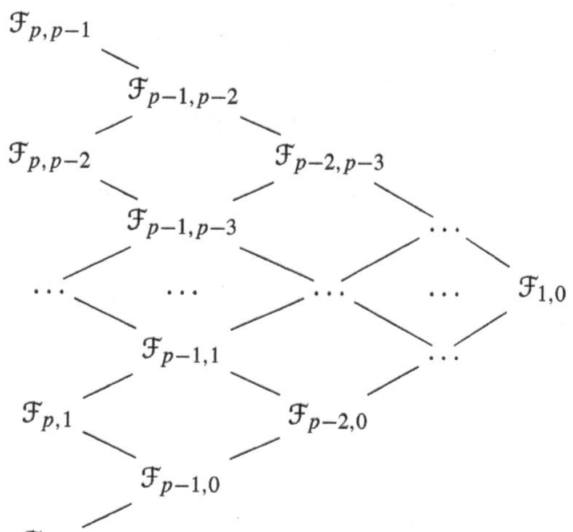


This should be compared with the corresponding diagram in Section 2.2.

The following is an immediate consequence of the definitions.

**Proposition 8.4.13** *If $M$ is a homogeneous $kE$-module then the vector bundles $\mathcal{F}_i(M)$ are homogeneous.* □

## 8.5 Twists and Syzygies

**Theorem 8.5.1** *Let $M$ be a finite-dimensional $kE$-module and let $1 \leq i \leq p-1$. Then there is a natural isomorphism*

$$\mathcal{F}_i(M)(-p+i) \cong \mathcal{F}_{p-i}(\Omega M).$$

*Proof.* Consider the diagram

$$\begin{array}{ccccccccc} 0 & \longrightarrow & \widetilde{\Omega M} & \longrightarrow & \widetilde{P_M} & \longrightarrow & \tilde{M} & \longrightarrow & 0 \\ & & \downarrow{\scriptstyle \theta_{\Omega M}} & & \downarrow{\scriptstyle \theta_{P_M}} & & \downarrow{\scriptstyle \theta_M} & & \\ 0 & \longrightarrow & \widetilde{\Omega M}(1) & \longrightarrow & \widetilde{P_M}(1) & \longrightarrow & \tilde{M}(1) & \longrightarrow & 0 \end{array} \tag{8.4}$$

where $P_M$ is a projective cover of $M$. Let

$$\delta\colon \mathsf{Ker}\,\theta_M \to \mathsf{Coker}\,\theta_{\Omega M}$$

be the switchback map. A simple diagram chase in conjunction with the fact that $\theta^p_{P_M} = 0$ yields that the restriction of $\delta$ to $\mathsf{Ker}\,\theta_M \cap \mathsf{Im}\,\theta^{i-1}_M$ lands in

$$\frac{\mathsf{Ker}\,\theta^{p-i}_{\Omega M}}{\mathsf{Ker}\,\theta^{p-i}_{\Omega M} \cap \mathsf{Im}\,\theta_{\Omega M}}(1).$$

Projecting the latter onto

$$\mathcal{F}_{p-i,p-i-1}(\Omega M)(1) = \frac{\mathsf{Ker}\,\theta^{p-i}_{\Omega M}}{\mathsf{Ker}\,\theta^{p-i-1}_{\Omega M} + \mathsf{Ker}\,\theta^{p-i}_{\Omega M} \cap \mathsf{Im}\,\theta_{\Omega M}}(1),$$

we get a map of bundles:

$$\delta'\colon \mathsf{Ker}\,\theta_M \cap \mathsf{Im}\,\theta^{i-1}_M \to \mathcal{F}_{p-i,p-i-1}(\Omega M)(1).$$

Since $\delta'$ evidently kills $\mathsf{Ker}\,\theta_M \cap \mathsf{Im}\,\theta^i_M$, we conclude that $\delta'$ factors through $\mathcal{F}_i(M)$. Hence, we have an induced map

$$\bar{\delta}\colon \mathcal{F}_i(M) \to \mathcal{F}_{p-i,p-i-1}(\Omega M)(1).$$

A simple block count shows that this is an isomorphism at each fiber. Hence by Theorem 6.2.1 this is an isomorphism of bundles. Thus using Lemma 8.4.8 (i.e., applying $\theta_{\Omega M}$ a further $p-i-1$ times), we have

$$\mathcal{F}_i(M) \cong \mathcal{F}_{p-i,p-i-1}(\Omega M)(1) \cong \mathcal{F}_{p-i}(\Omega M)(p-i).$$

Twisting by $\mathcal{O}(-p+i)$, we get the desired isomorphism.

Let $f\colon M \to N$ be a map of $kE$-modules. The naturality of the isomorphism

$$\mathcal{F}_i(M)(-p+i) \cong \mathcal{F}_{p-i}(\Omega M)$$

is equivalent to the commutativity of the diagram

$$\begin{array}{ccc} \mathcal{F}_i(M)(-p+i) & \xrightarrow{\mathcal{F}_i(f)(-p+i)} & \mathcal{F}_i(N)(-p+i) \\ \downarrow{\scriptstyle\cong} & & \downarrow{\scriptstyle\cong} \\ \mathcal{F}_{p-i}(\Omega M) & \xrightarrow{\mathcal{F}_{p-i}(\Omega f)} & \mathcal{F}_{p-i}(\Omega N). \end{array} \tag{8.5}$$

Note that although $\Omega M$ is only well defined up to projective summands, and $\Omega f$ is only well defined up to maps that factor through a projective, nonetheless $\mathcal{F}_{p-i}$ is zero on projective modules and the map $\mathcal{F}_{p-i}(\Omega f)$ is well defined. The commutativity follows from the construction of the map $\bar{\delta}$ and naturality of the isomorphism of Lemma 8.4.8. □

**Corollary 8.5.2** *Let $M$ be a finite-dimensional $kE$-module and let $1 \le i \le p-1$. Then $\mathcal{F}_i(\Omega^2 M) \cong \mathcal{F}_i(M)(-p)$.*

*Proof.* Apply the theorem twice. □

**Corollary 8.5.3** *We have $\mathcal{F}_1(\Omega^{2n}k) \cong \mathcal{O}(-np)$, and $\mathcal{F}_{p-1}(\Omega^{2n-1}k) \cong \mathcal{O}(1-np)$.*

*Proof.* This follows from the theorem and the corollary, using the isomorphism $\mathcal{F}_1(k) \cong \mathcal{O}$. □

**Remark 8.5.4** If $p = 2$ then Theorem 8.5.1 and Corollary 8.5.3 reduce to the statements that $\mathcal{F}_1(\Omega M) \cong \mathcal{F}_1(M)(-1)$ and $\mathcal{F}_1(\Omega^n k) \cong \mathcal{O}(-n)$.

**Theorem 8.5.5** *Let $M^*$ be the $k$-linear dual of $M$, as a $kE$-module. Then*

$$\mathcal{F}_i(M^*) \cong \mathcal{F}_i(M)^\vee(-i+1).$$

*Proof.* This follows from the more obvious isomorphism

$$\mathcal{F}_{i,i-1}(M^*) \cong \mathcal{F}_{i,0}(M)^\vee$$

together with Lemma 8.4.8. □

**Example 8.5.6** There is a locally split short exact sequence of modules of constant Jordan type for $\mathbb{Z}/2 \times \mathbb{Z}/2$ in characteristic two:

$$0 \to \Omega^2(k) \to \Omega(k) \oplus \Omega(k) \to k \to 0.$$

Applying $\mathcal{F}_1$ to this sequence gives the short exact sequence of sheaves

$$0 \to \mathcal{O}(-2) \to \mathcal{O}(-1) \oplus \mathcal{O}(-1) \to \mathcal{O} \to 0$$

on $\mathbb{P}^1$. This example shows that in Proposition 8.4.4, the functors $\mathcal{F}_i$ do not in general take locally split sequences of modules of constant Jordan type to split short exact sequences of vector bundles.

## 8.6 Chern Numbers of $\mathcal{F}_i(M)$

**Lemma 8.6.1** *Let $M$ be a $kE$-module of constant Jordan type.*

(i) *As elements of $\mathbb{Z}[h]/(h^r)$ we have*

$$\prod_{0\le j<i\le p} (1+jh)^{\operatorname{rank}\mathcal{F}_i(M)} c\left(\mathcal{F}_i(M), \frac{h}{1+jh}\right) = \prod_{0\le j<i\le p} c(\mathcal{F}_i(M)(j)) = 1.$$

(ii) *As elements of $\mathbb{Q}[h]/(h^r)$ we have*

$$\sum_{1\le i\le p} \left(\frac{1-e^{ih}}{1-e^h}\right) \mathsf{Ch}(\mathcal{F}_i(M)) = \sum_{0\le j<i\le p} \mathsf{Ch}(\mathcal{F}_i(M)(j)) = \dim M.$$

*Proof.* Combining Proposition 7.3.2 with Lemma 8.4.6, we have

$$\prod_{0\le j<i\le p} c(\mathcal{F}_{i,j}(M)) = c(\widetilde{M}) = 1,$$

$$\sum_{0\le j<i\le p} \mathsf{Ch}(\mathcal{F}_{i,j}(M)) = \mathsf{Ch}(\widetilde{M}) = \dim M.$$

Using Lemma 8.4.8 we can rewrite these as

$$\prod_{0\le j<i\le p} c(\mathcal{F}_i(M)(j)) = 1, \qquad \sum_{0\le j<i\le p} \mathsf{Ch}(\mathcal{F}_i(M)(j)) = \dim M.$$

The lemma now follows by applying the formula in Lemma 7.5.1 for the Chern polynomial and Chern character of the twists of a bundle. □

**Proposition 8.6.2** *Let $M$ be a $kE$-module of constant Jordan type. Then for $1 \le i \le p-1$ we have*

(i) $c(\mathcal{F}_{p-i}(\Omega(M), h)) = (1+(-p+i)h)^{\operatorname{rank}\mathcal{F}_i(M)} c\left(\mathcal{F}_i(M), \frac{h}{1+(-p+i)h}\right)$.
(ii) $\mathsf{Ch}(\mathcal{F}_{p-i}(\Omega(M))) = e^{(-p+i)h}\mathsf{Ch}(\mathcal{F}_i(M))$.
(iii) $c(\mathcal{F}_i(\Omega^2(M), h)) = (1-ph)^{\operatorname{rank}\mathcal{F}_i(M)} c\left(\mathcal{F}_i(M), \frac{h}{1-ph}\right)$.
(iv) $\mathsf{Ch}(\mathcal{F}_i(\Omega^2(M))) = e^{-ph}\mathsf{Ch}(\mathcal{F}_i(M))$.
(v) $c(\mathcal{F}_i(M^*), h) = (1-(i-1)h)^{\operatorname{rank}\mathcal{F}_i(M)} c\left(\mathcal{F}_i(M), \frac{-h}{1-(i-1)h}\right)$.
(vi) $\mathsf{Ch}(\mathcal{F}_i(M^*), h) = e^{-(i-1)h}\mathsf{Ch}(\mathcal{F}_i(M), -h)$.

*Proof.* Parts (i)-(iv) follow by combining Lemma 7.5.1 with Theorem 8.5.1 and Corollary 8.5.2. Parts (v) and (vi) follow by combining Lemmas 7.5.2 and 7.5.1 with Theorem 8.5.5:

$$\begin{aligned} c(\mathcal{F}_i(M^*), h) &= c(\mathcal{F}_i(M)^\vee(-i+1), h) \\ &= c(\mathcal{F}_i(M)(i-1), -h) \\ &= (1-(i-1)h)^{\operatorname{rank}\mathcal{F}_i(M)} c\left(\mathcal{F}_i(M), \frac{-h}{1-(i-1)h}\right) \end{aligned}$$

and

$$\begin{aligned} \mathsf{Ch}(\mathcal{F}_i(M^*), h) &= \mathsf{Ch}(\mathcal{F}_i(M)^\vee(-i+1), h) \\ &= \mathsf{Ch}(\mathcal{F}_i(M)(i-1), -h) \\ &= e^{-(i-1)h}\mathsf{Ch}(\mathcal{F}_i(M), -h). \end{aligned}$$

□

**Proposition 8.6.3** *Let $M$ be a $kE$-module of constant Jordan type. Then as elements of $\mathbb{Q}[h]/(h^r)$ we have*

$$\mathsf{Ch}(\mathcal{F}_p(\Omega^2(M))) = e^{-ph}\mathsf{Ch}(\mathcal{F}_p(M)) + \left(\frac{1-e^h}{1-e^{ph}}\right)(\dim\Omega^2(M) - e^{-ph}\dim M).$$

*Proof.* Applying Lemma 8.6.1 to $\Omega^2(M)$ and $M$, together with Lemma 7.5.1 and Corollary 8.5.2 we have

$$\begin{aligned} \left(\frac{1-e^{ph}}{1-e^h}\right)\mathsf{Ch}(\mathcal{F}_p(\Omega^2(M))) &= \sum_{j=0}^{p-1}\mathsf{Ch}(\mathcal{F}_p(\Omega^2(M)(j))) \\ &= \dim\Omega^2(M) - \sum_{0\le j<i<p}\mathsf{Ch}(\mathcal{F}_i(\Omega^2(M))(j)) \\ &= \dim\Omega^2(M) - \sum_{0\le j<i<p}\mathsf{Ch}(\mathcal{F}_i(M)(j-p)) \\ &= \dim\Omega^2(M) - e^{-ph}\sum_{0\le j<i<p}\mathsf{Ch}(\mathcal{F}_i(M)(j)) \\ &= \dim\Omega^2(M) - e^{-ph} \\ &\quad\times\left(\dim M - \sum_{j=0}^{p-1}\mathsf{Ch}(\mathcal{F}_p(M)(j))\right) \\ &= \dim\Omega^2(M) - e^{-ph} \\ &\quad\times\left(\dim M - \left(\frac{1-e^{ph}}{1-e^h}\right)\mathsf{Ch}(\mathcal{F}_p(M))\right). \end{aligned}$$

Now multiply both sides by $\dfrac{1-e^h}{1-e^{ph}}$. □

## 8.7 The Construction: $p = 2$

In this section and the next we give the construction used in the proof of Theorem 8.1.1 in the two cases $p = 2$ and $p$ odd. Then in Section 8.9 we give the details of the proof using this construction. Since $H^1(E, k)$ is the vector space dual of $J(kE)/J^2(kE)$, there are elements $y_1, \dots, y_r$ forming a vector space basis for $H^1(E, k)$ and corresponding to the linear functions $Y_1, \dots, Y_r$ on $J(kE)/J^2(kE)$ introduced in Section 8.2. Because of the difference in structure of the cohomology ring, we divide the discussion into two cases, according as $p = 2$ or $p$ is odd. In this section we deal with the case $p = 2$, and in the next section we explain the modifications needed for the slightly more difficult case where $p$ is odd.

So now we assume $p = 2$. In this case the cohomology ring $H^*(E, k)$ is the polynomial algebra $k[y_1, \dots, y_r]$. We define a $k$-algebra homomorphism

$$\rho\colon H^*(E, k) = k[y_1, \dots, y_r] \to k[Y_1, \dots, Y_r]$$

by $\rho(y_i) = Y_i$.

**Lemma 8.7.1** *If $\zeta \in H^n(E, k)$ is represented by a cocycle $\hat{\zeta}\colon \Omega^{n+j}k \to \Omega^j k$ (with $j \in \mathbb{Z}$) then the diagram*

$$\begin{array}{ccc} \mathcal{O}(-n-j) & \xrightarrow{\rho(\zeta)} & \mathcal{O}(-j) \\ \downarrow \cong & & \downarrow \cong \\ \mathcal{F}_1(\Omega^{n+j}k) & \xrightarrow{\mathcal{F}_1(\hat{\zeta})} & \mathcal{F}_1(\Omega^j k) \end{array}$$

*commutes.*

*Proof.* Consider $\hat{\zeta}\colon \Omega^n k \to k$. The commutative diagram (8.5) applied to $\hat{\zeta}$ and iterated $j$ times becomes

$$\begin{array}{ccc} \mathcal{F}_1(\Omega^n k)(-j) & \xrightarrow{\mathcal{F}_1(\hat{\zeta})(-j)} & \mathcal{F}_1(k)(-j) \\ \downarrow \cong & & \downarrow \cong \\ \mathcal{F}_1(\Omega^{n+j}k) & \xrightarrow{\mathcal{F}_1(\Omega^j \hat{\zeta})} & \mathcal{F}_1(\Omega^j k). \end{array}$$

Hence, it suffices to assume that $j = 0$. Additivity of the functor $\mathcal{F}_1$ allows us to assume that $\zeta$ is a monomial on generators $y_1, \dots, y_r$. Finally, since multiplication in cohomology corresponds to composition of the corresponding maps on Heller shifts of $k$, it suffices to prove our statement for a generator $\zeta = y_i$ of degree one.

In the case $j = 0$, $\zeta = y_i$, we need to show that the following diagram commutes (this is the diagram above twisted by $\mathcal{O}(1)$):

$$\begin{array}{ccc} \mathcal{F}_1(k) & \xrightarrow{\quad Y_i \quad} & \mathcal{F}_1(k)(1) \\ \Big\downarrow{\scriptstyle \simeq} & & \Big\| \\ \mathcal{F}_1(\Omega k)(1) & \xrightarrow{\mathcal{F}_1(\hat{y}_i)(1)} & \mathcal{F}_1(k)(1). \end{array} \tag{8.6}$$

Let $E_i$ be the subgroup of index two in $E$ such that $y_i$ is inflated from $E/E_i$ to $E$, namely the subgroup generated by all of $g_1, \ldots, g_r$ except $g_i$. Then $y_i$ represents the class of the extension

$$0 \to k \to M_i \to k \to 0$$

where $M_i$ is the permutation module on the cosets of $E_i$. This is a module of length two on which $X_1, \ldots, X_r$ act as zero except for $X_i$, which acts as a Jordan block of length two. We have a commutative diagram of $kE$-modules

$$\begin{array}{ccccccccc} 0 & \longrightarrow & \Omega k & \longrightarrow & P_0 & \longrightarrow & k & \longrightarrow & 0 \\ & & \Big\downarrow{\scriptstyle y_i} & & \Big\downarrow & & \Big\| & & \\ 0 & \longrightarrow & k & \longrightarrow & M_i & \longrightarrow & k & \longrightarrow & 0. \end{array} \tag{8.7}$$

The left vertical isomorphism $\overline{\delta}\colon \mathcal{F}_1(k) \xrightarrow{\sim} \mathcal{F}_1(\Omega k)(1)$ of the diagram (8.6) is given by the switchback map for the short exact sequence

$$0 \to \Omega k \to P_0 \to k \to 0$$

as in Diagram (8.4). Applying $\theta$ to the commutative diagram on free $\mathcal{O}$-modules induced by the module diagram (8.7), we get a commutative diagram

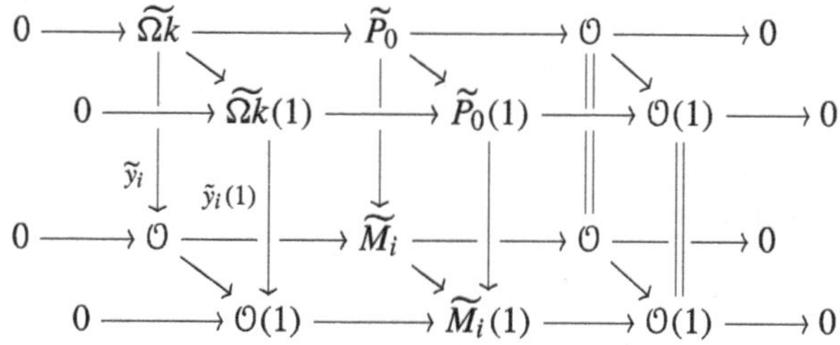

where all horizontal arrows going from the back to the front are given by the operator $\theta$ on the corresponding module. In the bottom layer of this diagram,

$$\begin{array}{ccccccccc} 0 & \longrightarrow & \mathcal{O} & \longrightarrow & \widetilde{M}_i & \longrightarrow & \mathcal{O} & \longrightarrow & 0 \\ & & \Big\downarrow{\scriptstyle \theta_k} & & \Big\downarrow{\scriptstyle \theta_{M_i}} & & \Big\downarrow{\scriptstyle \theta_k} & & \\ 0 & \longrightarrow & \mathcal{O}(1) & \longrightarrow & \widetilde{M}_i(1) & \longrightarrow & \mathcal{O}(1) & \longrightarrow & 0 \end{array}$$

the left- and right-hand maps $\theta_k$ are zero.

The map $\widetilde{y}_i : \widetilde{\Omega k} \to \mathcal{O}$ is induced by $y_i : \Omega k \to k$. To compute the composite $\mathcal{F}_1(\hat{y}_i)(1) \circ \overline{\delta}$ we first do the switchback map of the top layer and then push the result down via $\tilde{y}_i(1)$. Since the diagram is commutative, we can first push down via the identity map of the right vertical back arrow and then do the switchback of the bottom layer. Hence, the composite $\mathcal{F}_1(\hat{y}_i)(1) \circ \overline{\delta}$ is given by the switchback map $\overline{\delta} : \mathcal{O} \to \mathcal{O}(1)$ of the bottom layer. The action of $\theta_{M_i}$ is given by the matrix

$$\begin{pmatrix} 0 & 0 \\ Y_i & 0 \end{pmatrix}$$

(see Remark 8.2.1), and so the switchback map is multiplication by $Y_i$. □

## 8.8 The Construction: $p$ Odd

In this section we turn to construction used in the proof of Theorem 8.1.1 in the case where $p$ is odd. We write $\beta : H^1(E, k) \to H^2(E, k)$ for the Bockstein map, and we set $x_i = \beta(y_i)$. In terms of Massey products, this is given by $x_i = -\langle y_i, y_i, \ldots, y_i \rangle$ ($p$ terms).

The cohomology ring is a tensor product of an exterior algebra with a polynomial algebra:

$$H^*(E, k) \cong \Lambda(y_1, \ldots, y_r) \otimes_k k[x_1, \ldots, x_r],$$

see Section 1.8. We define a $k$-algebra homomorphism

$$\rho : k[x_1, \ldots, x_r] \to k[Y_1, \ldots, Y_r]$$

by $\rho(x_i) = Y_i^p$.

**Lemma 8.8.1** *Let $\zeta$ be a degree $n$ polynomial in $k[x_1, \ldots, x_r]$, regarded as an element of $H^{2n}(E, k)$. If $\zeta$ is represented by a cocycle $\hat{\zeta} : \Omega^{2(n+j)}k \to \Omega^{2j}k$ then the diagram*

$$\begin{array}{ccc} \mathcal{O}(-p(n+j)) & \xrightarrow{\rho(\zeta)} & \mathcal{O}(-pj) \\ \downarrow \cong & & \downarrow \cong \\ \mathcal{F}_1(\Omega^{2(n+j)}k) & \xrightarrow{\mathcal{F}_1(\hat{\zeta})} & \mathcal{F}_1(\Omega^{2j}k) \end{array}$$

*commutes.*

*Proof.* The proof is similar to the proof in the case $p = 2$, but more complicated. Again it suffices to treat the case where $\zeta = x_i$ and $j = 0$. In other words, we need to compute the composite $\mathcal{F}_1(\hat{x}_i) \circ f$ in the diagram

$$\begin{array}{ccc}
\mathcal{F}_1(k) & \xrightarrow{\;Y_i^p\;} & \mathcal{F}_1(k)(p) \\
{\scriptstyle f}\big\downarrow{\scriptstyle \simeq} & & \big\| \\
\mathcal{F}_1(\Omega^2 k)(p) & \xrightarrow{\;\mathcal{F}_1(\hat{x}_i)\;} & \mathcal{F}_1(k)(p),
\end{array}$$

where $f\colon \mathcal{F}_1(k) \to \mathcal{F}_1(\Omega^2 k)(p)$ is the isomorphism of Corollary 8.5.2. Let

$$0 \longrightarrow \Omega^2 k \longrightarrow P_1 \longrightarrow P_0 \longrightarrow k \longrightarrow 0$$

be a truncated projective resolution of $k$. Tracing through the proof of Theorem 8.5.1, we see that $f\colon \mathcal{F}_1(k) \to \mathcal{F}_1(\Omega^2 k)(p)$ is a composite of three maps:

(1) the switchback of the top two rows of the diagram (8.8) below which gives the isomorphism $\mathcal{F}(k) \to \mathcal{F}_{p-1,p-2}(\Omega k)(1)$,
(2) followed by the isomorphism

$$\theta_{\Omega k}^{p-2}\colon \mathcal{F}_{p-1,p-2}(\Omega k)(1) \to \mathcal{F}_{p-1}(\Omega k)(p-1)$$

of Lemma 8.4.8,
(3) followed by another switchback map, now for the bottom two rows of diagram (8.8), which gives the isomorphism $\mathcal{F}_{p-1}(\Omega k)(p-1) \simeq \mathcal{F}_1(\Omega^2 k)(p)$.

$$\begin{array}{ccccccccccc}
 & & & & 0 & \longrightarrow & \widetilde{\Omega k} & \longrightarrow & \widetilde{P_0} & \longrightarrow & \mathcal{O} \longrightarrow 0 \\
 & & & & & & \big\downarrow & & \big\downarrow & & \big\downarrow \\
 & & & & 0 & \longrightarrow & \widetilde{\Omega k}(1) & \longrightarrow & \widetilde{P_0}(1) & \longrightarrow & \mathcal{O}(1) \longrightarrow 0 \\
 & & & & & & \big\downarrow{\scriptstyle \theta_{\Omega k}^{p-2}} & & & & \\
0 & \longrightarrow & \widetilde{\Omega^2 k}(p-1) & \longrightarrow & \widetilde{P_1}(p-1) & \longrightarrow & \widetilde{\Omega k}(p-1) & \longrightarrow & 0 & & \\
 & & \big\downarrow & & \big\downarrow & & \big\downarrow & & & & \\
0 & \longrightarrow & \widetilde{\Omega^2 k}(p) & \longrightarrow & \widetilde{P_1}(p) & \longrightarrow & \widetilde{\Omega k}(p) & \longrightarrow & 0 & &
\end{array} \tag{8.8}$$

Let $E_i$ be the subgroup of index $p$ such that $x_i$ is inflated from $E/E_i$, namely the subgroup generated by all of $g_1, \ldots, g_r$ except for $g_i$. We let $M_i$ be the permutation module on the cosets of $E_i$. This is a module of length $p$ on which $X_1, \ldots, X_r$ act as zero except for $X_i$, which acts as a Jordan block of length $p$. Then $x_i$ represents the class of the 2-fold extension

$$0 \longrightarrow k \longrightarrow M_i \longrightarrow M_i \longrightarrow k \longrightarrow 0$$

where the middle map is multiplication by $X_i$. We construct a diagram analogous to (8.8) for this extension:

$$\begin{array}{ccccccccccccc}
 & & & & 0 & \to & \widetilde{N_i} & \to & \widetilde{M_i} & \to & \mathcal{O} & \to & 0 \\
 & & & & & & \downarrow & & \downarrow & & \downarrow & & \\
 & & & & 0 & \to & \widetilde{N_i}(1) & \to & \widetilde{M_i} & \to & \mathcal{O}(1) & \to & 0 \\
 & & & & & & \downarrow{\scriptstyle \theta_{N_i}^{p-2}} & & & & & & \\
0 & \to & \mathcal{O}(p-1) & \to & \widetilde{M_i}(p-1) & \to & \widetilde{N_i}(p-1) & \to & 0 & & & & \\
 & & \downarrow & & \downarrow & & \downarrow & & & & & & \\
0 & \to & \mathcal{O}(p) & \to & \widetilde{M_i}(p) & \to & \mathcal{O}(p) & \to & 0. & & & &
\end{array} \tag{8.9}$$

Here, $N_i = \mathrm{Im}\{X_i \colon M_i \to M_i\}$. Just as in the proof of Lemma 8.8.1, the module diagram

$$\begin{array}{ccccccccccc}
0 & \to & \Omega^2 k & \to & P_1 & \to & P_0 & \to & k & \to & 0 \\
 & & \downarrow{\scriptstyle x_i} & & \downarrow & & \downarrow & & \| & & \\
0 & \to & k & \to & M_i & \to & M_i & \to & k & \to & 0
\end{array}$$

induces a commutative diagram of vector bundles with (8.8) on top and (8.8) at the bottom. Arguing as in the proof of Lemma 8.7.1, we compute the composite $\mathcal{F}_1(\hat{x}_i) \circ f$ by first mapping the rightmost $\mathcal{O}$ of diagram (8.8) identically to the rightmost $\mathcal{O}$ of Diagram (8.8), and then applying our composite of a switchback, followed by $\theta_{N_i}^{p-2}$, followed by another switchback in the diagram (8.8). The maps $\theta_{M_i} \colon \widetilde{M_i} \to \widetilde{M_i}(1)$ and $\theta_{N_i} \colon \widetilde{N_i} \to \widetilde{N_i}(1)$ are simply multiplication by $Y_i$. Since the leftmost and rightmost vertical arrows in (8.8) are zero, to compute the composite of the three maps involved in diagram (8.8), we have to multiply first by $Y_i$, then by $Y_i^{p-2}$, then by $Y_i$ again. Hence, $\mathcal{F}_1(\hat{x}_i) \circ f = Y_i^p$. □

## 8.9 Proof of the Realisation Theorem

We are ready to prove the main theorem of this chapter, Theorem 8.1.1. We begin with a couple of lemmas. The following lemma will be applied to modules over $K[X_\alpha]/(X_\alpha^p)$, where projective and injective modules coincide.

**Lemma 8.9.1** *Let* $\phi = \begin{pmatrix} a & b \\ c & d \end{pmatrix} \colon A \oplus P \to B \oplus Q$ *be an injective map of modules, with* $a \colon A \to B$ *a split injection and* $P$ *an injective module. Then* $\phi$ *is a split injection.*

*Proof.* Let $B = B_1 \oplus B_2$, where $B_1 = \operatorname{Im} a$, and let

$$a = \begin{pmatrix} a_1 \\ 0 \end{pmatrix} : A \to B_1 \oplus B_2, \quad b = \begin{pmatrix} b_1 \\ b_2 \end{pmatrix} : P \to B_1 \oplus B_2.$$

With this notation, we have

$$\phi = \begin{pmatrix} a_1 & b_1 \\ 0 & b_2 \\ c & d \end{pmatrix} : A \oplus P \to B_1 \oplus B_2 \oplus Q.$$

Since $a_1$ is an isomorphism, there exist maps $u : B_1 \to Q$, $v : P \to A$ such that $c = ua_1$ and $b_1 = a_1 v$. Hence,

$$\begin{pmatrix} 1 & 0 & 0 \\ 0 & 1 & 0 \\ -u & 0 & 1 \end{pmatrix} \begin{pmatrix} a_1 & b_1 \\ 0 & b_2 \\ c & d \end{pmatrix} \begin{pmatrix} 1 & -v \\ 0 & 1 \end{pmatrix} = \begin{pmatrix} a_1 & 0 \\ 0 & b_2 \\ 0 & d - ub_1 \end{pmatrix} : A \oplus P \to B_1 \oplus B_2 \oplus Q.$$

By assumption, this map is injective. Hence, the map

$$\begin{pmatrix} b_2 \\ d - ub_1 \end{pmatrix} : P \to B_2 \oplus Q$$

is injective and therefore split. □

**Lemma 8.9.2** *Let $0 \to A \to B \to C \to 0$ be a short exact sequence of $kE$-modules, such that $A$ and $B$ have trivial stable constant Jordan type and $\mathcal{F}_1(A) \to \mathcal{F}_1(B)$ is injective. Then $C$ is of trivial stable constant Jordan type and $\mathcal{F}_1(C)$ is the cokernel of $\mathcal{F}_1(A) \to \mathcal{F}_1(B)$.*

*Proof.* Applying Lemma 8.9.1 to the restriction of the sequence to $K[X_\alpha]/(X_\alpha^p)$ for each $0 \neq \alpha \in \mathbb{A}^r(k)$, we conclude that this sequence is locally split. Hence, $C$ is a module of trivial stable constant Jordan type. By Proposition 8.4.4, applying $\mathcal{F}_1$ gives an exact sequence

$$0 \to \mathcal{F}_1(A) \to \mathcal{F}_1(B) \to \mathcal{F}_1(C) \to 0.$$

□

**Theorem 8.9.3** *Given any vector bundle $\mathcal{F}$ of rank $s$ on $\mathbb{P}^{r-1}$, there exists a finitely generated $kE$-module $M$ of stable constant Jordan type $[1]^s$ such that*

(i) *if $p = 2$, then $\mathcal{F}_1(M) \cong \mathcal{F}$.*
(ii) *if $p$ is odd, then $\mathcal{F}_1(M) \cong F^*(\mathcal{F})$, the inverse image of $\mathcal{F}$ along the Frobenius morphism $F : \mathbb{P}^{r-1} \to \mathbb{P}^{r-1}$.*

*Proof.* Given a vector bundle $\mathcal{F}$ on $\mathbb{P}^{r-1}$, using Corollary 6.12.5 of the Hilbert syzygy theorem we can form a resolution by sums of twists of the structure sheaf:

$$0 \to \bigoplus_{j=1}^{m_r} \mathcal{O}(a_{r,j}) \to \cdots \to \bigoplus_{j=1}^{m_1} \mathcal{O}(a_{1,j}) \to \bigoplus_{j=1}^{m_0} \mathcal{O}(a_{0,j}) \to \mathcal{F} \to 0. \tag{8.10}$$

Each of the maps in this resolution is a matrix whose entries are homogeneous polynomials in $Y_1, \dots, Y_r$ and each kernel is again a vector bundle. Replacing each $Y_i$ with $y_i$ if $p = 2$ and with $x_i$ if $p$ is odd gives matrices of cohomology elements which we may use to form a sequence of modules and homomorphisms, which takes the form

$$0 \to \bigoplus \Omega^{-\varepsilon a_{r,j}}(k) \to \cdots \to \bigoplus \Omega^{-\varepsilon a_{1,j}}(k) \to \bigoplus \Omega^{-\varepsilon a_{0,j}}(k) \qquad (8.11)$$

where $\varepsilon = 1$ if $p = 2$ and $\varepsilon = 2$ if $p$ is odd. We discussed these maps $x_i$ and $y_i$ in Section 1.8, and showed that they may be constructed in such a way that they strictly commute. So this sequence is a complex in the module category $\mathsf{mod}(kE)$. It therefore determines an object in the derived category $\mathsf{D}^b(kE)$. Using the functor $\mathcal{R}\colon \mathsf{D}^b(kE) \to \mathsf{stmod}(kE)$ (see Section 1.6) we get a module $M$ in $\mathsf{stmod}(kE)$. The claim is that $M$ has stable constant Jordan type $[1]^s$, and $\mathcal{F}_1(M)$ is isomorphic to $\mathcal{F}$ if $p = 2$ and to $F^*(\mathcal{F})$ if $p$ is odd.

If $p = 2$ it follows from Lemma 8.7.1 that applying $\mathcal{F}_1$ to (8.11) gives all but the augmentation part of the sequence (8.10). If $p$ is odd, it follows from Lemma 8.8.1 that it gives a modification of this sequence where each of the original matrices has been altered by replacing the variables $Y_i$ by their $p^{\text{th}}$ powers. This is the inverse image of the original resolution along the Frobenius map $F\colon \mathbb{P}^{r-1} \to \mathbb{P}^{r-1}$, and so it is a resolution of the inverse image bundle $F^*(\mathcal{F})$.

Let $M_i$ be the image in the stable category of the complex obtained by removing the terms of degrees from 0 to $i - 1$ from the complex (8.11). So for example $M_r$ is just the left-hand term in this complex and $M_0$ is equal to $M$. Since the functor $\mathcal{R}$ preserves triangles, we obtain triangles

$$M_{i+1} \to \bigoplus \Omega^{-\varepsilon a_{i,j}}(k) \to M_i$$

in $\mathsf{stmod}(kE)$. Thus we have constructed a diagram

$$\begin{array}{l} \bigoplus \Omega^{-\varepsilon a_{r,j}}(k) \to \bigoplus \Omega^{-\varepsilon a_{r-1,j}}(k) \longrightarrow \bigoplus \Omega^{-\varepsilon a_{r-2,j}}(k) \longrightarrow \cdots\cdots \longrightarrow \bigoplus \Omega^{-\varepsilon a_{0,j}}(k) \\ \qquad\qquad M_{r-1} \qquad\qquad M_{r-2} \qquad\qquad M_1 \qquad\qquad M_0 \end{array}$$

Applying $\mathcal{F}_1$ to this diagram, we see by downwards induction on $i$ and Lemma 8.9.2 that $\mathcal{F}_1(M_i)$ is the $i$th kernel in (8.10) if $p = 2$, and its Frobenius twist if $p$ is odd, so that

$$\mathcal{F}_1(M_{i+1}) \to \mathcal{F}_1(\bigoplus \Omega^{-\varepsilon a_{i,j}}(k))$$

is injective with cokernel $\mathcal{F}_1(M_i)$. Thus all the $M_i$ are modules of trivial stable constant Jordan type. Finally, applying this with $i = 0$, we see that $\mathcal{F}_1(M) \cong \mathcal{F}$ if $p = 2$ and $\mathcal{F}_1(M) \cong F^*(\mathcal{F})$ if $p$ is odd. □

**Remark 8.9.4** We shall return to this theorem in Chapter 11, where we shall interpret the construction of Section 8.8 in terms of the Orlov correspondence.

## 8.10 Functoriality

Let $\mathcal{F}$ and $\mathcal{F}'$ be vector bundles on $\mathbb{P}^{r-1}$ and $f : \mathcal{F} \to \mathcal{F}'$ be a sheaf homomorphism. Let $M$ and $M'$ be the modules constructed in the proof of Theorem 8.9.3 realising $\mathcal{F}$ and $\mathcal{F}'$ if $p = 2$ and realising $F^*(\mathcal{F})$ and $F^*(\mathcal{F}')$ if $p$ is odd.

By Corollary 6.12.5, we obtain a morphism of resolutions

$$\begin{array}{ccccccccccc} 0 & \longrightarrow & \bigoplus_{j=1}^{m_r} \mathcal{O}(a_{r,j}) & \longrightarrow \cdots \longrightarrow & \bigoplus_{j=1}^{m_1} \mathcal{O}(a_{1,j}) & \longrightarrow & \bigoplus_{j=1}^{m_0} \mathcal{O}(a_{0,j}) & \longrightarrow & \mathcal{F} & \longrightarrow & 0. \\ & & \downarrow & & \downarrow & & \downarrow & & \downarrow{\scriptstyle f} & & \\ 0 & \longrightarrow & \bigoplus_{j=1}^{m_r} \mathcal{O}(a'_{r,j}) & \longrightarrow \cdots \longrightarrow & \bigoplus_{j=1}^{m_1} \mathcal{O}(a'_{1,j}) & \longrightarrow & \bigoplus_{j=1}^{m_0} \mathcal{O}(a'_{0,j}) & \longrightarrow & \mathcal{F}' & \longrightarrow & 0. \end{array} \tag{8.12}$$

Each of the vertical maps in this diagram is a matrix whose entries are homogeneous polynomials in $Y_1, \ldots, Y_r$. As in the proof of Theorem 8.9.3, we replace $Y_i$ with $y_i$ ($p = 2$) or $x_i$ ($p$ odd) to obtain the vertical maps in the following commutative diagram.

$$\begin{array}{ccccccc} 0 & \longrightarrow & \bigoplus \Omega^{-\varepsilon a_{r,j}}(k) & \longrightarrow \ldots \longrightarrow & \bigoplus \Omega^{-\varepsilon a_{1,j}}(k) & \longrightarrow & \bigoplus \Omega^{-\varepsilon a_{0,j}}(k) \\ & & \downarrow & & \downarrow & & \downarrow \\ 0 & \longrightarrow & \bigoplus \Omega^{-\varepsilon a'_{r,j}}(k) & \longrightarrow \ldots \longrightarrow & \bigoplus \Omega^{-\varepsilon a'_{1,j}}(k) & \longrightarrow & \bigoplus \Omega^{-\varepsilon a'_{0,j}}(k) \end{array} \tag{8.13}$$

This gives us a morphism in $\mathsf{D}^b(kE)$. Applying the functor $\mathcal{R}\colon \mathsf{D}^b(kE) \to \mathsf{stmod}(kE)$, we obtain a module homomorphism from $M$ to $M'$.

A chain homotopy between maps of complexes (8.10) lying over $f : \mathcal{F} \to \mathcal{F}'$ is again given by matrices of polynomials, and therefore gives rise to a chain homotopy between maps in the Diagram (8.10). Therefore different choices of chain maps in (8.10) induce the same map in $\mathsf{D}^b(kE)$ and hence the same map in $\mathsf{stmod}(kE)$.

It follows that the construction in Theorem 8.9.3 gives a functor

$$\mathcal{G}\colon \mathsf{Vec}(\mathbb{P}^{r-1}) \to \mathsf{stmod}(kE)$$

whose image consists of modules of constant Jordan type.

## 8.11 Tensor Products

Similar techniques to those used in Section 8.5 prove the following theorems.

**Theorem 8.11.1** *Let $M$ and $N$ be $kE$-modules of constant Jordan type. Then*

$$\mathcal{F}_1(M \otimes_k N) \cong \bigoplus_{i=1}^{p-1} (\mathcal{F}_i(M) \otimes_{\mathcal{O}} \mathcal{F}_i(N))(i-1),$$

$$\mathcal{F}_2(M \otimes_k N) \cong \bigoplus_{i=1}^{p-2} (\mathcal{F}_i(M) \otimes_{\mathcal{O}} \mathcal{F}_{i+1}(N) \,\oplus\, \mathcal{F}_{i+1}(M) \otimes_{\mathcal{O}} \mathcal{F}_i(N))(i-1)$$

*and more generally for $1 \le \ell \le p-1$ we have*

$$\mathcal{F}_\ell(M \otimes_k N) \cong \bigoplus_{i=1}^{p-\ell} \bigoplus_{j=0}^{\ell-1} (\mathcal{F}_{i+j}(M) \otimes_{\mathcal{O}} \mathcal{F}_{i+\ell-1-j}(N))(i-1)$$

*cf.* (2.1). □

It is harder to write down a formula for $\mathcal{F}_p(M \otimes_k N)$, and we do not attempt it here. Symmetric and exterior powers may be treated in the same way. For example we have the following theorem.

**Theorem 8.11.2** *Let $p$ be odd, and let $M$ be a $kE$-module of constant Jordan type. Then*

$$\mathcal{F}_1(\Lambda^2(M)) \cong \bigoplus_{i=1}^{p-1} \Lambda^2(\mathcal{F}_i(M))(i-1),$$

$$\mathcal{F}_1(S^2(M)) \cong \bigoplus_{i=1}^{p-1} S^2(\mathcal{F}_i(M))(i-1),$$

$$\mathcal{F}_2(\Lambda^2(M)) \cong \mathcal{F}_2(S^2(M)) \cong \bigoplus_{i=1}^{p-2} (\mathcal{F}_i(M) \otimes_{\mathcal{O}} \mathcal{F}_{i+1}(M))(i-1)$$

*and more generally for $1 \le \ell \le p-1$ we have*

$$\mathcal{F}_\ell(\Lambda^2(M)) \cong \begin{cases} \displaystyle\bigoplus_{i=1}^{p-\ell} \Big(\Lambda^2(\mathcal{F}_i(M)) \oplus \bigoplus_{j=0}^{(\ell-1)/2} (\mathcal{F}_{i+j}(M) \otimes_{\mathcal{O}} \mathcal{F}_{i+\ell-1-j}(M))\Big)(i-1) & \ell \text{ odd} \\ \displaystyle\bigoplus_{i=1}^{p-\ell} \bigoplus_{j=0}^{(\ell-1)/2} (\mathcal{F}_{i+j}(M) \otimes_{\mathcal{O}} \mathcal{F}_{i+\ell-1-j}(M))(i-1) & \ell \text{ even.} \end{cases}$$

*The formula for $\mathcal{F}_\ell(S^2(M))$ is the same with $\Lambda^2$ replaced by $S^2$ on the right-hand side for $\ell$ odd.*

Higher symmetric and exterior powers, and more general Schur functors are much more complicated.

## 8.12 Negative Tate Cohomology

In this section we examine in detail the locally split sequences of modules of constant Jordan type coming from negative Tate cohomology that were briefly discussed in Example 5.3.5, and the short exact sequences of vector bundles coming from them (Proposition 8.4.4).

**Proposition 8.12.1** *Let $E$ be an elementary abelian $p$-group, and let $E'$ be a shifted subgroup of strictly smaller rank. If $\zeta \in \hat{H}^{-n}(E,k)$ with $n > 0$ then the restriction of $\zeta$ to $\hat{H}^{-n}(E',k)$ is equal to zero.*

*Proof.* By a change of basis, we may assume that $E'$ is a subgroup of $E$. Since $E$ is elementary abelian, there exists another subgroup $E''$ such that $E = E' \times E''$. It follows from the Künneth theorem for group cohomology that if $\eta' \in H^{n-1}(E',k)$ then there exists $\eta \in H^{n-1}(E,k)$ such that $\mathrm{res}_{E,E'}(\eta) = \eta'$. Taking transfer we get

$$\mathrm{Tr}_{E',E}(\eta') = \mathrm{Tr}_{E',E}\mathrm{res}_{E,E'}(\eta) = |E : E'|.\eta = 0.$$

So transfer from $E'$ to $E$ in cohomology is zero. Since

$$\mathrm{res}_{E,E'} \colon \hat{H}^{-n}(E,k) \to \hat{H}^{-n}(E',k)$$

is Tate dual to this transfer map (see Section 1.5), it follows that it is also the zero map. □

Now suppose that $E$ has rank $r \geq 2$. Let $j \in \mathbb{Z}$ and let $n > 0$. If

$$0 \neq \zeta \in \hat{H}^{-n}(E,k) \cong \mathrm{Ext}^1_{kE}(\Omega^{j-n-1}(k), \Omega^j(k))$$

then we have a corresponding short exact sequence

$$0 \to \Omega^j(k) \to M \to \Omega^{j-n-1}(k) \to 0. \tag{8.14}$$

Let us assume that if $p > 2$ then $j$ is even and $n$ is odd, so that both ends of this short exact sequence have stable constant Jordan type $[1]$. By Proposition 8.12.1, this sequence splits on restriction to every cyclic shifted subgroup of $E$. So it is a locally split short exact sequence of modules of constant Jordan type (see Definition 5.3.3), and $M$ has stable constant Jordan type $[1]^2$.

Applying the functor $\mathcal{F}_1$, we obtain a short exact sequence of vector bundles

$$0 \to \mathcal{F}_1(\Omega^j(k)) \to \mathcal{F}_1(M) \to \mathcal{F}_1(\Omega^{j-n-1}(k)) \to 0. \tag{8.15}$$

**Theorem 8.12.2** *If $r \geq 3$ then the sequence* (8.15) *splits.*

*Proof.* By Corollary 8.5.2 and the remark following it, the vector bundles at the ends of this sequence are twists of the structure sheaf. If $r \geq 3$ then by Corollary 6.9.2 for any values of $a$ and $b$ we have

$$\mathrm{Ext}^1_{\mathbb{P}^{r-1}}(\mathcal{O}(a), \mathcal{O}(b)) \cong H^1(\mathbb{P}^{r-1}, \mathcal{O}(b-a)) = 0.$$

It follows that the sequence splits. □

For $r = 2$ the situation is more complicated. If $p = 2$ then the sequence (8.15) never splits. The sequence (8.14) takes the form

$$0 \to \Omega^j(k) \xrightarrow{\binom{\eta_2}{-\eta_1}} \Omega^{j-a}(k) \oplus \Omega^{j-b}(k) \oplus (\text{projective}) \xrightarrow{(\eta_1,\eta_2)} \Omega^{j-n-1}(k) \to 0 \tag{8.16}$$

where $a + b = n + 1$. The ideal generated by $\eta_1$ and $\eta_2$ is the annihilator of $\zeta$ in $H^*(E, k)$. In the language of [47], the element $\zeta$ is the last survivor for the sequence $\eta_1, \eta_2$.

Finally we discuss the case $r = 2$, $p$ odd. If $\zeta \in \hat{H}^{-n}(E, k)$ has zero Tate pairing with every nilpotent element of $H^{n-1}(E, k)$ then the sequence (8.15) splits. On the other hand, if there is a nilpotent element with which $\zeta$ has non-zero pairing then the annihilator of $\zeta$ is again generated by a regular sequence $\eta_1, \eta_2$, and the situation is as described in the $p = 2$ case. The extension class representing $\eta$ takes the form (8.16) with $a$ and $b$ even.

As a concrete example, let $r = 2$ and $p$ be odd so that

$$H^*(E, k) = \Lambda(y_1, y_2) \otimes_k k[x_1, x_2]$$

and let $\zeta$ be an element which has a non-zero pairing with $y_1 y_2$. Then $\eta_1, \eta_2$ are $x_1, x_2$ and the sequence takes the form

$$0 \to \Omega^4(k) \xrightarrow{\binom{x_2}{-x_1}} \Omega^2(k) \oplus \Omega^2(k) \xrightarrow{(x_1,x_2)} k \to 0.$$

The sequence (8.15) takes the form

$$0 \to \mathcal{O}(-2p) \to \mathcal{O}(-p) \oplus \mathcal{O}(-p) \to \mathcal{O} \to 0.$$

## 8.13 The BGG Correspondence

The construction of the operator $\theta$ given in Section 8.2 is based on the Bernstein–Gelfand–Gelfand (or BGG) correspondence. Since there are a number of different versions of this correspondence, in this section we give a brief summary and put the construction in context. For the purposes of this book, it is not necessary to understand this section, but the reader may find it helpful. Further information on the BGG correspondence may be found in the papers [22, 51, 55, 91, 111, 121], and the appendix of [202].

The BGG correspondence for exterior algebras is as follows. Let $\Lambda = \Lambda(X_1, \dots, X_r)$ be an exterior algebra on generators $X_1, \dots, X_r$ in degree 1, and let $S = k[Y_1, \dots, Y_r]$ be a polynomial algebra on generators $Y_1, \dots, Y_r$ of degree 1. Then there is an equivalence of triangulated categories

$$\mathsf{D}^b_{\mathrm{gr}}(\Lambda) \simeq \mathsf{D}^b_{\mathrm{gr}}(S). \tag{8.17}$$

between the bounded derived categories of graded $\Lambda$-modules and graded $S$-modules. Objects in these categories have two gradings: an internal grading coming from the gradings on $\Lambda$ and $S$, and a homological grading in the complexes. In this equivalence, the internal shift of degree 1 in $\mathsf{D}^b_{\mathrm{gr}}(\Lambda)$ corresponds to the composite of the internal shift of degree $-1$ and a homological shift of degree 1 in $\mathsf{D}^b_{\mathrm{gr}}(S)$.

Inside $\mathsf{D}^b_{\mathrm{gr}}(\Lambda)$, $\mathsf{perf}_{\mathrm{gr}}(\Lambda)$ is the thick subcategory of perfect complexes, i.e., objects isomorphic to bounded complexes of finitely generated projective modules. The quotient $\mathsf{D}^b_{\mathrm{gr}}(\Lambda)/\mathsf{perf}_{\mathrm{gr}}(\Lambda)$ is equivalent to the stable module category $\mathsf{stmod}_{\mathrm{gr}}(\Lambda)$ of finitely generated graded $\Lambda$-modules, see Section 1.6.

Inside $\mathsf{D}^b_{\mathrm{gr}}(S)$, let $\mathsf{fl}_{\mathrm{gr}}(S)$ be the thick subcategory of objects isomorphic to bounded complexes of finite length graded $S$-modules. Then $\mathsf{D}^b_{\mathrm{gr}}(S)/\mathsf{fl}_{\mathrm{gr}}(S)$ is equivalent to $\mathsf{D}^b(\mathsf{Coh}(\mathbb{P}^{r-1}))$, the bounded derived category of coherent sheaves on projective space, see Theorem 3.9.12.

Under the equivalence (8.17), $\mathsf{perf}_{\mathrm{gr}}(\Lambda)$ corresponds to $\mathsf{fl}_{\mathrm{gr}}(S)$, and so we obtain the BGG equivalence of categories

$$\mathsf{stmod}_{\mathrm{gr}}(\Lambda) \simeq \mathsf{D}^b(\mathsf{Coh}(\mathbb{P}^{r-1})). \tag{8.18}$$

There is a modified version of BGG, which is essentially the construction described in [51], which goes as follows. Let $\mathsf{D}^b_{\mathrm{dg}}(S)$ be the bounded derived category of differential graded $S$-modules. This may be described as follows. We regard $S$ as a differential graded algebra with zero differential. An object in $\mathsf{D}^b_{\mathrm{dg}}(S)$ is a differential graded $S$-module, where the differential has degree 1 and where the underlying $S$-module is finitely generated. The arrows are obtained by taking homotopy classes of maps of differential graded modules, and inverting quasi-isomorphisms.

Let $\mathsf{D}^b(\Lambda)$ be the bounded derived category of ungraded $\Lambda$-modules. Then there is an equivalence of triangulated categories

$$\mathsf{D}^b(\Lambda) \simeq \mathsf{D}^b_{\mathrm{dg}}(S). \tag{8.19}$$

Again this takes the thick subcategory $\mathsf{perf}(\Lambda)$ of $\mathsf{D}^b(\Lambda)$ to the thick subcategory $\mathsf{fl}_{\mathrm{dg}}(S)$ of $\mathsf{D}^b_{\mathrm{dg}}(S)$ consisting of objects isomorphic to differential graded modules of finite length. So we obtain a BGG correspondence

$$\mathsf{stmod}(\Lambda) \simeq \mathsf{D}^b_{\mathrm{dg}}(S)/\mathsf{fl}_{\mathrm{dg}}(S). \tag{8.20}$$

There is an obvious functor

$$\mathsf{Tot}\colon \mathsf{D}^b_{\mathrm{gr}}(S) \to \mathsf{D}^b_{\mathrm{dg}}(S),$$

which takes a complex of graded $S$-modules and applies to total complex construction to obtain differential graded $S$-modules. Under the BGG correspondence, this corresponds to the forgetful functor

$$\mathsf{stmod}_{\mathrm{gr}}(\Lambda) \to \mathsf{stmod}(\Lambda)$$

which forgets the grading.

Finally, there is a functor from $\mathsf{D}^b_{\mathrm{dg}}(S)$ to finitely generated graded $S$-modules, which takes a differential graded $S$-module to its cohomology. This passes down to a functor

$$H^*\colon \mathsf{D}^b_{\mathrm{dg}}(S)/\mathsf{fl}_{\mathrm{dg}}(S) \to \mathsf{Coh}(\mathbb{P}^{r-1}).$$

Composing this with the equivalence (8.20) gives the functor $\mathcal{F}_1\colon \mathsf{stmod}(\Lambda) \to \mathsf{Coh}(\mathbb{P}^{r-1})$ of Section 8.4.

Putting all these together, we obtain the following diagram of categories and functors:

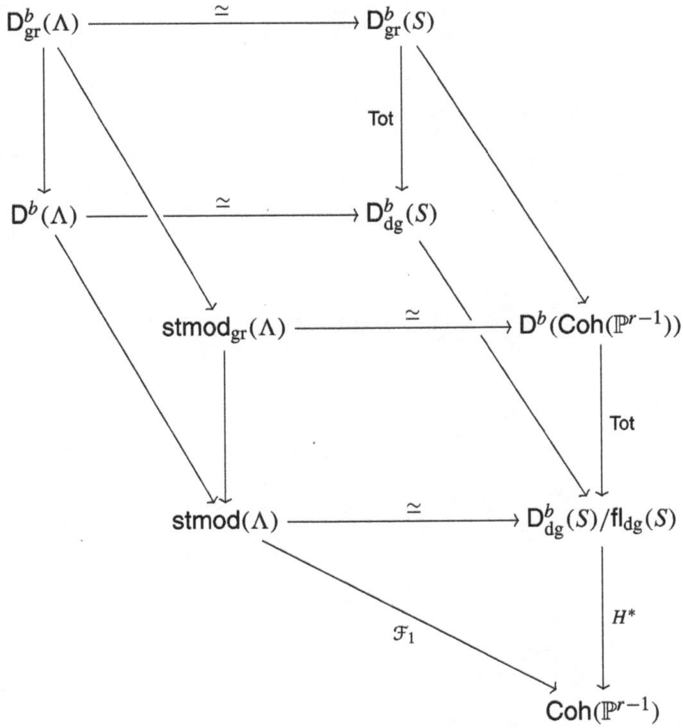


The appendix to the second and third editions of [202] gives an expanded version of the original paper. The version of the correspondence given there is particularly well suited to the study of modules of constant Jordan type for exterior algebras.

Consider a graded $\Lambda$-module $M$ with the property that the action of

$$X_\alpha = \lambda_1 X_1 + \cdots + \lambda_r X_r$$

on $M$ is exact except in degree zero for every $0 \neq \alpha = (\lambda_1, \ldots, \lambda_r) \in \mathbb{A}^r(k)$. The dimension of the homology of $M$ in degree zero for the action of $X_\alpha$ is then the alternating sum of the dimensions of the graded pieces, $\sum_i (-1)^i \dim_k(M_i)$, and is hence independent of $\alpha$. So such a module has constant Jordan type, and $\mathcal{F}_1(M)$ is a vector bundle.

**Theorem 8.13.1** *Every vector bundle on $\mathbb{P}^{r-1}$ arises as $\mathcal{F}_1(M)$ for a graded $\Lambda$-module $M$ satisfying the above condition. Furthermore, two such graded modules $M$ and $M'$ satisfy $\mathcal{F}_1(M) \cong \mathcal{F}_1(M')$ if and only if there are free modules $F$ and $F'$ such that $M \oplus F \cong M' \oplus F'$.*

We shall see some explicit examples of this theorem in the next chapter.

# 9

# Examples

In this chapter we look at some particular examples of vector bundles over projective space, and show how to realise them as $\mathcal{F}_i(M)$ with $M$ a module of constant Jordan type over an elementary abelian $p$-group and over exterior algebras. These examples include the tangent and cotangent bundles, the null correlation bundle, some bundles of Schwarzenberger, the Tango bundle, and the Horrocks–Mumford bundle. Some of these constructions will be used in Chapter 10.

The Schwarzenberger example shows that given a set of points $\{x_1, \dots, x_m\}$ in $\mathbb{P}^2$, there exists a module of stable constant Jordan type $[1]^2$ for $(\mathbb{Z}/2)^3$ whose restriction to a shifted subgroup of type $(\mathbb{Z}/2)^2$ is $\Omega^n k \oplus \Omega^{-n} k \oplus$ (projective), where $n$ is the number of $x_i$ that lie on the corresponding line $\mathbb{P}^1 \subseteq \mathbb{P}^2$.

## 9.1 Modules for $(\mathbb{Z}/2)^2$

Let $k$ be an algebraically closed field of characteristic two, $E = \langle g_1, g_2 \rangle \cong (\mathbb{Z}/2)^2$ and $X_1 = g_1 - 1$, $X_2 = g_2 - 1$ in $kE$. Thus $kE = k[X_1, X_2]/(X_1^2, X_2^2)$. Recall from Section 1.11 that we draw a diagram for $kE$ as a $kE$-module as follows.

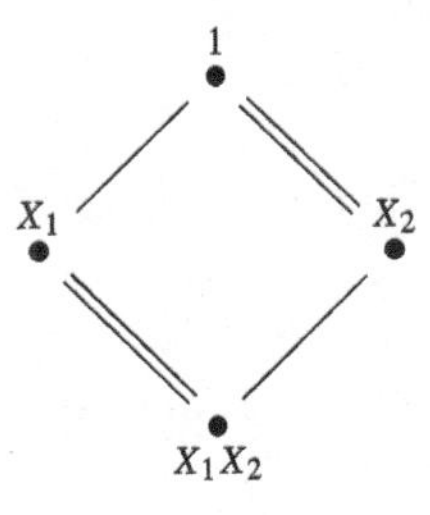


We wish to understand the structure of the sheaf $\widetilde{kE} = kE \otimes_k \mathcal{O}$ on $\mathbb{P}^1 = \operatorname{Proj} k[Y_1, Y_2]$ together with the operator $\theta = X_1 \otimes Y_1 + X_2 \otimes Y_2$ (see Section 8.2). We begin by examining the radical layers of $kE$, on the grounds that $\theta$ takes us down one radical layer. So we think of $\widetilde{kE}$ as

$$\begin{array}{ccc} \widetilde{kE/\mathrm{Rad}kE} & & \mathcal{O} \\ \big\downarrow \theta & & \big\downarrow \theta \\ \widetilde{\mathrm{Rad}kE/\mathrm{Rad}^2kE} & = & \mathcal{O} \oplus \mathcal{O} \\ \big\downarrow \theta & & \big\downarrow \theta \\ \widetilde{\mathrm{Rad}^2kE} & & \mathcal{O}. \end{array}$$

Since $\theta$ is really a map from $\widetilde{kE}$ to $\widetilde{kE}(1)$, the image of $\theta$ in the second radical layer is a copy of $\mathcal{O}(-1) \subset \mathcal{O} \oplus \mathcal{O}$ after untwisting. So we should use the Euler sequence

$$0 \to \mathcal{O}(-1) \to \mathcal{O} \oplus \mathcal{O} \to \mathcal{O}(1) \to 0$$

on $\mathbb{P}^1$ to rewrite the second radical layer $\mathcal{O} \oplus \mathcal{O}$ as an extension

$$\begin{array}{c} \mathcal{O}(1) \\ \big| \\ \mathcal{O}(-1) \end{array}$$

where the subsheaf $\mathcal{O}(-1)$ is the image of $\theta$, and is spanned by the element $X_1 \otimes Y_1 \oplus X_2 \otimes Y_2$ when regarded as an element of

$$\mathrm{Rad}kE/\mathrm{Rad}^2kE \otimes_k k[Y_1, Y_2].$$

Now $\theta^2 = 0$, so $\mathcal{O}(-1)$ is also the kernel of $\theta$ as a map from the second layer to the third layer. Thus we can write down the following diagram for $\widetilde{kE}$:

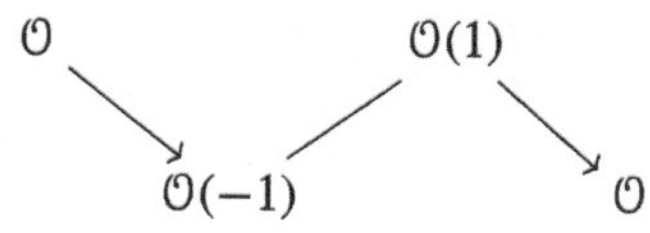

(9.1)

Here, the arrows going down right give the action of the operator $\theta$, while the lines going down left represent extensions of vector bundles. The radical layers are diagonal slices from top right to bottom left.

Now we turn to the syzygies of the trivial module. We have the following:

$\Omega(k)$ :

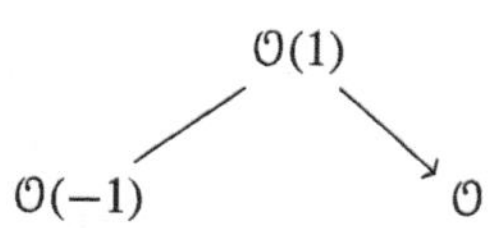

$\Omega^2(k)$ :

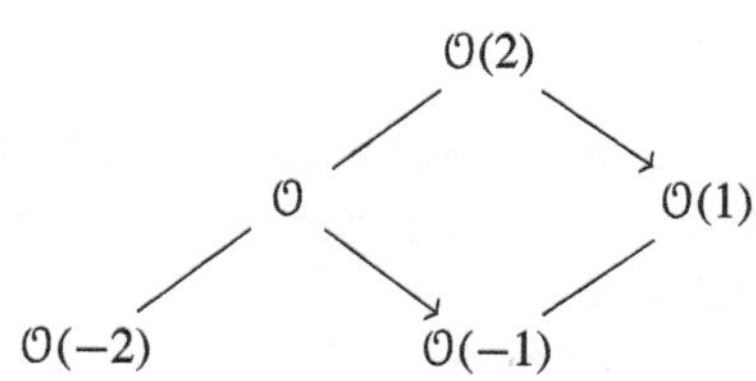

$\Omega^3(k)$ :

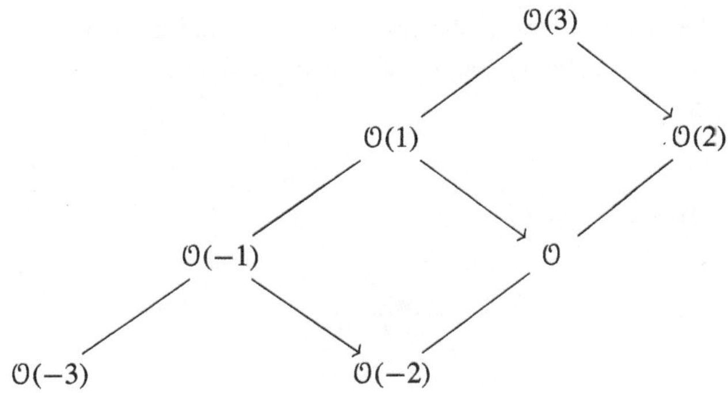


and so on. Dually, we have

$\Omega^{-1}(k)$ :

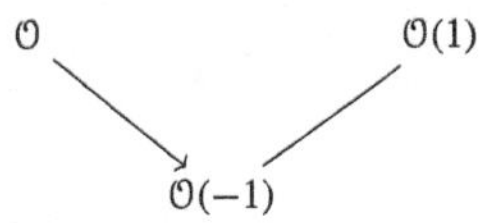


$\Omega^{-2}(k)$ :

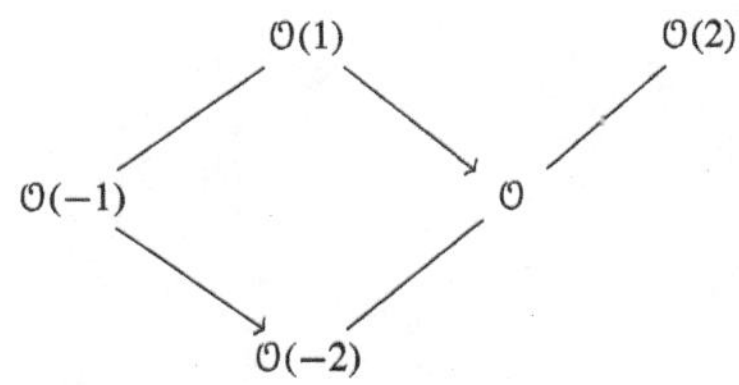


and so on.

## 9.2 Modules for $(\mathbb{Z}/p)^2$

The diagram corresponding to (9.1) for $k(\mathbb{Z}/3)^2$ with $k$ an algebraically closed field of characteristic three is as follows.

$k(\mathbb{Z}/3)^2$:

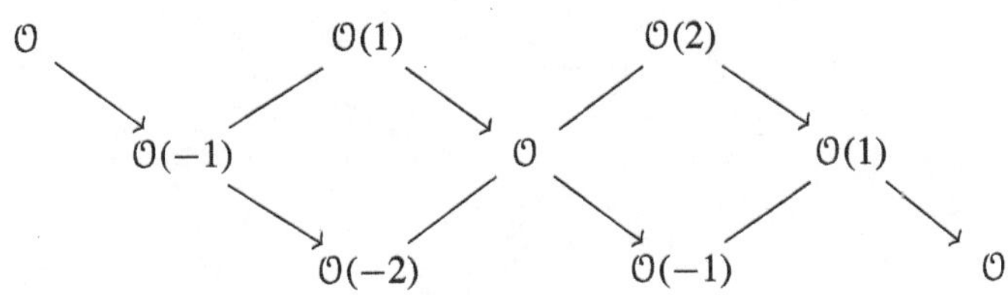


The diagrams for larger primes follow the same pattern. For example for $p = 5$ it is as follows:

$k(\mathbb{Z}/5)^2$:

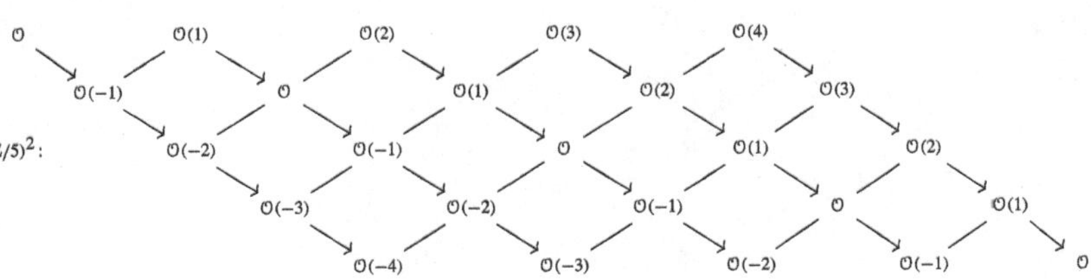

Again the radical layers are the diagonal slices going from top right to bottom left. In the case $p = 3$, the syzygies of the trivial module are as follows:

$\Omega(k)$ :

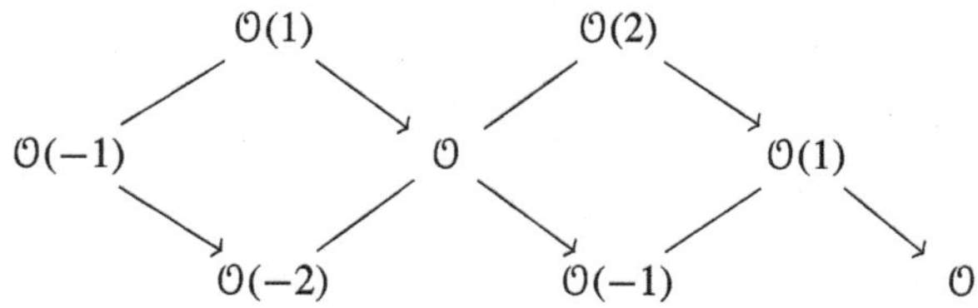


$\Omega^2(k)$ :

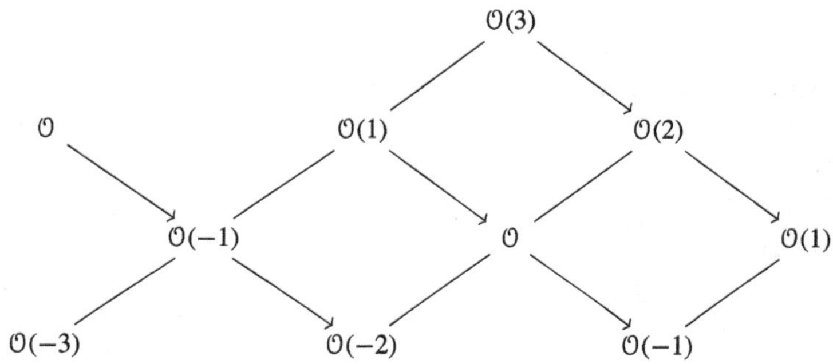


The next ones are more difficult:

$\Omega^3(k)$ :

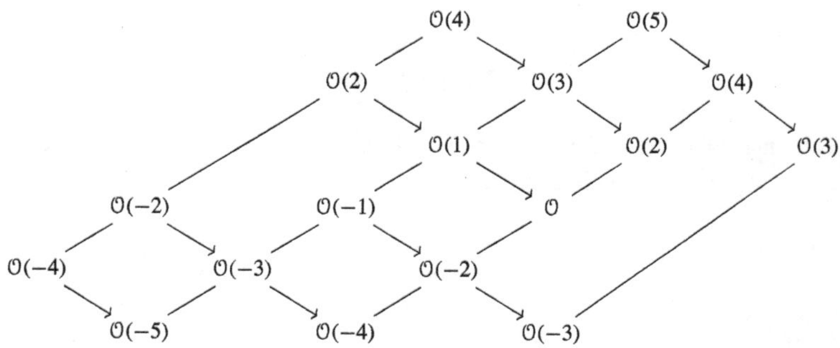

$\Omega^4(k)$ :

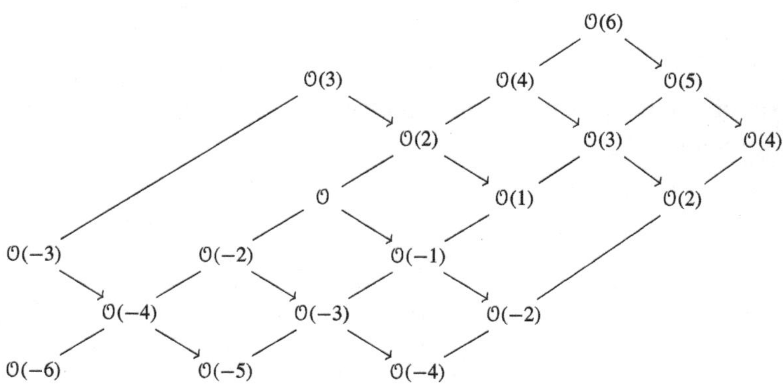

## 9.3 Larger Rank

We can make similar diagrams to a certain extent for larger rank. For example for $p = 2$ and rank 3 the group algebra has the following diagram:

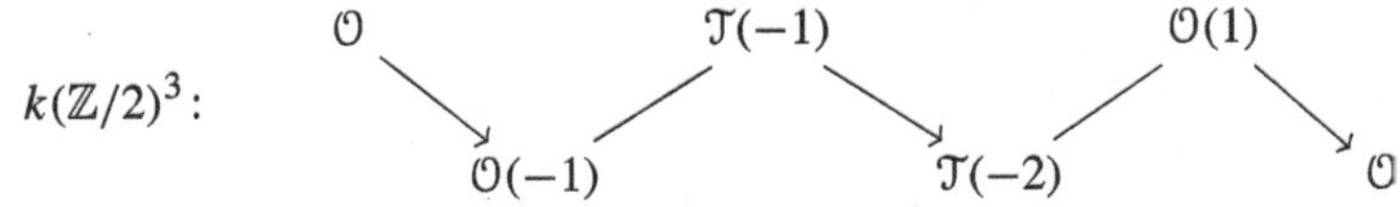


More generally, for $p = 2$ and larger rank the group algebra has the following diagram:

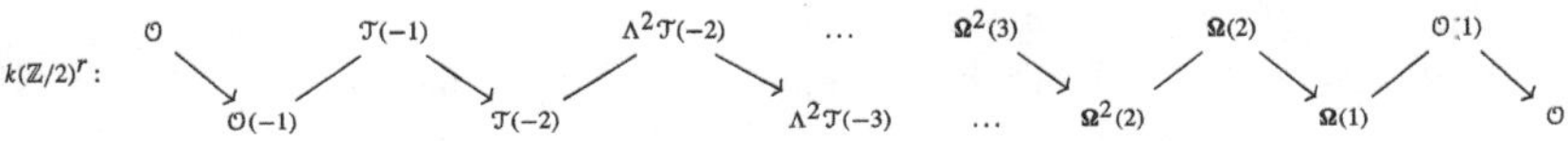


(recall from Section 6.10 that $\mathbf{\Omega}^{r-1-m}(r-m) \cong \Lambda^m\mathcal{T}(-m)$). This is in some sense an embodiment of the Koszul complex. The radical layers of the group algebra are diagonal slices in the diagram. The diagrams for $\Omega(k)$ and $\Omega^{-1}(k)$ for $p = 2$ are obtained from this diagram by removing the copy of $\mathcal{O}$ from the top left, respectively the bottom right.

**Theorem 9.3.1** *Let $E = (\mathbb{Z}/2)^r$ and $k$ an algebraically closed field of characteristic two. Then*

$$\mathcal{F}_2(kE) \cong \bigoplus_{j=0}^{r-1} \mathbf{\Omega}^j(j), \quad \mathcal{F}_2(\Omega(k)) \cong \bigoplus_{j=0}^{r-2} \mathbf{\Omega}^j(j), \quad \mathcal{F}_2(\Omega^{-1}(k)) \cong \bigoplus_{j=1}^{r-1} \mathbf{\Omega}^j(j).$$

For $p = 3$ and rank 3 the group algebra $k(\mathbb{Z}/3)^3$ has the following diagram:

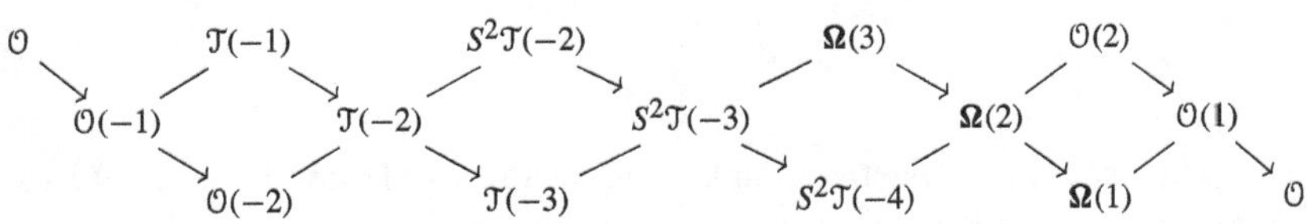


where $S^2\mathcal{T}$ denotes the symmetric square of the tangent bundle, which is a simple bundle of rank three on $\mathbb{P}^2$. Note that we have

$$S^2\mathcal{T}(-3) \cong S^2\mathbf{\Omega}(3).$$

For $p = 3$ and rank 4 the group algebra $k(\mathbb{Z}/3)^4$ has the following diagram:

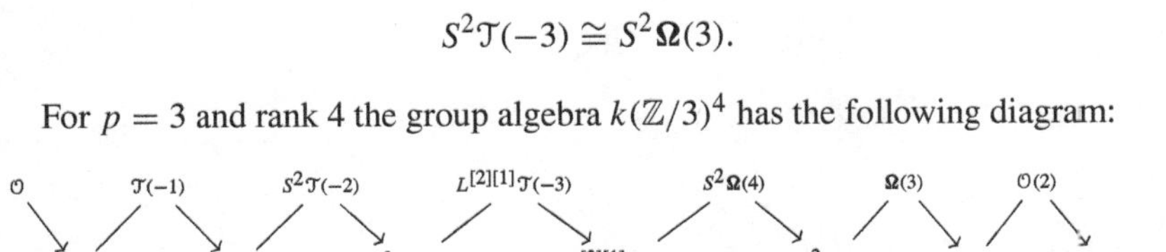

Here, $L^{[2][1]}$ is the simple functor in characteristic three obtained by taking the quotient of the symmetric cube by the Frobenius cube:

$$0 \to F^3(V) \to S^3(V) \to L^{[2][1]}(V) \to 0.$$

We have

$$L^{[2][1]}\mathcal{T}(-4) \cong L^{[2][1]}\boldsymbol{\Omega}(4).$$

It is easy to see how these patterns continue in the light of the description of the radical layers of $kE$ given in Section 1.17.

In the same vein, for $E$ of arbitrary rank and for any prime we have

$\mathsf{Soc}^2(kE)$ : 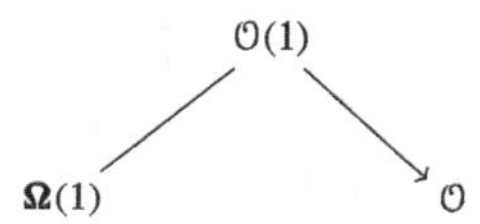


$kE/\mathsf{Rad}^2(kE)$ : 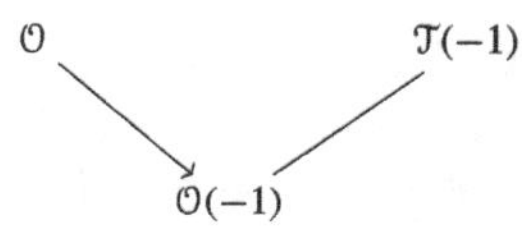


,

see Section 9.5.

## 9.4 Nilvarieties

The three-dimensional nilvariety of constant Jordan type given in Example 4.1.6, where the operator $\theta$ has matrix

$$\begin{pmatrix} 0 & Y_1 & 0 \\ Y_2 & 0 & Y_1 \\ 0 & -Y_2 & 0 \end{pmatrix},$$

may be thought of as a counterexample to the analogue of Dade's Lemma 1.9.5 for nilvarieties, because the dimension is not divisible by 9.

This example can be drawn in the fashion of the previous section as follows:

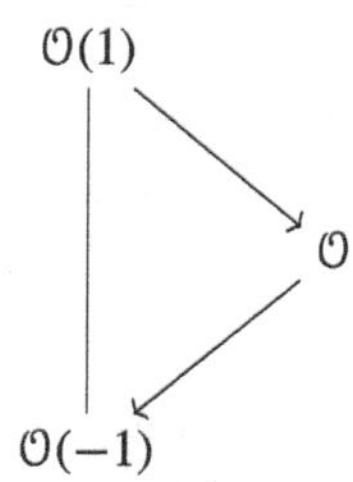

while the first example of dimension five described in Example 4.1.7 gives

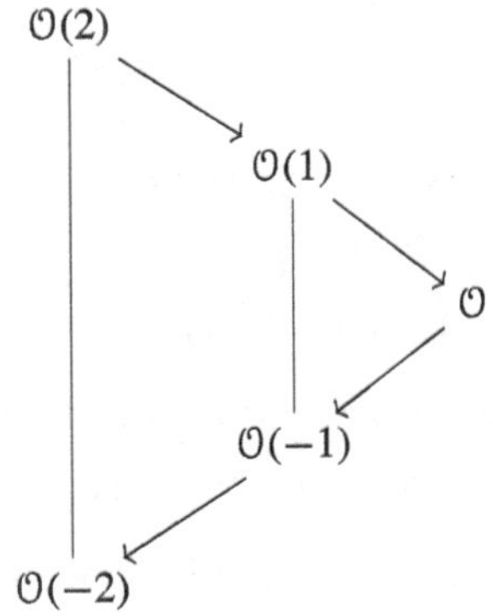


A closely related example is the nilvariety where the operator $\theta$ has matrix

$$\begin{pmatrix} 0 & 0 & Y_1 & 0 \\ Y_2 & 0 & 0 & Y_1 \\ 0 & 0 & 0 & 0 \\ 0 & 0 & -Y_2 & 0 \end{pmatrix}.$$

This may be thought of as a nilvariety with diagram

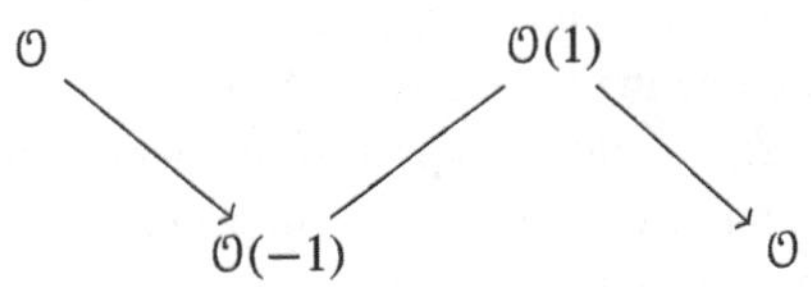


and forming a counterexample to the analogue of Dade's Endotrivial Theorem 5.4.2 for nilvarieties.

## 9.5 The Tangent and Cotangent Bundles

**Example 9.5.1** Let $M = kE/J^2(kE)$. Then $M$ has constant Jordan type $[2]\,[1]^{r-1}$. In the short exact sequence of vector bundles

$$0 \to \widetilde{M/\mathsf{Rad}M} \xrightarrow{\theta} \widetilde{\mathsf{Rad}M}(1) \to \mathcal{F}_1(M)(1) \to 0$$

the map $\theta$ (induced by $\theta_M$) is equal to the map defining the tangent bundle $\mathcal{T}$ of $\mathbb{P}^{r-1}$:

$$0 \to \mathcal{O} \to \mathcal{O}(1)^r \to \mathcal{T} \to 0.$$

It follows that $\mathcal{F}_1(M) \cong \mathcal{T}(-1)$. On the other hand we have $\mathcal{F}_{2,1}(M) \cong \mathcal{O}$, and hence $\mathcal{F}_2(M) \cong \mathcal{O}(-1)$.

The module $J^2(kE)$ is isomorphic to $\Omega(kE/J^2(kE))$. Applying Theorem 8.5.1, we find that

$$\mathcal{F}_{p-1}(J^2(kE)) \cong \mathcal{T}(-p), \qquad \mathcal{F}_{p-2}(J^2(kE)) \cong \mathcal{O}(-p+1).$$

**Example 9.5.2** Dually, let $M = \mathsf{Soc}^2(kE)$, also of constant Jordan type $[2]\,[1]^{r-1}$. Then in the short exact sequence

$$0 \to \mathcal{F}_1(M)(-1) \to \widetilde{M/\mathsf{Rad}M}(-1) \xrightarrow{\theta} \widetilde{\mathsf{Rad}M} \to 0$$

the map $\theta$ is equal to the map defining the cotangent bundle $\boldsymbol{\Omega}$ on $\mathbb{P}^{r-1}$:

$$0 \to \boldsymbol{\Omega} \to \mathcal{O}(-1)^r \to \mathcal{O} \to 0.$$

It follows that $\mathcal{F}_1(M) \cong \boldsymbol{\Omega}(1)$. On the other hand, we have $\mathcal{F}_2(M) \cong \mathcal{O}$.

The module $kE/\mathsf{Soc}^2(kE)$ is isomorphic to $\Omega^{-1}(\mathsf{Soc}^2(kE))$. Again applying Theorem 8.5.1, we find that

$$\mathcal{F}_{p-1}(kE/\mathsf{Soc}^2(kE)) \cong \boldsymbol{\Omega}(2), \qquad \mathcal{F}_{p-2}(kE/\mathsf{Soc}^2(kE)) \cong \mathcal{O}(2).$$

**Example 9.5.3** Suppose that $p \geq 3$, and let $M = \mathsf{Rad}(kE)/\mathsf{Rad}^3(kE)$, of contant Jordan type $[2]^r[1]^{r(r-1)/2}$. Then the short exact sequence

$$0 \to \mathcal{F}_{2,1}(M) \xrightarrow{\theta} \widetilde{\mathsf{Rad}M}(1) \to \mathcal{F}_1(M)(1) \to 0$$

can be viewed as follows. Consider the two-term complex $\mathcal{O} \to \mathcal{O}(1)^r$ where $\mathcal{O}$ is the term of degree zero and the homology is the tangent bundle $\mathcal{T}$ in degree one. Since we are in odd characteristic, the tensor square of this complex decomposes as the direct sum of the exterior square and the symmetric square. The exterior square is a complex of the form $0 \to \mathcal{O}(1)^r \to \mathcal{O}(2)^{r(r+1)/2}$ whose homology in degree two is the symmetric square $S^2(\mathcal{T})$. Untwisting, we get a short exact sequence

$$0 \to \mathcal{O}^r \to \mathcal{O}(1)^{r(r+1)/2} \to S^2(\mathcal{T})(-1) \to 0.$$

The first map in this sequence is the map corresponding to $\theta$, and so

$$\mathcal{F}_1(M) \cong S^2(\mathcal{T})(-2).$$

**Example 9.5.4** Suppose again that $p \geq 3$ and consider the $kE$-module of Loewy length three with underlying vector space $V^* \oplus (V \otimes_k V^*) \oplus V$ where $V = J(kE)/J^2(kE)$, and action given by

$$X\colon \begin{pmatrix} \phi \\ X' \otimes \phi' \\ X'' \end{pmatrix} \mapsto \begin{pmatrix} 0 \\ X \otimes \phi \\ \phi'(X)X' + \phi'(X')X \end{pmatrix}.$$

Here, $X$, $X'$, $X''$ are linear combinations of $X_1, \ldots, X_r$ and $\phi$, $\phi'$ are linear forms. It is easy to check that these matrices commute and cube to zero, and hence give a $kE$-module $M$. We have

$$V^* \otimes \mathcal{O}(-1) \xrightarrow{\theta} V \otimes V^* \otimes \mathcal{O} \to V \otimes \mathcal{O}(1)$$

and $\mathcal{F}_1(M) = \mathsf{Ker}\,\theta/\mathsf{Ker}\,\theta \cap \mathsf{Im}\,\theta$ as a subquotient of the middle term. This is the vector bundle of rank $(r-1)^2$ given by $\mathrm{Hom}(\mathcal{T},\mathcal{T}) \cong \mathcal{T} \otimes \mathbf{\Omega}$. The Jordan type of this module is $[3]\,[2]^{2(r-1)}[1]^{(r-1)^2}$.

On the other hand, if we want anticommuting matrices, we let

$$X\colon \begin{pmatrix} \phi \\ X' \otimes \phi' \\ X'' \end{pmatrix} \mapsto \begin{pmatrix} 0 \\ X \otimes \phi \\ \phi'(X)X' - \phi'(X')X \end{pmatrix}.$$

This then gives a module $M$ for the exterior algebra $\Lambda(X_1, \dots, X_r)$, this time with $\mathcal{F}_1(M) \cong \mathrm{Hom}(\mathcal{T},\mathcal{T})/(\text{multiples of the identity}) \cong (\mathcal{T} \otimes \mathbf{\Omega})/\mathcal{O}$. This time the Jordan type is $[2]^{2r}[1]^{r^2-2r}$.

## 9.6 The Null Correlation Bundle, $p = 2$

In Section 6.7, we described the null correlation bundle as an example of an indecomposable vector bundle of rank $(r-2)$ on $\mathbb{P}^{r-1}$ for $r$ even.

Let $p = 2$ and $r = 2m \geq 4$ be even, $E = (\mathbb{Z}/2)^r$, and let $M$ be the $(r+2)$-dimensional $kE$-module with basis elements $u, v_1, \dots, v_r, w$, with action given by: $X_i u = v_i$, $X_i v_j = w$ if $|i-j| = m$, $X_i v_j = 0$ otherwise. Then $\mathcal{F}_1(M)$ is the null correlation bundle of rank $r-2$ defined by the symplectic form

$$\langle \sum a_i X_i, \sum b_i X_i \rangle = \sum_{|i-j|=m} a_i b_j.$$

The monad (6.5) for the null correlation bundle,

$$0 \to \mathcal{O}(-1) \to \mathcal{O}^r \to \mathcal{O}(1) \to 0,$$

can be written in the form

$$0 \to \widetilde{\langle u \rangle}(-1) \xrightarrow{\theta} \widetilde{\langle v_1, \dots, v_r \rangle} \xrightarrow{\theta} \widetilde{\langle w \rangle}(1) \to 0.$$

It follows that for this module $M$, $\mathcal{F}_1(M)$ is isomorphic to the cohomology of this monad, namely the null correlation bundle.

The case $r = 4$ of the module $M$ was described in Example 5.2.11. In general, the matrix for the operator $\theta$ is given by

$$\left(\begin{array}{c|ccc|c} & & & & \\ \hline Y_1 & & & & \\ \vdots & & & & \\ Y_r & & & & \\ \hline & Y_r & \cdots & Y_1 & \end{array}\right)$$

where blank entries are taken to be zero as usual.

## 9.7 The Null Correlation Bundle, $p$ Odd

Realising the null correlation bundle when $p$ is odd is an instructive exercise in the techniques of Section 8.8; as in that section, we shall content ourselves with realising the Frobenius twist of the null correlation bundle.

Let $p$ be an odd prime, $r = 2m \geq 4$ be even, and $E = (\mathbb{Z}/p)^r$. Again we use the same symplectic form, but instead of using extensions of $k$ by $k$ as elements of $H^1(E, k)$ we are obliged to use elements of $H^2(E, k)$ as explained in Section 8.8. Therefore we take the sequence of modules and maps

$$\Omega^2 k \to k^r \to \Omega^{-2} k$$

where the first map is given by the column vector of elements $(x_1, \ldots, x_m, x_{m+1}, \ldots, x_r)^t$ in $H^2(E, k)$ and the second is given by the row vector $(x_{m+1}, \ldots, x_r, -x_1, \ldots, -x_m)$. Since the elements $x_i \in H^2(E, k)$ commute, the composite of these maps represents the zero element of $H^4(E, k)$ and so it factors through a projective module.

This allows us to perform the following construction in the stable module category: complete the map $\Omega^2 k \to k^r$ to a triangle

$$\Omega^2 k \to k^r \to X$$

in the stable module category $\mathsf{stmod}(kE)$. Then there exists a map $X \to \Omega^{-2} k$ whose composite with $k^r \to X$ is the given map $k^r \to \Omega^{-2} k$. Complete this to a triangle

$$Y \to X \to \Omega^{-2} k,$$

in $\mathsf{stmod}(kE)$, and then $Y$ will be the required module of stable constant Jordan type $[1]^{r-2}$. We illustrate this with the following diagram in $\mathsf{stmod}(kE)$:

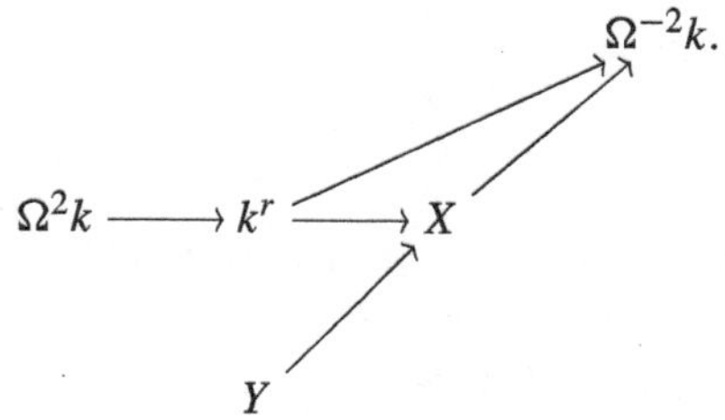


Note that as in Section 8.8, $\mathcal{F}_1(Y)$ is not isomorphic to the null correlation bundle but rather to the Frobenius twist of the null correlation bundle.

Let us make this construction more explicit in $\mathsf{mod}(kE)$. Recall that the beginning of the minimal resolution of $k$ as a $kE$-module is as follows:

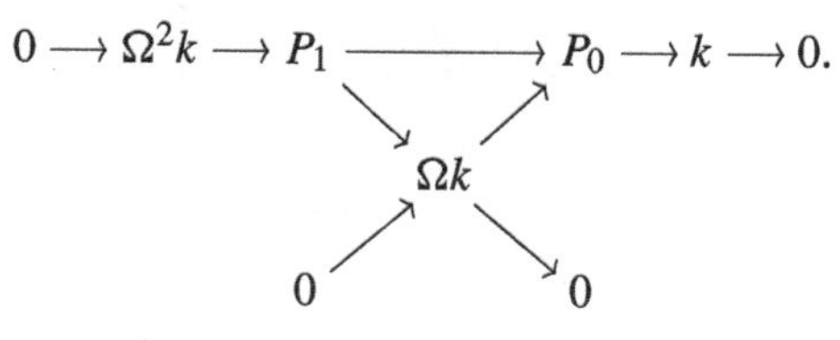

We turn $\Omega^2 k \to k^r$ into an injective map by adding $P_1$ to $k^r$. So $X$ is an extension of $\Omega k$ by $k^r$ as in the following diagram:

$$\begin{array}{ccccccccc}
 & & & & 0 & & 0 & & \\
 & & & & \downarrow & & \downarrow & & \\
 & & & & k^r & = & k^r & & \\
 & & & & \downarrow & & \downarrow & & \\
0 & \longrightarrow & \Omega^2 k & \longrightarrow & k^r \oplus P_1 & \longrightarrow & X & \longrightarrow & 0 \\
 & & \| & & \downarrow & & \downarrow & & \\
0 & \longrightarrow & \Omega^2 k & \longrightarrow & P_1 & \longrightarrow & \Omega k & \longrightarrow & 0 \\
 & & & & \downarrow & & \downarrow & & \\
 & & & & 0 & & 0 & &
\end{array}$$

Now the composite $\Omega^2 k \to k^r \to \Omega^{-2} k$ factors through some projective module, and therefore through the projective cover $P_{-1}$ of $\Omega^{-2} k$, which happens to be isomorphic to $P_1$ as both are direct sums of $r$ copies of the group algebra. So there exists a surjective map $k^r \oplus P_1 \to \Omega^{-2} k$ such that the composite

$$\Omega^2 k \to k^r \oplus P_1 \to \Omega^{-2} k$$

is the zero map in $\mathsf{mod}(kE)$. The required module $Y$ is the cohomology of this complex in the middle term. So it fits into a commutative diagram as follows:

$$\begin{array}{ccccccccc}
 & & 0 & & 0 & & & & \\
 & & \downarrow & & \downarrow & & & & \\
0 & \longrightarrow & \Omega^{-1} k & \longrightarrow & P_1 & \longrightarrow & \Omega^{-2} k & \longrightarrow & 0 \\
 & & \downarrow & & \downarrow & & \downarrow & & \\
0 & \longrightarrow & Y & \longrightarrow & X \oplus P_1 & \longrightarrow & \Omega^{-2} k & \longrightarrow & 0 \\
 & & \downarrow & & \downarrow & & & & \\
 & & X & = & X & & & & \\
 & & \downarrow & & \downarrow & & & & \\
 & & 0 & & 0 & & & &
\end{array}$$

From these diagrams we can see that the module $Y$ has a filtration with filtered quotients (from the top) $\Omega k$, $k^r$, $\Omega^{-1} k$. The extensions of $\Omega k$ by $k^r$ and of $k^r$ by $\Omega^{-1} k$ are given by the column vector $(x_1, \ldots, x_m, x_{m+1}, \ldots, x_r)^t$ and the row vector $(x_{m+1}, \ldots, x_r, -x_1, \ldots, -x_m)$, respectively. The reason the threefold extension exists is that the cup product of these two vectors is zero in $H^4(E, k)$.

**Proposition 9.7.1** *In odd characteristic, for the module Y in the above construction we have* $\mathcal{F}_1(Y) \cong F^*(\mathcal{F}_N)$, *the inverse image of the null correlation bundle along the Frobenius map on* $\mathbb{P}^{r-1}$.

**Remark 9.7.2** The constructions given in this section and the last work with any monad of the form

$$0 \to \mathcal{O}(-1)^{\oplus a} \to \mathcal{O}^{\oplus b} \to \mathcal{O}(1)^{\oplus c} \to 0.$$

In the next two sections we discuss more examples of monads of this form, giving rise to instanton bundles and Schwarzenberger's bundles.

## 9.8 Instanton Bundles

There is a generalisation of the null correlation bundle called an *instanton bundle* on $\mathbb{P}^{r-1}$ when $r$ is even. It is a bundle $\mathcal{F}_k$ appearing as the cohomology of a monad of the form

$$0 \to \mathcal{O}(-1)^{\oplus k} \xrightarrow{\beta} \mathcal{O}^{\oplus(2k+r-2)} \xrightarrow{\alpha} \mathcal{O}(1)^{\oplus k} \to 0$$

with $k \geq 1$. The positive integer $k$ is called the *quantum number* (or *instanton number* or *charge*) of the bundle. The usual definition of an instanton bundle is technical, and we shall not give it here, instead referring the reader to Okonek and Spindler [203]. It is proved in Proposition 2.4 of Hoffmann [157] that their definition is equivalent to the one given above together with the condition of trivial splitting type; the latter is an open condition and is omitted by some authors. There is a considerable literature on instanton bundles, see for example Ancona and Ottaviani [9, 10], Barth [34], Costa and Ottaviani [94, 95], Hartshorne [148], Hoffmann [157], Miro-Roig and Orus-Lacort [194], Okonek and Spindler [203], Rao [213] and Vedernikov [240]. In particular, in [95] it is shown that the moduli space of instanton bundles on $\mathbb{P}^{r-1}$ with instanton number $k$ is an affine variety. The paper of Atiyah, Drinfeld, Hitchin and Manin [15] explains the connection of complex instanton bundles over $\mathbb{P}^3$ with solutions of the Yang–Mills equations.

Instead of giving the definition, we shall give examples of instanton bundles for each even $r \geq 4$ and each $k \geq 1$, and describe some properties. The examples with $k = 1$ are just the null correlation bundles.

**Example 9.8.1** Let $\alpha$ be the $k \times (2k+r-2)$ matrix whose $(i, j)$ entry is $Y_{j-2i-1}$ if $1 \leq j - 2i - 1 \leq r$ and zero otherwise, and let $\beta$ be the $(2k+r-2) \times k$ matrix whose $(i, j)$ entry is $(-1)^{i+1}Y_{r+2+i-2j}$ if $1 \leq r+2+i-2j \leq r$ and zero otherwise.

It is easy to check that these matrices have zero product, and that $\alpha$ is surjective and $\beta$ is injective at all points of projective space. With these choices for $\alpha$ and $\beta$, we obtain an instanton bundle.

For example, if $r = 6$ and $k = 3$, the maps $\alpha$ and $\beta$ are given by

$$\alpha = \begin{pmatrix} Y_1 & Y_2 & Y_3 & Y_4 & Y_5 & Y_6 & & & & \\ & & Y_1 & Y_2 & Y_3 & Y_4 & Y_5 & Y_6 & & \\ & & & & Y_1 & Y_2 & Y_3 & Y_4 & Y_5 & Y_6 \end{pmatrix}$$

$$\beta = \begin{pmatrix} Y_6 & & \\ -Y_5 & & \\ Y_4 & Y_6 & \\ -Y_3 & -Y_5 & \\ Y_2 & Y_4 & Y_6 \\ -Y_1 & -Y_3 & -Y_5 \\ & Y_2 & Y_4 \\ & -Y_1 & -Y_3 \\ & & Y_2 \\ & & -Y_1 \end{pmatrix}.$$

**Proposition 9.8.2** *The Chern polynomial of an instanton bundle $\mathcal{F}_k$ on $\mathbb{P}^{r-1}$ ($r$ even) is equal to*

$$c(\mathcal{F}_k) = (1 - h^2)^{-k} = 1 + kh^2 + \binom{k+1}{2}h^4 + \binom{k+2}{3}h^6 + \cdots$$

*and the Chern character is given by*

$$\mathsf{Ch}(\mathcal{F}_k) = (2k + r - 2) - k(e^h + e^{-h}).$$

*Proof.* This follows from Lemma 7.3.5. □

Again for $p = 2$ we can take the monad describing an instanton bundle,

$$0 \to \mathcal{O}(-1)^{\oplus k} \to \mathcal{O}^{\oplus(2k+r-2)} \to \mathcal{O}(1)^{\oplus k} \to 0,$$

and apply the constructions of the last two sections to it. The modules $M_k$ obtained this way for $p = 2$ have constant Jordan type $[2]^k[1]^{r-2}$, and $\mathcal{F}_1(M_k) \cong \mathcal{F}_k$, the instanton bundle with quantum number $k$.

For example if $r = 6$ and $k = 3$, using the instanton bundle described in Example 9.8.1, the operator $\theta$ for this module is represented by the matrix

$$\left(\begin{array}{ccc|cccccccccc|ccc}
 & & & & & & & & & & & & & & & \\
 & & & & & & & & & & & & & & & \\
 & & & & & & & & & & & & & & & \\ \hline
Y_6 & & & & & & & & & & & & & & & \\
Y_5 & & & & & & & & & & & & & & & \\
Y_4 & Y_6 & & & & & & & & & & & & & & \\
Y_3 & Y_5 & & & & & & & & & & & & & & \\
Y_2 & Y_4 & Y_6 & & & & & & & & & & & & & \\
Y_1 & Y_3 & Y_5 & & & & & & & & & & & & & \\
 & Y_2 & Y_4 & & & & & & & & & & & & & \\
 & Y_1 & Y_3 & & & & & & & & & & & & & \\
 & & Y_2 & & & & & & & & & & & & & \\
 & & Y_1 & & & & & & & & & & & & & \\ \hline
 & & & Y_1 & Y_2 & Y_3 & Y_4 & Y_5 & Y_6 & & & & & & & \\
 & & & & & Y_1 & Y_2 & Y_3 & Y_4 & Y_5 & Y_6 & & & & & \\
 & & & & & & & Y_1 & Y_2 & Y_3 & Y_4 & Y_5 & Y_6 & & &
\end{array}\right)$$

A diagram for this module looks as follows (we have not used any notation to distinguish the actions of $X_1, \ldots, X_6$, but it is easy to see which is which from the above matrix).

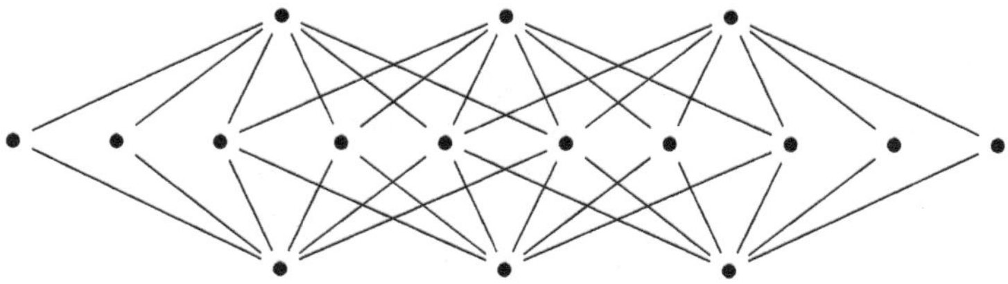

For $p$ odd, again we need to use the more complicated construction described in Section 9.7 (and of course put the signs back into the matrices). The modules we obtain this way have $\mathcal{F}_1(M)$ isomorphic to $F^*(\mathcal{F}_k)$, the inverse image of the instanton bundle along the Frobenius map.

## 9.9 Schwarzenberger's Bundles

The next example concerns the bundles $\mathcal{F}(x_1, \ldots, x_m)$. These were constructed by Schwarzenberger, and appear in the book of Okonek, Schneider and Spindler [202, Theorem 2.2.5]; see also [13]. Let $k$ be an algebraically closed field of characteristic 2 and let $E$ be an elementary abelian group of order 8. Let $\{x_1, \ldots, x_m\}$ be a finite set of points in $\mathbb{P}^2$. Then the following is a construction of a $kE$-module $M$ of dimension $4m + 2$ of constant Jordan type $[2]^{2m}[1]^2$ with the following property. Given a shifted subgroup $E'$ of $E$ of rank two, the restriction of $M$ to $E'$ is $\Omega^n k \oplus \Omega^{-n} k \oplus$ (projective), where $n$ is the number of points from the set $\{x_1, \ldots, x_m\}$ which lie on the corresponding line $\mathbb{P}^1 \subseteq \mathbb{P}^2$. The construction relies on the fact that there's an easier construction of the corresponding bundle than the one given in [202]. Namely, there's a monad of the form

$$\mathcal{O}(-1)^m \to \mathcal{O}^{2m+2} \to \mathcal{O}(1)^m$$

whose homology is $\mathcal{F}(x_1, \ldots, x_m)$. The maps in this monad are given as follows. The left-hand map is

$$\begin{pmatrix} w_1 & w_2 & \cdots & w_m \\ v_1 & 0 & & 0 \\ -u_1 & 0 & \cdots & 0 \\ 0 & v_2 & & 0 \\ 0 & -u_2 & & 0 \\ & & \cdots & \\ 0 & 0 & & v_m \\ 0 & 0 & & -u_m \\ 0 & 0 & & 0 \end{pmatrix}$$

while the right-hand map is

$$\begin{pmatrix} 0 & u_1 & v_1 & 0 & 0 & \dots & 0 & 0 & w_1 \\ 0 & 0 & 0 & u_2 & v_2 & & 0 & 0 & w_2 \\ & & & & & \dots & & & \\ & & & & & \dots & & & \\ 0 & 0 & 0 & 0 & 0 & & u_m & v_m & w_m \end{pmatrix}.$$

Here, for each $i$ with $1 \leq i \leq m$, the linear functions $u_i, v_i, w_i$ are chosen to be any basis for the space spanned by $Y_1, Y_2, Y_3$ with the property that $u_i$ and $v_i$ vanish on $x_i$ but $w_i$ does not.

**Example 9.9.1** Suppose that $x_1 = (1:0:0)$, $x_2 = (0:1:0)$, $x_3 = (0:0:1)$, $x_4 = (1:1:1)$ and $x_5 = (1:a:b)$. Then one choice of bases gives the matrices

$$\begin{pmatrix} Y_1 & Y_2 & Y_3 & Y_1 & Y_1 \\ Y_2 & 0 & 0 & 0 & 0 \\ -Y_3 & 0 & 0 & 0 & 0 \\ 0 & Y_3 & 0 & 0 & 0 \\ 0 & -Y_1 & 0 & 0 & 0 \\ 0 & 0 & Y_1 & 0 & 0 \\ 0 & 0 & -Y_1 & 0 & 0 \\ 0 & 0 & 0 & Y_1 - Y_2 & 0 \\ 0 & 0 & 0 & -Y_1 + Y_3 & 0 \\ 0 & 0 & 0 & 0 & aY_1 - Y_2 \\ 0 & 0 & 0 & 0 & -bY_1 + Y_3 \\ 0 & 0 & 0 & 0 & 0 \end{pmatrix} \quad \text{and}$$

$$\begin{pmatrix} 0 & Y_3 & Y_2 & 0 & 0 & 0 & 0 & 0 & 0 & 0 & 0 & Y_1 \\ 0 & 0 & 0 & Y_1 & Y_3 & 0 & 0 & 0 & 0 & 0 & 0 & Y_2 \\ 0 & 0 & 0 & 0 & 0 & Y_2 & Y_1 & 0 & 0 & 0 & 0 & Y_3 \\ 0 & 0 & 0 & 0 & 0 & 0 & 0 & Y_1 - Y_3 & Y_1 - Y_2 & 0 & 0 & Y_1 \\ 0 & 0 & 0 & 0 & 0 & 0 & 0 & 0 & 0 & bY_1 - Y_3 & aY_1 - Y_2 & Y_1 \end{pmatrix}.$$

So we form a $kE$-module $M$ with $J^3(M) = 0$ and Loewy layers of dimensions $m$, $2m+2$ and $m$, with these maps as the data for $\theta$. In other words, the monad above is

$$\widetilde{M/J(M)}(-1) \xrightarrow{\theta} \widetilde{J(M)/J^2(M)} \xrightarrow{\theta} \widetilde{J^2(M)}(1).$$

It is easy to see that $\mathcal{F}_1(M)$ is isomorphic to the given Schwarzenberger bundle, and that the restriction of $M$ to $E'$ has the required isomorphism type. The matrix for $\theta$ is as follows:

$$\left(\begin{array}{c||c||c} & & \\ \hline\hline \begin{array}{cccc} w_1 & w_2 & \cdots & w_m \\ v_1 & 0 & & 0 \\ -u_1 & 0 & \cdots & 0 \\ 0 & v_2 & & 0 \\ 0 & -u_2 & & 0 \\ & & \cdots & \\ 0 & 0 & & v_m \\ 0 & 0 & & -u_m \\ 0 & 0 & & 0 \end{array} & & \\ \hline\hline & \begin{array}{ccccccccc} 0 & u_1 & v_1 & 0 & 0 & \cdots & 0 & 0 & w_1 \\ 0 & 0 & 0 & u_2 & v_2 & & 0 & 0 & w_2 \\ & & & & & \cdots & & \cdot & \\ & & & & & \cdots & & & \\ 0 & 0 & 0 & 0 & 0 & & u_m & v_m & w_m \end{array} & \end{array}\right).$$

## 9.10 The Examples of Tango

Recall from §6.8 that Tango produced examples of indecomposable bundles $\mathcal{F}$ of rank $r-2$ on $\mathbb{P}^{r-1} = \mathbb{P}(V)$ fitting into exact sequences

$$0 \to \mathcal{O}(-2) \to \mathcal{O}(-1)^r \to \mathcal{O}^{2r-3} \to \mathcal{F} \to 0,$$

one for each subspace $W$ of codimension $2r-3$ in $\Lambda^2 V$ containing no non-trivial wedges. For $p = 2$ it is easy to construct modules of constant Jordan type for $(\mathbb{Z}/2)^r$ realising these bundles. Namely, thinking of $V$ as $J(kE)/J^2(kE)$, we can regard $W$ as a subspace of $J^2(kE)/J^3(kE) = \Lambda^2 V$. Let $\hat{W}$ be the inverse image of $W$ in $J^2(kE)$. Then $M = kE/\hat{W}$ is a $kE$-module of dimension $3r-2$ of constant Jordan type $[2]^r[1]^{r-2}$ satisfying

$$\mathcal{F}_1(M) \cong \mathcal{F}.$$

The sequence above defining $\mathcal{F}$ can be written as

$$0 \to \widetilde{M/\mathsf{Rad}(M)}(-2) \to \widetilde{\mathsf{Rad}(M)/\mathsf{Rad}^2(M)}(-1) \to \widetilde{\mathsf{Rad}^2(M)} \to \mathcal{F}_1(M) \to 0.$$

One can visualise the module $M$ as follows:

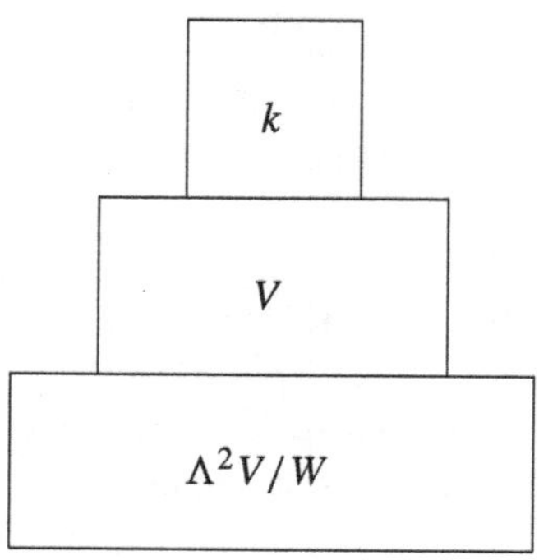

## 9.11 The Horrocks–Mumford Bundle

The Horrocks–Mumford bundle is an indecomposable vector bundle of rank two on $\mathbb{P}^4$ in any characteristic. In this section we construct this bundle and examine its automorphisms. Literature includes the original paper of Horrocks and Mumford [161] as well as the papers of Aure [20], Barth [33], Barth, Hulek and Moore [35], Barth and Moore [36], Boralevi [59], Borcea [61], Decker [99, 100], Decker and Schreyer [102, 103], Dowbor and Meltzer [106], Hulek [163, 164], Hulek and Vandeven [165], Kaji [172], Lee [186], as well as Example 7.3 of Eisenbud, Fløystad and Schreyer [111] and Section 2.3.3 of Okonek, Schneider and Spindler [202].

Let $V$ be a $k$-vector space of dimension five with basis $v_0, \dots, v_4$. We read these subscripts modulo five, so that for example $v_5 = v_0$. Let $v_0^*, \dots, v_4^*$ be the dual basis of $V^*$, and identify $V^*$ with $\Lambda^4 V$ via

$$v_i^* = v_{i+1} \wedge v_{i+2} \wedge v_{i+3} \wedge v_{i+4}.$$

Similarly, $\Lambda^2 V^*$ is identified with $\Lambda^3 V$ via

$$v_i^* \wedge v_{i+1}^* = v_{i+2} \wedge v_{i+3} \wedge v_{i+4} \tag{9.2}$$

$$v_i^* \wedge v_{i+2}^* = -v_{i+1} \wedge v_{i+3} \wedge v_{i+4}. \tag{9.3}$$

These identifications amount to fixing a determinant form on $V$.

Let $V'$ be another $k$-vector space of dimension five with basis $v_0', \dots, v_4'$ and define maps

$$f_+ \colon V' \to \Lambda^2 V \qquad f_+(v_i') = v_{i+2} \wedge v_{i-2}$$

$$f_- \colon V' \to \Lambda^2 V \qquad f_-(v_i') = v_{i+1} \wedge v_{i-1}.$$

Let $f_+^*, f_-^* \colon \Lambda^2 V^* \to (V')^*$ be the transposed maps.

**Lemma 9.11.1** *If $v' \in V'$, $v, v_+, v_- \in V$ satisfy $f_+(v') = v \wedge v_+$ and $f_-(v') = v \wedge v_-$ then $v' = 0$.*

*Proof.* We have

$$f_+(v') \wedge f_-(v') = v \wedge v_+ \wedge v \wedge v_- = 0.$$

If $v' = \sum_i a_i v_i'$ then we have

$$0 = f_+(v') \wedge f_-(v') = \sum_i a_i^2 v_{i+2} \wedge v_{i-2} \wedge v_{i+1} \wedge v_{i-1} = \sum_i a_i^2 v_i^*$$

and hence each $a_i = 0$. □

Now take the Koszul complex on $\mathbb{P}^4 = \mathbb{P}(V)$:

$$\begin{array}{ccccccccccccc} 0 \to \mathcal{O} \to V \otimes \mathcal{O}(1) & & \longrightarrow & & \Lambda^2 V \otimes \mathcal{O}(2) & & \longrightarrow & & \Lambda^2 V^* \otimes \mathcal{O}(3) & & \longrightarrow & & V^* \otimes \mathcal{O}(4) \to \mathcal{O}(5) \to 0 \\ & \searrow & & \nearrow & & \searrow & & \nearrow & & \searrow & & \nearrow & \\ & & \mathcal{T} & & & & \Lambda^2 \mathcal{T} & & & & \Lambda^3 \mathcal{T} & & \end{array}$$

Let $W$ be a two-dimensional vector space with basis $w_1, w_2$, and define

$$p\colon V' \otimes \mathcal{O}(2) \xrightarrow{\binom{f_+}{f_-}} W \otimes \Lambda^2 V \otimes \mathcal{O}(2) \to W \otimes \Lambda^2 \mathcal{T}$$

$$q\colon W \otimes \Lambda^2 \mathcal{T} \to W \otimes \Lambda^2 V^* \otimes \mathcal{O}(3) \xrightarrow{(-f_-^* \; f_+^*)} (V')^* \otimes \mathcal{O}(3).$$

Here, we are just regarding $W \otimes \Lambda^2 V$ as a direct sum of two copies of $\Lambda^2 V$, and the first map in the definition of $p$ then has components $f_+$ and $f_-$, tensored with the identity map on $\mathcal{O}(2)$. The point of using the space $W$ is that automorphisms of the Horrocks–Mumford bundle will make use of linear transformations on $W$. Similar remarks apply to the second map in the definition of $q$.

**Lemma 9.11.2** *We have $q \circ p = 0$.*

*Proof.* First note that the composite of the first map in the definition of $q$ with the second map in the definition of $p$ is equal to the differential in the Koszul complex. Let $\phi$ be a function of some degree in a graded module defining $\mathcal{O}(2)$, and we evaluate $q \circ p$ on an element of the form $v_i' \otimes \phi$. So we first apply $\binom{f_+}{f_-} \otimes 1$ to get

$$\begin{pmatrix} v_{i+2} \wedge v_{i-2} \\ v_{i+1} \wedge v_{i-1} \end{pmatrix} \otimes \phi.$$

Next we apply the Koszul differential to obtain

$$\begin{pmatrix} v_{i-1} \wedge v_{i+2} \wedge v_{i-2} \otimes Y_{i-1}\phi + v_i \wedge v_{i+2} \wedge v_{i-2} \otimes Y_i\phi + v_{i+1} \wedge v_{i+2} \wedge v_{i-2} \otimes Y_{i+1}\phi \\ v_{i-2} \wedge v_{i+1} \wedge v_{i-1} \otimes Y_{i-2}\phi + v_i \wedge v_{i+1} \wedge v_{i-1} \otimes Y_i\phi + v_{i+2} \wedge v_{i+1} \wedge v_{i-1} \otimes Y_{i+2}\phi \end{pmatrix}.$$

Applying the identification (9.2) of $\Lambda^2 V^*$ with $\Lambda^3 V$, we rewrite this as

$$\begin{pmatrix} v_i^* \wedge v_{i+1}^* \otimes Y_{i-1}\phi + v_{i+1}^* \wedge v_{i-1}^* \otimes Y_i\phi + v_{i-1}^* \wedge v_i^* \otimes Y_{i+1}\phi \\ v_i^* \wedge v_{i+2}^* \otimes Y_{i-2}\phi + v_{i+2}^* \wedge v_{i-2}^* \otimes Y_i\phi + v_{i-2}^* \wedge v_i^* \otimes Y_{i+2}\phi \end{pmatrix}.$$

Finally, applying $(-f_-^*, f_+^*) \otimes 1$ we get

$$0 - (v_i')^* \otimes Y_i\phi + 0 + 0 + (v_i')^* \otimes Y_i\phi + 0 = 0. \qquad \square$$

**Lemma 9.11.3** *We have*

(1) *$p$ is a locally split injection, and*
(2) *$q$ is a locally split surjection.*

*Proof.* (1) By Theorem 6.2.1 it suffices to prove that $p(-2)$ and hence $p$ is injective on fibres at every point in $\mathbb{P}^4 = \mathbb{P}(V)$. Choose a representative $0 \neq v \in V$ of $\bar{v} \in \mathbb{P}^4$. If $v'$ is in the kernel of $p$ at $\bar{v}$ then $\binom{f_+}{f_-}(v')$ is in the kernel of the differential in the Koszul complex and so there exist vectors $v_+$ and $v_-$ such that $f_+(v') = v \wedge v_+$ and $f_-(v') = v \wedge v_-$. By Lemma 9.11.1 this implies that $v' = 0$.

(2) The map $q$ is the dual of $p$. $\square$

Combining Lemmas 9.11.2 and 9.11.3 we have a monad

$$V' \otimes \mathcal{O}(2) \xrightarrow{p} W \otimes \Lambda^2 \mathcal{T} \xrightarrow{q} (V')^* \otimes \mathcal{O}(3). \tag{9.4}$$

**Definition 9.11.4** The Horrocks–Mumford bundle $\mathcal{F}_{\mathrm{HM}}$ is defined to be the cohomology of the monad (9.4).

**Proposition 9.11.5** *The Chern polynomial of the Horrocks–Mumford bundle* $\mathcal{F}_{\mathrm{HM}}$ *is equal to* $1 + 5h + 10h^2$.

*Proof.* The Chern polynomial of $\Lambda^2\mathcal{T}$ is $(1+2h)^{10}/(1+h)^5$. Applying Lemma 7.3.5 to the monad (9.4), we have

$$c(\mathcal{F}_{\mathrm{HM}}) = ((1+2h)^{10}/(1+h)^5)^2 \,/\, (1+2h)^5(1+3h)^5,$$

which is equal to $1 + 5h + 10h^2$ as an element of $\mathbb{Z}[h]/(h^5)$. □

**Theorem 9.11.6** *The Horrocks–Mumford bundle* $\mathcal{F}_{\mathrm{HM}}$ *is indecomposable.*

*Proof.* The Chern polynomial $1 + 5h + 10h^2$ does not factorise over the integers. See Lemma 7.3.9. □

**Remark 9.11.7** The minimal resolution of the Horrocks–Mumford bundle by direct sums of line bundles (cf. Corollary 6.12.5) is as follows:

$$0 \to \mathcal{O}(-5)^{\oplus 2} \to \mathcal{O}(-3)^{\oplus 20} \to \mathcal{O}(-2)^{\oplus 35} \to \mathcal{O}^{\oplus 4} \oplus \mathcal{O}(-1)^{\oplus 15} \to \mathcal{F}_{\mathrm{HM}} \to 0.$$

On the other hand, the construction above shows that there is a complex

$$0 \to W \otimes \mathcal{O} \to W \otimes V \otimes \mathcal{O}(1) \to (W \otimes \Lambda^2 V)/V' \otimes \mathcal{O}(2) \to (V')^* \otimes \mathcal{O}(3) \to 0$$

in which the maps are all linear, and which is exact except at the $\mathcal{O}(2)$ term, where the homology is $\mathcal{F}_{\mathrm{HM}}$. This complex has the form

$$0 \to \mathcal{O}^{\oplus 2} \to \mathcal{O}(1)^{\oplus 10} \to \mathcal{O}(2)^{\oplus 15} \to \mathcal{O}(3)^{\oplus 5} \to 0.$$

Writing $a, b, c, d, e$ for the functions of degree one in $V^*$ dual to the basis $v_0, \ldots, v_4$ of $V$, the matrices for the maps in this complex may be taken to be

$$\left(\begin{array}{c|c} a & \\ b & \\ c & \\ d & \\ e & \\ \hline & a \\ & b \\ & c \\ & d \\ & e \end{array}\right), \quad \left(\begin{array}{ccccc|ccccc} -c & & a & & & & & & & \\ & -d & & b & & & & & & \\ & & -e & & c & & & & & \\ d & & & -a & & & & & & \\ & e & & & -b & & & & & \\ \hline -b & a & & & & & & e & & -c \\ & -c & b & & & -d & & & a & \\ & & -d & c & & & -e & & & b \\ & & & -e & d & c & & -a & & \\ e & & & & -a & & d & & -b & \\ \hline & & & & & -b & a & & & \\ & & & & & & -c & b & & \\ & & & & & & & -d & c & \\ & & & & & & & & -e & d \\ & & & & & e & & & & -a \end{array}\right),$$

$$\left(\begin{array}{ccccc|ccccc|ccccc} d & & & c & & & & a & & & e & & & & b \\ & e & & & d & & & & b & & c & a & & & \\ e & & a & & & & & & & c & & d & b & & \\ & a & & b & & d & & & & & & & e & c & \\ & & b & & c & & e & & & & & & & a & d \end{array}\right).$$

## 9.12 Automorphisms of the Horrocks–Mumford Bundle

In this section, we describe the automorphism group of $\mathcal{F}_{HM}$, under the assumption that $k$ does not have characteristic 5.

Let $\varepsilon$ be a primitive fifth root of unity in $k$. We define automorphisms of $V$ using the following $5 \times 5$ matrices in $SL(5, k)$:

$$\alpha = \begin{pmatrix} \varepsilon & 0 & 0 & 0 & 0 \\ 0 & \varepsilon & 0 & 0 & 0 \\ 0 & 0 & \varepsilon & 0 & 0 \\ 0 & 0 & 0 & \varepsilon & 0 \\ 0 & 0 & 0 & 0 & \varepsilon \end{pmatrix} \qquad \beta = \begin{pmatrix} 1 & 0 & 0 & 0 & 0 \\ 0 & \varepsilon & 0 & 0 & 0 \\ 0 & 0 & \varepsilon^2 & 0 & 0 \\ 0 & 0 & 0 & \varepsilon^3 & 0 \\ 0 & 0 & 0 & 0 & \varepsilon^4 \end{pmatrix}$$

$$\gamma = \begin{pmatrix} 0 & 1 & 0 & 0 & 0 \\ 0 & 0 & 1 & 0 & 0 \\ 0 & 0 & 0 & 1 & 0 \\ 0 & 0 & 0 & 0 & 1 \\ 1 & 0 & 0 & 0 & 0 \end{pmatrix}$$

$$\zeta = \begin{pmatrix} 1 & 0 & 0 & 0 & 0 \\ 0 & \varepsilon^3 & 0 & 0 & 0 \\ 0 & 0 & \varepsilon^2 & 0 & 0 \\ 0 & 0 & 0 & \varepsilon^2 & 0 \\ 0 & 0 & 0 & 0 & \varepsilon^3 \end{pmatrix} \qquad \eta = -\frac{1}{\sqrt{5}} \begin{pmatrix} 1 & 1 & 1 & 1 & 1 \\ 1 & \varepsilon & \varepsilon^2 & \varepsilon^3 & \varepsilon^4 \\ 1 & \varepsilon^2 & \varepsilon^4 & \varepsilon & \varepsilon^3 \\ 1 & \varepsilon^3 & \varepsilon & \varepsilon^4 & \varepsilon^2 \\ 1 & \varepsilon^4 & \varepsilon^3 & \varepsilon^2 & \varepsilon \end{pmatrix}$$

$$\kappa = -\begin{pmatrix} 1 & 0 & 0 & 0 & 0 \\ 0 & 0 & 1 & 0 & 0 \\ 0 & 0 & 0 & 0 & 1 \\ 0 & 1 & 0 & 0 & 0 \\ 0 & 0 & 0 & 1 & 0 \end{pmatrix}.$$

We write $\alpha', \ldots, \kappa'$ for the automorphisms of $V'$ obtained by replacing each occurrence of $\varepsilon^i$ by $\varepsilon^{2i}$ in these matrices. Let $\alpha'', \ldots, \kappa''$ be the automorphisms of $W$ given as follows:

$$\alpha'' = \beta'' = \gamma'' = \begin{pmatrix} 1 & 0 \\ 0 & 1 \end{pmatrix} \quad \zeta'' = \begin{pmatrix} \varepsilon & 0 \\ 0 & \varepsilon^4 \end{pmatrix} \quad \eta'' = \frac{1}{\sqrt{5}} \begin{pmatrix} \varepsilon^2 - \varepsilon^3 & \varepsilon - \varepsilon^4 \\ \varepsilon - \varepsilon^4 & \varepsilon^3 - \varepsilon^2 \end{pmatrix}$$

$$\kappa'' = \begin{pmatrix} 0 & 1 \\ -1 & 0 \end{pmatrix}.$$

The elements $\alpha, \beta, \gamma$ satisfy the relations

$$\alpha^5 = \beta^5 = \gamma^5 = 1, \qquad \alpha\beta = \beta\alpha, \qquad \alpha\gamma = \gamma\alpha, \qquad \gamma\beta = \alpha\beta\gamma.$$

So the group $H = \langle \alpha, \beta, \gamma \rangle$ is an extraspecial group of order $5^{1+2}$ and exponent 5. Its centre is $Z = \langle \alpha \rangle$ of order 5, and we write $\bar{H}$ for $H/Z$.

The elements $\zeta$ and $\eta$ satisfy

$$\zeta^5 = 1, \qquad (\zeta\eta)^3 = 1, \qquad \eta^2 = -1$$

and so the group $L = \langle \zeta, \eta \rangle$ is isomorphic to $SL(2, 5)$. Furthermore, $\kappa$ is an element of $L$, and the subgroup $Q = \langle \eta, \kappa \rangle$ is a Sylow 2-subgroup of $L$, isomorphic to a quaternion group of order 8. The subgroup $R = \langle \zeta\eta \rangle$ is a Sylow 3-subgroup of $L$, of order 3.

The group $L$ normalises $H$,

$$\alpha\zeta = \zeta\alpha, \quad \alpha\eta = \eta\alpha, \quad \beta\zeta = \zeta\beta, \quad \beta\eta = \eta\gamma^{-1}, \quad \gamma\zeta = \zeta\alpha^3\beta\gamma, \quad \gamma\eta = \eta\beta,$$

and together they generate a semidirect product

$$N = \langle \alpha, \beta, \gamma, \zeta, \eta \rangle = H \rtimes L,$$

a group of order 15,000 and shape $5^{1+2}{:}\, SL(2, 5)$. The group $N$ is the normaliser in $SL(5, k)$ of $H$, and is the full automorphism group of $\mathfrak{F}_{\mathsf{HM}}$. Sylow 2-, 3- and 5-subgroups of $N$ are given by $Q$, $R$ and

$$S = \langle \alpha, \beta, \gamma, \zeta \rangle,$$

respectively.

The matrices $\alpha', \ldots, \kappa'$ and the matrices $\alpha'', \ldots, \kappa''$ satisfy the same relations as those given for $\alpha, \ldots, \kappa$ above, and so $V'$ and $W$ are also representations of $N$. In fact, $W$ is inflated from a representation of $\bar{N} = N/H \cong SL(2, 5)$. Corresponding to the isomorphism $\Lambda^2(W) \cong k$, the representation $W$ supports an invariant symplectic form given by the matrix

$$\begin{pmatrix} 0 & 1 \\ -1 & 0 \end{pmatrix}. \tag{9.5}$$

It follows that $W$ is self-dual.

Next we discuss the representation theory of $H$, and use it to explain why the normaliser of $H$ in $SL(5, k)$ is the group $N$ of shape $5^{1+2}{:}\, SL(2, 5)$. In an Appendix to this section, we give a listing of code in Magma [63] for finding the ordinary character table of $N$.

We continue to assume that $k$ does not have characteristic five. With this assumption, $H$ has 25 irreducible representations of dimension one with $Z$ in the kernel, and four faithful irreducible representations of dimension five. These four are distinguished by the scalar multiple of the identity matrix representing $\alpha$. We write $V_1, V_2, V_3, V_4$ for the irreducible representations of dimension five, where the scalar representing $\alpha$ on $V_i$ is equal to $\varepsilon^i$.

Let $\phi$ be an automorphism of $H$ that acts trivially on $Z$. If $\rho\colon H \to SL(5, k)$ is a homomorphism realising the representation $V_1$ then $\rho \circ \phi$ is another homomorphism realising an isomorphic representation. So there is a matrix $A_\phi$ such that for all $h \in H$ we have $\rho(\phi(h)) = A_\phi \rho(h) A_\phi^{-1}$. The matrix $A_\phi$ is well defined up to scalars, it can be chosen to be in $SL(5, k)$, and $A_\phi A_{\phi'}$ is a scalar multiple of $A_{\phi\phi'}$. Thus the matrices $A_\phi$ together with $\rho(H)$ generate the normaliser $N = N_{SL(5,k)}(H)$. The group $N$ surjects onto the group $\bar{N}$ of automorphisms of $H$ acting trivially on $Z$, which is an extension of the outer automorphisms $SL(2, 5)$ by the inner automorphisms $\bar{H}$. The extension of $SL(2, 5)$ by $\bar{H}$ splits because the centraliser of an involution in the centre of a Sylow 2-subgroup is a complement for $\bar{H}$. The kernel of $N \to \mathsf{Aut}(H)$ is $C_{SL(5,k)}(H)$, which consists just of the scalar matrices. There are five of these, and they are in $\rho(H)$. The group $N$ has order 15,000, and sits in an extension of the form

$$1 \to H \to N \to SL(2, 5) \to 1.$$

So we regard $H$ as a subgroup of $N$ and $V_1$ as a representation of $N$. Since

$$H^2(SL(2, 5), \mathbb{Z}/5) = 0$$

the extension splits, so that $N$ is isomorphic to the semidirect product $H \rtimes SL(2, 5)$.

$$\begin{array}{ccccccccc}
 & & & & 1 & & 1 & & \\
 & & & & \downarrow & & \downarrow & & \\
1 & \longrightarrow & Z & \longrightarrow & H & \longrightarrow & \bar{H} & \longrightarrow & 1 \\
 & & \| & & \downarrow & & \downarrow & & \\
1 & \longrightarrow & Z & \longrightarrow & N & \longrightarrow & \bar{N} & \longrightarrow & 1 \\
 & & & & \downarrow \uparrow & & \downarrow \uparrow & & \\
 & & & & SL(2,5) & = & SL(2,5) & & \\
 & & & & \downarrow & & \downarrow & & \\
 & & & & 1 & & 1 & &
\end{array}$$

So $V_i$ extends uniquely to a representation of $N$ for $1 \le i \le 4$ which we continue to denote $V_i$. Thus we have

$$V_1 \cong V, \quad V_2 \cong V', \quad V_3 \cong (V')^* \cong \Lambda^4(V'), \quad V_4 \cong V^* \cong \Lambda^4(V).$$

Next, we apply this discussion to the construction of the Horrocks–Mumford bundle given in Section 9.11. We begin by explaining the roles of the maps $\binom{f_+}{f_-} : V' \to W \otimes \Lambda^2 V$ and $(-f_-, f_+)\colon W \otimes \Lambda^2 V^* \to (V')^*$ used in the constructions there.

**Lemma 9.12.1** *There is an isomorphism of $kN$-modules $W \otimes V_2 \cong \Lambda^2 V_1$.*

*Proof.* The maps $f_+$ and $f_-$ of the previous section give rise to a map

$$W \otimes V_2 \xrightarrow{(f_-, -f_+)} \Lambda^2 V_1 \tag{9.6}$$

$$\begin{pmatrix} v'_i \\ v'_j \end{pmatrix} \quad \mapsto \quad v_{i+1} \wedge v_{i-1} - v_{j+2} \wedge v_{j-2}. \tag{9.7}$$

We check that

$$(f_-, -f_+) \circ (\alpha'' \otimes \alpha') = \Lambda^2(\alpha) \circ (f_-, -f_+),$$

and similarly for $\beta, \ldots, \kappa$. Therefore, $(f_1, -f_+)$ is a $kN$-module homomorphism. It is non-zero, and $\Lambda^2 V_1$ is irreducible, so it is an isomorphism. □

Using the isomorphism $W \cong W^*$ given by the symplectic form (9.5), the adjoint of the map given in Lemma 9.12.1 is the map

$$\begin{pmatrix} f_+ \\ f_- \end{pmatrix} : V_2 \to W \otimes \Lambda^2 V_1$$

appearing in Section 9.11.

Dually, we have the following.

**Lemma 9.12.2** *There is an isomorphism*

$$W \otimes V_3 \xrightarrow{(f_+^*, f_-^*)} \Lambda^2 V_4.$$

The adjoint of this is the map

$$(-f_-^*, f_+^*) \colon V_3 \to W \otimes \Lambda^2 V_4$$

used in Section 9.11.

We have thus seen that the maps used in the construction of $\mathcal{F}_{\mathrm{HM}}$ commute with the automorphisms given by the group $N$. Thus $N$ acts as automorphisms of the Horrocks–Mumford bundle.

## Appendix: Characters of $N$

Recall the shape of the group $N$ of order 15,000:

$$\begin{array}{l} N \\ \big| \; SL(2,5) \\ H \\ \big| \; (\mathbb{Z}/5)^2 \\ Z \\ \big| \; \mathbb{Z}/5 \\ 1\,. \end{array}$$

In this Appendix, we provide a listing of code in the Magma computer algebra language [63] to construct the group $N$ as a subgroup of $SL(5, p)$, provided $p$ is congruent to 1 modulo 5, and produce its character table. The value of $p$ may be changed from 11 to any other prime congruent to 1 modulo 5 by changing the first line. For primes not congruent to 1 modulo 5, it is necessary first to extend the ground field so that it contains fifth roots of unity.

```
p:=11;
G:=SL(5,p);
H:=SylowSubgroup(G,5);
Z:=Centre(H);
K:=H;
while Order(K) ne 5 do z:=Random(Z); K:=sub<H|z>; end while;
while IsAbelian(K) do
 while Order(K) ne 25 do x:=Random(H); K:=sub<H|z,x>; end while;
 while Order(K) ne 125 do y:=Random(H); K:=sub<H|z,x,y>; end while;
end while;
N:=Normaliser(G,K);
X:=CharacterTable(N);
```

The character degrees are as follows: 1 (trivial module $k$), 2 (twice, $W$ and $W'$), 3 (twice, $U$ and $U'$), 4 (twice, $Y = W \otimes W'$ and $Y'$), 5 (five times, $V_1, \ldots, V_4, Z$), 6 ($T$), 10 (eight times, $W \otimes V_i$ and $W' \otimes V_i$ for $1 \le i \le 4$), 15 (eight times, $U \otimes V_i$ and $U' \otimes V_i$ for $1 \le i \le 4$), 20 (eight times, $Y \otimes V_i$ and $Y' \otimes V_i$ for $1 \le i \le 4$), 24 (five times), 25 (four times, $Z \otimes V_i$ for $1 \le i \le 4$) and 30 (four times, $T \otimes V_i$ for $1 \le i \le 4$), for a total of 50 irreducible characters. Here are some tensor products of characters of low degree:

| $k$ | $W$ | $W'$ | $U$ | $U'$ | $Y$ | $Y'$ | $Z$ |
|---|---|---|---|---|---|---|---|
| $W$ | $k\backslash U$ | $Y$ | $W' \oplus Y'$ | $T$ | $W' \oplus T$ | $U \oplus Z$ | $Y' \oplus T$ |
| $W'$ | | $k\backslash U'$ | $T$ | $W' \oplus Y'$ | $W \oplus T$ | $U' \oplus Z$ | $Y' \oplus T$ |
| $U$ | | | $U\backslash k \oplus Z$ | $Y \oplus Z$ | $U' \oplus Y \oplus Z$ | $W \oplus Y' \oplus T$ | $U \oplus U' \oplus Y \oplus Z$ |
| $U'$ | | | | $U'\backslash k \oplus Z$ | $U \oplus Y \oplus Z$ | $W' \oplus Y' \oplus T$ | $U \oplus U' \oplus Y \oplus Z$ |

Here, the backslash separates the exterior square from the symmetric square in the case of tensoring a character with itself. Further tensor products are given by $\Lambda^2(V_1) = W \otimes V_2$, $S^2(V_1) = U' \otimes V_2$, $\Lambda^3(V_1) = W \otimes V_3$, $S^3(V_1) = (W \oplus Z) \otimes V_3$, $\Lambda^4(V_1) = V_4$, $S^4(V_1) = (U \oplus U' \oplus Y \oplus Y') \otimes V_4$, $V_1 \otimes V_2 = (W' \oplus U) \otimes V_3$, $V_1 \otimes V_3 = (W \oplus U') \otimes V_4$, and $V_1 \otimes V_4 = k \oplus$ irreducible 24.

*Second proof of Lemma 9.11.2 in characteristic* $\geq 7$. Recall that we have

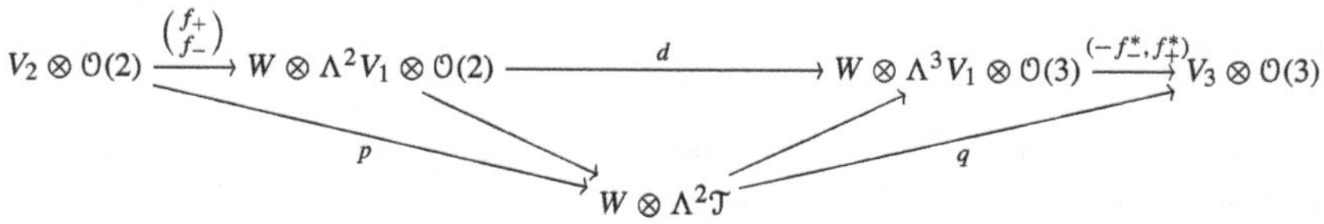

All of the maps commute with the action of $N$, and so the composite $q \circ p$ is represented by a $kN$-module homomomorphism $V_1 \otimes V_2 \to V_3$. But $V_1 \otimes V_2 \cong (W' \oplus U) \otimes V_3$ is a sum of two irreducible characters, and does not admit any non-zero $kN$-module homomorphisms to $V_3$. □

## 9.13 Realising the Horrocks–Mumford Bundle

We introduced the Horrocks–Mumford bundle $\mathcal{F}_{\mathsf{HM}}$ in Section 9.11. For the exterior algebra on five variables we can produce a module $M$ of dimension 32 and constant Jordan type $[2]^{15}[1]^2$ with $\mathcal{F}_1(M) \cong \mathcal{F}_{\mathsf{HM}}$ in any characteristic, by using the matrices given in that section, see Figure 9.1. This figure shows the action of $\theta = aX_1 + bX_2 + cX_3 + dX_4 + eX_5$.

In this section, we give a module theoretic construction for $\mathcal{F}_{\mathsf{HM}}$ which is similar to the construction of the functors $\mathcal{F}_i$ but different.

**Remark 9.13.1** In Chapter 8, we proved that given any vector bundle $\mathcal{F}$ of rank $s$ on $\mathbb{P}^{r-1}$, there is a finitely generated $kE$-module $M$ of stable constant Jordan type $[1]^s$ such that:

(i) if $p = 2$ then $\mathcal{F}_1(M) \cong \mathcal{F}$, and
(ii) if $p$ is odd, then $\mathcal{F}_1(M) \cong F^*(\mathcal{F})$, where $F\colon \mathbb{P}^{r-1} \to \mathbb{P}^{r-1}$ is the Frobenius morphism.

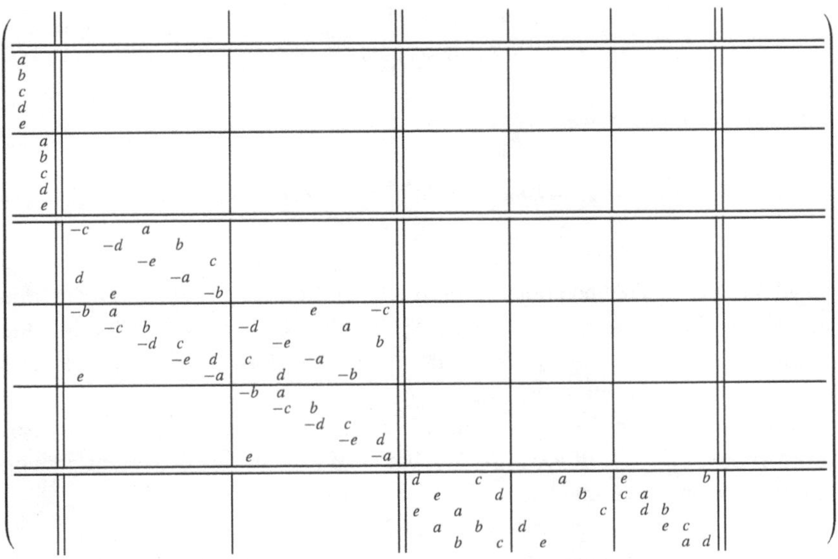

Figure 9.1 Matrix for $\theta$ for the Horrocks–Mumford bundle example.

However, we also proved in Section 10.2 that for a module $M$ of stable constant Jordan type $[1]^s$, the Chern numbers $c_i(\mathcal{F}_1(M))$ are divisible by $p$ for $1 \leq i \leq p-2$. So although $F^*(\mathcal{F}_{\mathsf{HM}})$ can be constructed by this general construction, no twist of $\mathcal{F}_{\mathsf{HM}}$ itself can be $\mathcal{F}_1(M)$ for a module $M$ of constant Jordan type $[1]^2$, since $c_1(\mathcal{F}_{\mathsf{HM}}(i)) = 2i+5$ and $c_2(\mathcal{F}_{\mathsf{HM}}(i)) = i^2+5i+10$ cannot both be divisible by $p$ for $p \geq 7$. This demonstrates that we are forced to use a different construction.

We begin with a description of extensions of simple modules over semidirect products of the form $kP \rtimes G$ in preparation for the construction. Let $P$ be a finite $p$-group, $k$ be a field of characteristic $p$, $G$ be a finite $p'$-group acting as algebra automorphisms on $kP$, and let $A = kP \rtimes G$, a finite-dimensional algebra over $k$. Let $V$ be the $A$-module given by $J(kP)/J^2(kP)$. Note that $P$ acts trivially on $V$, so that it is really just a $kG$-module. The simple $A$-modules all have $P$ in the kernel, so they are the same as the simple $kG$-modules. The following proposition says that $\mathrm{Ext}^1$ between these simple modules can be computed just using character theory.

**Proposition 9.13.2** *If $S_1$ and $S_2$ are $A$-modules with $P$ in the kernel then*

$$\mathrm{Ext}^1_A(S_1, S_2) \cong \mathrm{Hom}_{kG}(V \otimes_k S_1, S_2).$$

*Proof.* As $kP$-modules, $S_1$ and $S_2$ are trivial, so

$$\begin{aligned}\mathrm{Ext}^1_{kP}(S_1, S_2) &\cong \mathrm{Ext}^1_{kP}(k, k) \otimes_k \mathrm{Hom}_k(S_1, S_2)\\ &\cong V^* \otimes_k \mathrm{Hom}_k(S_1, S_2)\\ &\cong \mathrm{Hom}_k(V \otimes_k S_1, S_2).\end{aligned}$$

Since $G$ is a $p'$-group, we have

$$\mathrm{Ext}^1_A(S_1, S_2) \cong \mathrm{Ext}^1_{kP}(S_1, S_2)^H \cong \mathrm{Hom}_{kG}(V \otimes_k S_1, S_2). \qquad \square$$

More explicitly, given a $kG$-module homomorphism $\rho\colon V \otimes_k S_1 \to S_2$, there is an adjoint $kG$-module homomorphism $\rho'\colon V \to \mathrm{Hom}_k(S_1, S_2)$. We can describe the corresponding extension of $S_1$ by $S_2$ as follows. As a module for $kG$, the extension splits as $S_1 \oplus S_2$. The action of $P$ has $J^2(kP)$ in the kernel, and $v \in J(kP)/J^2(kP)$ acts as the matrix $\left(\begin{smallmatrix} 0 & 0 \\ \rho'(v) & 0 \end{smallmatrix}\right)$.

Note that if $G$ is not a $p'$-group then $\mathrm{Hom}_{kG}(V \otimes_k S_1, S_2)$ is canonically isomorphic to a subspace of $\mathrm{Ext}^1_A(S_1, S_2)$, but there may also be extensions coming from extensions of $kG$-modules.

Let us apply this in the case of $kE \rtimes N$-modules, where $E = (\mathbb{Z}/p)^5$ and $N = 5^{1+2}\colon SL(2,5)$ acting on $X_1, \ldots, X_5$ via the matrices given in Section 9.12. Recall from Sections 9.11 and 9.12 that we have a diagram of sheaves on $\mathbb{P}^4$,

$$\begin{array}{ccccccc}
V_2 \otimes \mathcal{O}(2) & \xrightarrow{\binom{f_+}{f_-}} & W \otimes \Lambda^2 V_1 \otimes \mathcal{O}(2) & \xrightarrow{\quad d \quad} & W \otimes \Lambda^3 V_1 \otimes \mathcal{O}(3) & \xrightarrow{(-f_-^*, f_+^*)} & V_3 \otimes \mathcal{O}(3), \\
& \searrow^{p} & \searrow & & \nearrow & \nearrow_{q} & \\
& & & W \otimes \Lambda^2 \mathcal{T} & & &
\end{array} \tag{9.8}$$

and that the Horrocks–Mumford bundle $\mathcal{F}_{\mathsf{HM}}$ is defined to be $\mathsf{Ker}\,(q)/\mathsf{Im}\,(p)$ as a subquotient of $W \otimes \Lambda^2\mathcal{T}$.

Using Proposition 9.13.2, we realise the Koszul differential

$$d \colon W \otimes \Lambda^2 V_1 \otimes \mathcal{O}(2) \to W \otimes \Lambda^3 V_1 \otimes \mathcal{O}(3)$$

by a homomorphism $V_1 \otimes (W \otimes \Lambda^2 V_1) \to W \otimes \Lambda^3 V_1$ and hence by an extension of $kE \rtimes N$-modules

$$0 \to W \otimes \Lambda^3 V_1 \to M \to W \otimes \Lambda^2 V_1 \to 0.$$

Thus $M$ is a $kE$-module of Loewy length two. Write $N_1$ for the kernel of $(-f_-, f_+)\colon W \otimes \Lambda^3 V_1 \twoheadrightarrow V_3$ and $N_2$ for the cokernel of $\binom{f_+}{f_-} \colon V_2 \rightarrowtail W \otimes \Lambda^2 V_1$. Then we have a complex of $kE \rtimes N$-modules

$$N_1 \overset{\phi}{\rightarrowtail} M \overset{\psi}{\twoheadrightarrow} N_2.$$

This gives rise to a diagram of sheaves on $\mathbb{P}^4$:

$$\begin{array}{ccc}
& \widetilde{M} & \xrightarrow{\ \widetilde{\psi}\ } \widetilde{N}_2. \\
& \big\downarrow{\scriptstyle \theta} & \\
\widetilde{N}_1(1) & \xrightarrow{\widetilde{\phi}(1)} & \widetilde{M}(1)
\end{array} \tag{9.9}$$

**Theorem 9.13.3** *In Diagram* (9.9), *we have*

$$\theta(\mathsf{Ker}\,\widetilde{\psi}) \subseteq \mathsf{Im}\,(\widetilde{\phi}(1))$$

*and*

$$\mathsf{Im}\,(\theta) \cap \mathsf{Im}\,(\widetilde{\phi}(1)) \,/\, \theta(\mathsf{Ker}\,\widetilde{\psi}) \cong \mathcal{F}_{\mathsf{HM}}(-2),$$

*the Horrocks–Mumford bundle untwisted twice.*

*Proof.* In Diagram (9.9), $\widetilde{\mathsf{Soc}(M)}$ is in the kernel of $\widetilde{\psi}$ and $\theta$, so we may quotient it out. Similarly, the images of $\widetilde{\phi}(1)$ and $\theta$ are contained in $\widetilde{\mathsf{Soc}(M)}(1)$ so we obtain a diagram

$$\begin{array}{ccccccccc}
0 & \longrightarrow & \widetilde{V}_2 & \longrightarrow & \widetilde{W \otimes \Lambda^2 V_1} & \xrightarrow{\ \widetilde{\psi}\ } & \widetilde{N}_2 & \longrightarrow & 0 \\
& & & & \big\downarrow{\scriptstyle \theta} & & & & \\
0 & \longrightarrow & \widetilde{N}_1(1) & \xrightarrow{\widetilde{\phi}(1)} & \widetilde{W \otimes \Lambda^3 V_1}(1) & \longrightarrow & \widetilde{V}_3(1) & \longrightarrow & 0
\end{array}$$

where the rows are exact. Twisting this twice and comparing with Diagram (9.8) proves the theorem. □

## 9.14 The Horrocks Parent Bundle and the Tango Bundle

In this section, we describe the Horrocks parent bundle of rank three on $\mathbb{P}^5$ in any characteristic. Further information on this bundle can be found in Horrocks [160] as well as the papers [11, 101, 236]. Let $V$ be a vector space of dimension six over $k$, and choose a decomposition $V = U \oplus U^*$ as a sum of a vector space of dimension three and its dual. Choose a basis $v_1, v_2, v_3$ of $U$, and let $v_4, v_5, v_6$ be the dual basis of $U^*$. We have an action of $SL(3, k)$ on $U$ as matrices with determinant one, and on $U^*$ as the transpose inverses of these matrices. This gives an action of $G = SL(3, k) \rtimes \mathbb{Z}/2$ on $V$. Here, the $\mathbb{Z}/2$ acts on $SL(3, k)$ through transpose inverse on matrices. The group $G$ will be the full automorphism group of the Horrocks parent bundle.

Using the determinant form given by these bases, we have isomorphisms $U \cong \Lambda^2(U^*)$ and $U^* \cong \Lambda^2(U)$ given by

$$v_1 \longleftrightarrow v_5 \wedge v_6, \qquad v_2 \longleftrightarrow v_6 \wedge v_4, \qquad v_3 \longleftrightarrow v_4 \wedge v_5 \tag{9.10}$$

and

$$v_4 \longleftrightarrow v_2 \wedge v_3, \qquad v_5 \longleftrightarrow v_3 \wedge v_1, \qquad v_6 \longleftrightarrow v_1 \wedge v_2. \tag{9.11}$$

So we have

$$\Lambda^2(V) = \Lambda^2(U) \oplus \Lambda^2(U^*) \oplus (U \otimes U^*) \cong U^* \oplus U \oplus (U \otimes U^*) \cong V \oplus (U \otimes U^*).$$

The isomorphisms (9.10) and (9.11) combine to give maps

$$V \xrightarrow{\phi} \Lambda^2(V) \xrightarrow{\psi} V$$

with $\psi\phi = \mathrm{Id}_V$.

There is a non-degenerate $G$-invariant symplectic form $\omega$ on $V$ given by

$$\omega((u, \rho), (u', \rho')) = \rho'(u) - \rho(u')$$

for $u, u' \in U$ and $\rho, \rho' \in U^*$. With respect to the basis $v_1, \dots, v_6$ of $V$ this has matrix $\left(\begin{smallmatrix} 0 & I \\ -I & 0 \end{smallmatrix}\right)$. This symplectic form induces a $G$-invariant isomorphism $V \cong V^*$, under which the image of $\omega$ is $\omega^* = v_1 \wedge v_4 + v_2 \wedge v_5 + v_3 \wedge v_6$.

The form $\omega$ induces a non-degenerate $G$-invariant symmetric bilinear form on $\Lambda^2 V$ via

$$((v \wedge v'), (v'' \wedge v''')) \mapsto \omega(v, v')\omega(v'', v''') + \omega(v, v'')\omega(v''', v') + \omega(v, v''')\omega(v', v'')$$

and hence an isomorphism $\Lambda^2 V \cong (\Lambda^2 V)^* = \Lambda^2 V^*$. The following lemma is easy to check.

**Lemma 9.14.1** (i) *$\phi$ sends every element $v \in V$ to a sum of wedges of elements $v' \wedge v''$ satisfying $\omega(v, v') = \omega(v, v'') = 0$.*

(ii) *For all $v, v', v''$ in $V$ we have $\omega(v, \psi(v' \wedge v'')) = \omega(v', \psi(v'' \wedge v)) = \omega(v'', \psi(v \wedge v'))$. In particular, we have $\omega(v, \psi(v, v')) = 0$.* □

These maps allow us to define a complex of sheaves on $\mathbb{P}^5 = \mathbb{P}(V)$:

$$0 \to \mathcal{O}(-2) \to (k \oplus V) \otimes \mathcal{O}(-1) \to \Lambda^2(V) \otimes \mathcal{O} \to (V \oplus k) \otimes \mathcal{O}(1) \to \mathcal{O}(2) \to 0. \tag{9.12}$$

The first two maps in the complex are given by the Koszul differential combined with the map $\phi$. The third and fourth maps are related to these by taking the duals using the isomorphisms above, and then conjugating by the matrix $\begin{pmatrix} 0 & I \\ -I & 0 \end{pmatrix}$. More explicitly, these maps are as follows:

(i) $\mathcal{O}(-2) \to (k \oplus V) \otimes \mathcal{O}(-1)$
$v : \lambda \mapsto (0, \lambda v)$

(ii) $(k \oplus V) \otimes \mathcal{O}(-1) \to \Lambda^2 V \otimes \mathcal{O}$
$v : (\lambda, v') \mapsto \lambda\phi(v) + v \wedge v'$

(iii) $\Lambda^2 V \otimes \mathcal{O} \to (V \oplus k) \otimes \mathcal{O}(1)$

$$v : v' \wedge v'' \mapsto (\omega(v, v')v'' + \omega(v', v'')v + \omega(v'', v)v', \omega(v, \psi(v' \wedge v'')))$$

(iv) $(V \oplus k) \otimes \mathcal{O}(1) \to \mathcal{O}(2)$
$v : (v', \lambda) \mapsto \omega(v, v')$.

**Theorem 9.14.2** *The complex* (9.12) *only has homology in the middle term, and is exact elsewhere.*

*Proof.* It is easy to check using the explicit formulae above together with Lemma 9.14.1 that the composites are zero. Alternatively, it can be verified by multiplying the matrices given below for the maps. More abstractly, it comes from the fact that the construction is based on the exterior square of a null correlation bundle for the symplectic form, see [160].

Exactness at the second term amounts to showing that if $0 \neq v \in V$, and $(\lambda, w) \in k \oplus V$ satisfies

$$\lambda\phi(v) + v \wedge w = 0 \tag{9.13}$$

then $\lambda = 0$ and $w$ is a multiple of $v$. To see this, let $v = (v', v'')$, $w = (w', w'')$ with $v', w' \in U$ and $v'', w'' \in U^*$. Projecting the equation (9.13) onto the summand $U \otimes U^*$ of $\Lambda^2(V)$, the term $\lambda\phi(v)$ vanishes and we are left with $v' \wedge w'' + v'' \wedge w' = 0$ in $\Lambda^2(V)$. As elements of $U \otimes U^*$ this gives $v' \otimes w'' = w' \otimes v''$. Since $v \neq 0$ this implies that for some scalar $\mu \in k$ we have $w' = \mu v'$ and $w'' = \mu v''$. Thus $w = \mu v$. Plugging this back into (9.13), we see that $\lambda\phi(v) = 0$. Since $\phi$ is injective, we conclude that $\lambda = 0$.

For exactness at the fourth term, we first check that (0, 1) is in the image. Given $0 \neq v = \lambda_1 v_1 + \cdots + \lambda_6 v_6$, some $\lambda_i$ is non-zero. Since there are elements of $G$

sending any $v_i$ to any other, we may assume that $i = 1$ without loss of generality. Consider the element

$$-\lambda_1 v_2 \wedge v_3 - \lambda_6 v_2 \wedge v_4 + \lambda_5 v_3 \wedge v_4.$$

A short computation using the definitions shows that $v$ sends this element to $(0, \lambda_1)$, and hence dividing by $\lambda_1$, $(0, 1)$ is also in the image.

Having established this, we can ignore the second term in the image, and prove that given $v' \in V$ with $\omega(v, v') = 0$, there exists an element in $\Lambda^2 V$ taken by $v$ to something with $v'$ in its first component. If $v'$ is not a multiple of $v$, then by nondegeneracy of $\omega$ we may choose $v'' \in V$ with $\omega(v'', v) = 1$ and $\omega(v', v") = 0$. Then

$$v\colon v' \wedge v'' \mapsto (v', \omega(v, \psi(v' \wedge v'')))$$

does the trick. If $v' = v$, we choose $v''$ and $v'''$ with $\omega(v, v'') = 0$, $\omega(v, v''') = 0$, $\omega(v'', v''') = 1$. Then

$$v\colon v'' \wedge v''' \mapsto (v, \omega(v, \psi(v'' \wedge v''')))$$

does the trick. So we are done. □

**Definition 9.14.3** The Horrocks parent bundle is the homology at the middle term of the complex (9.12).

**Remark 9.14.4** From the construction, we see that the dual of the Horrocks parent bundle is the pullback of this bundle through the automorphism $\begin{pmatrix} 0 & I \\ -I & 0 \end{pmatrix}$ of $\mathbb{P}^5$.

Writing $a, b, c, d, e, f$ for the functions of degree one in $V^*$ dual to the basis $v_1, \ldots, v_6$ of $V$, the matrices for the maps in this complex may be taken to be

$$\left(\begin{array}{c} 0 \\ \hline a \\ b \\ c \\ d \\ e \\ f \end{array}\right), \quad \left(\begin{array}{c|cccccc} a & & & & & f & -e \\ b & & & & -f & & d \\ c & & & & e & -d & \\ d & & c & -b & & & \\ e & -c & & a & & & \\ f & b & -a & & & & \\ \hline & d & & & -a & & \\ & e & & & & -a & \\ & f & & & & & -a \\ & & d & & -b & & \\ & & e & & & -b & \\ & & f & & & & -b \\ & & & d & -c & & \\ & & & e & & -c & \\ & & & f & & & -c \end{array}\right),$$

$$\left(\begin{array}{cccccc|ccccccccc} & & & & f & -e & & b & c & & -a & & & & -a \\ & & & -f & & d & -b & & & a & & c & & & -b \\ & & & e & -d & & -c & & & & -c & & a & b & \\ & c & -b & & & & & & & -e & d & & -f & & d \\ -c & & a & & & & e & -d & & & & & & -f & e \\ b & -a & & & & & f & & -d & & f & -e & & & \\ \hline d & e & f & -a & -b & -c & & & & & & & & & \end{array}\right),$$

$$\left(\begin{array}{cccccc|c} d & e & f & a & b & c & 0 \end{array}\right).$$

**Proposition 9.14.5** *The Horrocks parent bundle has rank three and Chern polynomial* $1 + 3h^2$.

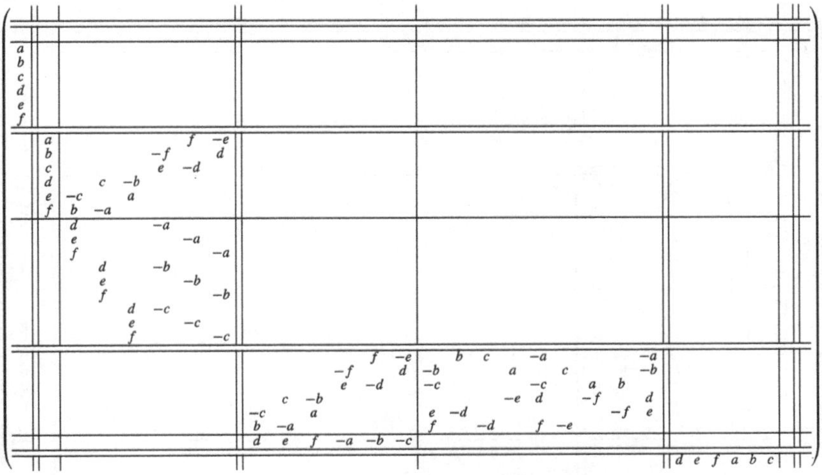

Figure 9.2 Matrix for $\theta$ for the Parent bundle example.

*Proof.* Using the defining sequence (9.12), we see that the rank is $1 - 7 + 15 - 7 + 1 = 3$ and the Chern polynomial is

$$(1 - 2h)(1 - h)^{-7}(1 + h)^{-7}(1 + 2h)$$

which is equal to $1 + 3h^2$ as an element of $\mathbb{Z}[h]/(h^6)$. □

**Theorem 9.14.6** *If $k$ does not have characteristic two, then the Horrocks parent bundle is irreducible.*

*Proof.* Since it has rank three, if it is not irreducible, then either it or its dual contains a line bundle $\mathcal{O}(j)$. The quotient is then a vector bundle of rank two with Chern polynomial $(1 + 3h^2)/(1 + jh)$, and hence $j = 0$.

Now restrict to the projective subspace $\mathbb{P}(U)$. The restriction of (9.12) is a direct sum of two twists of the (exact) Koszul complex and a monad:

$$0 \longrightarrow \mathcal{O}(-2) \longrightarrow U \otimes \mathcal{O}(-1) \longrightarrow U^* \otimes \mathcal{O} \longrightarrow \mathcal{O}(1) \longrightarrow 0$$

$$0 \longrightarrow \mathcal{O}(-1) \longrightarrow U \otimes \mathcal{O} \longrightarrow U^* \otimes \mathcal{O}(1) \longrightarrow \mathcal{O}(2) \longrightarrow 0$$

$$0 \longrightarrow U^* \otimes \mathcal{O}(-1) \longrightarrow U \otimes U^* \otimes \mathcal{O} \longrightarrow U \otimes \mathcal{O}(1) \longrightarrow 0.$$

The maps in the monad are $u\colon \rho \to u \otimes \rho$ and $u\colon u' \otimes \rho \to \rho(u')u - \rho(u)u'$ for $u, u' \in U$ and $\rho \in U^*$.

It follows that the restriction of the Horrocks parent bundle to $\mathbb{P}(U)$ is the homology of this monad, which is the quotient of $\mathcal{T} \otimes \mathbf{\Omega} = \mathcal{H}om(\mathcal{T}, \mathcal{T})$ by the copy of $\mathcal{O}$ given by multiples of the identity. If $k$ does not have characteristic two, then

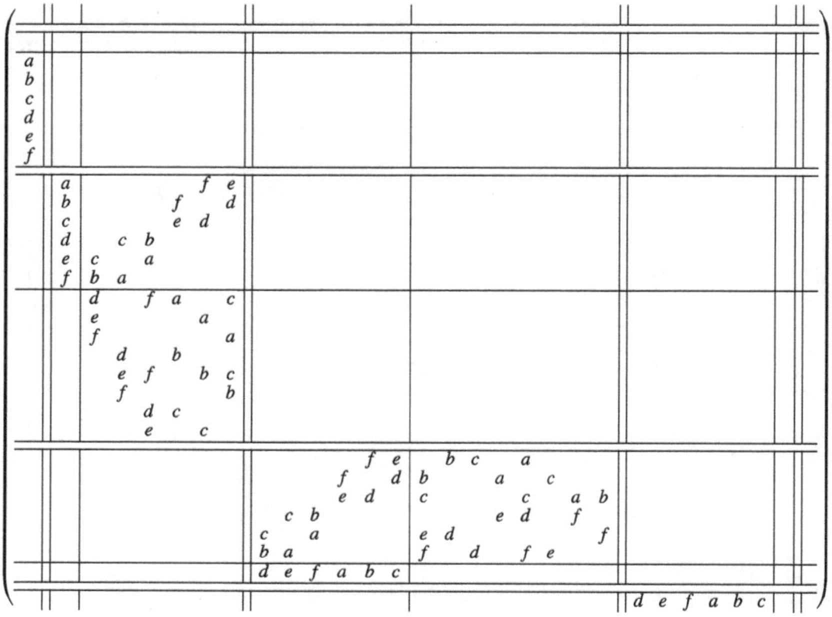

Figure 9.3 Matrix for $\theta$ for the Tango bundle example.

$\mathcal{H}om(\mathcal{T}, \mathcal{T})$ decomposes as a direct sum of $\mathcal{O}$ and a self-dual bundle of rank three which does not have a subbundle isomorphic to $\mathcal{O}$. □

**Theorem 9.14.7** *If $k$ has characteristic two, then the Horrocks parent bundle contains a copy of $\mathcal{O}$, and the quotient is an irreducible vector bundle of rank two on $\mathbb{P}^5$ with Chern polynomial $1 + 3h^2$.*

*Proof.* Under the map (iii) above, $v$ sends the symplectic form $\omega^* \in \Lambda^2 V$ to $2v$. This is zero in characteristic two. Since $\omega^*$ is not in the image of the map (ii) for any $v \in V$, it follows that this defines a trivial subbundle of the Horrocks parent bundle. The quotient is irreducible because its Chern polynomial does not factorise as a product of two linear factors. □

**Definition 9.14.8** The Tango bundle is the vector bundle of rank two on $\mathbb{P}^5$ in characteristic two described in the theorem. It is thus the homology in the middle of a complex of the form

$$0 \to \mathcal{O}(-2) \to (k \oplus V) \otimes \mathcal{O}(-1) \to \Lambda^2(V)/\langle\omega^*\rangle \otimes \mathcal{O} \to (V \oplus k) \otimes \mathcal{O}(1) \to \mathcal{O}(2) \to 0$$

which is exact elsewhere.

For the exterior algebra on six variables, in odd characteristic, we can produce a module $M$ of dimension 31 and constant Jordan type $[2]^{14}[1]^3$ with $\mathcal{F}_1(M)$ isomorphic to the Horrocks parent bundle on $\mathbb{P}^5$, by using the matrices given in Section 9.14, see Figure 9.2 on page 229. This matrix shows the action of $\theta = aX_1 + bX_2 + cX_3 + dX_4 + eX_5 + fX_6$.

In characteristic two, the exterior algebra is isomorphic to the group algebra of an elementary abelian 2-group. So again using the matrices from Section 9.14, we obtain a module for $E = (\mathbb{Z}/2)^6$ of dimension 30 and stable constant Jordan type $[2]^{14}[1]^2$ in characteristic two such that $\mathcal{F}_1(M)$ is isomorphic to the Tango bundle on $\mathbb{P}^5$. See Figure 9.3 on page 230.

# 10

# Restrictions Coming from Chern Numbers

The purpose of this chapter is to use the theory of Chern numbers and the Hirzebruch–Riemann–Roch theorem to obtain information about modules of constant Jordan type and the vector bundles associated with them.

The method works more generally for studying linear varieties of matrices of constant rank, and we begin with an easy example of how to use Chern numbers to study this situation.

## 10.1 Matrices of Constant Rank

In this section, we use the theory of Chern numbers to prove theorems of Sylvester and Westwick on matrices of constant rank. We then use this to obtain information about modules of constant Jordan type.

Recall from Section 7.1 that the Chow ring of $\mathbb{P}^{r-1}$ is

$$A^*(\mathbb{P}^{r-1}) \cong \mathbb{Z}[h]/(h^r).$$

If $\mathcal{F}$ is a vector bundle on $\mathbb{P}^{r-1}$, we write

$$c(\mathcal{F}) = \sum_{j \geq 0} c_j(\mathcal{F})h^j \in A^*(\mathbb{P}^{r-1})$$

for the Chern polynomial, where $c_0(\mathcal{F}) = 1$ and the $c_i(\mathcal{F}) \in \mathbb{Z}$ $(1 \leq i \leq r-1)$ are the Chern numbers of $\mathcal{F}$. If $0 \to \mathcal{F} \to \mathcal{F}' \to \mathcal{F}'' \to 0$ is a short exact sequence of vector bundles then we have $c(\mathcal{F}') = c(\mathcal{F})c(\mathcal{F}'')$.

**Theorem 10.1.1** (Sylvester [231]) *Let $A_1, \dots, A_r$ span a linear variety of $n \times n$ matrices of constant rank $n-1$ with $n > 1$. Then $r \leq 3$, and if $r = 3$ then $n$ is odd.*

*Proof.* We use $A_1, \dots, A_r$ to construct an operator on vector bundles on $\mathbb{P}^{r-1}$,

$$\theta \colon \mathcal{O}^{\oplus n} \to \mathcal{O}(1)^{\oplus n}$$

as described in Section 8.2. Since $A_1, \dots, A_r$ span a linear variety of constant rank $n-1$, it follows using Theorem 6.2.1 that $\theta$ is a map of vector bundles whose kernel and cokernel are vector bundles of rank one. By Theorem 6.3.1 the kernel and cokernel are twists of the structure sheaf. So we obtain an exact sequence of vector bundles

$$0 \to \mathcal{O}(a) \to \mathcal{O}^{\oplus n} \xrightarrow{\theta} \mathcal{O}(1)^{\oplus n} \to \mathcal{O}(b) \to 0$$

for some $a, b \in \mathbb{Z}$. Applying Chern polynomials, we obtain

$$c(\mathcal{O}(a))c(\mathcal{O}(1)^{\oplus n}) = c(\mathcal{O}^{\oplus n})c(\mathcal{O}(b))$$

as elements of $\mathbb{Z}[h]/(h^r)$. Thus

$$(1 + ah)(1 + h)^n \equiv 1 + bh \pmod{h^r}.$$

If $r \geq 2$ then equating coefficients of $h$ gives $b = n + a$. If $r \geq 3$ then equating coefficients of $h^2$ gives $an + \binom{n}{2} = 0$, so that $a = -(n-1)/2$ and it follows that $n$ is odd. If $r \geq 4$ then equating coefficients of $h^3$ gives $a\binom{n}{2} + \binom{n}{3} = 0$. It follows that $n = 1$, contradicting the hypotheses. □

To see that the theorem is sharp, we exhibit the following examples. For $r = 2$ we have the matrices

$$Y_1A_1 + Y_2A_2 = \begin{pmatrix} Y_2 & & & & & & 0 \\ Y_1 & Y_2 & & & & & 0 \\ & Y_1 & \ddots & & & & \vdots \\ & & \ddots & Y_2 & & & \vdots \\ & & & Y_1 & Y_2 & & 0 \\ & & & & Y_1 & & 0 \end{pmatrix}$$

while for $r = 3$ and $n$ odd we have the matrices

$$Y_1A_1+Y_2A_2+Y_3A_3 = \left(\begin{array}{ccccccccccc} & -Y_3 & & & & & -Y_1 & & & & \\ Y_3 & & & & & & -Y_2 & -Y_1 & & & \\ & & & & & & -Y_3 & -Y_2 & -Y_1 & & \\ & & & & & & & \ddots & \ddots & \ddots & \\ & & & & & & & & -Y_3 & -Y_2 & -Y_1 \\ & & & & & & & & & -Y_3 & -Y_2 \\ & & & & & & & & & & \\ Y_1 & Y_2 & Y_3 & & & & & & & & \\ & Y_1 & Y_2 & Y_3 & & & & & & & \\ & & \ddots & \ddots & \ddots & & & & & & \\ & & & Y_1 & Y_2 & Y_3 & & & & & \\ & & & & Y_1 & Y_2 & & & & & \end{array}\right)$$

(Westwick [245, Theorem 4.1]).

Using the same method together with a determinant argument, we have the following generalisation of Theorem 10.1.1.

**Theorem 10.1.2** (Westwick [246]) *Let $A_1, \ldots, A_r$ span a linear variety of $n \times m$ matrices of constant rank $s > 0$. Then $r \leq m + n - 2s + 1$.*

*Proof.* By transposing if necessary, we may assume that $m \le n$. Again we use $A_1, \ldots, A_r$ to construct an operator on vector bundles on $\mathbb{P}^{r-1}$

$$\theta\colon \mathcal{O}^{\oplus m} \to \mathcal{O}(1)^{\oplus n}.$$

This is a map of vector bundles whose kernel and cokernel are vector bundles of rank $m-s$ and $n-s$ respectively. So we obtain an exact sequence of vector bundles

$$0 \to \mathcal{F} \to \mathcal{O}^{\oplus m} \to \mathcal{O}(1)^{\oplus n} \to \mathcal{F}' \to 0.$$

We obtain

$$c(\mathcal{F})(1+h)^n = c(\mathcal{F}')$$

as elements of $\mathbb{Z}[h]/(h^r)$, with $\deg c(\mathcal{F}) \le m-s$ and $\deg c(\mathcal{F}') \le n-s$. If $r \ge m+n-2s+2$ then comparing coefficients of $h^{n-s+1}, \ldots, h^{(m-s)+(n-s)+1}$ gives us a system of $m-s+1$ equations of the form

$$\begin{array}{llll}
\binom{n}{n-s+1} & + \binom{n}{n-s} c_1(\mathcal{F}) & + \cdots + \binom{n}{n-m+1} c_{m-s}(\mathcal{F}) & = 0 \\
\binom{n}{n-s+2} & + \binom{n}{n-s+1} c_1(\mathcal{F}) & + \cdots + \binom{n}{n-m+2} c_{m-s}(\mathcal{F}) & = 0 \\
\quad\cdots & \quad\cdots & \quad\cdots & \cdots \\
\binom{n}{n+m-2s+1} & + \binom{n}{n+m-2s} c_1(\mathcal{F}) & + \cdots + \binom{n}{n-s+1} c_{m-s}(\mathcal{F}) & = 0.
\end{array}$$

Now the determinant of the coefficients in this set of equations is equal to

$$\pm\frac{\binom{n}{n-s+1}\binom{n+1}{n-s+1}\cdots\binom{n+m-s}{n-s+1}}{\binom{n-s+1}{n-s+1}\binom{n-s+2}{n-s+1}\cdots\binom{n+m-2s+1}{n-s+1}}$$

(see Muir and Metzler [195], Chapter XX, Section 733, page 682), and this is clearly non-zero. Therefore the system of equations is inconsistent. Since we assumed that $r \ge m+n-2s+2$ in order to obtain a square matrix of coefficients, this proves that $r \le m+n-2s+1$. □

**Example 10.1.3** Example 5.1.4 shows that there are $10 \times 10$ matrices of constant rank 8 with $r = 4$, but it is not known whether there are examples of spaces of dimension five consisting of $n \times n$ matrices of constant rank $n-2$ (Westwick [248]).

**Theorem 10.1.4** *Let $p > 2$ and let $M$ be a $k(\mathbb{Z}/p)^r$-module of constant Jordan type $[2]^s[1]^t$. If $s > 0$ then $t \ge r-1$.*

*Proof.* By Proposition 5.1.17, we have $\mathsf{Rad}^2(M) = 0$. Let $A_1, \ldots, A_r$ be the matrices for the actions of

$$X_1, \ldots, X_r \colon M/\mathsf{Rad}(M) \to \mathsf{Rad}(M).$$

This is a linear variety of $n \times m$ matrices of constant rank $s$, where $m + n = \dim_k(M) = 2s + t$. Applying Theorem 10.1.2, we obtain $r \le m + n - 2s + 1 = t + 1$. □

## 10.2 Congruences on Chern Numbers

In this section we obtain congruences for Chern numbers that will be applied to give information about modules of constant Jordan type.

**Lemma 10.2.1** *For a vector bundle $\mathcal{F}$ of rank $s$ on $\mathbb{P}^{r-1}$ we have*

$$c(\mathcal{F})c(\mathcal{F}(1)) \cdots c(\mathcal{F}(p-1)) \equiv 1 - sh^{p-1} \pmod{(p, h^p)}$$

*as an element of $\mathbb{Z}[h]/(h^r)$.*

*Proof.* We give two proofs, the first in terms of the Chern roots, and the second directly from the definition of Chern classes (see Definitions 7.1.6 and 7.2.6).

We write

$$c(\mathcal{F}) = \prod_{j=1}^{s}(1 + \alpha_j h),$$

where the $\alpha_j$ are the Chern roots. Then by formula (7.3) we have

$$c(\mathcal{F}(i)) = \prod_{j=1}^{s}(1 + (\alpha_j + i)h).$$

Thus we have

$$c(\mathcal{F})c(\mathcal{F}(1)) \cdots c(\mathcal{F}(p-1)) = \prod_{j=1}^{s}(1+\alpha_j h)(1+(\alpha_j+1)h) \cdots (1+(\alpha_j+p-1)h).$$

Now by Fermat's little theorem, we have the identity

$$x(x+y) \cdots (x+(p-1)y) \equiv x^p - xy^{p-1} \pmod{p} \tag{10.1}$$

and so putting $x = 1 + \alpha_j h$, $y = h$ we obtain

$$\begin{aligned} c(\mathcal{F})c(\mathcal{F}(1)) \cdots c(\mathcal{F}(p-1)) &\equiv \prod_{j=1}^{s}((1+\alpha_j h)^p - (1+\alpha_j h)h^{p-1}) \pmod{p} \\ &\equiv \prod_{j=1}^{s}(1 - h^{p-1} + (\alpha_j^p - \alpha_j)h^p) \pmod{p} \\ &\equiv 1 - sh^{p-1} \pmod{(p, h^p)}. \end{aligned}$$

*A priori*, this is a congruence between polynomials with algebraic integer coefficients. But if two rational integers are congruent modulo $p$ as algebraic integers then they are also congruent modulo $p$ as rational integers. This is because their difference, divided by $p$, is both an algebraic integer and a rational number, therefore an integer. □

The second proof gives slightly more information, so formulate it as a separate statement. The first proof can probably be made to give the same information, but it is less clear.

**Lemma 10.2.2** *For a vector bundle $\mathcal{F}$ of rank $s$ on $\mathbb{P}^{r-1}$ we have*

$$c(\mathcal{F})c(\mathcal{F}(1))\cdots c(\mathcal{F}(p-1)) \equiv 1 - sh^{p-1} \pmod{(p, h^{2p-2})}$$

*as an element of $\mathbb{Z}[h]/(h^r)$.*

*Proof.* The argument is similar to the proof of Lemma 10.2.1, but avoids algebraic integers. Recall that if $\mathcal{F} = M^\sim$ where $M$ is a finitely generated $R$-module with Poincaré series

$$p_M(t) = \frac{\sum_j a_j t^j}{(1-t)^r}$$

then

$$c(\mathcal{F}) = \prod_j (1 + jh)^{a_j} \in \mathbb{Z}[h]/(h^r).$$

So again using (10.1) we have

$$\begin{aligned} c(\mathcal{F})c(\mathcal{F}(1))\ldots c(\mathcal{F}(p-1)) &= \prod_j ((1+jh)(1+(j+1)h)\ldots \\ &\quad (1+(j+p-1)h))^{a_j} \\ &\equiv \prod_j ((1+jh)^p - (1+jh)h^{p-1})^{a_j} \pmod{p} \\ &\equiv \prod_j (1 - h^{p-1} + (j^p - j)h^p)^{a_j} \pmod{p} \\ &\equiv \prod_j (1 - h^{p-1})^{a_j} \pmod{p} \\ &\equiv 1 - \sum_j a_j h^{p-1} \pmod{(p, h^{2p-2})}. \end{aligned}$$

Recalling from Definition 7.1.2 that

$$s = \operatorname{rank}(\mathcal{F}) = \sum_j a_j,$$

this proves the lemma. □

## 10.3 Restrictions on Stable Jordan Type, $p$ Odd

The main theorem of this section, Theorem 10.3.1 complements the Realisation Theorem 8.1.1.

**Theorem 10.3.1** (Benson and Pevtsova [53]) *Suppose that $M$ has stable constant Jordan type $[1]^s$. Then $p$ divides the Chern numbers $c_m(\mathcal{F}_1(M))$ for $1 \leq m \leq p-2$.*

*Proof.* The trivial vector bundle $\widetilde{M}$ has a filtration with filtered quotients (not in order) $\mathcal{F}_1(M), \mathcal{F}_p(M), \mathcal{F}_p(M)(1), \ldots, \mathcal{F}_p(M)(p-1)$. So we have

$$\begin{aligned}1 &= c(\widetilde{M})\\ &= c(\mathcal{F}_1(M))c(\mathcal{F}_p(M))c(\mathcal{F}_p(M)(1))\cdots c(\mathcal{F}_p(M)(p-1))\\ &\equiv c(\mathcal{F}_1(M)) \pmod{(p, h^{p-1})}\end{aligned}$$

by Lemma 10.2.1. It follows that the coefficients $c_m(\mathcal{F}_1(M))$ are divisible by $p$ for $1 \leq m \leq p-2$. □

**Remark 10.3.2** For $p = 2$ this theorem says nothing. But for $p$ odd, it at least forces $c_1(\mathcal{F}_1(M))$ to be divisible by $p$. As an explicit example, the twists of the Horrocks–Mumford bundle $\mathcal{F}_{\mathsf{HM}}(i)$ have $c_1 = 2i + 5$ and $c_2 = i^2 + 5i + 10$. For $p \geq 7$ these cannot both be divisible by $p$, and so there is no module $M$ of stable constant Jordan type $[1]^2$ and integer $i$ such that $\mathcal{F}_1(M) \cong \mathcal{F}_{\mathsf{HM}}(i)$.

**Remark 10.3.3** In Section 9.7, we constructed for $p$ odd and $r$ even a $kE$-module $Y$ of stable constant Jordan type $[1]^{r-2}$ with $\mathcal{F}_1(Y) \cong F^*(\mathcal{F}_N)$, the inverse image of the null correlation bundle along the Frobenius map on $\mathbb{P}^{r-1}$.

The Chern polynomial of the null correlation bundle was computed in Proposition 7.3.6 to be

$$(1-h)^{-1}(1+h)^{-1} = 1 + h^2 + h^4 + \cdots + h^{r-2} \in \mathbb{Z}[h]/(h^r).$$

Since the coefficient of $h^2$ is not divisibe by $p$, it follows from Theorem 10.3.1 that the null correlation bundle is not of the form $\mathcal{F}_1(M)$ for any $kE$-module $M$ of stable constant Jordan type $[1]^{r-2}$ as long as $p \geq 5$.

The same is true for $p = 3$ but needs slightly more argument, along the lines of the method for Theorem 5.12.1. If $M$ is such a module then $\Lambda^{r-1}M$ is projective by Dade's lemma, so the dimension of $M$ is congruent to $r-2$ modulo $p^r$. Since $r > 1$, it follows that $\mathcal{F}_p(M)$ is a bundle of rank divisible by $p$. So in the application of Lemma 10.2.1 in the proof of Theorem 10.3.1, the constant $s$ is divisible by $p$ and so the proof shows that $c_m(\mathcal{F}_1(M))$ is divisible by $p$ for $1 \leq m \leq p-1$. If $p = 3$ then $p - 1 = 2$ and looking at the coefficient of $h^2$ again shows that the null correlation bundle has to be pulled back through the Frobenius map before it can be realised in this way.

## 10.4 Eliminating More Stable Jordan Types

In this section, we come back to the question of what stable types can occur for modules of constant Jordan type. The new tool is Lemma 10.2.1, and the main theorem of this section is as follows.

**Theorem 10.4.1** (Benson [43]) *If a module for* $(\mathbb{Z}/p)^r$ *has stable constant Jordan type* $[a_1]\dots[a_t]$ *with*

$$\sum_{i=1}^{t} a_i \le \min(r-1, p-2)$$

*then* $a_1 = \cdots = a_t = 1$.

*Proof.* Let $M$ be a module with stable constant Jordan type $[a_1]\dots[a_t]$. Since $\widetilde{M}$ is a direct sum of copies of $\mathcal{O}$, we have $c(\widetilde{M}) = 1$. Furthermore, by Lemmas 8.4.6 and 8.4.8, $\widetilde{M}$ has a filtration in which the filtered quotients are isomorphic to $\mathcal{F}_i(M)(j)$ for $0 \le j < i \le p$. Let $a = \sum_{i=1}^{t} a_i$. By hypothesis, $a \le p-2$, so using Lemma 10.2.1 we have

$$\prod_{j=1}^{p-2} c(\mathcal{F}_j(M))c(\mathcal{F}_j(M)(1))\dots c(\mathcal{F}_j(M)(j-1)) \equiv 1 \pmod{(p, h^{p-1})}.$$

The left-hand side is a polynomial of degree $a \le p-2$. Furthermore, we have $a \le r-1$ and $h^{r-1} \ne 0$. So this congruence may be read as an equation in the polynomial ring $\mathbb{F}_p[h]$. The units in this ring are the constant polynomials, so all factors in the product are equal to one. If $j \ge 2$, both $c(\mathcal{F}_j(M))$ and $c(\mathcal{F}_j(M)(1))$ then have to be congruent to 1 modulo $p$. Since

$$c_1(\mathcal{F}_j(M)(1)) = c_1(\mathcal{F}_j(M)) + \operatorname{rank}(\mathcal{F}_j(M))$$

(see Lemma 7.5.1), this implies that the rank of $\mathcal{F}_j(M)$ is divisible by $p$. Since $a < p$, this implies that $\mathcal{F}_j(M) = 0$ for $j \ge 2$, so each $a_i = 1$. □

This proves a weak version of Conjecture 5.15.2 provided $p$ is large:

**Corollary 10.4.2** *If $M$ is a $k(\mathbb{Z}/p)^r$-module of stable constant Jordan type* $[2]\,[1]^j$ *with $p > r$ then $j \ge r-2$.* □

We can get slightly more information in the case where $j = r-2$ by examining the coefficients more carefully. We illustrate this in the case $r = 3$, $j = 1$.

**Theorem 10.4.3** *If $M$ is a module of stable constant Jordan type* $[2]\,[1]$ *for* $(\mathbb{Z}/p)^3$ *with $p \ge 5$ then $p \equiv 1 \pmod 3$.*

*Proof.* Using Lemma 7.5.1 as before, we have

$$(1+\alpha h)(1+(\alpha+1)h)(1+\beta h) \equiv 1 \pmod{(p, h^3)}$$

where $\alpha = c_1(\mathcal{F}_2(M))$ and $\beta = c_1(\mathcal{F}_1(M))$. Comparing coefficients of $h$ and $h^2$, we obtain

$$\begin{aligned} 2\alpha+\beta+1 &\equiv 0 \\ \alpha(\alpha+1)+(2\alpha+1)\beta &\equiv 0 \pmod{p}. \end{aligned}$$

Substituting $\beta = -(2\alpha+1)$ into the second equation gives

$$3\alpha^2+3\alpha+1 \equiv 0 \pmod{p}.$$

This has a solution if and only if $-3$ is a square modulo $p$. Since $p \geq 5$, this is equivalent to the condition that $p \equiv 1 \pmod 3$ by quadratic reciprocity. □

If $p \equiv 1 \pmod 3$ then the Chern class information for a module of stable constant Jordan type [2] [1] for $(\mathbb{Z}/p)^3$ is consistent. For example, if $p = 7$ then the two solutions of the equations in the proof of the theorem are $\alpha \equiv 1$, $\beta \equiv 4$ and $\alpha \equiv 5$, $\beta \equiv 3 \pmod 7$.

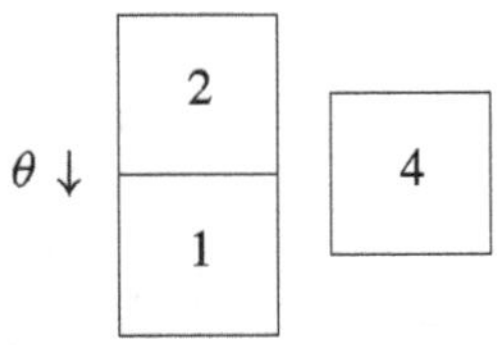


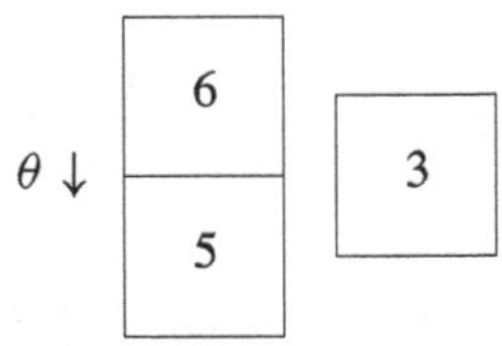


These two possibilities correspond to the equations

$$\begin{aligned} (1+h)(1+2h)(1+4h) &\equiv 1+h^3 \pmod 7 \\ (1+3h)(1+5h)(1+6h) &\equiv 1-h^3 \pmod 7. \end{aligned}$$

It is not clear whether such a module exists if $p = 7$, or more generally if $p \equiv 1 \pmod 3$, but if it does then its restriction to $(\mathbb{Z}/p)^2$ is not isomorphic to any of the modules of type [2] [1] that we know how to construct (see Section 9.5 for some examples), because these examples all have either $\alpha \equiv -1$, $\beta \equiv 1$ or $\alpha \equiv 0$, $\beta \equiv -1 \pmod p$, and in particular $3\alpha^2+3\alpha+1 \not\equiv 0 \pmod p$.

The general theorem obtained by the method of Theorem 10.4.3 is as follows.

**Theorem 10.4.4** *If $M$ is a $kE$-module of stable constant Jordan type $[2]\,[1]^{r-2}$ for $(\mathbb{Z}/p)^r$ with $p > r$ and $r$ odd then $p \equiv 1$ modulo some prime factor of $r$.*

*Proof.* Using Lemma 7.5.1 as before, we have

$$(1+\alpha h)(1+(\alpha+1)h)c(\mathcal{F}_1(M)) \equiv 1 \pmod{p, h^r}$$

where $\alpha = c_1(\mathcal{F}_2(M))$. So the left-hand side is congruent to $1 - ah^r \pmod p$ for some integer $a$. By the factor theorem, both $\alpha$ and $\alpha+1$ are $r$th roots of $a$ modulo

$p$, and so $p$ does not divide $\alpha$, $\alpha + 1$ or $a$. Furthermore, writing $\alpha^{-1}$ for an inverse of $\alpha$ modulo $p$, it follows that $1 + \alpha^{-1}$ is an $r$th root of unity modulo $p$. Since it is not equal to one, it follows that for some prime factor $q$ of $r$, there is a primitive $q$th root of unity modulo $p$ and so $p \equiv 1 \pmod{q}$. □

## 10.5 Restrictions on Jordan Type for $p = 2$

For $p = 2$ it does not make sense to try to restrict the stable Jordan type, but there are interesting restrictions on Jordan type. Some of these come from Dade's lemma, but we shall deduce them instead from the theory of Chern numbers. As soon as the rank is at least four, we shall have to make use of the Hirzebruch–Riemann–Roch theorem to get some of the information. We obtain complete information up to rank four, but it already becomes apparent that the pattern for higher rank will be complicated.

Throughout this section, $M$ will be a module of constant Jordan type $[2]^n[1]^m$ for $E \cong (\mathbb{Z}/2)^r$. We shall write $c_i$ for the Chern numbers of $\mathcal{F}_2(M)$ and $d_i$ for the Chern numbers of $\mathcal{F}_1(M)$.

We begin with some lemmas giving information for $m$ large.

**Lemma 10.5.1** *If $m \geq r - 1$ then for each $n \geq 0$ there exists a module of constant Jordan type $[2]^n[1]^m$ for $(\mathbb{Z}/2)^r$.*

*Proof.* Example 5.6.16 gives modules of constant Jordan type $[2]^n[1]^{r-1}$ for every value of $n$. Direct sums of these with copies of the trivial module of constant Jordan type $[1]$ show that as long as $m \geq r - 1$, there are no restrictions on the Jordan type. □

The following should be compared with Theorem 10.1.4.

**Lemma 10.5.2** *If $r$ is even and $m = r - 2$ then for each even value of $n \geq 0$ there exists a module of constant Jordan type $[2]^n[1]^{r-2}$ for $(\mathbb{Z}/2)^r$.*

*Proof.* This follows from the construction in Section 9.6 of modules $M_k$ such that $\mathcal{F}_1(M_k)$ is an instanton bundle of charge $k$ on $\mathbb{P}^{r-1}$. The modules $M_k$ constructed there have type $[2]^{2k}[1]^{r-2}$. □

For $r$ odd we get much weaker information.

**Lemma 10.5.3** *If $r$ is odd and $m = r - 2$ then there are modules of constant Jordan types $[2]^r[1]^{r-2}$ and $[2]^{2^{r-1}-2r+2}[1]^{r-2}$ for $(\mathbb{Z}/2)^r$.*

*Proof.* This follows from the construction in Section 9.10 of modules $M$ of constant Jordan type $[2]^r[1]^{r-2}$ associated with Tango's rank $r-2$ examples of simple vector bundles on $\mathbb{P}^r$ for any $r$. These modules cannot compete with the instanton examples used to prove Lemma 10.5.2 when $r$ is even, but they're all we've got for $r$ odd. The modules $\Omega(M)$ have constant Jordan type $[2]^{2^{r-1}-2r+2}[1]^{r-2}$. □

The method we shall adopt in order to get restrictions on Jordan type is as follows. Recall from Theorem 8.4.9 that the trivial sheaf $\widetilde{M} = M \otimes_k \mathcal{O}$ has a filtration in which the quotients are $\mathcal{F}_1(M)$, $\mathcal{F}_2(M)$ and $\mathcal{F}_2(M)(1)$. Using Lemma 7.5.1 for the Chern numbers of the twisted bundle $\mathcal{F}_2(M)(1)$, we obtain the following equation:

$$(1 + d_1h + d_2h^2 + d_3h^3 + \cdots)(1 + c_1h + c_2h^2 + c_3h^3 + \cdots)$$
$$(1+(c_1+n)h+(c_2+(n-1)c_1+\tbinom{n}{2})h^2+(c_3+(n-2)c_2+\tbinom{n-1}{2}c_1+\tbinom{n}{3})h^3+\cdots) = 1.$$

This may be used inductively to express the $d_i$ in terms of the $c_i$. The result is the following:

$$\begin{aligned}
d_1 &= -2c_1 - n \\
d_2 &= -2c_2 + 3c_1^2 + (2n+1)c_1 + \tbinom{n+1}{2} \\
d_3 &= -2c_3 + 6c_1c_2 - 4c_1^3 + (n+1)(2c_2 - 3c_1^2) - (n+1)^2c_1 - \tbinom{n+2}{3} \\
d_4 &= -2c_4 + 6c_1c_3 + 3c_2^2 - 12c_1^2c_2 + 5c_1^4 + (2n+3)(c_3 - 3c_1c_2 + 2c_1^3) \\
&\quad + \tfrac{3n^2+9n+8}{2}c_1^2 - (n^2+3n+3)c_2 + \tfrac{(n+1)(n+2)(2n+3)}{6}c_1 + \tbinom{n+3}{4} \\
&\cdots
\end{aligned}$$

From these formulas, information then comes from Remark 7.3.8. Namely, for $m < j < r$ we get an equation between the $c_i$ from the fact that $d_j = 0$.

**Rank Two.** For $r = 2$, there is only one restriction on Jordan type. Namely, if $m = 0$ then $n$ is even. This restriction comes from Dade's lemma, but to deduce it using Chern numbers, we argue as follows. We have $d_1 = 0$, so

$$c_1(\mathcal{F}_2(M)) + c_1(\mathcal{F}_2(M)(1)) = 0.$$

Since $c_1(\mathcal{F}_2(M)(1)) = c_1(\mathcal{F}_2(M)) + n$, this gives

$$n = -2c_1.$$

It follows that $n$ is even. Since $kE$ has constant Jordan type $[2]^2$, Lemma 10.5.1 shows that this is the only restriction.

**Rank Three.** For $r = 3$ and $m = 0$, again Dade's lemma shows that $n$ is divisible by 4. To show this using Chern classes, we note that again $n = -2c_1(\mathcal{F}_2(M))$. The equation

$$(1 - \tfrac{n}{2}h + c_2h^2)(1 + \tfrac{n}{2}h + (c_2 + (n-1)(-\tfrac{n}{2}) + \tbinom{n}{2})h^2) = 1$$

shows that $c_2 = \frac{n^2}{8}$, and hence $n$ is divisible by 4.

For $r = 3$ and $m = 1$, Theorem 5.4.2 shows that $n$ is congruent to $-1$ or 0 modulo four. To show this using Chern classes, we have

$$(1 + c_1h + c_2h^2)(1 + (c_1 + n)h + (c_2 + (n-1)c_1 + \tbinom{n}{2})h^2)(1 + d_1h) = 1$$

in $\mathbb{Z}/(h^3)$. Comparing coefficients of $h$ gives

$$d_1 = -2c_1 - n,$$

and then comparing coefficients of $h^2$ gives

$$c_1(c_1 + n) + 2c_2 + (n-1)c_1 + \tbinom{n}{2} + (-2c_1 - n)(2c_1 + n) = 0.$$

Reading this equation modulo two, we get

$$\tbinom{n}{2} + n^2 \equiv 0 \pmod 2.$$

Thus $n$ is congruent to $-1$ or 0 modulo four.

To see that these are the only restrictions for $r = 3$, we note that the modules $kE$, $k$ and $\Omega(k)$ have constant Jordan type $[2]^4$, $[1]$ and $[2]^3[1]$ respectively, and then use Lemma 10.5.1.

Here is a diagram of these restrictions for rank three. Dots indicate possible Jordan types, and the pattern repeats to the right with period 4.

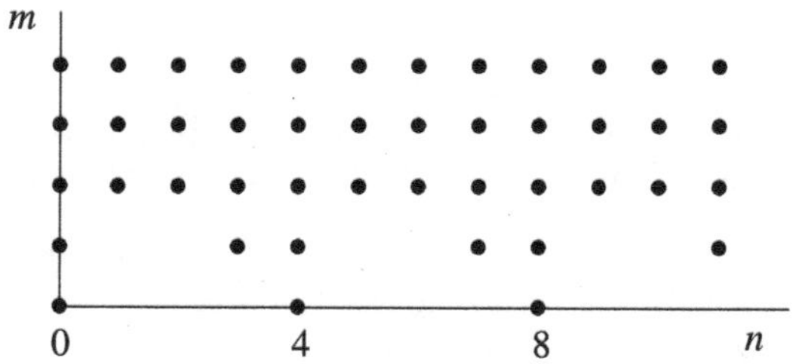


**Rank Four.** For $r = 4$ and $m = 0$, we shall see that the Dade's lemma restriction that $n$ is divisible by 8 can be proved using Chern classes, with the use of the Hirzebruch–Riemann–Roch theorem.

The method of the previous two cases shows that the Chern polynomial of $\mathcal{F}_2(M)$ has the form

$$1 - \frac{n}{2}h + \frac{n^2}{8}h^2 - \frac{n^3 - 4n}{48}h^3.$$

Without resorting to the Hirzebruch–Riemann–Roch theorem, all we get from this is the restriction that $n$ is divisible by 4. But it follows from Hirzebruch–Riemann–Roch that for any vector bundle on $\mathbb{P}^3$, we have

$$c_1c_2 + c_3 \equiv 0 \pmod 2,$$

see Example 7.7.11. Thus

$$\frac{n^3 - 4n}{48} + \binom{-\frac{n}{2}}{}\binom{\frac{n^2}{8}}{} \equiv 0 \pmod 2,$$

which forces $n$ to be divisible by 8.

For $r = 4$ and $m = 1$, Theorem 5.4.2 shows that $n$ is congruent to 0 or $-1$ modulo 8. This information may also be obtained from Hirzebruch–Riemann–Roch, but we defer the proof of this until Section 10.8.

For $r = 4$ and $m = 2$, we find that $d_1 \equiv n \pmod 2$ and then $d_2 \equiv \binom{n}{2} + n^2 \pmod 2$. The Hirzebruch–Riemann–Roch Theorem implies that $d_1 d_2$ is even, which in turn implies that $n$ is not congruent to one modulo four.

Continuing with $r = 4$ and $m = 2$, if $n \equiv 3 \pmod 8$, then working mod 8 and $h^4$ we have

$$\begin{aligned}(1 + c_1 h + c_2 h^2 + c_3 h^3)(1 + (c_1 + 3)h + (c_2 + 2c_1 + 3)h^2 \\ + (c_3 + c_2 + c_1 + 1)h^3)(1 + d_1 h + d_2 h^2) \equiv 1.\end{aligned}$$

Equating coefficients of $h$ and $h^2$ gives

$$\begin{aligned}d_1 &\equiv -2c_1 - 3, \\ d_2 &\equiv -2c_2 + 3c_1^2 - c_1 - 2.\end{aligned}$$

Equating coefficients of $h^3$ then gives

$$2c_3 + 2c_1c_2 + 4c_1^3 + 4c_1^2 + 2 \equiv 0.$$

By Hirzebruch–Riemann–Roch, $c_3 + c_1c_2$ is even and so we obtain a contradiction. Thus $n$ cannot be 3 modulo 8.

So the limitations we have found for rank four are:

(i) If $m = 0$ then $n$ is divisible by 8.
(ii) If $m = 1$ then $n$ is congruent to 0 or $-1$ modulo 8.
(iii) If $m = 2$ then $n$ is not congruent to 1, 3 or 5 modulo 8.

To see that these form a complete set of conditions, $kE$ has type $[2]^8$, $k$ has type $[1]$, and $\mathsf{Rad}(kE)$ has type $[2]^7[1]$. Sums of these together with Lemmas 10.5.1 and 10.5.2 show that all possibilities for $n$ allowed by the above conditions do actually occur.

Here is a diagram of the restrictions for rank four. Dots indicate possible Jordan types, and the pattern repeats to the right with period 8.

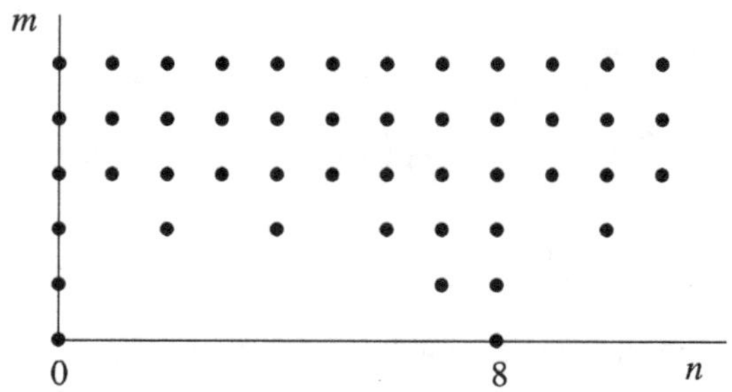


We shall go some way towards explaining this pattern in Sections 10.6–10.8.

## 10.6 Applying Hirzebruch–Riemann–Roch for $p = 2$: The Case $m = 0$

We continue with the notation of the previous section. At this stage, it becomes apparent that there should be a better way to organise the information, especially the congruences coming from the Hirzebruch–Riemann–Roch Theorem 7.8.7. Here is how to recover the information coming from Dade's lemma this way, for $p = 2$.

**Lemma 10.6.1** *If $m = 0$ then the Chern character of $\mathcal{F}_2(M)$ is given by*

$$\mathsf{Ch}(\mathcal{F}_2(M)) = 2n/(1 + e^h).$$

*Proof.* We have a short exact sequence

$$0 \to \mathcal{F}_2(M) \to \widetilde{M} \to \mathcal{F}_2(M)(1) \to 0$$

and $M$ has dimension $2n$, so we have

$$\mathsf{Ch}(\mathcal{F}_2(M)) + \mathsf{Ch}(\mathcal{F}_2(M)(1)) = 2n.$$

By Lemma 7.5.1 we have

$$\mathsf{Ch}(\mathcal{F}_2(M)(1)) = e^h \mathsf{Ch}(\mathcal{F}_2(M)).$$

Hence

$$(1 + e^h)\mathsf{Ch}(\mathcal{F}_2(M)) = 2n. \qquad \square$$

**Lemma 10.6.2** *The coefficient of $h^{r-1}$ in the power series expansion of*

$$\left(\frac{h}{1 - e^{-h}}\right)^r \frac{1}{1 + e^h}$$

*is equal to* $\dfrac{1}{2^r}$.

*Proof.* By Cauchy's integral formula, the coefficient is

$$\frac{1}{2\pi i}\oint \frac{dh}{(1-e^{-h})^r(1+e^h)}$$

where the contour of integration is a small anticlockwise circle around the origin. Substituting $z = 1 - e^{-h}$ (see Remark 7.8.8), this is equal to

$$\frac{1}{2\pi i}\oint \frac{dz}{z^r(2-z)} = \frac{1}{2\pi i}\oint \left(\frac{2^{r-1}+2^{r-2}z+\cdots+z^{r-1}}{z^r}+\frac{1}{2-z}\right)\frac{dz}{2^r} = \frac{1}{2^r}.$$

□

**Theorem 10.6.3** *If $m = 0$ then $n$ is divisible by $2^{r-1}$.*

*Proof.* Applying the Hirzebruch–Riemann–Roch Theorem 7.8.7 to $\mathcal{F}_2(M)$ and using Lemma 10.6.1 we find that $\chi(\mathcal{F}_2(M))$ is the coefficient of $h^{r-1}$ in the power series

$$\left(\frac{h}{1-e^{-h}}\right)^r \frac{2n}{1+e^h}.$$

By Lemma 10.6.2 this is equal to

$$\frac{2n}{2^r}.$$

Since $\chi(\mathcal{F}_2(M))$ is an integer this shows that $n$ is divisible by $2^{r-1}$. □

**Remark 10.6.4** In contrast with Dade's lemma, this method works for nilvarieties of constant Jordan type just as well as for modules of constant Jordan type. So we obtain the following theorem, which holds in any characteristic.

**Theorem 10.6.5** *Let $k$ be an algebraically closed field, and let $M$ be a nilvariety of rank $r$ over $k$ of constant Jordan type $[2]^n$. Then $n$ is divisible by $2^{r-1}$.*

**Remark 10.6.6** This theorem should not be so surprising, because if $A_1, \ldots, A_r$ span a nilvariety of rank $r$ consisting of matrices that square to zero then for $1 \le i < j \le r$ we have $(A_i + A_j)^2 = 0$, so that

$$A_j A_i = -A_i A_j.$$

So the matrices generate a representation of an exterior algebra. Dade's lemma holds in any characteristic for exterior algebras, see for example Theorem 3.1 of Aramova, Avramov and Herzog [12].

**Remark 10.6.7** In contrast with this theorem, Example 4.1.6 shows that for $p$ odd, if a nilvariety of rank $r$ has constant Jordan type $[p]^n$, it does not follow that $n$ is divisible by $p^{r-1}$. We investigate this further in Section 10.9.

## 10.7 Bypassing Hirzebruch–Riemann–Roch

We have seen that the integrality conditions on Chern numbers coming from the Hirzebruch–Riemann–Roch theorem are consequences of the integrality of the coefficients in the Poincaré seires of a corresponding graded module. We ought therefore to be able to obtain the results of the last section directly using Poincaré series.

*Second proof of Theorem 10.6.5.* Let $M$ be a $kE$-module of constant Jordan type $[2]^n$, let $R = k[Y_1, \dots, Y_r]$, and let $N$ be a finitely generated graded $R$-module such that $N^\sim \cong \mathcal{F}_2(M)$. Let

$$p_N(t) = \frac{\sum_i a_i t^i}{(1-t)^r}$$

with $\sum_i a_i t^i \in \mathbb{Z}[t, t^{-1}]$. Then since $\widetilde{M}$ has a filtration with filtered quotients $\mathcal{F}_2(M)$ and $\mathcal{F}_2(M)(1)$, we have

$$(1+t)p_N(t) = \frac{2n}{(1-t)^r} + q(t)$$

with $q(t) \in \mathbb{Z}[t, t^{-1}]$. Substitute $t = -1$ to get

$$0 = \frac{2n}{2^r} + q(-1).$$

Since $q(-1) \in \mathbb{Z}$, it follows that $n$ is divisible by $2^{r-1}$. □

## 10.8 Applying and Bypassing Hirzebruch–Riemann–Roch for $p = 2$: The Case $1 \le m \le r-3$

Recall from Section 6.3 that if $\mathcal{F}$ is a vector bundle of rank $m$ on $\mathbb{P}^{r-1}$ with $1 \le m \le r-3$ then there are three possibilities:

(i) $\mathcal{F}$ is a direct sum of line bundles,

(ii) $r \le 6$ and $\mathcal{F}$ has a summand isomorphic to one of the examples mentioned in Section 6.3, or

(iii) $\mathcal{F}$ has a summand which is a new indecomposable vector bundle of low rank on projective space.

In this section we investigate case (i). Let $k$ be an algebraically closed field of characteristic two and $E \cong (\mathbb{Z}/2)^r$. Let $M$ be a $kE$-module of constant Jordan type $[2]^n[1]^m$.

**Lemma 10.8.1** *If $\mathcal{F}_1(M)$ is a direct sum of line bundles $\mathcal{O}(a_1) \oplus \cdots \oplus \mathcal{O}(a_m)$ then the Chern character of $\mathcal{F}_2(M)$ is given by*

$$\mathsf{Ch}(\mathcal{F}_2(M)) = \frac{2n + m - e^{a_1 h} - \cdots - e^{a_m h}}{1 + e^h}.$$

*Proof.* The vector bundle $\widetilde{M}$ has a filtration with filtered quotients $\mathcal{F}_2(M)$, $\mathcal{F}_2(M)(1)$ and $\mathcal{F}_1(M)$. So

$$(1 + e^h)\mathsf{Ch}(\mathcal{F}_2(M)) + e^{a_1 h} + \cdots + e^{a_m h} = 2n + m. \qquad \square$$

**Lemma 10.8.2** *The coefficient of $h^{r-1}$ in the power series expansion of*

$$\left(\frac{h}{1 - e^{-h}}\right)^r \frac{2n + m - e^{a_1 h} - \cdots - e^{a_m h}}{1 + e^h}$$

*differs from*

$$\frac{2n + m - (-1)^{a_1} - \cdots - (-1)^{a_m}}{2^r}$$

*by an integer.*

*Proof.* As usual, by Cauchy's integral formula, the coefficient is

$$\frac{1}{2\pi i} \oint \frac{(2n + m - e^{a_1 h} - \cdots - e^{a_m h})dh}{(1 - e^{-h})^r (1 + e^h)}.$$

Substituting $z = 1 - e^{-h}$ (see Remark 7.8.8), this is equal to

$$\frac{1}{2\pi i} \oint \frac{2n + m - \sum_{i=1}^m (1 - z)^{-a_i}}{z^r (2 - z)}\, dz.$$

This is the coefficient of $z^{r-1}$ in the power series

$$\frac{2n + m - \sum_{i=1}^m (1 - z)^{-a_i}}{2 - z}$$
$$= (\tfrac{1}{2} + \tfrac{1}{2^2} z + \tfrac{1}{2^3} z^2 + \cdots)(2n + m - \textstyle\sum_{i=1}^m (1 + a_i z + \binom{a_i+1}{2} z^2 + \binom{a_i+2}{3} z^3 + \cdots)).$$

So this coefficient is equal to

$$\frac{2n + m - \sum_{i=1}^m (1 + 2a_i + 2^2\binom{a_i+1}{2} + 2^3\binom{a_i+2}{3} + \cdots + 2^{r-1}\binom{a_i+r-2}{r-1})}{2^r}. \tag{10.2}$$

Now the power series expansion

$$(1 - 2)^{-a_i} = 1 + 2a_i + 2^2\binom{a_i+1}{2} + \cdots \tag{10.3}$$

converges 2-adically, and so

$$1 + 2a_i + 2^2\binom{a_i + 1}{2} + 2^3\binom{a_i + 2}{3} + \cdots$$
$$+ 2^{r-1}\binom{a_i + r - 2}{r - 1} \equiv (-1)^{a_i} \pmod{2^r}.$$

It follows that the expression (10.2) differs by an integer from

$$\frac{2n + m - \sum_{i=1}^m (-1)^{a_i}}{2^r}$$

as required. $\square$

**Theorem 10.8.3** *If $\mathcal{F}_1(M) \cong \mathcal{O}(a_1) \oplus \cdots \oplus \mathcal{O}(a_m)$ then $n$ is congruent to one of the integers $0, -1, \ldots, -m$ modulo $2^{r-1}$. More precisely, it is congruent to minus the number of $a_i$ that are odd.*

As in the case of Theorem 10.6.5, we provide two proofs, the first using the Hirzebruch–Riemann–Roch theorem and the second avoiding it. The relationship between these two proofs is quite instructive, but the second one is clearly easier.

*First proof.* Applying the Hirzebruch–Riemann–Roch Theorem 7.8.7 and Lemma 10.8.1, we find that $\chi(\mathcal{F}_2(M))$ is the coefficient of $h^{r-1}$ in the power series

$$\left(\frac{h}{1-e^{-h}}\right)^r \left(\frac{2n+m-e^{a_1h}-\cdots-e^{a_mh}}{1+e^h}\right).$$

By Lemma 10.8.2 this is differs from

$$\frac{2n+m-\sum_{i=1}^m(-1)^{a_i}}{2^r}$$

by an integer. Thus $2n+m-\sum_{i=1}^m(-1)^{a_i}$ is divisible by $2^r$. It follows that $n$ is congruent modulo $2^{r-1}$ to minus the number of $a_i$ that are odd. □

*Second proof.* Let $R = k[Y_1, \ldots, Y_r]$, and let $N_1, N_2$ be finitely generated graded $R$-modules with $N_1^\sim \cong \mathcal{F}_1(M)$ and $N_2^\sim \cong \mathcal{F}_2(M)$. Since $\mathcal{F}_1(M) \cong \mathcal{O}(a_1) \oplus \cdots \oplus \mathcal{O}(a_m)$, we may take for $N_1$ a sum of shifts of $R$ by degrees $a_1, \ldots, a_m$. The Poincaré series of $N_1$ and $N_2$ are then given by

$$p_{N_1}(t) = \frac{t^{a_1}+\cdots+t^{a_m}}{(1-t)^r}, \qquad p_{N_2}(t) = \frac{f(t)}{(1-t)^r}$$

with $f(t) \in \mathbb{Z}[t, t^{-1}]$. Since $\widetilde{M}$ is a trivial bundle of rank $2n+m$ having a filtration with filtered quotients $\mathcal{F}_2(M)$, $\mathcal{F}_2(M)(1)$ and $\mathcal{F}_1(M)$, we have

$$(1+t)p_{N_2}(t) + \frac{t^{a_1}+\cdots+t^{a_m}}{(1-t)^r} = \frac{2n+m}{(1-t)^r} + q(t)$$

where $q(t) \in \mathbb{Z}[t, t^{-1}]$. Substitute $t = -1$ to get

$$\frac{(-1)^{a_1}+\cdots+(-1)^{a_m}}{2^r} = \frac{2n+m}{2^r} + q(-1).$$

Since $q(-1) \in \mathbb{Z}$, it follows that $2n+m-\sum_{i=1}^m(-1)^{a_i}$ is divisible by $2^r$, and so $n$ is congruent modulo $2^{r-1}$ to minus the number of $a_i$ that are odd. □

Note that all the possibilities allowed by Theorem 10.8.3 actually occur. For example we can take

$$M = \Omega^{-a_1}(k) \oplus \cdots \oplus \Omega^{-a_m}(k).$$

If $m = 1$ then we are in the case of Theorem 5.4.2, and that theorem shows that $n$ is congruent to 0 or $-1$ modulo $2^{r-1}$. In this case $\mathcal{F}_1(M)$ has to be a line bundle, and so Theorem 10.8.3 gives the same congruence for $n$.

If $2 \leq m \leq r - 3$, then vector bundles that are not sums of line bundles are hard to come by. In particular, no examples are known with $r > 6$ (cf. Section 6.3). So we have the following theorem.

**Theorem 10.8.4** *Let $k$ be an algebraically closed field of characteristic two and $E \cong (\mathbb{Z}/2)^r$. If $M$ is a module of constant Jordan type $[2]^n[1]^m$ and $n$ is not congruent to one of the integers $0, -1, \dots, -m$ modulo $2^{r-1}$ then $\mathcal{F}_1(M)$ is a vector bundle on $\mathbb{P}^{r-1}$ which is not a sum of line bundles. In particular, if $r > 6$ and $m \leq r - 3$ then $\mathcal{F}_1(M)$ is a new vector bundle of low rank on projective space.* □

This theorem explains why, in the case $r = 4$ in Section 10.5, we had to resort to examples where $\mathcal{F}_1(M)$ is not a direct sum of line bundles in order to obtain existence for Jordan type $[2]^n[1]^2$ with $n$ congruent to 2 or 4 modulo 8.

For $r = 6$, in Section 9.14 we constructed a module of constant Jordan type $[2]^{14}[1]^2$ associated with the Tango bundle. This does not satisfy the congruences of the theorem.

Combining the information given in this section with Lemma 10.5.1, the case that remains for discussion is the case $m = r - 2$. In this case, we have Lemmas 10.5.2 and 10.5.3, which give somewhat incomplete information.

## 10.9 Nilvarieties of Constant Jordan Type $[p]^n$ for $p \geq 3$

In this section we investigate nilvarieties of rank $r$ and constant Jordan type $[p]^n$ for $p \geq 3$. In the case of $kE$-modules of constant Jordan type, Dade's lemma implies that $p^{r-1}|n$. However, we already know from Example 4.1.6 that there are nilvarieties of rank two and constant Jordan type $[p]$, so the divisibility statement coming from Dade's lemma does not apply to nilvarieties. In this section we obtain a lower bound on the power of $p$ dividing $n$. This lower bound is exact for $p = 3$, but the situation is unclear for $p \geq 5$.

Again we have a choice of using Hirzebruch–Riemann–Roch or working directly with Poincaré series. We choose the latter approach, as it turns out to be more straightforward.

**Theorem 10.9.1** *Let $p \geq 3$ and let $M$ be a rank $r$ nilvariety of constant Jordan type $[p]^n$. Then*

$$p^{\left\lfloor \frac{r-1}{p-1} \right\rfloor} | n$$

*where $\left\lfloor \frac{r-1}{p-1} \right\rfloor$ denotes the largest integer less than or equal to $\frac{r-1}{p-1}$.*

*Proof.* Let $R = k[Y_1, \ldots, Y_r]$, and let $N$ be a finitely generated graded $R$-module with $\mathcal{F}_p(M) \cong N^\sim$. Write the Poincaré series of $N$ as

$$p_N(t) = \frac{f(t)}{(1-t)^r}.$$

The vector bundle $\widetilde{M}$ has a filtration with filtered quotients

$$\mathcal{F}_p(M),\ \mathcal{F}_p(M)(1), \ldots, \mathcal{F}_p(M)(p-1),$$

so we have

$$(1 + t + \cdots + t^{p-1})p_N(t) = \frac{np}{(1-t)^r} + q(t) \qquad (10.4)$$

where $q(t) \in \mathbb{Z}[t, t^{-1}]$. Let $\varepsilon$ be a primitive $p$th root of unity in $\mathbb{C}$, so that

$$1 + \varepsilon + \cdots + \varepsilon^{p-1} = 0.$$

Then $\mathbb{Z}[\varepsilon]$ is the ring of cyclotomic integers in the cyclotomic number field $\mathbb{Q}[\varepsilon]$. This ring is a Dedekind domain in which the prime ideal lying above $(p)$ is $(1-\varepsilon)$, and we have $(1-\varepsilon)^{p-1} = (p)$.

Substituting $t = \varepsilon$ in equation (10.4), we obtain the equation

$$\frac{np}{(1-\varepsilon)^r} + q(\varepsilon) = 0$$

in $\mathbb{Q}[\varepsilon]$. Since $q(\varepsilon) \in \mathbb{Z}[\varepsilon]$, it follows that $np$ is in the $r$th power of the ideal $(1-\varepsilon)$. Since it is a rational integer, it follows that it is divisible by $p^a$, where $a$ is the smallest integer greater than or equal to $\frac{r}{p-1}$, namely $\left\lfloor \frac{r-1}{p-1} \right\rfloor + 1$. So $n$ is divisible by $p^{\left\lfloor \frac{r-1}{p-1} \right\rfloor}$. □

The following theorem shows that Theorem 10.9.1 is sharp in the case $p = 3$.

**Theorem 10.9.2** *Let $k$ be an algebraically closed field of characteristic three. Then there exists a nilvariety $M$ of rank $r$ of constant Jordan type $[3]^n$ with $n = 3^{\left[\frac{r-1}{2}\right]}$.*

*Proof.* It suffices to prove this with $r$ even. In this case, we may take a tensor product of $r/2$ copies of Example 4.1.6. In characteristic three, a tensor product of two Jordan blocks of length three is a sum of three Jordan blocks of length three. □

**Example 10.9.3** If $r = 4$, the tensor product of two copies of Example 4.1.6 gives a nilvariety of constant Jordan type where the operator $\theta$ acts as the matrix

$$\begin{pmatrix} 0 & Y_2 & 0 & Y_4 & 0 & 0 & 0 & 0 & 0 \\ Y_1 & 0 & Y_2 & 0 & Y_4 & 0 & 0 & 0 & 0 \\ 0 & -Y_1 & 0 & 0 & 0 & Y_4 & 0 & 0 & 0 \\ Y_3 & 0 & 0 & 0 & Y_2 & 0 & Y_4 & 0 & 0 \\ 0 & Y_3 & 0 & Y_1 & 0 & Y_2 & 0 & Y_4 & 0 \\ 0 & 0 & Y_3 & 0 & -Y_1 & 0 & 0 & 0 & Y_4 \\ 0 & 0 & 0 & -Y_3 & 0 & 0 & 0 & Y_2 & 0 \\ 0 & 0 & 0 & 0 & -Y_3 & 0 & Y_1 & 0 & Y_2 \\ 0 & 0 & 0 & 0 & 0 & -Y_3 & 0 & -Y_1 & 0 \end{pmatrix}.$$

If $k$ has characteristic 3 then the Jordan type is $[3]^3$ while if $k$ has characteristic at least 5 or characteristic zero, the Jordan type is $[5]\,[3]\,[1]$.

## 10.10 Nilvarieties with a Single Jordan Block

In contrast with Theorem 10.9.2, for $p \geq 5$ the situation is less clear. Theorem 10.9.1 allows for the existence of nilvarieties of rank $p - 1$ and constant Jordan type $[p]$, whereas we have the following theorem saying that this cannot occur.

**Theorem 10.10.1** (Causa, Re and Teodorescu [85]) *Let $M$ be a nilvariety of rank $r$ and constant Jordan type $[m]$. Then $r \leq 2$, and if $r = 2$ then $m$ is odd.*

*Proof.* Suppose that $M$ is a nilvariety of constant Jordan type $[m]$. Then $\mathcal{F}_m(M)$ is a line bundle, so by Theorem 6.3.1 we have $\mathcal{F}_m(M) \cong \mathcal{O}(a)$ for some integer $a$. The bundle $\widetilde{M} \cong \mathcal{O}^{\oplus m}$ has a filtration with filtered quotients

$$\mathcal{O}(a), \mathcal{O}(a+1), \dots, \mathcal{O}(a+m-1)$$

and so

$$0 = c_1(\widetilde{M}) = a + (a+1) + \cdots + (a+m-1) = ma + m(m-1)/2.$$

Thus $a = -(m-1)/2$ and so $m$ is odd.

If $r \geq 3$ then for all $i$ and $j$ we have

$$\mathrm{Ext}^1_{\mathcal{O}_{\mathbb{P}^{r-1}}}(\mathcal{O}(i), \mathcal{O}(j)) \cong H^1(\mathbb{P}^{r-1}, \mathcal{O}(j-i)) = 0$$

and so the filtration splits, giving

$$\mathcal{O}^{\oplus m} \cong \mathcal{O}(a) \oplus \mathcal{O}(a+1) \oplus \cdots \oplus \mathcal{O}(a+m-1),$$

which contradicts Theorem 6.1.6. □

**Theorem 10.10.2** (Causa, Re and Teodorescu [85]) *For each odd value of $m \geq 3$ there exists a nilvariety of rank two and constant Jordan type $[m]$ of the form*

$$Y_1A_1 + Y_2A_2 = \begin{pmatrix} 0 & Y_1 & & & & & \\ \alpha_1 Y_2 & 0 & Y_1 & & & 0 & \\ & \alpha_2 Y_2 & 0 & Y_1 & & & \\ & & \alpha_3 Y_2 & \ddots & \ddots & & \\ & 0 & & \ddots & 0 & Y_1 \\ & & & & \alpha_{m-1} Y_2 & 0 \end{pmatrix}.$$

*The set of values of $(\alpha_1, \ldots, \alpha_{m-1}) \in k^{m-1}$ that give such nilvarieties form a non-empty dense open subset of a complete intersection of dimension $(m-1)/2$ (i.e., projectively of dimension $(m-3)/2$).*

*Proof.* The given matrices form a nilvariety of constant Jordan type $[m]$ if and only if the $\alpha_i$ are all non-zero, and

$$(Y_1A_1 + Y_2A_2)^m = 0. \tag{10.5}$$

We claim that condition (10.5) is equivalent to the condition that $(Y_1A_1 + Y_2A_2)^m$ annihilates the first basis vector $v_1$ of $k^m$. One direction is clear; for the other direction, the observation that $(Y_1A_1 + Y_2A_2)^i v_1$ is $Y_2^i$ times the $i$th basis vector $v_i$ plus a linear combination of previous basis vectors $v_1, \ldots, v_{i-1}$ shows inductively that it annihilates all basis vectors.

The condition that $(Y_1A_1 + Y_2A_2)^m v_1 = 0$ gives $(m-1)/2$ equations. The $i$th equation says that the $(2i)$th coordinate is zero, and is of the form

$$(\text{polynomial in } \alpha_1, \ldots, \alpha_{(m+2i-1)/2}) + \alpha_{2i+2} \cdots \alpha_{(m+2i+1)/2} = 0.$$

These equations describe a complete intersection in $k^{m-1}$ of dimension $(m-1)/2$. The open subset defined by the non-vanishing of $\alpha_1, \ldots \alpha_{m-1}$ is non-empty and dense. □

**Example 10.10.3** If $m = 5$ the general solution to the equations of Theorem 10.10.2 is of the form

$$A_1 = \begin{pmatrix} 0 & 1 & 0 & 0 & 0 \\ 0 & 0 & 1 & 0 & 0 \\ 0 & 0 & 0 & 1 & 0 \\ 0 & 0 & 0 & 0 & 1 \\ 0 & 0 & 0 & 0 & 0 \end{pmatrix} \qquad A_2 = \lambda \begin{pmatrix} 0 & 0 & 0 & 0 & 0 \\ a & 0 & 0 & 0 & 0 \\ 0 & 1 & 0 & 0 & 0 \\ 0 & 0 & -(a+1)^2 & 0 & 0 \\ 0 & 0 & 0 & a(a+1) & 0 \end{pmatrix}$$

$(\lambda \neq 0, a \neq 0, -1)$.

**Example 10.10.4** We can do slightly better than rank two in Theorem 10.10.1 if we look for a nilvariety of constant Jordan type $[m]^2$. The following is an example of constant Jordan type $[5]^2$ and rank three, with $k$ an algebraically closed field of characteristic not equal to two.

$$\begin{pmatrix} 0 & Y_2 & Y_3 & 0 & 0 & 0 & 0 & 0 & 0 & 0 \\ 2Y_1 & 0 & Y_2 & -4Y_3 & 0 & 0 & 0 & 0 & 0 & 0 \\ 0 & -Y_1 & 0 & Y_2 & Y_3 & 0 & 0 & 0 & 0 & 0 \\ 0 & 0 & Y_1 & 0 & Y_2 & Y_3 & 0 & 0 & 0 & 0 \\ 0 & 0 & 0 & -2Y_1 & 0 & 0 & Y_3 & 0 & 0 & 0 \\ 0 & 0 & 0 & 0 & 0 & 0 & Y_2 & Y_3 & 0 & 0 \\ 0 & 0 & 0 & 0 & 0 & 2Y_1 & 0 & Y_2 & -4Y_3 & 0 \\ 0 & 0 & 0 & 0 & 0 & 0 & -Y_1 & 0 & Y_2 & Y_3 \\ 0 & 0 & 0 & 0 & 0 & 0 & 0 & Y_1 & 0 & Y_2 \\ 0 & 0 & 0 & 0 & 0 & 0 & 0 & 0 & -2Y_1 & 0 \end{pmatrix}$$

So now the question is, how sharp is Theorem 10.9.1 for $p \geq 5$? For example, does there exist a nilvariety of constant Jordan type $[p]^n$ and rank $p-1$ with $n$ not divisible by $p$? The following extended version of Theorem 10.10.1 shows that for $p$ large we cannot expect $n$ to be small.

**Theorem 10.10.5** *Let $M$ be a nilvariety of rank $r \geq 3$ and constant Jordan type $[m]^n$. Then $\mathfrak{F}_m(M)$ is a vector bundle of rank $n$ on $\mathbb{P}^{r-1}$ which is not a direct sum of line bundles. In particular, if $n \leq r-2$ and $r \geq 7$ this gives a new vector bundle of low rank on projective space (cf. Section 6.3).*

*Proof.* Suppose to the contrary that $\mathfrak{F}_m(M)$ is a direct sum of line bundles. Arguing as in the proof of Theorem 10.10.1, the filtration of $\widetilde{M}$ splits and we obtain

$$\mathcal{O}^{\oplus nm} \cong \mathfrak{F}_m(M) \oplus \mathfrak{F}_m(M)(1) \cdots \oplus \mathfrak{F}_m(M)(m-1)$$

contradicting Theorem 6.1.6. □

## 10.11 Babylonian Towers

Recall from Remark 6.3.4 the Babylonian tower theorem for vector bundles on projective spaces. It states that a vector bundle on $\mathbb{P}^{r-1}$ which extends to a vector bundle on $\mathbb{P}^{r+n-1}$ for all $n \geq 0$ has to be a direct sum of line bundles. In this section we examine the corresponding situation for modules of constant Jordan type.

**Definition 10.11.1** We say that a $k(\mathbb{Z}/p)^r$-module $M$ of constant Jordan type *stably extends* to $(\mathbb{Z}/p)^{r+n}$ if there exists a $k(\mathbb{Z}/p)^{r+n}$-module $M'$ of constant Jordan type such that $M'{\downarrow_{k(\mathbb{Z}/p)^r}} \cong M \oplus$ (projective).

**Theorem 10.11.2** *Let $p$ be an odd prime, and suppose that $M$ is a $k(\mathbb{Z}/p)^r$-module of stable constant Jordan type $[a_1]\dots[a_t]$ which stably extends to*

*$k(\mathbb{Z}/p)^{r+n}$ for all $n \geq 0$, and suppose that $\sum_{i=1}^{t} a_i \leq p-2$. Then $a_i = 1$ for $1 \leq i \leq t$, so that the stable constant Jordan type of $M$ is $[1]^t$. Furthermore, $\mathcal{F}_1(M)$ is a direct sum of line bundles $\bigoplus_{i=1}^{t} \mathcal{O}(pn_i)$.*

*Proof.* The fact that we have $a_i = 1$ for $1 \leq i \leq t$ is proved in Theorem 10.4.1. By the Babylonian tower theorem for vector bundles (Remark 6.3.4), $\mathcal{F}_1(M)$ is a direct sum of line bundles $\bigoplus_{i=1}^{t} \mathcal{O}(m_i)$. It remains to prove that each $m_i$ is divisibe by $p$. If $n$ is large enough so that $r+n > t$, then the Chern polynomial $c(\mathcal{F}_1(M))$ is a polynomial of degree at most $t$ in $\mathbb{Z}[h]/(h^{r+n})$. So the factorisation

$$c(\mathcal{F}_1(M)) = \prod_{i=1}^{t}(1+m_ih)$$

may be regarded as an equation in $\mathbb{Z}[[h]] = \varprojlim_n A^*(\mathbb{P}^{r+n-1})$. By Theorem 10.3.1, we have $p|c_i(\mathcal{F}_1(M))$ for $1 \leq i \leq p-2$. Since $t \leq p-2$, this implies that

$$p|c_t(\mathcal{F}_1(M)) = \prod_{i=1}^{t} m_i,$$

and so for some value of $i$, $m_i$ is divisible by $p$. Thus

$$(1+m_ih)^{-1}c(\mathcal{F}_1(M)) \in \mathbb{Z}[[h]]$$

is a polynomial of degree at most $t-1$, all of whose coefficients of positive powers of $h$ are divisible by $p$, and so $\prod_{j \neq i} m_j$ is divisible by $p$. Continuing this way, stripping off one factor at a time, we see that all the $m_j$ are divisible by $p$, as required. □

**Conjecture 10.11.3** (Babylonian tower conjecture for modules of constant Jordan type) Let $p$ be an odd prime, and suppose that $M$ is a $k(\mathbb{Z}/p)^r$-module of stable constant Jordan type $[a_1]\dots[a_t]$ which stably extends to $k(\mathbb{Z}/p)^{r+n}$ for all $n \geq 0$, and suppose that $\sum_{i=1}^{t} a_i \leq p-2$. Then $M$ decomposes as a direct sum of modules of the form $\Omega^{-n_i}(k)$ and projective modules (cf. Corollary 8.5.3).

# 11

# Orlov's Correspondence

## 11.1 Introduction

This chapter is a lightly modified version of the paper [45]. The goal is to describe an equivalence between the derived category of an elementary abelian $p$-group of rank $r$, and the category of (graded) reduced matrix factorisations of the polynomial $y_1 X_1^p + \cdots + y_r X_r^p$. This gives another interpretation of the Realisation Theorem 8.1.1 for vector bundles on projective space from modular representations of constant Jordan type.

Orlov [204] established an equivalence of singularity categories which he called "reduction of dimension". This has been further investigated by many authors, including for example Baranovsky and Pecharich [29], Burke and Stevenson [69], Burke and Walker [70], Chen [87]. Our purpose here is to investigate the consequences for representation theory of an elementary abelian $p$-group $E$ over a field $k$ of characteristic $p$ as explicitly as possible. In this context, we explain the Orlov correspondence and how it lifts to the derived category. We use the Orlov correspondence to give a new interpretation of the construction given in Section 8.8 for proving Theorem 8.1.1, realising vector bundles from modules of constant Jordan type, see Remark 11.12.4.

I should point out that Theorem 11.3.3, the main theorem of this chapter, is a special case of Theorem 7.5 of Burke and Stevenson [69]; even the functors realising the equivalences in the theorem are the same. The proof presented here uses a minimum of heavy machinery, taking advantage of the special situation in hand to reduce to an explicit computation involving the "bidirectional Koszul complex," introduced in Section 11.6.

Let $R_0$ be the polynomial ring $k[X_1, \ldots, X_r]$, and let

$$R = R_0[y_1, \ldots, y_r].$$

We make this a graded ring with the $X_i$ in degree zero and the $y_i$ in degree one. Let $f$ be the degree one polynomial

$$f = y_1 X_1^p + \cdots + y_r X_r^p \in R,$$

and let

$$A = R/(f), \qquad B = R/(X_1^p, \dots, X_r^p).$$

Let $E = \langle g_1, \dots, g_r \rangle \cong (\mathbb{Z}/p)^r$ be an elementary abelian $p$-group of rank $r$, and let $k$ be a field of characteristic $p$. The map sending $X_i$ to $g_i - 1 \in kE$ gives the identification

$$kE = k[X_1, \dots, X_r]/(X_1^p, \dots, X_r^p)$$

so that

$$B = k[y_1, \dots, y_r] \otimes_k kE.$$

The elements $y_1, \dots, y_r$ will eventually be identified with the generators for the polynomial part of the cohomology ring $H^*(E, k)$. So if $p$ is odd, the $y_i$ here correspond to the $x_i$ in Section 1.8 while if $p = 2$ they correspond to the elements $y_i^2$ given there. We thus have the following diagram of graded rings and homomorphisms:

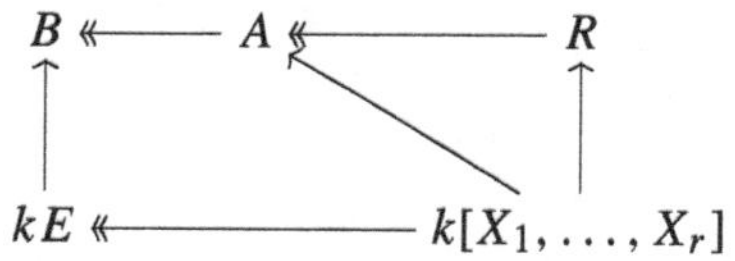

The ring $kE$ is a complete intersection and $A$ is a hypersurface. Our goal is to investigate the Orlov correspondence as an equivalence of triangulated categories

$$\mathsf{D}^b(kE) \underset{\Phi}{\overset{\Psi}{\rightleftarrows}} \mathsf{D}_{\mathrm{sg}}(A) \simeq \underline{\mathsf{RMF}}(f) \simeq \underline{\mathsf{MCM}}(A).$$

Here, $\mathsf{D}_{\mathrm{sg}}(A)$ is the singularity category of $A$, namely the Verdier quotient of $\mathsf{D}^b(A)$ by the thick subcategory of perfect complexes. This is equivalent to $\underline{\mathsf{RMF}}(f)$, the category of reduced matrix factorisations of $f$, which in turn is equivalent to the stable category $\underline{\mathsf{MCM}}(A)$ of maximal Cohen–Macaulay $A$-modules. See Theorem 11.9.3. This descends to an equivalence

$$\mathsf{stmod}(kE) \underset{\bar\Phi}{\overset{\bar\Psi}{\rightleftarrows}} \underline{\underline{\mathsf{RMF}}}(f) \simeq \underline{\underline{\mathsf{MCM}}}(A)$$

where $\underline{\underline{\mathsf{MCM}}}(A)$ is the Verdier quotient of $\underline{\mathsf{MCM}}(A)$ by the thick subcategory of maximal Cohen–Macaulay approximations of modules sitting in a finite set of degrees, and $\underline{\underline{\mathsf{RMF}}}(f)$ is the corresponding quotient of $\underline{\mathsf{RMF}}(f)$.

The functor $\Psi$ is easy to describe. Given a bounded complex of $kE$-modules $M_*$, we tensor over $k$ with $k[y_1, \dots, y_r]$ to get a bounded complex of $B$-modules, pull back to get a bounded complex of $A$-modules, and then apply the quotient functor $\mathsf{D}^b(A) \to \mathsf{D}_{\mathrm{sg}}(A)$. For the functor $\Phi$, a reduced matrix factorisation of $f$ is given by a pair of free $R$-modules $F$ and $F'$ together with maps $F'[-1] \xrightarrow{V} F \xrightarrow{U} F'$ with $UV$ and $VU$ equal to multiplication by $f$. Given such data, we apply $B \otimes_R -$ to obtain free $B$-modules $B \otimes_R F$, $B \otimes_R F'$ and then take graded pieces to get free $kE$-modules:

$$\cdots \xrightarrow{V} (B \otimes_R F)_0 \xrightarrow{U} (B \otimes_R F')_0 \xrightarrow{V} (B \otimes_R F)_1 \xrightarrow{U} (B \otimes_R F')_1 \xrightarrow{V} \cdots.$$

This complex only goes finitely far to the left, since $F$ and $F'$ are finitely generated, but it goes infinitely far to the right. For technical reasons, we then need to index this complex so that the term $(B \otimes_R F)_0$ is in homological degree $r$. The resulting complex is not exact, but it is exact sufficiently far to the right. So it is a semi-injective resolution of a well-defined object in $\mathsf{D}^b(kE)$. We shall show that this defines a functor $\Phi\colon \underline{\mathsf{RMF}}(f) \to \mathsf{D}^b(kE)$ such that $\Phi\Psi$ and $\Psi\Phi$ are naturally isomorphic to the identity functors.

Our main tools are a bidirectional Koszul complex, which replaces Orlov's use of local duality, and a theorem of Schoutens [223] relating the singularity category to the singular set.

**Grading Conventions.** We grade everything homologically, so that the differential decreases degree. When we talk of complexes of graded modules, there are two subscripts. The first subscript gives the homological degree and the second gives the internal degree. If $C$ is a complex of graded modules with components $C_{i,j}$ then we write $C[n]$ for the homological shift: $C[n]_{i,j} = C_{i+n,j}$, and $C(n)$ for the internal shift: $C(n)_{i,j} = C_{i,j+n}$.

## 11.2 Maximal Cohen–Macaulay Modules

Recall from Eisenbud [110] that there is an equivalence of categories between the stable category of maximal Cohen–Macaulay $A$-modules and matrix factorisations of the polynomial $f$. This correspondence is also explained in detail in Chapter 7 of Yoshino [251]. Our exposition is broadly taken from there.

We write $\mathsf{mod}(A)$ for the abelian category of finitely generated $A$-modules and module homomorphisms. A *maximal Cohen–Macaulay* $A$-module is a module whose depth is equal to $2r - 1$, which is the Krull dimension of $A$. We write $\mathsf{MCM}(A)$ for the full subcategory of $\mathsf{mod}(A)$ consisting of maximal Cohen–Macaulay $A$-modules. This is an exact category with enough projectives and enough injectives, and the projective and injective objects coincide. Such a category is called a *Frobenius category*, see Section I.2 of Happel [144].

The *stable category* $\underline{\mathsf{MCM}}(A)$ of maximal Cohen–Macaulay modules has the same objects as $\mathsf{MCM}(A)$, but the homomorphisms are given by

$$\underline{\mathrm{Hom}}_A(M, N) = \mathrm{Hom}_A(M, N)/P\mathrm{Hom}_A(M, N)$$

where $P\mathrm{Hom}_A(M, N)$ is the subspace of $\mathrm{Hom}_A(M, N)$ consisting of homomorphisms that factor through a free $A$-module. Note that $\underline{\mathsf{MCM}}(A)$ is a triangulated category.

So let $M$ be a maximal Cohen–Macaulay $A$-module. By the Auslander–Buchsbaum formula, as an $R$-module $M$ has projective dimension one. So it has a resolution of the form

$$0 \to F \xrightarrow{U} F' \to M \to 0$$

where $F$ and $F'$ are free $R$-modules, which necessarily have the same number of generators. Here, the generators of a free module are allowed to be in different degrees. We can therefore think of $U$ as a square matrix with entries in $R$. Since $fM = 0$, we have $fF' \subseteq UF$. So for any $x \in F'$ there exists a unique $y \in F$ such that $fx = Uy$. This defines for us an $R$-module homomorphism $V\colon F' \to F[1]$ sending $x$ to $y$. It satisfies $UV = f.I$, and therefore also $VU = f.I$ since $R$ is an integral domain.

**Definition 11.2.1** A *matrix factorisation* $(U, V)$ of $f$ consists of a pair of square matrices $U$ and $V$ of the same size, with entries in $R$, satisfying

$$UV = VU = f.I$$

where $I$ denotes the identity matrix. This can be regarded as a pair of maps between free $R$-modules

$$U\colon F \to F', \qquad V\colon F'[-1] \to F.$$

A *map* of matrix factorisations is given by a pair of maps of free modules making the two obvious squares commute. This gives matrix factorisations the structure of an exact category which we denote $\mathsf{MF}(f)$.

Given a matrix factorisation of $f$, let $\bar{F} = A \otimes_R F$, $\bar{F}' = A \otimes_R F'$ to give a complex

$$\cdots \to \bar{F}[-1] \xrightarrow{U} \bar{F}'[-1] \xrightarrow{V} \bar{F} \xrightarrow{U} \bar{F}'. \tag{11.1}$$

In fact, this is an exact sequence, since if $x \in F$ with $U\bar{x} = 0$ then there exists $y \in F'$ such that $Ux = fy$, so $fx = VUx = Vfy = fVy$ and hence $x = Vy$ (and similarly if $x \in F'$ with $V\bar{x} = 0$). It follows that (11.1) is a free resolution of the cokernel of $U$.

The matrix factorisation $(U, V) = (1, f)$ (with $F = F' = R$) has $\mathsf{Coker}\,(U) = 0$, so to get a correspondence between maximal Cohen–Macaulay modules and matrix factorisations we need to work modulo copies of this one. Furthermore, the matrix factorisation $(U, V) = (f, 1)$ (with $F = R[-1]$, $F' = R$) has $\mathsf{Coker}\,(U) \cong A$. So for the stable category of maximal Cohen–Macaulay modules we need to work modulo this one as well.

**Definition 11.2.2** The category $\underline{\mathsf{MF}}(f) = \mathsf{MF}(f)/\{(1, f)\}$ has the same objects as $\mathsf{MF}(f)$, but the homomorphisms are obtained by quotienting out those that factor through a direct sum of copies of $(1, f)$.

The category $\underline{\mathsf{RMF}}(f) = \mathsf{MF}(f)/\{(1, f), (f, 1)\}$ of *reduced matrix factorisations* has the same objects as $\mathsf{MF}(f)$ but the homomorphisms are obtained by quotienting out those that factor through a direct sum of copies of $(1, f)$ and $(f, 1)$.

**Theorem 11.2.3** (Eisenbud [110]) *The processes described above induce equivalences of categories between* $\underline{\mathsf{MF}}(f)$ *and* $\mathsf{MCM}(A)$ *and between* $\underline{\mathsf{RMF}}(f)$ *and* $\underline{\mathsf{MCM}}(A)$.

Given a matrix factorisation $F'[-1] \xrightarrow{V} F \xrightarrow{U} F'$, the shift $\tau$ in the triangulated category $\underline{\mathsf{RMF}}(f)$ takes it to $F \xrightarrow{-U} F' \xrightarrow{-V} F[1]$. Doing the same again gives $F' \xrightarrow{V} F[1] \xrightarrow{U} F'[1]$. So the square of the shift is equal to the internal degree shift for graded $A$-modules:

$$\tau^2 = [1].$$

Now given any $A$-module $N$, its $(2r-1)$st syzygy $\Omega^{2r-1}N$ is a maximal Cohen–Macaulay module, and hence has a periodic resolution of the form (11.1). So the minimal resolution of $N$ eventually agrees with the minimal resolution of a maximal Cohen–Macaulay module $N'$. The pitchfork lemma of Auslander and Buchweitz [21] allows us to extend to a map of resolutions, so that we obtain a map $N' \to N$, well defined in the stable category. The module $N'$ is called a *maximal Cohen–Macaulay approximation* to $N$. Adding a free module to $N'$ to make this map surjective, we obtain a short exact sequence

$$0 \to N'' \to N' \to N \to 0$$

where $N''$ has finite injective dimension.

This allows us to identify the maximal Cohen–Macaulay modules with the singularity category.

**Theorem 11.2.4** (Buchweitz [66]) *The inclusion of the maximal Cohen–Macaulay modules in the derived category induces an equivalence between* $\underline{\mathsf{MCM}}(A)$ *and* $\mathsf{D}_{\mathrm{sg}}(A)$.

## 11.3 The Orlov Correspondence

In this section we give a version of the Orlov correspondence for a complete intersection of dimension zero. Let $C = k[X_1, \dots, X_r]/(f_1, \dots, f_r)$ where $f_1, \dots, f_r$ is a regular sequence contained in the square of the maximal ideal $(X_1, \dots, X_r)$.

**Example 11.3.1** Let

$$E = \langle g_1, \dots, g_r \rangle \cong (\mathbb{Z}/p)^r$$

and let $k$ be a field of characteristic $p$. Let $kE$ be the group algebra of $E$ over $k$, and let

$$X_i = g_i - 1 \in kE.$$

Then $kE = k[X_1, \ldots, X_r]/(X_1^p, \ldots, X_r^p)$ is a complete intersection of codimension $r$ and dimension zero.

Let $R_0 = k[X_1, \ldots, X_r]$ and let $R = k[y_1, \ldots, y_r] \otimes_k R_0$. We regard $R$ as a graded polynomial ring with the $y_i$ in degree one and the $X_i$ in degree zero. Let

$$f = y_1 f_1 + \cdots + y_r f_r \in R,$$

an element of degree one. Let

$$A = R/(f)$$

and

$$B = R/(f_1, \ldots, f_r) = k[y_1, \ldots, y_r] \otimes_k C.$$

We have a diagram

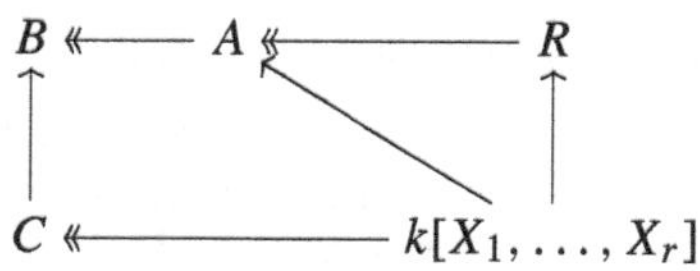


Taking Proj of these graded rings, we get the diagram in Section 2 of Orlov [204]:

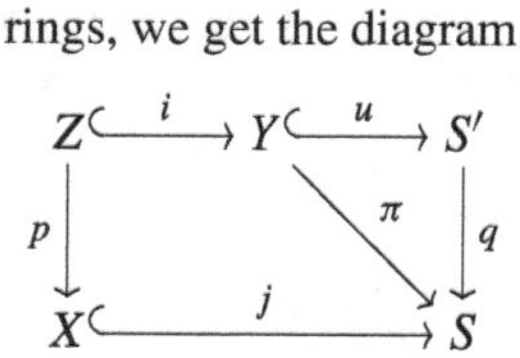


**Theorem 11.3.2** (Orlov) *The functor* $\mathbb{R}i_* p^* \colon \mathsf{D}^b(X) \to \mathsf{D}^b(Y)$ *descends to an equivalence of categories* $\mathsf{stmod}(C) = \mathsf{D}_{\mathrm{sg}}(X) \to \mathsf{D}_{\mathrm{sg}}(Y)$. *The right adjoint*

$$\mathbb{R}p_* i^\flat = \mathbb{R}p_* \mathbb{L}i^*(- \otimes \omega_{Z/Y})[-r+1] \tag{11.2}$$

*gives the inverse equivalence.*

Our goal is to prove the following lift of the Orlov correspondence to the derived category.

**Theorem 11.3.3** *There is an equivalence of categories* $\mathsf{D}^b(C) \simeq \mathsf{D}_{\mathrm{sg}}(A)$ *lifting the equivalence* $\mathsf{D}_{\mathrm{sg}}(X) \simeq \mathsf{D}_{\mathrm{sg}}(Y)$ *of Orlov.*

**Remark 11.3.4** In Orlov's version, he makes use of local duality as described in Chapter III, Corollary 7.3 of Hartshorne [146] to identify the right adjoint (11.2). In our version, this is replaced by the self-duality (11.14) of a complex ${}_A\hat{\boldsymbol{\Delta}}_A$ giving a Tate resolution of $A$ as an $A$-$A$-bimodule. The discrepancy between the shift of $-r+1$ in (11.2) and the shift of $-r$ in (11.14) is explained by the fact that in our situation the sheaf $\omega_{Z/Y}$ is just $\mathcal{O}(-1)$.

In the following corollary, we spell out the consequences for the modular representation theory of elementary abelian $p$-groups.

**Corollary 11.3.5** *Let $E$ be an elementary abelian $p$-group of rank $r$. Then the following triangulated categories are equivalent:*

(1) *The derived category $\mathsf{D}^b(kE)$.*
(2) *The singularity category of graded $A$-modules $\mathsf{D}_{\mathrm{sg}}(A)$ where*

$$A = R/(f), \quad R = k[y_1, \dots, y_r, X_1, \dots, X_r], \quad f = y_1X_1^p + \dots + y_rX_r^p,$$

*the $X_i$ have degree zero and the $y_i$ have degree one.*
(3) *The stable category of maximal Cohen–Macaulay graded $A$-modules.*
(4) *The category of reduced graded matrix factorisations of $f$ over $R$.*

We shall see that under the correspondence given by this Corollary, the image of the trivial $kE$-module $k$ is a $2^{r-1} \times 2^{r-1}$ matrix factorisation given by taking the even and odd terms in a bidirectional Koszul complex. The perfect complexes correspond to the maximal Cohen–Macaulay approximations to the $A$-modules which are non-zero only in finitely many degrees, so that the equivalence descends to Orlov's equivalence

$$\mathsf{stmod}(kE) \simeq \mathsf{D}_{\mathrm{sg}}(\mathsf{Proj}\,(A)).$$

The elements $y_i$ correspond to the basis for the primitive elements in $H^2(E,k)$ obtained by applying the Bockstein map to the basis of $H^1(E,k)$ dual to $X_1, \dots, X_r$.

## 11.4 The Functors

First we describe the functor $\Phi\colon \mathsf{D}^b(C) \to \mathsf{D}_{\mathrm{sg}}(A)$. If $M_*$ is a bounded complex of finitely generated $C$-modules then the tensor product

$$k[y_1, \dots, y_r] \otimes_k M_*$$

is a bounded complex of finitely generated $B$-modules, which may then be regarded as a bounded complex of finitely generated $A$-modules. Passing down to the singularity category $\mathsf{D}_{\mathrm{sg}}(A)$, we obtain $\Phi(M_*)$. Thinking of $\mathsf{D}_{\mathrm{sg}}(A)$ as equivalent to $\underline{\mathsf{MCM}}(A)$, we can view $\Phi(M_*)$ as a maximal Cohen–Macaulay approximation to $k[y_1, \dots, y_r] \otimes_k M_*$.

Next we describe the functor $\Psi\colon \mathsf{D}_{\mathrm{sg}}(A) \to \mathsf{D}^b(C)$. An object $N$ in $\mathsf{D}_{\mathrm{sg}}(A)$ can be thought of as a maximal Cohen–Macaulay $A$-module. It is therefore represented by a reduced graded matrix factorisation of the polynomial $f$ over $R$. Namely, we have a pair of finitely generated free $R$-modules $F$ and $F'$ and maps

$$F \xrightarrow{U} F' \xrightarrow{V} F(1)$$

such that $UV$ and $VU$ are both equal to $f$ times the identity map. Then we have the following sequence of free $C$-modules:

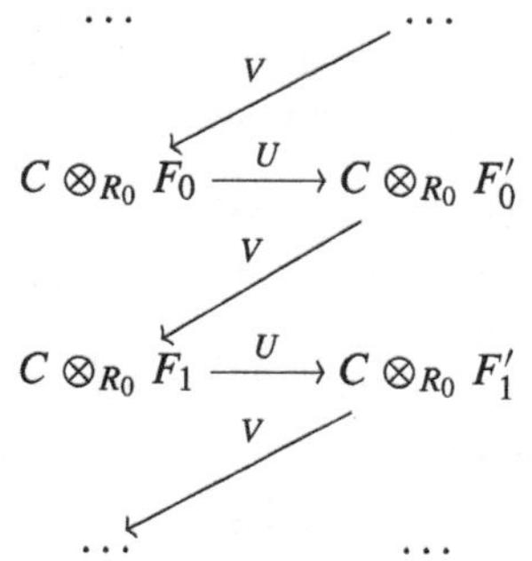

Since the free modules are finitely generated, this is zero far enough up the page. We shall see in Lemma 11.4.1 below, that the resulting complex only has homology in a finite number of degrees. It is therefore a complex of free $C$-modules, bounded to the left, and whose homology is totally bounded. It is therefore a semi-injective resolution of a well-defined object in $\mathsf{D}^b(C)$. We shift in degree so that the term $F_0 \otimes_{R_0} C$ appears in degree $-r$, and this is the object $\Psi(N)$ in $\mathsf{D}^b(C)$.

Another way of viewing the object $\Psi(N)$ is to take a complete resolution of $N$ as an $A$-module:

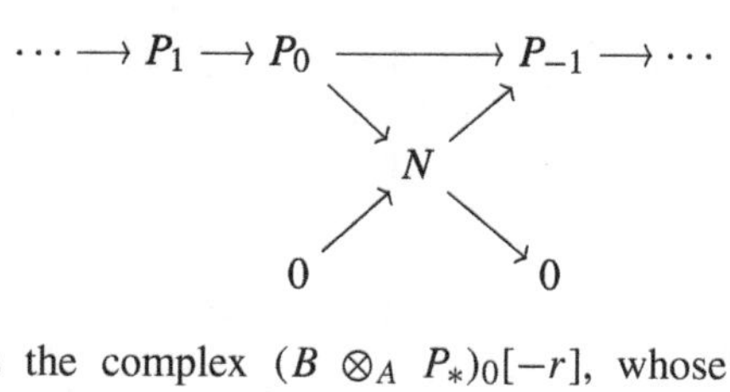

and then $\Psi(N)$ is the complex $(B \otimes_A P_*)_0[-r]$, whose degree $n$ term is $(B \otimes_A P_*)_{(n-r,0)}$.

**Lemma 11.4.1** *If $M$ is a maximal Cohen–Macaulay $A$-module then for all $j \geq 0$, $\mathrm{Tor}_j^A(B, M)$ is non-zero only in finitely many degrees.*

*Proof.* This follows from the fact that $B$ is locally (but not globally) a complete intersection as an $A$-module. More explicitly, for $1 \leq i \leq r$, in the ring $R[y_i^{-1}]$ we have the following equation:

$$X_i^p = y_i^{-1}(f - y_1 f_1 - \cdots \overset{i}{\uparrow} \cdots - y_r f_r).$$

It follows that

$$B[y_i^{-1}] = A[y_i^{-1}]/(f_1, \cdots \overset{i}{\uparrow} \cdots, f_r)$$

is a complete intersection of codimension $r - 1$ over $A[y_i^{-1}]$. Using a Koszul complex, it follows that for all $A$-modules $M$ and all $j \geq r$ we have

$$\mathrm{Tor}_j^A(B, M)[y_i^{-1}] = \mathrm{Tor}_j^{A[y_i^{-1}]}(B[y_i^{-1}], M[y_i^{-1}]) = 0.$$

If $M$ is a maximal Cohen–Macaulay module then the minimal resolution of $M$ is periodic and so $\mathrm{Tor}_j^A(B, M)[y_i^{-1}] = 0$ for all $j \geq 0$. Since $\mathrm{Tor}_j^A(B, M)$ is finitely generated, it is annihilated by a high enough power of each $y_i$ and is hence it is non-zero only in finitely many degrees. □

## 11.5 An Example

Before delving into proofs, let us examine an example in detail. Let $E = (\mathbb{Z}/p)^2 = \langle g_1, g_2 \rangle$, an elementary abelian group of rank two, and let $k$ be a field of characteristic $p$. Then setting $X_1 = g_1 - 1$, $X_2 = g_2 - 1$, we have

$$C = kE = k[X_1, X_2]/(X_1^p, X_2^p)$$
$$A = k[y_1, y_2, X_1, X_2]/(y_1X_1^p + y_2X_2^p).$$

Let us compute $\Phi(k)$, where $k$ is the trivial $kE$-module. This means we should resolve $k[y_1, y_2]$ as an $A$-module, and look at the corresponding matrix factorisation. This minimal resolution has the form

$$A(-1) \oplus A(-1) \xrightarrow{\begin{pmatrix} y_2X_2^{p-1} & -y_1X_1^{p-1} \\ X_1 & X_2 \end{pmatrix}} A \oplus A(-1) \xrightarrow{\begin{pmatrix} X_2 & y_1X_1^{p-1} \\ -X_1 & y_2X_2^{p-1} \end{pmatrix}} A \oplus A \xrightarrow{(X_1, X_2)}$$
$$A \to k[y_1, y_2].$$

This pair of $2 \times 2$ matrices gives a matrix factorisation of the polynomial $y_1X_1^p + y_2X_2^p$, and it is the matrix factorisation corresponding to the trivial $kE$-module.

Applying the functor $\Psi$ to this matrix factorisation, we obtain the minimal injective resolution of the trivial $kE$-module, shifted in degree by two. The elements $y_1$ and $y_2$ give the action of the polynomial generators of $H^*(E, k)$ of degree two on the minimal resolution.

Similarly, we compute $\Phi(kE)$ using the following resolution:

$$A(-1)\oplus A(-1) \xrightarrow{\begin{pmatrix} y_2 & -y_1 \\ X_1^p & X_2^p \end{pmatrix}} A\oplus A(-1) \xrightarrow{\begin{pmatrix} X_2^p & y_1 \\ -X_1^p & y_2 \end{pmatrix}} A\oplus A \xrightarrow{(X_1^p, X_2^p)}$$
$$A \to kE \otimes_k k[y_1, y_2].$$

This should be compared with the resolution of $k[X_1, X_2]$:

$$A(-2)\oplus A(-2) \xrightarrow{\begin{pmatrix} y_2 & -y_1 \\ X_1^p & X_2^p \end{pmatrix}} A(-1)\oplus A(-2) \xrightarrow{\begin{pmatrix} X_2^p & y_1 \\ -X_1^p & y_2 \end{pmatrix}} A(-1)$$
$$\oplus A(-1) \xrightarrow{(y_2, -y_1)} A \to k[X_1, X_2].$$

This is eventually the same resolution as above, but shifted two places to the left. Thus $\Phi(kE)$ is a maximal Cohen–Macaulay approximation to a module concentrated in a single degree.

## 11.6 The Bidirectional Koszul Complex

In this section, we construct the bidirectional Koszul complex. This first appears in a paper of Tate [237], and reappears in many places. This allows us to describe the minimal resolution of $B$ as an $A$-module. This computes the value of the functor $\Phi$ on the free $C$-module of rank one.

Let $\Lambda_n$ $(0 \le n \le r)$ be the free $R$-module of rank $\binom{r}{n}$ on generators

$$e_{j_1} \wedge \cdots \wedge e_{j_n}$$

with $1 \le j_1 < \cdots < j_n \le r$. We use the convention that the wedge is alternating, in the sense that $e_i \wedge e_j = -e_j \wedge e_i$ and $e_i \wedge e_i = 0$, to give meaning to wedge products with indices out of order or repeated indices.

We give $\Lambda_*$ a differential $d\colon \Lambda_n \to \Lambda_{n-1}$ described by

$$d(e_{j_1} \wedge \cdots \wedge e_{j_n}) = \sum_i (-1)^{i-1} f_{j_i}(e_{j_1} \wedge \cdots \overset{j_i}{\uparrow} \cdots \wedge e_{j_n}) \tag{11.3}$$

where the vertical arrow indicates a missing term. We also give $\Lambda_*$ a differential in the other direction $\delta\colon \Lambda_n \to \Lambda_{n+1}(1)$ described by

$$\delta(e_{j_1} \wedge \cdots \wedge e_{j_n}) = \sum_j y_j e_j \wedge (e_{j_1} \wedge \cdots \wedge e_{j_n}). \tag{11.4}$$

We call the graded $R$-module $\Lambda_*$ with these two differentials the *bidirectional Koszul complex* with respect to the pair of sequences $f_1, \ldots, f_r$ and $y_1, \ldots, y_r$:

$$\Lambda_r \underset{\underset{(1)}{\delta}}{\overset{d}{\rightleftarrows}} \Lambda_{r-1} \rightleftarrows \quad \cdots \quad \rightleftarrows \Lambda_1 \underset{\underset{(1)}{\delta}}{\overset{d}{\rightleftarrows}} \Lambda_0\,.$$

**Lemma 11.6.1** *The map $d\delta + \delta d\colon \Lambda_n \to \Lambda_n(1)$ is equal to multiplication by $f = \sum_i y_i f_i$.*

*Proof.* We have

$$\delta d(e_{j_1} \wedge \cdots \wedge e_{j_n}) = \sum_{i,j} (-1)^{i-1} y_j f_j e_j \wedge (e_{j_1} \wedge \cdots \overset{j_i}{\uparrow} \cdots \wedge e_{j_n}),$$

whereas

$$d\delta(e_{j_1} \wedge \cdots \wedge e_{j_n}) = \sum_{i,j} (-1)^{i-1} y_j f_j e_j \wedge (e_{j_1} \wedge \cdots \overset{j_i}{\uparrow} \cdots \wedge e_{j_n})$$
$$+ \sum_i y_i f_i (e_{j_1} \wedge \cdots \wedge e_{j_n}). \qquad \square$$

Thus, taking even and odd parts of $\Lambda_*$, we see that

$$\bigoplus_{n=0}^{\lfloor \frac{r}{2} \rfloor} \Lambda_{2n}(n-1) \xrightarrow{d+\delta} \bigoplus_{n=1}^{\lfloor \frac{r+1}{2} \rfloor} \Lambda_{2n-1}(n-1) \xrightarrow{d+\delta} \bigoplus_{n=0}^{\lfloor \frac{r}{2} \rfloor} \Lambda_{2n}(n) \tag{11.5}$$

is a matrix factorisation of $f$, called the *Koszul factorisation*. Note that the free $R$-modules in this matrix factorisation all have rank $2^{r-1}$, because this is the sum of the even binomial coefficients as well as the sum of the odd binomial coefficients. For notation, we write

$$K_0(-1) \xrightarrow{d+\delta} K_1 \xrightarrow{d+\delta} K_0 \tag{11.6}$$

for the Koszul factorisation, and we write $K$ for the cokernel of $d+\delta\colon K_1 \to K_0$.

For example, if $r = 4$ we get the Koszul factorisation

$$\Lambda_4(1) \oplus \Lambda_2 \oplus \Lambda_0(-1) \xrightarrow{\begin{pmatrix} d & \delta & 0 \\ 0 & d & \delta \end{pmatrix}} \Lambda_3(1) \oplus \Lambda_1 \xrightarrow{\begin{pmatrix} \delta & 0 \\ d & \delta \\ 0 & d \end{pmatrix}} \Lambda_4(2) \oplus \Lambda_2(1) \oplus \Lambda_0.$$

The minimal resolution of the $A$-module $B$ is obtained by applying $A \otimes_R -$ to

$$\cdots \to \Lambda_4 \oplus \Lambda_2(-1) \oplus \Lambda_0(-2) \to \Lambda_3 \oplus \Lambda_1(-1) \xrightarrow{d+\delta} \Lambda_2 \oplus \Lambda_0(-1) \xrightarrow{d+\delta} \Lambda_1$$
$$\xrightarrow{d} \Lambda_0. \tag{11.7}$$

This takes $r$ steps to settle down to the Koszul factorisation, but in large degrees (i.e., far enough to the left) it agrees with

$$\cdots \to K_0(-2) \xrightarrow{d+\delta} K_1(-1) \xrightarrow{d+\delta} K_0(-1) \xrightarrow{d+\delta} K_1 \xrightarrow{d+\delta} K_0. \tag{11.8}$$

For notation, we write $\bar{\Lambda}_i = A \otimes_R \Lambda_i$, so that the minimal resolution of $B$ as an $A$-module takes the form

$$\cdots \to \bar{\Lambda}_4 \oplus \bar{\Lambda}_2(-1) \oplus \bar{\Lambda}_0(-2) \to \bar{\Lambda}_3 \oplus \bar{\Lambda}_1(-1) \xrightarrow{d+\delta} \bar{\Lambda}_2 \oplus \bar{\Lambda}_0(-1) \xrightarrow{d+\delta} \bar{\Lambda}_1$$
$$\xrightarrow{d} \bar{\Lambda}_0. \tag{11.9}$$

Let us write $\Delta_{i,j}$ for this complex. As usual, $i$ denotes the homological degree and $j$ the internal degree. Thus $\Delta \to B$ is a free resolution.

A complete resolution of $B$ as an $A$-module is also easy to write down at this stage. Namely, we just continue (11.8) to the right in the obvious way. Let us write $\hat{\Delta}_{i,j}$ for this complete resolution. Then we notice a self-duality up to shift:

$$\operatorname{Hom}_A(\hat{\Delta}, A) \cong \hat{\Delta}[-r] \tag{11.10}$$

and a periodicity

$$\hat{\Delta}[2] \cong \hat{\Delta}(1).$$

Next we observe that the minimal resolution of $A/(y_1, \dots, y_r) = k[X_1, \dots, X_r]$ as an $A$-module takes the form

$$\cdots \to \bar{\Lambda}_{r-3}(-3) \oplus \bar{\Lambda}_{r-1}(-2) \xrightarrow{d+\delta} \bar{\Lambda}_{r-2}(-2) \oplus \bar{\Lambda}_r(-1) \xrightarrow{d+\delta} \bar{\Lambda}_{r-1}(-1) \xrightarrow{\delta} \bar{\Lambda}_r. \tag{11.11}$$

This again takes $r$ steps to settle down to the Koszul factorisation, but in large degrees (i.e., far enough to the left) it agrees with the result of applying $A \otimes_R -$ to

$$\cdots \to K_1(\lfloor \tfrac{-r-2}{2} \rfloor) \xrightarrow{d+\delta} K_0(\lfloor \tfrac{-r-1}{2} \rfloor) \xrightarrow{d+\delta} K_1(\lfloor \tfrac{-r}{2} \rfloor) \xrightarrow{d+\delta} K_0(\lfloor \tfrac{-r+1}{2} \rfloor). \tag{11.12}$$

**Theorem 11.6.2** *The minimal resolutions over $A$ of $A/(y_1, \dots, y_r) = k[X_1, \dots, X_r]$ after $r$ steps and of $B$ after $2r$ steps are equal.*

*Proof.* Compare (11.12) with (11.8). □

Now let $M_*$ be a bounded complex of $C$-modules, regarded as an object in $\mathsf{D}^b(C)$, and let $X_{i,j}$ be a free resolution of $k[y_1, \dots, y_r] \otimes_k M_*$ as an $A$-module.

Thus for large positive homological degree $i$, this is a periodic complex corresponding to a matrix factorisation of $f$, namely $\Phi(M_*)$. The maps $\Delta_{*,*} \to B$ and $X_{*,*} \to k[y_1, \dots, y_r] \otimes_k M_*$ induce homotopy equivalences

$$B \otimes_A X_{*,*} \leftarrow \Delta_{*,*} \otimes_A X_{*,*} \to \Delta_{*,*} \otimes_A (k[y_1, \dots, y_r] \otimes_k M_*).$$

Now $f_1, \dots, f_r$ annihilate $M_*$, and so they act as zero in $\Delta_{*,*} \otimes_A (k[y_1, \dots, y_r] \otimes_k M_*)$. So the operator $d$ in the complex $\Delta_{*,*}$ acts as zero in the tensor product, which therefore decomposes as a direct sum of pieces, each living in a finite set of degrees. To be more explicit, it decomposes as a sum of the following pieces tensored over $A$ with $(k[y_1, \dots, y_r] \otimes_k M_*)$:

$$\begin{array}{llll}
 & & & \bar{\Lambda}_0 \\
 & & \bar{\Lambda}_0(-1) \xrightarrow{\delta} & \bar{\Lambda}_1 \\
 & \bar{\Lambda}_0(-2) \xrightarrow{\delta} & \bar{\Lambda}_1(-1) \xrightarrow{\delta} & \bar{\Lambda}_2 \\
\bar{\Lambda}_0(-3) \xrightarrow{\delta} & \bar{\Lambda}_1(-2) \xrightarrow{\delta} & \bar{\Lambda}_2(-1) \xrightarrow{\delta} & \bar{\Lambda}_3 \\
\dots & \dots & &
\end{array}$$

Eventually, this just consists of copies of the Koszul complex for parameters $y_1, \dots, y_r$ on $k[y_1, \dots, y_r] \otimes_k M_*$, shifted in degree by $(2n, -n+r)$. This Koszul complex is quasi-isomorphic to $M_*$ shifted $(2n, -n+r)$.

It follows that if we take a *complete* resolution over $A$ of $k[y_1, \dots, y_r] \otimes_k M_*$, apply $B \otimes_A -$ to it, and take the part with internal degree zero, we obtain a complex which is quasi-isomorphic to $M_*$ shifted in degree by $r$. This process is exactly the functor $\Psi$ applied to $\Phi(M_*)$. To summarise, we have proved the following:

**Theorem 11.6.3** *The composite functor* $\Psi \circ \Phi\colon \mathsf{D}^b(C) \to \mathsf{D}^b(C)$ *is naturally isomorphic to the identity functor.* □

If we restrict just to $C$-modules rather than complexes, we have the following formulation:

**Theorem 11.6.4** *Let $M$ be a $C$-module. Then for $i \geq 0$ we have*

$$\begin{aligned}
\mathrm{Tor}^A_{i+r,j}(B, k[y_1, \dots, y_r] \otimes_k M) &\cong \mathrm{Tor}^A_{i,j}(k[X_1, \dots, X_r], k[y_1, \dots, y_r] \otimes_k M) \\
&\cong \begin{cases} M & i = 2j \\ 0 & \textit{otherwise.} \end{cases}
\end{aligned}$$

*Proof.* This follows from Theorems 11.6.2 and 11.6.3. □

## 11.7 A Bimodule Resolution

A similar bidirectional Koszul complex can be used to describe the minimal resolution of $A$ as an $A$-$A$-bimodule. This works more generally for any hypersurface (or indeed with suitable modifications for any complete intersection, see Section 3 of Wolffhardt [249]), so we introduce it in that context. Let $S = k[u_1, \dots, u_n]$ where each $u_i$ is a homogeneous variable of non-negative degree. Let $\phi(u_1, \ldots, u_n) \in S$ be a homogeneous polynomial of positive degree and let $H = S/(\phi)$ be the corresponding hypersurface. We write

$$\begin{aligned} S \otimes_k S &= k[u'_1, \ldots, u'_n, u''_1, \ldots, u''_n] \\ H \otimes_k H &= S \otimes_k S/(\phi(u'_1, \ldots, u'_n), \phi(u''_1, \ldots, u''_n)). \end{aligned}$$

Then we can form the bidirectional Koszul complex on the two sequences

$$u'_1 - u''_1, \ldots, u'_n - u''_n$$

and

$$\begin{gathered} (\phi(u'_1, u'_2 \ldots, u'_n) - \phi(u''_1, u'_2, \ldots, u'_n))/(u'_1 - u''_1), \\ (\phi(u''_1, u'_2, u'_3, \ldots, u'_n) - \phi(u''_1, u''_2, u'_3, \ldots, u'_n))/(u'_2 - u''_2), \\ \ldots \\ (\phi(u''_1, \ldots, u''_{n-1}, u'_n) - \phi(u''_1, \ldots, u''_{n-1}, u''_n))/(u'_n - u''_n). \end{gathered}$$

Note that the latter is indeed a sequence of polynomials, and that the sum of the products of corresponding terms in these two sequences gives

$$\phi(u'_1, \ldots, u'_n) - \phi(u''_1, \ldots, u''_n).$$

We therefore obtain a matrix factorisation of this difference over $S \otimes_k S$ looking much like (11.5). The construction corresponding to (11.9) in this situation gives a resolution of the module $S = (S \otimes_k S)/(u'_1 - u''_1, \ldots, u'_r - u''_r)$ over the hypersurface $(S \otimes_k S)/(\phi' - \phi'')$. Since $\phi'$ is a not a zerodivisor on either the module or the hypersurface, we can mod it out, retaining exactness, to obtain a resolution of $S/\phi = H$ as a module over $(S \otimes_k S)/(\phi' - \phi'', \phi') = H \otimes_k H$ (i.e., as an $H$-$H$-bimodule). We write ${}_H\mathbf{\Delta}_H$ for this resolution. It is eventually periodic with period two. The corresponding complete resolution is periodic with period two, and we denote it by ${}_H\hat{\mathbf{\Delta}}_H$.

Applying this in the particular case of $A$ as an $A$-$A$-bimodule, we have

$$\begin{aligned} A \otimes_k A = k[&y'_1, .., y'_r, y''_1, .., y''_r, X'_1, .., X'_r, X''_1, .., X''_r] \\ &/(y'_1 f'_1 + \cdots + y'_r f'_r, y''_1 f''_1 + \cdots + y''_r f''_r). \end{aligned} \tag{11.13}$$

Let ${}_A\hat{\mathbf{\Delta}}_A$ be the corresponding complete resolution. Exactly as in (11.10) we have a self-duality up to shift:

$$\mathrm{Hom}_{A \otimes_k A}({}_A\hat{\mathbf{\Delta}}_A, A \otimes_k A) \cong {}_A\hat{\mathbf{\Delta}}_A[-r] \tag{11.14}$$

and a periodicity

$${}_A\hat{\mathbf{\Delta}}_A[2] \cong {}_A\hat{\mathbf{\Delta}}_A(1). \tag{11.15}$$

Now regarding $B$ as an $A$-$B$-bimodule via the map $A \to B$, we have a free resolution given by ${}_A\mathbf{\Delta}_A \otimes_A {}_AB_B$. We write ${}_A\mathbf{\Delta}_B$ for this resolution, and ${}_A\hat{\mathbf{\Delta}}_B$ for the corresponding complete resolution ${}_A\hat{\mathbf{\Delta}}_A \otimes_A {}_AB_B$. Similarly, if we regard $B$ as a

$B$-$A$-bimodule, we have a free resolution ${}_B\mathbf{\Delta}_A = {}_B B_A \otimes_A {}_A\mathbf{\Delta}_A$ and a complete resolution ${}_B\hat{\mathbf{\Delta}}_A = {}_B B_A \otimes_A {}_A\hat{\mathbf{\Delta}}_A$. The duality (11.14) gives

$$\mathrm{Hom}_{A\otimes_k B}({}_A\hat{\mathbf{\Delta}}_B, A\otimes_k B) \cong {}_B\hat{\mathbf{\Delta}}_A[-r] \tag{11.16}$$

and the periodicity (11.15) gives

$${}_A\hat{\mathbf{\Delta}}_B[2] \cong {}_A\hat{\mathbf{\Delta}}_B(1). \tag{11.17}$$

Finally, for any left $A$-module $N$ we have a free resolution ${}_A\mathbf{\Delta}_A \otimes_A N$ and a complete resolution ${}_A\hat{\mathbf{\Delta}}_A \otimes_A N$.

## 11.8 The Adjunction

**Proposition 11.8.1** *The functor $\Psi$ is right adjoint to $\Phi$.*

*Proof.* Let $M_*$ be a bounded complex of $C$-modules. If $N$ is a maximal Cohen–Macaulay $A$-module, then ${}_A\hat{\mathbf{\Delta}}_A \otimes_A N$ is a complete resolution of $N$ as an $A$-module. Then

$$\Psi(N) = ({}_B B_A \otimes_A ({}_A\hat{\mathbf{\Delta}}_A \otimes_A N))_0[-r] \cong ({}_B\hat{\mathbf{\Delta}}_A \otimes_A N)_0[-r].$$

Write $\underline{\mathrm{Hom}}$ for homomorphisms of complexes modulo homotopy. Since $({}_B\hat{\mathbf{\Delta}}_A \otimes_A N)_0[-r]$ is semi-injective, homomorphisms in $\mathsf{D}^b(C)$ from an object to it are just homotopy classes of maps of complexes. So using the duality (11.16), we have

$$\begin{aligned}
\mathrm{Hom}_{\mathsf{D}^b(C)}(M_*, \Psi(N)) &= \underline{\mathrm{Hom}}^n_C(M_*, ({}_B\hat{\mathbf{\Delta}}_A \otimes_A N)_0[-r]) \\
&\cong \underline{\mathrm{Hom}}_B(k[y_1,\dots,y_r]\otimes_k M_*, {}_B\hat{\mathbf{\Delta}}_A \otimes_A N[-r]) \\
&\cong \underline{\mathrm{Hom}}_B(k[y_1,\dots,y_r]\otimes_k M_*, \mathrm{Hom}_A({}_A\hat{\mathbf{\Delta}}_B, N)) \\
&\cong \underline{\mathrm{Hom}}_A({}_A\hat{\mathbf{\Delta}}_B \otimes_B (k[y_1,\dots,y_r]\otimes_k M_*), N) \\
&\cong \underline{\mathrm{Hom}}_A(\Phi(M), N).
\end{aligned}$$

In the last line, we are using the fact that

$${}_A\hat{\mathbf{\Delta}}_B \otimes_B (k[y_1,\dots,y_r]\otimes_k M_*) = {}_A\hat{\mathbf{\Delta}}_A \otimes_A (k[y_1,\dots,y_r]\otimes_k M_*)$$

is a complete resolution over $A$ of a maximal Cohen–Macaulay approximation to the complex $k[y_1,\dots,y_r]\otimes_k M_*$. □

## 11.9 The Equivalence

**Proposition 11.9.1** *The category $\mathsf{D}_{\mathrm{sg}}(A)$ is generated by the $A$-module $k[y_1,\dots,y_r]$.*

*Proof.* The singular locus of $A$ is defined by the equations $\frac{\partial f}{\partial y_i} = 0$ and $\frac{\partial f}{\partial X_i} = 0$. The former give the equations $f_1 = 0,\dots,f_r = 0$; since $C$ is a zero-dimensional complete intersection, these equations define the same variety as $X_1 = 0,\dots,X_r = 0$. The latter give the equations $\sum_j y_j \frac{\partial f_j}{\partial X_i} = 0$. Since

$f_1, \ldots, f_r$ are in the square of the maximal ideal, $\frac{\partial f_j}{\partial X_i}$ has zero constant term and so no new conditions are imposed by these equations. So we have shown that the prime ideals in the singular locus are those containing $(X_1, \ldots, X_r)$.

By Theorem 1.7.7, the singularity category $\mathsf{D}_{\mathrm{sg}}(A)$ is generated by the modules $A/\mathfrak{p}$ with $\mathfrak{p} \supseteq (X_1, \ldots, X_r)$, namely the quotients of $k[y_1, \ldots, y_r]$ by prime ideals. These in turn are generated by the single object $k[y_1, \ldots, y_r]$, by the Hilbert syzygy theorem. □

**Theorem 11.9.2** *The composite functor* $\Phi \circ \Psi \colon \mathsf{D}_{\mathrm{sg}}(A) \to \mathsf{D}_{\mathrm{sg}}(A)$ *is naturally isomorphic to the identity functor.*

*Proof.* Consider the adjunction of Proposition 11.8.1. By Theorem 11.6.3, the unit of this adjunction gives an isomorphism $k \to \Psi\Phi(k)$, where $k$ is the residue field of $C$, regarded as an object in $\mathsf{D}^b(C)$ by putting it in degree zero. It follows that $\Phi\Psi(\Phi(k)) = \Phi(\Psi\Phi(k)) \cong \Phi(k)$. An easy diagram chase shows that this isomorphism is given by the counit of the adjunction. It follows that the counit of the adjunction is an isomorphism for every object in the thick subcategory of $\mathsf{D}_{\mathrm{sg}}(A)$ generated by $\Phi(k)$. Since $\Phi(k) \cong k[y_1, \ldots, y_r]$, by Proposition 11.9.1 this is the whole of $\mathsf{D}_{\mathrm{sg}}(A)$. □

**Theorem 11.9.3** *The functors* $\Phi \colon \mathsf{D}^b(C) \to \mathsf{D}_{\mathrm{sg}}(A)$ *and* $\Psi \colon \mathsf{D}_{\mathrm{sg}}(A) \to \mathsf{D}^b(C)$ *are inverse equivalences of categories.*

*Proof.* This follows from Theorems 11.6.3 and 11.9.2. □

**Theorem 11.9.4** *The equivalence* $\mathsf{D}^b(C) \underset{\Psi}{\overset{\Phi}{\rightleftarrows}} \mathsf{D}_{\mathrm{sg}}(A)$ *descends to an equivalence*

$$\mathsf{stmod}(C) \underset{\bar{\Psi}}{\overset{\bar{\Phi}}{\rightleftarrows}} \underline{\underline{\mathsf{MCM}}}(A)\,.$$

*Proof.* It follows from Theorem 11.6.2 that $\Phi(C)$ is a shift of $k[X_1, \ldots, X_r]$. The theorem now follows, because $\mathsf{stmod}(C)$ is the quotient of $\mathsf{D}^b(C)$ by the thick subcategory generated by $C$ and $\underline{\underline{\mathsf{MCM}}}(A)$ is the quotient of $\underline{\mathsf{MCM}}(A)$ by the thick subcategory generated by $k[X_1, \ldots, X_r]$. □

## 11.10 The Trivial Module

So, in the case $C = kE$, where does the trivial $kE$-module $k$ go to under the correspondence of Theorem 11.9.3? To answer this, we must find the minimal resolution of $k[y_1, \ldots, y_r]$ as an $A$-module. This is again given in terms of a bidirectional Koszul complex, this time for the pair of sequences $X_1, \ldots, X_r$ and $y_1X_1^{p-1}, \ldots, y_rX_r^{p-1}$. The sum of the products of corresponding terms in these sequences again gives $f = \sum_i y_i X_i^p$, and so we obtain a matrix factorisation of $f$ by taking the even and odd parts of this bidirectional Koszul complex. Adding up the even and odd binomial coefficients, we see that this is a $2^{r-1} \times 2^{r-1}$ matrix factorisation of $f$. The minimal resolution of $k[y_1, \dots, y_r]$ as an $A$-module is given by (11.9) with respect to this version of the bidirectional Koszul complex.

In a similar way, we can find the image of any module of the form $kE/(X_1^{a_1}, \dots, X_r^{a_r})$ under the correspondence of Theorem 11.9.3 by doing the same process with a bidirectional Koszul complex for the pair of sequences $X_1^{a_1}, \dots, X_r^{a_r}$ and $y_1 X_1^{p-a_1}, \dots, y_r X_r^{p-a_r}$.

Let us look at some examples. First we look at the case $r = 1$. In this case $E$ is cyclic of order $p$, $kE = k[X]/(X^p)$, $A = k[y, X]/(yX^p)$ and $B = k[y, X]/(X^p)$. The indecomposable $kE$-modules are the Jordan blocks $J_n = k[X]/(X^n)$ $(1 \le n \le p)$. Resolving $J_n \otimes_k k[y]$ we get

$$\cdots \to A[-1] \xrightarrow{X^n} A[-1] \xrightarrow{yX^{p-n}} A \xrightarrow{X^n} A.$$

So the matrix factorisation corresponding to $J_n$ is given by the $1 \times 1$ matrices

$$A[-1] \xrightarrow{(yX^{p-n})} A \xrightarrow{(X^n)} A.$$

Next let $r = 2$, so that $A = k[y_1, y_2, X_1, X_2]/(y_1 X_1^p + y_2 X_2^p)$. To find the matrix factorisation corresponding to the trivial module, we resolve the $A$-module $k[y_1, y_2] = A/(X_1, X_2)$. Using the construction in Section 11.6, we obtain the following minimal resolution:

$$A[-1] \oplus A[-1] \xrightarrow{\begin{pmatrix} y_2 X_2^{p-1} & -y_1 X_1^{p-1} \\ X_1 & X_2 \end{pmatrix}} A \oplus A[-1] \xrightarrow{\begin{pmatrix} X_2 & y_1 X_1^{p-1} \\ -X_1 & y_2 X_2^{p-1} \end{pmatrix}}$$
$$A \oplus A \xrightarrow{\begin{pmatrix} X_1 & X_2 \end{pmatrix}} A. \tag{11.18}$$

The two square matrices in this resolution alternate, and so the matrix factorisation corresponding to the trivial module is as follows:

$$A \oplus A[-1] \xrightarrow{\begin{pmatrix} X_2 & y_1 X_1^{p-1} \\ -X_1 & y_2 X_2^{p-1} \end{pmatrix}} A \oplus A \xrightarrow{\begin{pmatrix} y_2 X_2^{p-1} & -y_1 X_1^{p-1} \\ X_1 & X_2 \end{pmatrix}} A[1] \oplus A.$$

In rank three, the minimal resolution of $k[y_1, y_2, y_3]$ as an $A$-module takes the following form.

$$A[-1]^{\oplus 3} \oplus A[-2] \xrightarrow{\begin{pmatrix} 0 & X_3 & -X_2 & y_1 X_1^{p-1} \\ -X_3 & 0 & X_1 & y_2 X_2^{p-1} \\ X_2 & -X_1 & 0 & y_3 X_3^{p-1} \\ y_1 X_1^{p-1} & y_2 X_2^{p-2} & y_3 X_3^{p-1} & 0 \end{pmatrix}} A[-1]^{\oplus 3} \oplus A \xrightarrow{\begin{pmatrix} 0 & -y_3 X_3^{p-1} & y_2 X_2^{p-1} & X_1 \\ y_3 X_3^{p-1} & 0 & -y_1 X_1^{p-1} & X_2 \\ -y_2 X_2^{p-1} & y_1 X_1^{p-1} & 0 & X_3 \\ X_1 & X_2 & X_3 & 0 \end{pmatrix}}$$
$$A^{\oplus 3} \oplus A[-1] \xrightarrow{\begin{pmatrix} 0 & X_3 & -X_2 & y_1 X_1^{p-1} \\ -X_3 & 0 & X_1 & y_2 X_2^{p-1} \\ X_2 & -X_1 & 0 & y_3 X_3^{p-1} \end{pmatrix}} A^{\oplus 3} \xrightarrow{\begin{pmatrix} X_1 & X_2 & X_3 \end{pmatrix}} A$$

The left-hand pair of matrices therefore gives the matrix factorisation of $f$ corresponding to the trivial $kE$-module.

## 11.11 Computer Algebra

Here is some code in the computer algebra language Macaulay2 for computing the functor $\mathsf{mod}(kE) \hookrightarrow \mathsf{D}^b(kE) \xrightarrow{\Phi} \underline{\mathsf{RMF}}(f)$. It is given here for the trivial module for a group of rank two with $p = 7$, but the code is easy to modify. The last two commands print out the sixth and seventh matrices in the minimal resolution, which in this case is easily far enough to give a matrix factorisation.

```
p=7
R = ZZ/p[X1,X2,y1,y2]
f = X1^p * y1 + X2^p * y2
A = R/(f)
U = cokernel matrix {{X1,X2}}
F = resolution (U,LengthLimit=>8)
F.dd_6
F.dd_7
```

To modify the code to work for other $kE$-modules $M$, the fifth line should be changed to give a presentation of

$$U = k[y_1, \dots, y_r] \otimes_k M$$

as an $A$-module. Don't forget the relations saying that $X_i^p$ annihilates $M$. For example, if $M = \Omega(k)$ for the same group of rank two above, then $U$ has two generators and three relations; it is the cokernel of the matrix

$$\begin{pmatrix} X_1^p & X_2 & 0 \\ 0 & X_1 & X_2^p \end{pmatrix} : A^{\oplus 3} \to A^{\oplus 2}.$$

So the fifth line should be changed to

```
U = cokernel matrix {{X1^p,X2,0},{0,X1,X2^p}}.
```

## 11.12 Cohomology

For this section, we stick with the case $C = kE$. The elements $y_1, \dots, y_r \in A$ act on maximal Cohen–Macaulay modules $N$ as maps $N \to N(1)$. Now $N(1)$ is isomorphic to $\Omega^{-2}(N)$, and $\Omega^{-1}$ is the shift functor in the triangulated category $\underline{\mathsf{MCM}}(A)$. It follows that under equivalence of categories of Theorem 11.9.3, these elements correspond to maps in $\mathsf{D}^b(kE)$ from $M_*$ to $M_*[2]$. We claim that these elements act as the polynomial part of the cohomology ring, namely the subring generated by the Bocksteins of the elements of degree one.

Recall that $kE$ is a Hopf algebra, either via the group theoretic diagonal map defined by $\Delta(g_i) = g_i \otimes g_i$, $\Delta(X_i) = X_i \otimes 1 + 1 \otimes X_i + X_i \otimes X_i$ or via the restricted Lie algebra diagonal map defined by $\tilde{\Delta}(X_i) = X_i \otimes 1 + 1 \otimes X_i$. In both cases, we make $A$ into a right $kE$-comodule via $\Delta, \tilde{\Delta}\colon A \to A \otimes kE$ defined by the same formula on $X_i$ and via $\Delta(y_i) = \tilde{\Delta}(y_i) = y_i \otimes 1$. It is easy to check that this is a coaction: $(1 \otimes \Delta) \circ \Delta = (\Delta \otimes 1) \circ \Delta$ and $(1 \otimes \tilde{\Delta}) \circ \tilde{\Delta} = (\tilde{\Delta} \otimes 1) \circ \tilde{\Delta}$. We denote the corresponding tensor products $N \otimes_k M$ and $N \tilde{\otimes}_k M$, where $M$ is a $kE$-module and $N$ and the tensor product are $A$-modules.

**Lemma 11.12.1** *If $N$ is a maximal Cohen–Macaulay $A$-module and $M$ is a $kE$-module then $N \otimes_k M$ and $N \tilde{\otimes}_k M$ are maximal Cohen–Macaulay $A$-modules.*

*Proof.* The $kE$-module $M$ has a finite filtration in which the filtered quotients are copies of the trivial $kE$-module $k$. So the tensor products $N \otimes_k M$ and $N \tilde{\otimes}_k M$ both have finite filtrations in which the quotients are isomorphic to $N$. The lemma now follows from the fact that every extension of maximal Cohen–Macaulay modules is maximal Cohen–Macaulay. □

Recall that the maximal Cohen–Macaulay module $K$ corresponding to the Koszul factorisation (11.6) is the image of the trivial $kE$-module $k$ under $\Phi$.

**Proposition 11.12.2** *If $M$ is a $kE$-module then $\Phi(M)$ may be taken to be the maximal Cohen–Macaulay module $K \otimes_k M$ (or $K \tilde{\otimes}_k M$).*

*Proof.* It follows from Lemma 11.12.1 that $K \otimes_k M$, resp. $K \tilde{\otimes}_k M$ is a maximal Cohen–Macaulay approximation to $k[y_1, \dots, y_r] \otimes_k M$. □

Let $V$ be the linear space spanned by $X_1, \dots, X_r$. Then we may use the polynomial $f$ to identify the space spanned by $y_1, \dots, y_r$ with the Frobenius twist of the dual $F(V^*)$. There is an action of $GL(V)$ on $kE$ induced by linear substitutions of the $X_i$, and this induces an action on the linear space $F(V^*)$ spanned by the $y_i$.

**Theorem 11.12.3** (i) *The maps*

$$\Psi(y_1), \dots, \Psi(y_r)\colon k \to k[2]$$

*form a vector space basis for the image of the Bockstein map $H^1(E, k) \to H^2(E, k)$.*

(ii) *For any $M$ in $\mathsf{stmod}(kE)$, the induced map*

$$\bar{\Psi}(y_i)\colon M \to \Omega^{-2}(M)$$

*is equal to the map*

$$\bar{\Psi}(y_i) \otimes 1\colon k \otimes_k M \to \Omega^{-2}(k) \otimes M$$

*and also to the map*

$$\bar{\Psi}(y_i)\,\tilde{\otimes}\,1\colon k\,\tilde{\otimes}_k\,M \to \Omega^{-2}(k)\,\tilde{\otimes}\,M.$$

*Proof.* (i) Consider the action of $GL(V)$ on $H^2(E,k)$. For $p$ odd, we have

$$H^2(E,k) \cong F(V^*) \oplus \Lambda^2(V^*).$$

For $p = 2$, we have $H^2(E,k) \cong S^2(V)$; in this case $S^2(V^*)$ has two composition factors as a $GL(V)$-module, given by the non-split short exact sequence

$$0 \to F(V^*) \to S^2(V^*) \to \Lambda^2(V^*) \to 0.$$

In both cases, there is a unique $GL(V)$-invariant subspace of $H^2(E,k)$ isomorphic to $F(V^*)$, and this is the image of the Bockstein map. It therefore suffices to prove that the $\bar{\Psi}(y_i)$ are not all equal to zero. To see this, take the Koszul complex for $K$ with respect to the parameters $y_1,\dots,y_r$. Since $K$ is a maximal Cohen–Macaulay approximation to $k[y_1,\dots,y_r]$, the homology of this complex disappears after applying $\bar{\Psi}$. If $\bar{\Psi}(y_i)$ were zero, this could not be the case. It follows that $\bar{\Psi}(y_i) \neq 0$ and the theorem is proved.

(ii) This follows from Proposition 11.12.2. □

**Remark 11.12.4** Theorem 11.12.3 gives an interpretation of the realisation theorem of Benson and Pevtsova, Theorem 8.1.1. Namely, given any $k[y_1,\dots,y_r]$-module $N$, inflating it and applying $\bar{\Psi}$ to a maximal Cohen–Macaulay approximation to the resulting $A$-module gives the same $kE$-module $M$ as the process described in Section 8.8. In particular, it follows from the analysis in Section 8.9 that $\mathcal{F}_1(M)$ is the Frobenius twist of the sheaf on $\mathbb{P}^{r-1}$ corresponding to $N$ and $\mathcal{F}_i(M) = 0$ for $1 < i < p$. So if $N$ is locally free over $k[y_1,\dots,y_r]$ then $M$ has constant Jordan type.

## 11.13 Twisted Versions of $kE$

In this section we indicate how to modify the constructions of this chapter to deal with the twisted versions of $kE$ discussed in Section 1.18. This is similar to the modifications of the Koszul complex used by Avramov, Gasharov and Peeva in §2 of [23]. Recall that $\mathbf{q} \in \mathrm{Mat}_r(k)$ is a matrix satisfying $q_{ij}q_{ji} = 1$ for $1 \le i, j \le r$ and $q_{ii} = 1$ for $1 \le i \le r$, $\mathbf{n} = (n_1,\dots,n_r)$ is an $r$-tuple of strictly positive integers, and $C_{\mathbf{q},\mathbf{n}}$ is the algebra with generators $X_1,\dots,X_r$ satisfying $X_iX_j = q_{ij}X_jX_i$ for $1 \le i, j \le r$ and $X_i^{p^{n_i}} = 0$ for $1 \le i \le r$; these algebras arise in discussions of blocks with normal abelian defect group, abelian inertial quotient and one isomorphism class of simple module.

In this case, we let $R$ be the twisted polynomial ring with generators $y_1,\dots,y_r,X_1,\dots,X_r$ satisfying the following relations:

$$y_iy_j = q_{ji}^{p^{n_i+n_j}}\,y_jy_i; \quad X_iX_j = q_{ij}X_jX_i; \quad X_jy_i = y_iX_j \qquad (11.19)$$

with the variables $X_i$ in degree zero and the variables $y_i$ in degree one. Thus $y_1 X_1^{p^{n_1}}, \dots, y_r X_r^{p^{n_r}}$ are commuting elements of $R$, and we set $f = y_1 X_1^{p^{n_1}} + \cdots + y_r X_r^{p^{n_r}}$.

In the bidirectional Koszul complex, the sense in which the wedges are alternating is modified to

$$e_i \wedge e_j = -q_{ij}^{p^{n_i+n_j}} e_j \wedge e_i.$$

The definition of the differential $d$ in (11.3) then becomes

$$d(e_{j_1} \wedge \cdots \wedge e_{j_n}) = \sum_i (-1)^{i-1} \left( \prod_{t=1}^{i-1} q_{j_t j_i}^{p^{n_{j_t}+n_{j_i}}} \right) (e_{j_1} \wedge \dots \overset{j_i}{\uparrow} \cdots \wedge e_{j_n})$$

while the definition of $\delta$ in (11.4) needs no alteration.

With these definitions, we obtain the following theorem.

**Theorem 11.13.1** *The following triangulated categories are equivalent:*

(1) *The derived category $D^b(C_{\mathbf{q},\mathbf{n}})$.*
(2) *The singularity category of graded $A$-modules $\mathsf{D}_{\mathrm{sg}}(A)$ where $A = R/f$, $R$ is the twisted polynomial ring with generators $y_1, \dots, y_r$ in degree one, $X_1, \dots, X_r$ in degree zero, and commutation relations* (11.19), *and $f = y_1 X_1^{p^{n_1}} + \cdots + y_r X_r^{p^{n_r}}$.*
(3) *The stable category of maximal Cohen–Macaulay graded $A$-modules.*
(4) *The category of reduced graded matrix factorisations of $f$ over $R$.*

*Proof.* The proofs are the same as in the untwisted case, with the modifications outlined above. □

# 12

# Phenomenology of Modules over Elementary Abelian $p$-Groups

## 12.1 Introduction

This chapter is about the following question. How small can a module be, and still have an interesting variety? Methods for producing such modules are similar to those used in discussions of constant Jordan type in previous chapters. We consider various measures of size, including dimension and Loewy length.

**Example 12.1.1** My interest in this question was kindled by the discovery that the 42-dimensional Specht module $S^{(3^3)}$ for the symmetric group $\mathfrak{S}_9$ in characteristic 3, when restricted to the elementary abelian subgroup $(\mathbb{Z}/3)^3$, has variety given by the equation

$$Y_1^2Y_2^2 + Y_1^2Y_3^2 + Y_2^2Y_3^2 = 0$$

in affine space $\mathbb{A}^3$. This was computed by Jon Carlson using the computer algebra package Magma [63] around 2005, and an abstract argument appears in the paper of Lim [187].

The question is, in odd characteristic, how can a module of such small dimension have such an interesting variety? All the methods we knew, at the time when this was computed, produced much larger modules. We shall see later in this chapter, that there is in fact a module of dimension 36 for $(\mathbb{Z}/3)^3$ with the same variety.

Let $k$ be an algebraically closed field of characteristic $p$, and let $E = (\mathbb{Z}/p)^r$ be an elementary abelian $p$-group of rank $r$. We use the notation from Sections 1.3 and 1.9. In particular, $E = \langle g_1, \dots, g_r\rangle$, $X_i = g_i - 1$, so that

$$kE = k[X_1, \dots, X_r]/(X_1^p, \dots, X_r^p).$$

The affine space $V_E^{\mathrm{rk}} \cong \mathbb{A}^r(k)$ has $X_1, \dots, X_r$ as a basis, and its coordinate ring is $k[V_E^{\mathrm{rk}}] = k[Y_1, \dots, Y_r]$, regarded as a graded ring with each $Y_i$ in degree one. Thus $Y_i$ is the function on $V_E^{\mathrm{rk}}$ defined by $Y_i(X_j) = \delta_{i,j}$ (Kronecker delta). If $\alpha = (\lambda_1, \dots, \lambda_r) \in V_E^{\mathrm{rk}}$ then we write $X_\alpha$ for $\lambda_1 X_1 + \cdots + \lambda_r X_r$.

We can think of $Y_1, \ldots, Y_r$ as corresponding to the elements $y_1, \ldots, y_r$ in $H^1(E,k) = \mathrm{Ext}^1_{kE}(k,k)$, see Section 1.8. If $p = 2$, these generated the polynomial ring $H^*(E,k)$, and so we may identify $H^*(E,k)$ with $k[V_E^{\mathrm{rk}}]$. If $p$ is odd, the $y_i$ anticommute, and their Bocksteins $x_i$ generate the polynomial part of the cohomology ring. Since the Bockstein map is semilinear with respect to the Frobenius map, this means that the $x_i$ correspond with $Y_i^p$, as we saw in Chapter 8.

**Example 12.1.2** Let $W$ be a linear subspace of $V_E^{\mathrm{rk}}$ of dimension $c$. Choose a basis $X_{\beta_1}, \ldots, X_{\beta_c}$ for $W$, and let

$$kE' = k[X_{\beta_1}, \ldots, X_{\beta_c}]/(X^p_{\beta_1}, \ldots, X^p_{\beta_c}).$$

This is a subalgebra flatly embedded in $kE$. Let $M(W)$ be the $kE$-module $kE \otimes_{kE'} k$. Then $\dim_k(M(W)) = p^{r-c}$, the Loewy length of $M(W)$ is $(p-1)(r-c)+1$, and $V_E^{\mathrm{rk}}(M(W)) = W$.

Let $U$ be a linear subspace of $V_E^{\mathrm{rk}}$ of dimension $r-c$ such that $U \cap W = \{0\}$. Choose a basis $\alpha_1, \ldots, \alpha_{r-c}$ for $U$, and set

$$kE'' = k[X_{\alpha_1}, \ldots, X_{\alpha_{r-c}}]/(X^p_{\alpha_1}, \ldots, X^p_{\alpha_{r-c}}). \tag{12.1}$$

Then $M(W)\!\downarrow_{kE''}$ is a free module of rank one.

Similarly in the exterior algebra case, if $W$ is a linear subspace of $V_\Lambda^{\mathrm{rk}}$ of dimension $c$ with basis $X_{\beta_1}, \ldots, X_{\beta_c}$, let

$$\Lambda' = \Lambda(X_{\beta_1}, \ldots, X_{\beta_c})$$

and let $M(W)$ be the $\Lambda$-module $\Lambda \otimes_{\Lambda'} k$. Then $\dim_k(M(W)) = 2^{r-c}$, the Loewy length of $M(W)$ is $r-c+1$, and $V_\Lambda^{\mathrm{rk}}(M(W)) = W$.

These modules have as small a dimension as possible among modules with rank variety $W$, as the following theorem shows.

**Theorem 12.1.3** (Carlson [78, §4]) *If $M$ is a $kE$-module such that $V_E^{\mathrm{rk}}(M)$ has dimension $m$ and degree $d$ then* $\dim_k M/p^{r-m}$ *is an integer at least as big as $d$.*

*Proof.* By Bézout's theorem, a generic subspace $V' \subseteq V_E^{\mathrm{rk}}$ of dimension $r-m+1$ intersects $V_E^{\mathrm{rk}}(M)$ in $d$ distinct lines through the origin. So by parts (4) and (6) of Proposition 1.9.11, $M(V) \otimes_k M$ decomposes as a direct sum of $d$ modules, each of whose rank varieties is a single line through the origin. For such a summand, the restriction to a subgroup of rank $r-1$ not containing this line is projective, so the dimension is divisible by $p^{r-1}$. So $\dim_k(M(V) \otimes_k M) = p^{m-1}\dim_k(M)$ is $p^{r-1}$ times an integer at least $d$. □

Essentially the same proof gives the following.

**Theorem 12.1.4** *If $M$ is a $kE$-module such that $V_E^{\mathrm{rk}}(M)$ has an irreducible component of dimension $m$ and degree $d$ then* $\dim_k(M) \geq d.p^{r-m}$.

**Remarks 12.1.5** The conclusion no longer states that $\dim_k(M)$ is divisible by $p^{r-m}$, because of the possible existence of other irreducible components of $V_E^{\text{rk}}(M)$ of larger dimension than $m$.

The adjustment of the proof is to notice that $V'$ intersects the given irreducible component of $V_E^{\text{rk}}(M)$ in $d$ distinct lines through the origin, none of which is in any other irreducible component.

The corresponding theorem for exterior algebras is as follows.

**Theorem 12.1.6** *Let* $\Lambda = \Lambda(X_1, \ldots, X_r)$.

(i) *If* $M$ *is a* $\Lambda$*-module such that* $V_\Lambda^{\text{rk}}(M)$ *has dimension* $m$ *and degree* $d$*, then* $\dim_k M/2^{r-m}$ *is an integer at least as big as* $d$.

(ii) *If* $M$ *is a* $\Lambda$*-module such that* $V_\Lambda^{\text{rk}}(M)$ *has an irreducible component of dimension* $m$ *and degree* $d$ *then* $\dim_k(M) \geq d.2^{r-m}$.

Given this situation, it is natural to regard the rank variety of a $kE$-module or of a $\Lambda$-module as "interesting" if it is not a finite union of linear subspaces of $V_E^{\text{rk}}$, respectively $V_\Lambda^{\text{rk}}$.

One surprising aspect of the discussion is that it seems to be much easier to produce small modules for elementary abelian groups with interesting varieties in characteristic two than in odd characteristic. In a lot of ways, in characteristics other than two, exterior algebras are much closer in behaviour to elementary abelian 2-groups than elementary abelian $p$-groups are. For this reason, we include these in our discussions whenever possible.

## 12.2 Module Constructions

We begin by looking at Carlson's "favourite example" [73, pp. 418–9], for $E = (\mathbb{Z}/2)^3$ over a field $k$ of characteristic two. The representing matrices in this example are given by

$$g_1 \mapsto \begin{pmatrix} 1 & 0 & 0 & 0 \\ 0 & 1 & 0 & 0 \\ 1 & 0 & 1 & 0 \\ 0 & 1 & 0 & 1 \end{pmatrix} \quad g_2 \mapsto \begin{pmatrix} 1 & 0 & 0 & 0 \\ 0 & 1 & 0 & 0 \\ a & 0 & 1 & 0 \\ 0 & b & 0 & 1 \end{pmatrix} \quad g_3 \mapsto \begin{pmatrix} 1 & 0 & 0 & 0 \\ 0 & 1 & 0 & 0 \\ 0 & c & 1 & 0 \\ 1 & 0 & 0 & 1 \end{pmatrix}.$$

This module is an extension of $k \oplus k$ by $k \oplus k$, so it is determined by a $2 \times 2$ matrix of elements of $\operatorname{Ext}^1_{kE}(k, k)$. This matrix is

$$\begin{pmatrix} y_1 + ay_2 & y_3 \\ cy_3 & y_1 + by_2 \end{pmatrix}.$$

In this way he obtained a three-parameter family of modules of dimension four whose varieties are the quadrics

$$(Y_1 + aY_2)(Y_1 + bY_2) = cY_3^2. \tag{12.2}$$

To verify this, we note that

$$X_1 \mapsto \begin{pmatrix} 0&0&0&0\\ 0&0&0&0\\ 1&0&0&0\\ 0&1&0&0 \end{pmatrix} \quad X_2 \mapsto \begin{pmatrix} 0&0&0&0\\ 0&0&0&0\\ a&0&0&0\\ 0&b&0&0 \end{pmatrix} \quad X_3 \mapsto \begin{pmatrix} 0&0&0&0\\ 0&0&0&0\\ 0&c&0&0\\ 1&0&0&0 \end{pmatrix}$$

and so

$$Y_1X_1 + Y_2X_2 + Y_3X_3 \mapsto \begin{pmatrix} 0&0&0&0\\ 0&0&0&0\\ Y_1+aY_2 & Y_3 & 0&0\\ cY_3 & Y_1+bY_2 & 0&0 \end{pmatrix}.$$

The rank of this is less than half the dimension of the module precisely when the determinant of the bottom left $2 \times 2$ block vanishes.

We can modify Carlson's example by making the group a bit bigger, $E = (\mathbb{Z}/2)^4$, and using the matrix

$$\begin{pmatrix} y_1 & y_3 \\ y_2 & y_4 \end{pmatrix}$$

to get a module of dimension four whose variety is the ruled surface $Y_2Y_3 = Y_1Y_4$. More explicitly, the matrices for this module are given by

$$g_1 \mapsto \begin{pmatrix} 1&0&0&0\\ 0&1&0&0\\ 1&0&1&0\\ 0&0&0&1 \end{pmatrix} \quad g_2 \mapsto \begin{pmatrix} 1&0&0&0\\ 0&1&0&0\\ 0&0&1&0\\ 1&0&0&1 \end{pmatrix} \quad g_3 \mapsto \begin{pmatrix} 1&0&0&0\\ 0&1&0&0\\ 0&1&1&0\\ 0&0&0&1 \end{pmatrix}$$

$$g_4 \mapsto \begin{pmatrix} 1&0&0&0\\ 0&1&0&0\\ 0&0&1&0\\ 0&1&0&1 \end{pmatrix}.$$

The module $M$ defined by these matrices sits in a triangle in $\mathsf{stmod}(kE)$ of the form

$$\Omega(k) \oplus \Omega(k) \xrightarrow{\begin{pmatrix} y_1 & y_3 \\ y_2 & y_4 \end{pmatrix}} k \oplus k \to M \to k \oplus k.$$

Restricting to shifted subgroups of $kE$ of rank three gives the modules in Carlson's family.

This construction generalises in an obvious way to give a $2n$-dimensional module for an elementary abelian group of rank $n^2$ whose variety is given by the vanishing of the $n \times n$ generic determinant in $n^2$ variables. This variety has dimension $n^2 - 1 = r - 1$ and degree $n$, so this module hits the Bézout bound from Theorem 12.1.3.

## 12.3 Odd Primes Are More Difficult

The mysterious question at this point is this. If we try to make the same sort of example for $p$ odd, it is impossible to produce a module of dimension $2p$ over an elementary abelian group of order $p^3$ whose variety is given by equation (12.2). So the bound from Theorem 12.1.3 is not sharp in this case. More generally, we have the following theorem:

**Theorem 12.3.1** *If $p$ is an odd prime, $E \cong (\mathbb{Z}/p)^r$ and $M$ is a $kE$-module of dimension $2p$ then $V_E^{\mathrm{rk}}(M)$ is the whole of affine space $\mathbb{A}^r(k)$, a linear subspace of dimension $r-1$, or the union of two such subspaces intersecting in a linear subspace of dimension $r-2$.*

*Proof.* First we note that $V_E^{\mathrm{rk}}(M)$ has to have dimension $r-1$ or $r$ by Theorem 12.1.3, because $\dim_k(M) = 2p < p^2$.

Now suppose that $V_E^{\mathrm{rk}}(M)$ is not the whole of $\mathbb{A}^r(k)$. Then some linear combination of $X_1, \dots, X_r$ acts as the matrix

$$\begin{pmatrix} 0 & 0 & & & & 0 & 0 \\ 0 & 0 & & & & 0 & 0 \\ 1 & 0 & & & & & \\ 0 & 1 & & & & & \\ & & \ddots & & & & \\ 0 & 0 & & 1 & 0 & 0 & 0 \\ 0 & 0 & & 0 & 1 & 0 & 0 \end{pmatrix}.$$

The action of each $X_i$ commutes with this matrix, and is hence of the form

$$\begin{pmatrix} A_{1,i} & 0 & \cdots & 0 \\ A_{2,i} & A_{1,i} & & 0 \\ \vdots & & \ddots & \vdots \\ A_{p,i} & A_{p-1,i} & \cdots & A_{1,i} \end{pmatrix}$$

where each $A_{n,i}$ is a $2 \times 2$ matrix.

The matrices $A_{1,i}$ all commute and are nilpotent, and so after a change of basis within the $2 \times 2$ blocks if necessary, we have

$$A_{1,i} = \begin{pmatrix} 0 & 0 \\ \lambda_i & 0 \end{pmatrix}.$$

Now we have two cases. The first case is where some $\lambda_i$ is non-zero. Then the fact that $A_{1,i}A_{n,j} = A_{n,j}A_{1,i}$ forces all the $A_{n,j}$ to be lower triangular. So $M$ has a submodule $M_1$, spanned by the even-numbered basis vectors, of dimension $p$. Then $V_E^{\mathrm{rk}}(M_1)$ and $V_E^{\mathrm{rk}}(M/M_1)$ are both hyperplanes. So $V_E^{\mathrm{rk}}(M)$ is a subvariety of codimension one contained in the union of these two hyperplanes (which may be equal), and is therefore equal to this union.

The second case is where all the $\lambda_i$ are equal to zero. In this case we have $A_{2,i}A_{2,j} = A_{2,j}A_{2,i}$, and so after a change of basis within the $2 \times 2$ blocks if necessary, the matrices $A_{2,i}$ are simultaneously lower triangular:

$$A_{2,i} = \begin{pmatrix} \mu_i & 0 \\ \kappa_i & \nu_i \end{pmatrix}.$$

Then $V_E^{\mathrm{rk}}(M)$ is given by the vanishing of the $(2p-2) \times (2p-2)$ minor in the lower left of the matrix, and therefore has the form

$$(\mu_1 Y_1 + \cdots + \mu_r Y_r)^{p-1}(\nu_1 Y_1 + \cdots + \nu_p Y_p)^{p-1} = 0.$$

It is therefore either a single hyperplane or a union of two hyperplanes. □

## 12.4 Relative Cohomology

In this section, we present a variation on Carlson's $L_\zeta$ modules from Section 1.10, making use of relative cohomology.

Let $W$ be a linear subspace of $V_E^{\mathrm{rk}}$ of dimension $c$, and let $X_{\beta_1}, \dots, X_{\beta_c}$ be a basis for $W$. Then we have a shifted subgroup $E'$ generated by $g_{\beta_1} = 1 + X_{\beta_1}, \dots, g_{\beta_c} = 1 + X_{\beta_c}$, and

$$kE' = k[X_{\beta_1}, \dots, X_{\beta_c}]/(X_{\beta_1}^p, \dots, X_{\beta_c}^p)$$

is flatly embedded in $kE$. If $U$ is a linear subspace of dimension $r-c$ complementary to $W$, then choosing a basis $X_{\alpha_1}, \dots, X_{\alpha_{r-c}}$ for $U$, we have another shifted subgroup $E''$ generated by $g_{\alpha_1} = 1 + X_{\alpha_1}, \dots, g_{\alpha_{r-c}} = 1 + X_{\alpha_{r-c}}$ and a flat embedding

$$kE'' = k[X_{\alpha_1}, \dots, X_{\alpha_{r-c}}]/(X_{\alpha_1}^p, \dots, X_{\alpha_{r-c}}^p) \to kE.$$

Then

$$kE = k[X_{\beta_1}, \dots, X_{\beta_c}, X_{\alpha_1}, \dots, X_{\alpha_{r-c}})/(X_{\beta_1}^p, \dots, X_{\beta_c}^p, X_{\alpha_1}^p, \dots, X_{\alpha_{r-c}}^p).$$

Letting $E_0$ be the shifted subgroup of $kE$ generated by $E'$ and $E''$, we have

$$kE = kE_0 = kE' \otimes_k kE''.$$

The cohomology of $E$ relative to $E'$ is given by

$$H^*(kE, kE', k) \cong H^*(E_0/E', k) \cong H^*(E'', k).$$

The relative syzygies are the same as the syzygies for the quotient $\Omega^i_{E_0/E'}(k)$. There are natural maps $\Omega^i_E(k) \to \Omega^i_{E_0/E'}(k)$.

Now consider a matrix $(\zeta_{ij})$ of elements of $H^*(E, k)$ satisfying Condition 1.10.5, and the corresponding triangle

$$\bigoplus_i \Omega^{a_i+1}(k) \to L_{(f_{ij})} \to \bigoplus_j \Omega^{b_j}(k) \xrightarrow{(\hat{f}_{ij})} \bigoplus_i \Omega^{a_i}(k)$$

in $\mathsf{stmod}(kE)$. Since we want to replace syzygies by relative syzygies, we had better work in $\mathsf{mod}(kE)$ rather than $\mathsf{stmod}(kE)$. So consider the extension

$$0 \to \bigoplus_i \Omega^{a_i+1}(k) \to L_{(f_{ij})} \to \bigoplus_j \Omega^{b_j}(k) \to 0 \tag{12.3}$$

corresponding to

$$(\zeta_{ij}) \in \mathrm{Ext}^1_{kE}(\bigoplus_j \Omega^{b_j}(k), \bigoplus_i \Omega^{a_i+1}(k)) \cong \underline{\mathrm{Hom}}_{kE}\big(\bigoplus_j \Omega^{b_j}(k), \bigoplus_i \Omega^{a_i}(k)\big).$$

Here, the module $L_{(f_{ij})}$ is a particular representative of the object in $\mathsf{stmod}(kE)$ with the same name, and it may have projective summands at this stage. That may change when we replace syzygies by relative syzygies in the extension, as we are about to do.

First we discuss the case $p = 2$.

Let $i$ index a row of the matrix $(f_{ij})$. Let $W$ be a subspace of $V_E^{\mathrm{rk}}$ contained in the kernels of the functions $f_{ij}(X_1, \dots, X_r)$ for $1 \le j \le n$. Then the cohomology elements $f_{ij}(x_1, \dots, x_r)$ $(1 \le j \le n)$ are inflated from $E_0/E'$, and hence may be regarded as elements of relative cohomology $H^*(kE, kE', k)$. We may make use of this under the following condition:

**Condition 12.4.1** For each value of $j$, either $f_{ij} = 0$ or $b_j = 0$.

Under this condition, whenever the extension $f_{ij}$ is non-zero, we have $\Omega^{b_j}(k) = k$. So we can replace these extensions with the corresponding ones for $\Omega^{a_i+1}_{E_0/E'}(k)$ representing the elements of relative cohomology.

Similarly, let $j$ index a column of the matrix $(f_{ij})$ and let $W$ be a subspace of $V_E^{\mathrm{rk}}$ contained in the kernels of the functions $f_{ij}(X_1, \dots, X_r)$ for $1 \le i \le n$. Then the cohomology elements $f_{ij}(x_1, \dots, x_r)$ $(1 \le i \le n)$ are inflated from $E_0/E'$, and hence may be regarded as elements of relative cohomology $H^*(kE, kE', k)$. We may make use of this under the following condition:

**Condition 12.4.2** For each value of $i$, either $f_{ij} = 0$ or $a_i = -1$.

Under this condition, whenever the extension $f_{ij}$ is non-zero, we have $\Omega^{a_i+1}(k) = k$. So we can replace these extensions with the corresponding ones for $\Omega^{b_j}_{E_0/E'}(k)$ representing the elements of relative cohomology.

For $p$ odd, we have to take into account the Frobenius map, because we are using $f_{ij}(y_1, \dots, y_r)$ instead of $f_{ij}(x_1, \dots, x_r)$. So in this case we should assume that $W$ is a subspace of $V_E^{\mathrm{rk}}$ contained in the kernels of the functions $f_{ij}(X_1^p, \dots, X_r^p)$ for $1 \le j \le n$ or for $1 \le i \le n$ in the two cases above. Apart from this, everything else works the same way.

**Theorem 12.4.3** *Let $L'_{(\zeta_{ij})}$ be the module obtained by the replacements described above in the short exact sequence* (12.3). *Then $V^{\mathrm{rk}}_E(L'_{(\zeta_{ij})}) = V^{\mathrm{rk}}_E(L_{(\zeta_{ij})})$, namely the hypersurface defined by the equation* $\det(f_{ij}) = 0$.

*Proof.* Outside of the set $W$, the computation is exactly the same as in Theorem 1.10.6. Inside $W$, the matrix has a zero row or a zero column, so the determinant vanishes. The map in the triangle representing the matrix is not an isomorphism because the relative syzygy restricts to a direct sum of more than one copy of the trivial module, so $W \subseteq V^{\mathrm{rk}}_E(L'_{(\zeta_{ij})})$. □

Of course, the same argument applies if we replace more than one of the summands in this manner.

## 12.5 Small Modules for Quadrics, $p = 2$

In this section, following Reid's thesis [214], we examine small modules for an elementary abelian group of even rank $E = (\mathbb{Z}/2)^r$ ($r = 2m$) in characteristic two, with rank variety given by the irreducible quadric

$$Y_1Y_2 + Y_3Y_4 + \cdots + Y_{2m-1}Y_{2m}. \tag{12.4}$$

The same discussion applies to an exterior algebra $\Lambda$ on $r = 2m$ generators in any characteristic.

**Conjecture 12.5.1** If $M$ is a $\Lambda$-module with rank variety $V^{\mathrm{rk}}_\Lambda(M)$ given by the hypersurface determined by the equation (12.4) then $\dim_k(M) \geq 2^m$.

Related theorems and conjectures appear in the literature on matrix factorisations. See for example [67, 178].

There are a number of ways of producing modules of dimension exactly $2^m$ with rank variety given by the hypersurface determined by (12.4).

The first is to use the construction given in [178]. Namely, we write down a pair $A, B$ of $2^{m-1} \times 2^{m-1}$ matrices of linear combinations of the $Y_i$ whose product $AB$ is a power of (12.4) times the identity (hence so is $BA$). Then the determinants of $A$ and $B$ are also powers of (12.4). Denote by $A_i$ the specialisation of $A$ with $Y_i = 1$ and $Y_j = 0$ for $j \neq i$. Then the representation

$$g_i \mapsto \begin{pmatrix} I & 0 \\ A_i & I \end{pmatrix}$$

has Loewy length two and rank variety (12.4). To produce such matrices $A$ and $B$ we proceed by induction on $m$. For $m = 2$, the matrices

$$A = \begin{pmatrix} Y_1 & Y_3 \\ Y_2 & Y_4 \end{pmatrix}, \qquad B = \begin{pmatrix} Y_4 & -Y_3 \\ -Y_2 & Y_1 \end{pmatrix}$$

clearly work. For the inductive step, if $A'$ and $B'$ are $2^{m-2} \times 2^{m-2}$ matrices with this property for $Y_1Y_2 + \cdots + Y_{2m-3}Y_{2m-2}$ then we take

$$A = \begin{pmatrix} Y_{2m-1}I & A' \\ B' & -Y_{2m}I \end{pmatrix}, \qquad B = \begin{pmatrix} Y_{2m}I & A' \\ B' & -Y_{2m-1}I \end{pmatrix}.$$

For example, if $m = 3$ the matrices $A$ and $B$ are given by

$$A = \begin{pmatrix} Y_5 & 0 & Y_1 & Y_3 \\ 0 & Y_5 & Y_2 & Y_4 \\ Y_4 & -Y_3 & -Y_6 & 0 \\ -Y_2 & Y_1 & 0 & -Y_6 \end{pmatrix} \qquad B = \begin{pmatrix} Y_6 & 0 & Y_1 & Y_3 \\ 0 & Y_6 & Y_2 & Y_4 \\ Y_4 & -Y_3 & -Y_5 & 0 \\ -Y_2 & Y_1 & 0 & -Y_5 \end{pmatrix}.$$

The second method for producing a module of this dimension is using the technique of Section 12.4. Consider the subgroup $E' = \langle g_1, g_3, \ldots, g_{2m-1} \rangle$ generated by the odd indexed generators of $E$. The corresponding subspace $W$ of $V_E^{\mathrm{rk}}$ spanned by $X_1, X_3, \ldots, X_{2m-1}$ has the property that the function (12.4) vanishes on it. The element

$$y_1y_2 + \cdots + y_{2m-1}y_{2m} \in \operatorname{Ext}^1_{kE}(\Omega k, k) \cong H^2(E, k)$$

therefore descends to an element of $\operatorname{Ext}^1_{kE}(\Omega_{E/E'}(k), k)$. So we obtain a short exact sequence

$$0 \to k \to M \to \Omega_{E/E'}(k) \to 0.$$

It follows from Theorem 12.4.3 that $V_E^{\mathrm{rk}}(M)$ is the hypersurface defined by the element (12.4). The dimension of this module is one more than the dimension of $\Omega_{E/E'}(k)$, so we have $\dim_k(M) = 2^m$.

Here is a diagram for the case $m = 3$, with $\dim_k(M) = 8$.

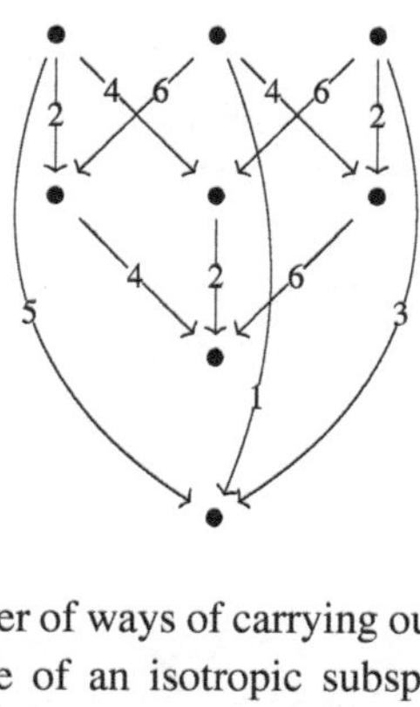

There are an infinite number of ways of carrying out this second recipe. Namely, there is one for each choice of an isotropic subspace of dimension $m$ for the quadratic form on $V_E^{\mathrm{rk}}$, and none of them are isomorphic. The duals of these modules also have the same property and the same dimension, and are not isomorphic to any of these.

## 12.6 Small Modules for Quadrics, $p$ Odd

It is known that given any closed homogeneous subvariety of affine $r$-space, there is a finitely generated $kE$-module with that as its rank variety. The general construction uses Carlson's $L_\zeta$ technology, as in Section 1.10. But the modules produced this way are big. For example, if we were to use this method for the quadric (12.2), we would need an element of cohomology $\zeta$ of degree four. The corresponding module $L_\zeta$ is the kernel of a representative cocycle $\hat{\zeta}\colon \Omega^4 k \to k$. This gives a module of dimension $6p^3$. We shall see that by judicious modification of this technique, we can reduce the figure to $2p^2$, but it seems hard to see how to get smaller than that.

For $p$ odd, we can't use $\mathrm{Ext}^1_{kE}(k,k)$ in this way for obvious reasons, so we have to use $\mathrm{Ext}^2_{kE}(k,k) \cong \mathrm{Ext}^1_{kE}(\Omega k, k)$. Using the corresponding matrix of elements

$$\begin{pmatrix} x_1 & x_3 \\ x_2 & x_4 \end{pmatrix},$$

we obtain an extension of $k\oplus k$ by $\Omega k\oplus\Omega k$ whose variety is given by the irreducible quadric $Y_1Y_4 + Y_2Y_3 = 0$.

But we can do better than this. If a row of the matrix does not use all of the variables, we can use a relative syzygy of $k$ instead of the absolute syzygy as in Section 12.4. The point is that the variety of the subgroup used for the relative syzygy should be contained in the variety defined by the polynomials appearing in the row or column.

To illustrate, here's a small module with interesting variety, for $p$ odd. We set $p = 3$ to give the example, but the construction works in general. Let $E = (\mathbb{Z}/3)^4 = \langle g_1, g_2, g_3, g_4\rangle$ and $k$ be a field of characteristic three. Represent the action of the elements $X_1 = g_1 - 1, \dots, X_4 = g_4 - 1$ by single edges, double edges, wavy edges and dotted edges respectively. Consider the following module.

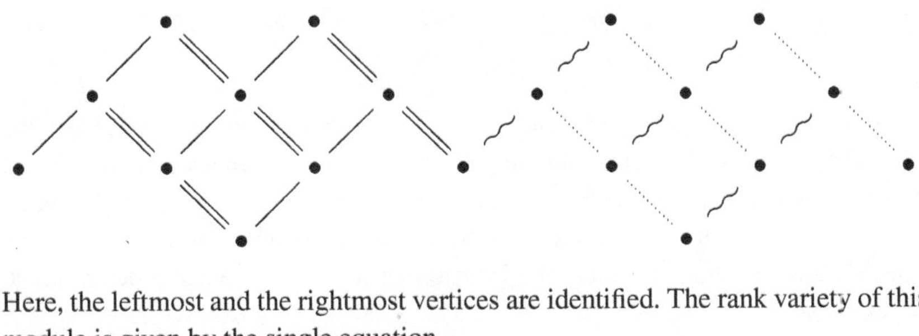

Here, the leftmost and the rightmost vertices are identified. The rank variety of this module is given by the single equation

$$Y_2Y_3 = Y_1Y_4.$$

Restricting to a shifted subgroup of $kE$ of rank three gives modules with any irreducible quadric as the variety.

Think of this module as an extension of two relative syzygies of the form $\Omega_{E/E'}k$ (with different subgroups $E'$) by two copies of $k$, corresponding to the matrix

$$\begin{pmatrix} x_1 & x_3 \\ x_2 & x_4 \end{pmatrix} \in \mathrm{Mat}_2(H^2(E,k)).$$

The crucial properties of this matrix are: each column only involves two of the variables so that it can be regarded as an element of cohomology relative to a subgroup $(\mathbb{Z}/p)^2$. Thus Condition 12.4.2 is satisfied, and so the computation of the rank variety follows from Theorem 12.4.3, giving the hypersurface determined by $Y_1^pY_4^p - Y_2^pY_3^p$. Since the coefficients are in $\mathbb{F}_p$, this determines the same variety as $Y_1Y_4 - Y_2Y_3$. The module $M$ above sits in a short exact sequence

$$0 \to k \oplus k \to M \to \Omega_{E/\langle g_3,g_4\rangle}(k) \oplus \Omega_{E/\langle g_1,g_2\rangle}(k) \to 0.$$

## 12.7 Trying to Understand the Specht Module $S^{(3^3)}$

In this section, we produce small modules with the same variety as the Specht module $S^{(3^3)}$ of Example 12.1.1, but whose dimension is, surprisingly, even smaller.

To do this, we apply the technique of Section 12.4 with the matrix

$$\begin{pmatrix} x_3 & x_1 & & \\ -x_1 & x_3 & x_3 & \\ & x_3 & x_3 & x_2 \\ & & -x_2 & x_3 \end{pmatrix}$$

to obtain a module $M$ of dimension 36 for $(\mathbb{Z}/3)^3$ whose variety is

$$Y_1^2Y_2^2 + Y_1^2Y_3^2 + Y_2^2Y_3^2 = 0$$

and which sits in a short exact sequence

$$0 \to k \oplus k \oplus k \oplus k \to M \to \Omega_{E/\langle g_2\rangle}(k) \oplus \Omega_{E/\langle g_2\rangle}(k) \oplus$$
$$\Omega_{E/\langle g_1\rangle}(k) \oplus \Omega_{E/\langle g_1\rangle}(k) \to 0.$$

Comparing with the Specht module $S^{(3^3)}$ for $\Sigma_9$ restricted to $(\mathbb{Z}/3)^3$, the mysterious things about this module are that it is 36 dimensional instead of 42 dimensional. This is, surprisingly, even smaller than the unexpectedly small Specht module. But this module also doesn't have enough symmetry to be the Specht module. There are more symmetric matrices with the same determinant, for example

$$\begin{pmatrix} x_3 & x_1 & x_2 & \\ & & x_3 & x_1 \\ x_1 & & & x_2 \\ & x_2 & & x_3 \end{pmatrix}$$

but this produces a module of dimension 54, which is too big.

This matrix looks a bit more symmetric:

$$\begin{pmatrix} x_1+x_2+x_3 & x_1-x_2 & x_2-x_3 & x_3-x_1 \\ x_1-x_2 & & x_2 & \\ x_2-x_3 & & & x_3 \\ x_3-x_1 & x_1 & & \end{pmatrix}$$

but it still produces a module of dimension 36 sitting in a short exact sequence

$$0 \to k \oplus k \oplus k \oplus k \to M \to \Omega_{E/\langle g_1g_2g_3\rangle}(k) \oplus \Omega_{E/\langle g_3\rangle}(k) \oplus \Omega_{E/\langle g_1\rangle}(k) \\ \oplus \Omega_{E/\langle g_2\rangle}(k) \to 0.$$

How to understand a module of dimension 42 with the correct variety is still a mystery.

## 12.8 Modules with Small Loewy Length

This section is based on joint work with my former Ph.D. student, Fergus Reid [54], which in turn grew out of his thesis [214].

Let $E \cong (\mathbb{Z}/p)^r$ be an elementary abelian $p$-group of rank $r$ and let $k$ be an algebraically closed field of characteristic $p$. If $M$ is a finite-dimensional $kE$-module then its rank variety $V_E^{\mathrm{rk}}(M)$ is a closed homogeneous subset of affine $r$-space $V_E^{\mathrm{rk}} = \mathbb{A}^r(k)$ of dimension equal to the complexity of $M$. If $M$ has complexity $c$ then its Loewy length is at least $(p-1)(r-c)+1$. It is the purpose of this section to investigate what happens when equality holds. The case of Loewy length $p$ and complexity $r-1$ was investigated in Reid's thesis, and this section generalises those results to arbitrary codimension. The main theorem of this section is as follows.

**Theorem 12.8.1** *Let $E \cong (\mathbb{Z}/p)^r$, $k$ be an algebraically closed field of characteristic $p$, and $M$ be a finite-dimensional $kE$-module of complexity $c$. Then $M$ has Loewy length at least $(p-1)(r-c)+1$. If $M$ has exactly this Loewy length then the following hold:*

(1) *If $p$ is odd then $V_E^{\mathrm{rk}}(M)$ is a union of a finite number of linear subspaces of $V_E^{\mathrm{rk}} = \mathbb{A}^r(k)$ of dimension $c$. Every such subvariety occurs for some $M$ of Loewy length $(p-1)(r-c)+1$.*
(2) *If $p=2$ then every irreducible component of $V_E^{\mathrm{rk}}(M)$ is equidimensional of dimension $c$. Given a closed homogeneous subvariety $V \subseteq \mathbb{A}^r(k)$ which is a finite union of set theoretic complete intersections of dimension $c$, there exists such a $kE$-module $M$ of Loewy length $r-c+1$ with $V_E^{\mathrm{rk}}(M) = V$.*

If $p=2$, the group algebra $kE$ is isomorphic to an exterior algebra on $r$ generators. The case $p=2$ of Theorem 12.8.1 generalises in an obvious way to modules in any characteristic for an exterior algebra as follows.

**Theorem 12.8.2** *Let $\Lambda = \Lambda(X_1, \dots, X_r)$ be an exterior algebra on variables $X_1, \dots, X_r$ over an algebraically closed field $k$, and $M$ be a finite-dimensional $\Lambda$-module of complexity $c$. Then $M$ has Loewy length at least $r - c + 1$. If $M$ has exactly this Loewy length then $V_\Lambda^{\mathrm{rk}}(M)$ is equidimensional of dimension $c$. Given a closed homogeneous subvariety $V \subseteq \mathbb{A}^r(k)$ which is a finite union of set theoretic complete intersections of dimension $c$, there exists such a $\Lambda$-module $M$ of Loewy length $r - c + 1$ with $V_\Lambda^{\mathrm{rk}}(M) = V$.*

**Remark 12.8.3** In the case $p$ odd, Theorem 12.8.1 describes exactly which subvarieties can occur for modules with complexity $c$ and Loewy length $(p-1)(r-c)+1$. For $p = 2$, and more generally for exterior algebras in any characteristic in Theorem 12.8.2 there is a gap, as described by the following series of observations.

The twisted cubic defined by the homogeneous radical ideal

$$(Y_1Y_3 - Y_2^2, Y_2Y_4 - Y_3^2, Y_1Y_4 - Y_2Y_3) \subseteq k[Y_1, \dots, Y_4]$$

is not a complete intersection in the sense that this ideal is not a complete intersection; but as a variety it is a *set theoretic* complete intersection, because it can be expressed as an intersection (with multiplicity two) of the hypersurfaces

$$Y_1Y_3 - Y_2^2 = 0 \quad \text{and} \quad Y_3(Y_2Y_4 - Y_3^2) - Y_4(Y_1Y_4 - Y_2Y_3) = 0.$$

But not every equidimensional closed homogeneous subvariety of $\mathbb{A}^r(k)$ is a set theoretic complete intersection. For example, the ideal $(Y_1, Y_2) \cap (Y_3, Y_4) \subseteq k[Y_1, \dots, Y_4]$ describes the union of two 2-dimensional linear subspaces of $\mathbb{A}^4$ intersecting in the origin. It is shown in Example 9.17 of [167] that this has codimension two but arithmetic rank three, and in particular it is not a set theoretic complete intersection. Nonetheless, it is obviously a union of two complete intersections, and hence the theorem guarantees that there is a module of Loewy length three for $(\mathbb{Z}/2)^4$ in characteristic two (or an exterior algebra in four variables in any characteristic) with this variety.

There are also irreducible examples. Example 3.4.2 in Hartshorne [145], after homogenisation, is the closed homogeneous subset of $\mathbb{A}^5(k)$ of dimension three defined parametrically by coordinates

$$(s^3, s^2t, stu, su(u-s), u^2(u-s)).$$

The paper describes this over the complex numbers, but the argument given there shows that over any algebraically closed field $k$, this is not a set theoretic complete intersection. The argument is that set theoretic complete intersections have the property that locally, the self-intersections have codimension one. In this example, the corresponding map $\mathbb{P}^2 \to \mathbb{P}^4$ is injective except for the points $(1:0:0)$ and $(1:0:1)$, which both go to $(1:0:0:0:0)$, so this point is a self-intersection of codimension two. We do not know whether there is a module of Loewy length three for $(\mathbb{Z}/2)^5$ (or for an exterior algebra on five generators in other characteristics) with this variety.

We thank Srikanth Iyengar and Anurag Singh for referring us to the examples in [167] and Burt Totaro for pointing us to Hartshorne's paper.

The rest of this section is devoted to the proof of Theorems 12.8.1 and 12.8.2. The following will be used in the proof of Theorem 12.8.20.

**Proposition 12.8.4** *Let $X_{\alpha_1}, \dots, X_{\alpha_c}$ be a basis for a subspace $U$ of $V_E^{\mathrm{rk}}$ of dimension $r-c$ of $V_E^{\mathrm{rk}}$ and let $kE'' \subseteq kE$ be as defined in* (12.1). *Let $M$ be a $kE$-module of dimension $p^{r-c}$ such that $M{\downarrow_{kE''}}$ is free of rank one. Then $V_E^{\mathrm{rk}}(M)$ is a linear subspace $W$ of $V_E^{\mathrm{rk}}$ of dimension $c$ complementary to $U$.*

*Proof.* Since $kE$ is commutative, its action on $M$ is as $kE''$-module homomorphisms. We have $\mathrm{End}_{kE''}(kE'') \cong kE''$, with the isomorphism being given by multiplication by elements of $kE''$. So there is a function $f\colon \{X_1, \dots, X_r\} \to kE''$ such that the action of $X_i$ on $M$ is the same as the action of $f(X_i)$. The elements $X_i - f(X_i)$ then act as zero on $M$. The images of these elements in $J(kE)/J^2(kE)$ span a subspace $W$ of dimension at least $c$ intersecting $U$ in the origin, and hence $W$ has dimension exactly $c$ and is a complement to $U$. It now follows from Lemma 6.4 of Carlson [74] that $V_E^{\mathrm{rk}}(M) = W$. □

**Lemma 12.8.5** *Suppose that $M$ is a module with $\mathsf{Rad}^m(M) = 0$. Then for $0 \le i \le m$ we have $\mathsf{Rad}^i(M) \subseteq \mathsf{Soc}^{m-i}(M)$.* □

**Lemma 12.8.6** *If $M_1$ and $M_2$ are $kE$-modules of Loewy length $\ell_1$ and $\ell_2$ then $M_1 \otimes_k M_2$ has Loewy length at most $\ell_1 + \ell_2 - 1$.*

*Proof.* Tensor products of semisimple $kE$-modules are semisimple. So defining

$$N_i = \sum_{i_1+i_2=i} J^{i_1}(kE)M_1 \otimes_k J^{i_2}(kE)M_2 \subseteq M_1 \otimes_k M_2,$$

it follows that $N_0 = M_1 \otimes_k M_2$, $N_i/N_{i+1}$ is semisimple, and we have $N_{\ell_1+\ell_2-1} = 0$ because if $i_1 + i_2 = \ell_1 + \ell_2 - 1$ then either $i_1 \ge \ell_1$ or $i_2 \ge \ell_2$. □

**Definition 12.8.7** We say that a closed homogenous subvariety $V$ of $V_E^{\mathrm{rk}}$ is *equidimensional* if all irreducible components of $V$ have the same dimension. A (homogeneous) hypersurface is equidimensional of codimension one. Equivalently, it is defined by a single homogeneous polynomial equation.

**Theorem 12.8.8** *Let $E \cong (\mathbb{Z}/p)^r$, $k$ an algebraically closed field of characteristic $p$ and $M$ a finite-dimensional $kE$-module of complexity $c$. Then the Loewy length of $M$ is at least $(p-1)(r-c)+1$. If equality occurs then $V_E^{\mathrm{rk}}(M)$ is equidimensional, and for $0 \le i \le (p-1)(r-c)+1$ we have*

$$\mathsf{Rad}^i(M) = J^i(kE')M,$$

*where* $kE'' = k[X_{\alpha_1}, \dots, X_{\alpha_{r-c}}]/(X_{\alpha_1}^p, \dots, X_{\alpha_{r-c}}^p)$ *is a flatly embedded subalgebra such that* $M{\downarrow_{kE''}}$ *is free.*

*Proof.* Since $V_E^{\mathrm{rk}}(M)$ has dimension $c$, there exists a linear subspace $U$ of $V_E^{\mathrm{rk}}$ of dimension $r-c$ whose intersection with $V_E^{\mathrm{rk}}(M)$ is exactly the origin. Let $\alpha_1, \dots, \alpha_{r-c}$ be a basis for $U$. Then

$$kE'' = k[X_{\alpha_1}, \dots, X_{\alpha_{r-c}}]/(X_{\alpha_1}^p, \dots, X_{\alpha_{r-c}}^p)$$

is a subalgebra flatly embedded in $kE$, and $M{\downarrow_{kE''}}$ is a free module. So the radical length of $M{\downarrow_{kE''}}$ is the radical length of $kE''$, which is equal to $(p-1)(r-c)+1$. It follows that the radical length of $M$ has to be at least this big.

Now suppose that equality occurs. Then letting $m = (p-1)(r-c)+1$, by Lemma 12.8.5 we have

$$J^i(kE'')M \subseteq \mathsf{Rad}^i(M) \subseteq \mathsf{Soc}^{m-i}(M) \subseteq J^i(kE'')M$$

and so $\mathsf{Rad}^i(M) = J^i(kE'')M$.

If $V_E^{\mathrm{rk}}(M)$ has an irreducible component $V$ of dimension less than $c$ then there exists a linear subspace $W$ of dimension $r-c+2$ such that $V \cap W$ has dimension one, and $W$ intersects other irreducible components of $V_E^{\mathrm{rk}}(M)$ in $\{0\}$. Let $M(W)$ be the module constructed in Example 12.1.2, with dimension $p^{c-2}$ and Loewy length $(p-1)(c-2)+1$. Then $V_E^{\mathrm{rk}}(W \otimes_k M)$ decomposes as a union of $V \cap W$ and another subvariety intersecting $V \cap W$ in $\{0\}$. So by a theorem of Carlson [75], $W \otimes_k M$ has a direct summand $M'$ with $V_E^{\mathrm{rk}}(M') = V \cap W$. The Loewy length of $M'$ is at most the Loewy length of $W \otimes_k M$, which by Lemma 12.8.6 is at most

$$((p-1)(c-2)+1) + ((p-1)(r-c)+1) - 1 = (p-1)(r-2)+1.$$

Since $M'$ has complexity one, it follows that it is a counterexample to the inequality in the first part of the theorem. This contradiction shows that $V_E^{\mathrm{rk}}(M)$ cannot have such an irreducible component. □

Next, we describe the corresponding theorem for exterior algebras. The analogue of Theorem 12.8.8 is the following, and the proof is the same. See for example Theorem 3.2 and Remark 3.4 (3) of [12].

**Theorem 12.8.9** *Let* $\Lambda = \Lambda(X_1, \dots, X_r)$ *be an exterior algebra on variables* $X_1, \dots, X_r$ *over an algebraically closed field* $k$, *and* $M$ *a finite-dimensional* $\Lambda$*-module of complexity* $c$. *Then the Loewy length of* $M$ *is at least* $r-c+1$. *If equality occurs then* $V_\Lambda^{\mathrm{rk}}(M)$ *is equidimensional, and for* $0 \le i \le r-c+1$ *we have* $\mathsf{Rad}^i(M) = J^i(\Lambda'')M$, *where* $\Lambda'' = \Lambda(X_{\alpha_1}, \dots, X_{\alpha_{r-c}})$ *is an exterior subalgebra of rank* $r-c$ *flatly embedded in* $\Lambda$ *in such a way that* $M{\downarrow_{\Lambda''}}$ *is free.* □

We are now ready to deal with the special case discussed in Reid's thesis [214], namely modules of complexity $c = r-1$. This depends on the following result from the theory of matrix factorisations:

**Theorem 12.8.10** *Let $f$ be a homogeneous polynomial of degree $d$ in $k[Y_1, \ldots, Y_r]$. Then there exist $m > 0$ and $m \times m$ matrices $A_1, \ldots, A_d \in \mathrm{Mat}_m(k[Y_1, \ldots, Y_r])$ such that each $A_i$ is a matrix whose entries are either zero or polynomials of degree one, and $A_1 \ldots A_d = f.I$, where $I$ is the $m \times m$ identity matrix.*

*Proof.* This follows from Theorem 1.2 of Herzog, Ulrich and Backelin [152]. □

**Remark 12.8.11** In the theorem, it follows that for $1 \le i \le d$ we also have

$$A_i A_{i+1} \ldots A_d A_1 \ldots A_{i-1} = f.I.$$

**Corollary 12.8.12** *Let $f$ be a homogeneous polynomial of degree $d$ in $k[Y_1, \ldots, Y_r]$. Then there exists $m > 0$ and a matrix $A \in \mathrm{Mat}_m(k[Y_1, \ldots, Y_r])$ whose entries are either zero or polynomials of degree one, such that the hypersurface determined by $\det(A)$ is the same as that determined by $f$.*

*Proof.* If $f$ is irreducible, this follows from Theorem 12.8.10 by taking $A = A_1$, since

$$\det(A_1) \ldots \det(A_d) = \det(f.I) = f^m$$

and so $\det(A)$ is a power of $f$. For general $f$ it follows by taking direct sums. □

**Remark 12.8.13** In the corollary, one may insist that $\det(A)$ is a power of $f$, but the matrices may be larger, and there is no advantage to doing this.

**Theorem 12.8.14** *Let $k$ be a field and $\Lambda = \Lambda(X_1, \ldots, X_r)$ be an exterior algebra over $k$. If $M$ is a $\Lambda$-module of Loewy length two then $V_\Lambda^{\mathrm{rk}}(M)$ is either a hypersurface or the whole of $V_\Lambda^{\mathrm{rk}}$. Conversely, given any homogeneous hypersurface $V$ of $V_\Lambda^{\mathrm{rk}}$, there exists a $\Lambda$-module $M$ of Loewy length two with $V_\Lambda^{\mathrm{rk}}(M) = V$.*

*Proof.* The first statement follows from Theorem 12.8.8. For the converse, we'll make use of Corollary 12.8.12. Interpret $k[Y_1, \ldots, Y_r]$ as the coordinate ring of $\mathbb{A}^r(k)$. Thus $Y_1, \ldots, Y_r$ are dual to $X_1, \ldots, X_r$. For notation, if $\alpha = (\alpha_1, \ldots, \alpha_r) \in \mathbb{A}^r(k)$ and $f \in k[Y_1, \ldots, Y_r]$, we write $f(\alpha)$ to mean $f(\alpha_1, \ldots, \alpha_r)$. We can write $V \subseteq V_\Lambda^{\mathrm{rk}}$ as $V_\Lambda^{\mathrm{rk}}(f)$, the zeros of a homogeneous polynomial $f(Y_1, \ldots, Y_r) \in k[Y_1, \ldots, Y_r]$. By Corollary 12.8.12, there exists a matrix $A \in \mathrm{Mat}_m(k[Y_1, \ldots, Y_r])$ whose entries are either zero or polynomials of degree one, such that the hypersurface determined by $\det(A)$ is equal to $V$. Let $M$ be the module given by

$$X_\alpha \mapsto \begin{pmatrix} 0 & 0 \\ A(\alpha) & 0 \end{pmatrix}.$$

Then $V_\Lambda^{\mathrm{rk}}(M)$ is the set of $\alpha$ such that this matrix has rank less than $\frac{1}{2}\dim_k(M)$. The only relevant minor is the bottom left square, and so the condition is that $\alpha$ is a zero of $\det(A)(\alpha) = f(\alpha)$. Thus $V_\Lambda^{\mathrm{rk}}(M) = V$. □

**Corollary 12.8.15** *Let $k$ be a field of characteristic two and let $E$ be an elementary abelian 2-group. Given any homogeneous hypersurfaces $V \subseteq V_E^{\mathrm{rk}}$, there exists a $kE$-module $M$ of Loewy length two with $V_E^{\mathrm{rk}}(M) = V$.* □

Next, we turn to odd primes. The following completes the proof of Theorem 12.8.1 in the case $c = r - 1$. A proof is given in Theorem 12.8.20 for $c$ arbitrary, and it does not use this special case, so logically it is not necessary to read this proof. On the other hand it is easier, and good preparation for reading the proof of Theorem 12.8.20.

**Theorem 12.8.16** *Let $k$ be an algebraically closed field of characteristic $p \geq 3$ and let $E$ be an elementary abelian $p$-group. Let $M$ be a $kE$-module of Loewy length $p$ with $V_E^{\mathrm{rk}}(M)$ a proper subvariety of $V_E^{\mathrm{rk}}$. Then $V_E^{\mathrm{rk}}(M)$ is a finite union of linear hypersurfaces in $V_E^{\mathrm{rk}}$.*

*Proof.* Choose $\beta \in V_E^{\mathrm{rk}} \setminus V_E^{\mathrm{rk}}(M)$. Then the action of $X_\beta$ on $M$ is projective, and setting $kE' = k[X_\beta]/(X_\beta^p)$, it follows from Theorem 12.8.8 that for $0 \leq i \leq p$ we have $\mathsf{Rad}^i(M) = X_\beta^i.M$. So $M$ has a basis with respect to which the matrix for $X_\beta$ is

$$\begin{pmatrix} 0 & 0 & \cdots & 0 & 0 \\ I & 0 & & 0 & 0 \\ 0 & I & & 0 & 0 \\ \vdots & & \ddots & & \vdots \\ 0 & 0 & \cdots & I & 0 \end{pmatrix}.$$

Here, there are $p$ blocks, and 0 and $I$ represent zero and identity square matrices of size $\dim_k(M)/p$. Since $\mathsf{Rad}^i(M) = X_\beta^i.M$, the generators $X_1, \ldots, X_r$ act as matrices of the form

$$\begin{pmatrix} 0 & 0 & \cdots & 0 & 0 \\ * & 0 & & 0 & 0 \\ * & * & & 0 & 0 \\ \vdots & & \ddots & & \vdots \\ * & * & \cdots & * & 0 \end{pmatrix}.$$

Furthermore, since the $X_j$ commute with $X_\beta$, this forces the matrices representing the $X_j$ to take the form

$$\begin{pmatrix} 0 & 0 & \cdots & & 0 & 0 \\ A_j & 0 & & & 0 & 0 \\ B_j & A_j & & & 0 & 0 \\ \vdots & \ddots & \ddots & & & \vdots \\ & & \ddots & A_j & 0 & 0 \\ * & & \cdots & B_j & A_j & 0 \end{pmatrix}.$$

Since $p \geq 3$, the fact that these matrices commute with each other now forces $A_1, \dots, A_r$ to commute with each other. Commuting matrices over an algebraically closed field may be simultaneously lower triangularised. So after a change of basis within the blocks above, we may assume that $A_1, \dots, A_r$ are lower triangular.

We are now in a position to compute $V_E^{\mathrm{rk}}(M)$. There is only one minor of size $\frac{p-1}{p}\dim_k(M)$ that can be non-zero, namely the bottom left part of the matrices above, obtained by removing one block from the top of each column and one block from the right of each row. If $\alpha = (\lambda_1, \dots, \lambda_r)$, then this minor for $X_\alpha$ has determinant

$$\det(\lambda_1 A_1 + \cdots + \lambda_r A_r)^{p-1}.$$

Since $A_1, \dots, A_r$ are lower triangular, this is a product of linear forms in $\lambda_1, \dots, \lambda_r$. It follows that $V_E^{\mathrm{rk}}(M)$ is the union of linear hypersurfaces determined by this product. □

Next, we handle the case of exterior algebras for arbitrary complexity, thereby completing the proof of Theorem 12.8.2 and of part (2) of Theorem 12.8.1.

**Theorem 12.8.17** *Let $\Lambda = \Lambda(X_1, \dots, X_r)$ be an exterior algebra over an algebraically closed field $k$. If $V \subseteq V_\Lambda^{\mathrm{rk}}$ is a finite union of set theoretic complete intersections of dimension $c$, then there exists a $\Lambda$-module $M$ of Loewy length $r - c + 1$ with $V_\Lambda^{\mathrm{rk}}(M) = V$.*

*Proof.* Since the Loewy length of $M_1 \oplus M_2$ is equal to the larger of the Loewy lengths of $M_1$ and $M_2$, using Proposition 1.9.11 (3), it suffices to prove the theorem in the case where $V$ is a complete intersection. Let $f_1, \dots, f_{r-c}$ be a regular sequence of homogeneous elements in $k[Y_1, \dots, Y_r]$ such that the variety of the ideal $(f_1, \dots, f_{r-c})$ is equal to $V$. For each $f_i$, Theorem 12.8.14 produces a module $M_i$ of Loewy length two with $V_\Lambda^{\mathrm{rk}}(M_i) = V(f_i)$, the hypersurface defined by $f_i$. Set $M = M_1 \otimes_k \cdots \otimes_k M_{r-c}$. Using Proposition 1.9.11 (4), we have

$$V_\Lambda^{\mathrm{rk}}(M) = V_\Lambda^{\mathrm{rk}}(M_1) \cap \cdots \cap V_\Lambda^{\mathrm{rk}}(M_{r-c}) = V(f_1) \cap \cdots \cap V(f_{r-c}) = V.$$

Furthermore, by Lemma 12.8.6 and induction on the codimension of $V$, the Loewy length of $M$ is equal to $r - c + 1$. □

**Remark 12.8.18** If $k$ does not have characteristic two, it is necessary to consider $\Lambda$ as a graded algebra with the $X_i$ in degree one, in order for it to be a Hopf algebra so that we can tensor modules together. Fortunately, the modules $M_i$ in the proof of the theorem may be regarded as graded $\Lambda$-modules.

**Corollary 12.8.19** *Let $k$ be an algebraically closed field of characteristic 2 and $E = (\mathbb{Z}/2)^r$. If $V \subseteq V_E^{\mathrm{rk}}$ is a finite union of set theoretic complete intersections of dimension $c$, then there exists a $kE$-module $M$ of Loewy length $r - c + 1$ with $V_E^{\mathrm{rk}}(M) = V$.* □

Finally, we complete the proof of Theorem 12.8.1 by proving that if $p$ is odd and $M$ has minimal Loewy length for its complexity, then its variety is a finite union of linear subspaces of the same dimension. The proof is a generalisation of the proof of Theorem 12.8.16, but the technical details are harder to follow.

**Theorem 12.8.20** *Let $k$ be an algebraically closed field of characteristic $p \geq 3$ and let $E = (\mathbb{Z}/p)^r$ be an elementary abelian $p$-group. Let $M$ be a $kE$-module with complexity $c$ and Loewy length $(p-1)(r-c)+1$. Then $V_E^{\mathrm{rk}}(M)$ is a finite union of linear subspaces of $V_E^{\mathrm{rk}}$ of dimension $c$. Every such subvariety occurs this way.*

*Proof.* The fact that every such variety is realised by a module with the right Loewy length is easy to see by looking at direct sums of modules of the form $M(W)$ described in Example 12.1.2. So we assume that $M$ is a module with complexity $c$ and Loewy length $(p-1)(r-c)+1$ and we examine its variety.

Let $kE = k[X_1, \dots, X_r]/(X_1^p, \dots, X_r^p)$. As in the proof of Theorem 12.8.8, choose a linear subspace $U$ of dimension $r-c$ with $U \cap V_E^{\mathrm{rk}}(M) = \{0\}$ and choose a basis $\alpha_1, \dots, \alpha_{r-c}$ for $U$. Set

$$kE'' = k[X_{\alpha_1}, \dots, X_{\alpha_{r-c}}]/(X_{\alpha_1}^p, \dots, X_{\alpha_{r-c}}^p).$$

Then $M{\downarrow_{kE''}}$ is a free module.

Set $M_0 = J(kE')M$, which by Theorem 12.8.8 is equal to $\mathsf{Rad}(M)$. For $1 \leq i \leq r-c$, set

$$M_i = X_{\alpha_1}M + \cdots + X_{\alpha_i}^2 M + \cdots + X_{\alpha_{r-c}}M,$$
$$M_i' = X_{\alpha_1}M + \cdots + X_{\alpha_i}^3 M + \cdots + X_{\alpha_{r-c}}M.$$

For $1 \leq i < j \leq r-c$, set

$$M_{ij} = X_{\alpha_1}M + \cdots + X_{\alpha_i}^2 M + \cdots + X_{\alpha_j}^2 M + \cdots + X_{\alpha_{r-c}}M$$

(just two terms with squares). Then $X_{\alpha_i}$ induces isomorphisms

$$M/M_0 \to M_0/M_i \to M_i/M_i'.$$

By Theorem 12.8.8, $X_1, \dots, X_r$ send $M$ into $M_0$ and $M_0$ into $M_i$, and hence also induce maps

$$M/M_0 \to M_0/M_i \to M_i/M_i'.$$

Since they commute with $X_{\alpha_i}$, if we identify these spaces using $X_{\alpha_i}$ then the two maps induced by each $X_\ell$ are the same. Since the $X_\ell$ commute with each other, we may choose a basis for $M/M_0$ with respect to which these maps are represented by lower triangular matrices, which we denote $A_{j,\ell}$. Furthermore, we have a commutative square of isomorphisms

$$\begin{array}{ccc} M/M_0 & \xrightarrow{X_{\alpha_{j'}}} & M_0/M_{j'} \\ {\scriptstyle X_{\alpha_j}}\downarrow & & \downarrow{\scriptstyle X_{\alpha_j}} \\ M_0/M_j & \xrightarrow[X_{\alpha_{j'}}]{} & (M_j + M_{j'})/M_{j,j'}. \end{array}$$

The vertical maps induced by an $X_\ell$ commute with $X_{\alpha_{j'}}$ so they are both the same, and the horizontal maps induced by another $X_{\ell'}$ commute with $X_{\alpha_j}$ so they are both the same. Therefore the matrices $A_{j,\ell}$ and $A_{j',\ell'}$ commute. It follows that we may choose a single basis for $M/M_0$ so that all the matrices $A_{j,\ell}$ ($1 \le j \le r-c$, $1 \le \ell \le r$) are lower triangular. Write $m_1, \dots, m_n$ for such a basis. So the image of $m_b$ under $A_{j,\ell}$ is contained in the linear span of $m_b, \dots, m_n$. Then a basis for $M$ is given by the elements

$$X_{\alpha_1}^{a_1} \dots X_{\alpha_{r-c}}^{a_{r-c}}.m_b \tag{12.5}$$

with $0 \le a_i \le p-1$ and $1 \le b \le n$.

Next, we observe that the actions of $X_1, \dots, X_r$ on $M$ are as $kE''$-module endomorphisms. We have $\mathrm{End}_{kE''}(M) \cong \mathrm{Mat}_n(kE'')$, and by Theorem 12.8.8, the actions of $X_1, \dots, X_r$ are by elements of $\mathrm{Mat}_n(J(kE''))$. Furthermore, if $x \in \mathrm{Mat}_n(J^2(kE''))$ then we may write $x = \sum_i y_i z_i$ with $y_i, z_i \in \mathrm{Mat}_n(J(kE''))$ and $y_i^p = 0$, $z_i^p = 0$ (to see this, it suffices to consider the case where $x$ has only one non-zero entry, and where that entry is a product of elements of $J(kE'')$). So by Proposition 2.2 of Friedlander and Pevtsova [124], if we adjust the action of $X_1, \dots, X_r$ by elements of $\mathrm{Mat}_n(J^2(kE''))$ we do not change $V_E^{\mathrm{rk}}(M)$. We may make such an adjustment so that the action of $X_1, \dots, X_r$ on $M$ is determined by the equations

$$X_\ell.m_b = \sum_j A_{j,\ell} X_{\alpha_j}.m_b.$$

These equations do determine the action on all basis elements (12.5) since $X_\ell$ commutes with $X_{\alpha_1}^{a_1} \dots X_{\alpha_{r-c}}^{a_{r-c}}$.

We make a filtration of $M$ by submodules

$$M = N_1 \supseteq \cdots \supseteq N_{n+1} = \{0\}$$

where $N_i$ is the span of the basis elements (12.5) of $M$ with $b \geq i$. The subquotient $N_i/N_{i+1}$ $(1 \leq i \leq n)$ has a basis consisting of the images of the basis elements (12.5) of $M$ with $b = i$. It is an example of a module satisfying the hypothesis of the theorem, of dimension $p^{r-c}$, and where the matrices $A_{j,\ell}$ described above have only one entry. Proposition 12.8.4 now applies, and we deduce that $V_E^{\mathrm{rk}}(N_i/N_{i+1})$ is a linear subspace $W_i \subseteq V_E^{\mathrm{rk}}$ of dimension $c$. It follows using Proposition 1.9.11 (5) that $V_E^{\mathrm{rk}}(M) \subseteq W_1 \cup \cdots \cup W_n$. By Theorem 12.8.8, $V_E^{\mathrm{rk}}(M)$ is equidimensional of dimension $c$, so it follows that $V_E^{\mathrm{rk}}(M)$ is a union of some subset of $\{W_1, \ldots, W_n\}$. □

## 12.9 Small Modules for $(\mathbb{Z}/p)^2$

Let $E = (\mathbb{Z}/p)^2 = \langle g, h\rangle$ be an elementary abelian $p$-group of rank two, with $p$ odd, and let $k$ be an algebraically closed field of characteristic $p$. Set $x = g - 1$ and $y = h - 1$ in $kE$, so that $kE = k[x, y]/(x^p, y^p)$.

If $M$ is a finite-dimensional indecomposable periodic $kE$-module then $\dim_k M$ is of the form $np$ for some $n > 0$. By Proposition 1.9.11 (6) (Carlson's connectedness theorem), the rank variety of $M$ consists of a single line through the origin. By changing the group basis for $kE$ if necessary, we can assume that this line is a coordinate axis. In other words, we can assume that the restriction of $M$ to $k[x]/(x^p)$ is not projective, and thus $x$ acts on $M$ with at least one Jordan block of length less than $p$. Our main theorem then states that the action of $x$ on $M$ also has some Jordan block of length $p$ then $\dim_k M > p^{3/2}$:

**Theorem 12.9.1** *Let $E = (\mathbb{Z}/p)^2$ and $k$ be a field of characteristic $p$. If there exists an indecomposable $kE$-module $M$ of dimension $np$ satisfying $0 < \dim_k x^{p-1}M < n$ and $\dim_k y^{p-1}M = n$ then $n > \sqrt{p}$.*

Of course, a module $M$ satisfying the condition of the theorem is necessarily periodic.

We prove in the next section that the bound is close to sharp, in the sense that for all large enough primes $p$ there is a module satisfying the conditions of the theorem and whose dimension satisfies

$$p^{\frac{3}{2}} < \dim_k(M) < p^{\frac{3}{2}} + \sqrt{2}\, p^{\frac{5}{4}}.$$

The rest of this section is devoted to the proof of Theorem 12.9.1. So we suppose that $M$ is an indecomposable $kE$-module of dimension $np$ such that $0 < \dim_k x^{p-1}M < n$ and $\dim_k y^{p-1}M = n$. We also assume that $n < \sqrt{p}$, and that $n$ is as small as possible subject to the existence of such an $M$. Our goal is to obtain a contradiction.

Notice that the condition $0 < \dim_k x^{p-1}M < n$ states that the action of $x$ on $M$ has a mixture of Jordan blocks of length $p$ and length less that $p$, while the condition $\dim_k y^{p-1}M = n$ states that all Jordan blocks for the action of $y$ on $M$ have length exactly $p$. Thus $y^{p-1}M = \mathsf{Ker}\,(y, M)$.

**Lemma 12.9.2** *If $N$ is a non-zero submodule of $M$ such that*

$$\dim_k N = p.\dim_k y^{p-1}N \quad \textit{and} \quad \dim_k x^{p-1}N > 0$$

*then $N = M$.*

*Proof.* If $N$ is a proper submodule of $M$ of dimension $mp$, there are two cases to consider. If $\dim_k x^{p-1}N < m$ then $N$ is a smaller counterexample to the theorem. So by minimality of $n$, this cannot occur. If, on the other hand, $\dim_k x^{p-1}N = m$ then the rank variety $V_E^r(N)$ does not contain the line corresponding to $x$, and so $V_E^r(N) \cap V_E^r(M) = \{0\}$. Thus $V_E^r(M/N) = V_E^r(M) \cup V_E^r(N)$, and Carlson's connectedness theorem [75] (or Theorem 8.2 of [74]) shows that $M/N$ is a direct sum $M_1 \oplus M_2$ with $V_E^r(M_1) = V_E^r(M)$ and $V_E^r(M_2) = V_E^r(N)$. Thus $\mathrm{Ext}^1_{kE}(M_1, N) = 0$, which implies that the inverse image of $M_1$ in $M$ decomposes as $M_1' \oplus N$ with $M_1' \cong M_1$. Hence $M$ decomposes as a direct sum of $M_1'$ and the inverse image of $M_2$ in $M$. Since $M$ was assumed to be indecomposable, this is a contradiction. □

Choose $v \in M$ such that $x^{p-1}v \neq 0$. Replacing $v$ by a suitable power of $y$ times $v$ if necessary we may assume also that $yx^{p-1}v = 0$. Define integers $t_0, \dots, t_{p-1}$ between 0 and $p-1$ so that $y^{p-1-t_i}x^iv \neq 0$ but $y^{p-t_i}x^iv = 0$. Multiplying this last equation by $x$ and using the equation $yx^{p-1}v = 0$, we see that

$$0 \le t_0 \le t_1 \le \cdots \le t_{p-1} = p-1.$$

Since $y^{p-1}M = \mathsf{Ker}\,(y, M)$, we may choose elements $w_1, \dots, w_p$ so that

$$y^{t_i}w_i = x^iv \qquad \text{and} \qquad xw_i = y^{t_{i+1}-t_i}w_{i+1} \quad (0 \le i \le p-2). \tag{12.6}$$

To choose such elements, start with $w_{p-1}$ and use downwards induction on $i$. We then have

$$x^jw_i = y^{t_{i+j}-t_i}w_{i+j}, \qquad (1 \le i+j \le p-1). \tag{12.7}$$

Let $N$ be the linear span in $M$ of the elements $y^iw_j$ for $0 \le i \le p-1$ and $0 \le j \le n$. Since $\dim_k M = np$, there is a non-trivial linear relation between these elements. Applying a power of $y$ to this relation if necessary, we see that there is a non-trivial linear relation between the elements $y^{p-1}w_j$ for $0 \le j \le n$. Thus for some $m \le n$, $y^{p-1}w_m$ is in the linear span of the $y_{p-1}w_j$ for $j < m$. Since multiplication by $y^{p-1}$ is an isomorphism from $N/yN$ to $y^{p-1}N$, it follows that $w_m$ is in the linear span of the remaining $y^iw_j$ with $0 \le i \le p-1$ and $0 \le j \le m$. Applying $y$ to this relation and using induction, we see that $w_m$ is in the linear span of the $y^i w_j$ with $0 \le i \le p-1$ and $0 \le j \le m-1$. It follows that this linear span is closed under the action of $x$.

Thus we have shown that $N$ is the linear span in $M$ of the elements $y^i w_j$ with $0 \le i \le p-1$ and $0 \le j \le m-1$, and $N$ is a $kE$-submodule of $M$ of dimension $mp$ with $m \le n$, and $\dim_k y^{p-1}N = m$. Applying Lemma 12.9.2 we see that $N = M$.

Under the action of $x$, the set of $y^i w_j$ with $0 < j \le m$ and $t_{j+1} > t_j + i$ together with the $y^i w_0$ with $0 \le i \le p-1$ forms a set of generators for $M$.

Next we examine the exact form of the relation expressing $w_n$ in terms of the $y^i w_j$ with $0 \le i \le p-1$ and $0 \le j \le n-1$. For this purpose we filter $M$ by *level*. We define the level of $y^i w_j$ to be $\ell(y^i w_j) = i + j - t_i$. The level of a linear combination of these basis elements is the smallest level of a basis element appearing with non-zero coefficient. Among the basis elements appearing with non-zero coefficient in $w_n$, look at the one with the smallest value of $j$, and define $a = n - j$ for this $j$.

**Lemma 12.9.3** *We have* $\ell(w_n) = n - t_n$.

*Proof.* The statement $\ell(w_n) = n - t_n$ is equivalent to the statement that $\ell(xw_{n-1}) = \ell(w_{n-1}) + 1$.

If $\ell(w_n) \le n - t_n - 1$ then choose $q$ so that $a(q-1) < p < aq$. Since $a \le n < \sqrt{p}$, we have $q > \sqrt{p}$ and so $a < q$. We also have $\ell(xw_{n-1}) \le \ell(w_{n-1})$ and so

$$\ell(x^{aq} w_{n-1}) \le aq - q < a(q-1) < p.$$

This implies that $x^{aq} w_{n-1} \ne 0$, which contradicts $x^p w_{n-1} = 0$.

On the other hand, if $\ell(w_n) \ge n - t_n + 1$, write $p = nq' + r$, with $0 < r < n$. Thus $nq' < p < n(q'+1)$. Since $n < \sqrt{p}$ we have $q' + 1 > \sqrt{p} > n$ and so $q' \ge n$. Then

$$\ell(x^{p-1}v) = \ell(x^{nq'+r-1}v) \ge (n+1)q' + r = p + q' \ge p + n,$$

which contradicts $x^{p-1}v \ne 0$. □

Next we do a dimension count. It is easy to check that the elements $x^k y^i w_j$ with $t_j > t_{j-1} + i$ and $0 \le k \le p-1$ together with the elements $x^k y^i w_0$ with $0 \le k < n - a$ and $0 \le i \le p-1$ span $M$. The number of such elements is at most $np$. It follows that $x^{p-1} y^i w_j \ne 0$ provided $t_j > t_{j-1} + i$ and $0 \le k \le p-1$.

It now follows that for $n - a \le j \le n-1$ we have $t_{j+1} = t_j + 1$, since multiples of $p$ generate mod $a$. Then examining $x^{p-1} w_0$ we see that $t_0 = 0$, $t_j \le j$ for $1 \le j \le n - a - 1$ and $t_j = j$ for $n - a \le j \le n - 1$.

Now dualise $M$ (using the Lie dual rather than the group dual) to get that all the $t_j$ are equal to $j$.

If $a < n$ then some linear combination of the $w_j$ is sent to zero by $x$, so generates a $kE$-submodule of dimension $p$ on which $y$ acts freely. The quotient is then a smaller counterexample to the theorem provided $a < n - 1$. If $a = n - 1$ then the module decomposes, again contradicting the hypotheses.

If, on the other hand, $a = n$ then $x$ acts freely on $M$, contradicting the hypotheses of the theorem.

## 12.10 The Bound is Close to Sharp

In this section, we show that the bound given in Theorem 12.9.1 is close to sharp. We begin with a number theoretic lemma.

**Lemma 12.10.1** *Given a prime $p \geq 5$, there exist positive integers $n$ with*

$$p^{\frac{1}{2}} < n < p^{\frac{1}{2}} + \sqrt{2}\, p^{\frac{1}{4}}$$

*and $q < n$ such that*

$$nq + q < p < nq + n.$$

*For $p$ large enough, $n$ and $q$ can also be made to satisfy*

$$nq + \frac{n}{2} < p < nq + q + \frac{n}{2}.$$

*Proof.* If $p = 5$ we can take $n = 3$ and $q = 1$. So assume that $p > 5$. Select integers $m$ so that

$$m^2 > p > (m-1)^2 \tag{12.8}$$

and $b$ so that

$$m^2 - (b-1)^2 > p \geq m^2 - b^2. \tag{12.9}$$

Thus $m^2 > m^2 - (b-1)^2 > (m-1)^2$, and so $1 < b < \sqrt{2m-1} + 1$. The second inequality in (12.9) is strict because $p$ is a prime bigger than 5, so that $b$ cannot equal $m - 1$. Set $n = m + b - 1$ and $q = m - b$. Thus $nq + q = m^2 - b^2 < p$ and $nq + n = m^2 - (b-1)^2 > p$. Finally, $n = m + b - 1 < m + \sqrt{2m} < p^{\frac{1}{2}} + \sqrt{2}\, p^{\frac{1}{4}}$.

The inequalities in the second statement follows from the fact that $b < \sqrt{2m-1} + 1$, so that provided $p$ is large, it follows that $m$ is large compared with $b$. Both $n$ and $q$ are therefore close to $m$, so $n/2 < q$ and $q + n/2 > n$. □

**Theorem 12.10.2** *For all large enough primes $p$, there exists a $k(\mathbb{Z}/p)^2$-module $M$ satisfying the conditions of Theorem 12.9.1, with*

$$p^{\frac{3}{2}} < \dim_k(M) < p^{\frac{3}{2}} + \sqrt{2} p^{\frac{5}{4}}.$$

*Proof.* Given a (large enough) prime $p$, and integers $n$ and $q$ satisfying the conditions of the lemma, we build a module $M$ of dimension $np$ as follows. Set $\alpha = n + nq - p$ and $\beta = p - nq - q$. The inequalities in the lemma show that we have $0 < \alpha < \frac{n}{2}$ and $0 < \beta < \frac{n}{2}$.

Take two generators $v$ and $w$, with relations:

$$y^\beta w = x^{n-\alpha} v \qquad x^\alpha w = y^{n+1-\beta} v.$$

Using these relations we have $x^n v = x^\alpha y^\beta w = y^\beta y^{n+1-\beta} v = y^{n+1} v$. So

$$\begin{aligned} x^{p-1} w &= x^{n-2\alpha-1}(x^n)^q x^\alpha w = x^{n-2\alpha-1}(x^n)^q y^{n+1-\beta} v \\ &= x^{n-2\alpha-1} y^{n-\beta+1}(y^{n+1})^q v = x^{n-2\alpha-1} y^{n-2\beta+1} y^p v = 0. \end{aligned}$$

We also have

$$x^{p-1} v = x^{n-\alpha-1}(x^n)^q v = x^{n-\alpha-1}(y^{n+1})^q v = x^{n-\alpha-1} y^{p-\beta} v \neq 0$$

while $x^p v = x^{n-\alpha} y^{p-\beta} v = y^{p-\beta} y^\beta v = 0$. So writing $w_i = x^i v$ for $0 \le i < n-\alpha$ and $w_i = x^{i-(n-\alpha)} w$ for $n-\alpha \le i < n$, the module $M$ has a basis consisting of the elements $y^j w_i$ with $0 \le i < n$ and $0 \le j < p$.

To see that $M$ is indecomposable, we argue as follows. Since $x^{p-1} v \neq 0$ and $x^{p-1} w = 0$, any endomorphism sends $w$ to a multiple of $w$ plus an element of $\mathrm{Rad}(M)$. So if $M$ decomposes then without loss of generality one of the summands is generated by $w$. But then the other summand is generated by $v$ plus a multiple of $w$. But $x^{n-\alpha} v$ is in the submodule generated by $w$, so the summands have non-zero intersection, which is a contradiction. □

# Appendix A

## Modules for $\mathbb{Z}/p$

Here are tables of the tensor products of modules for $\mathbb{Z}/p$ in characteristic $p$ given by the formulas in Section 2.3 for small values of $p$. In these tables we use bracket notation for Jordan types as explained in Section 4.4, so that for example $[3]^2[2]$ is the notation for $J_3 \oplus J_3 \oplus J_2$.

| $p=3$ | [1] | [2] | [3] |
|---|---|---|---|
| [1] | [1] | [2] | [3] |
| [2] | [2] | [3][1] | $[3]^2$ |
| [3] | [3] | $[3]^2$ | $[3]^3$ |

| $p=5$ | [1] | [2] | [3] | [4] | [5] |
|---|---|---|---|---|---|
| [1] | [1] | [2] | [3] | [4] | [5] |
| [2] | [2] | [3][1] | [4][2] | [5][3] | $[5]^2$ |
| [3] | [3] | [4][2] | [5][3][1] | $[5]^2[2]$ | $[5]^3$ |
| [4] | [4] | [5][3] | $[5]^2[2]$ | $[5]^3[1]$ | $[5]^4$ |
| [5] | [5] | $[5]^2$ | $[5]^3$ | $[5]^4$ | $[5]^5$ |

| $p=7$ | [1] | [2] | [3] | [4] | [5] | [6] | [7] |
|---|---|---|---|---|---|---|---|
| [1] | [1] | [2] | [3] | [4] | [5] | [6] | [7] |
| [2] | [2] | [3][1] | [4][2] | [5][3] | [6][4] | [7][5] | $[7]^2$ |
| [3] | [3] | [4][2] | [5][3][1] | [6][4][2] | [7][5][3] | $[7]^2[4]$ | $[7]^3$ |
| [4] | [4] | [5][3] | [6][4][2] | [7][5][3][1] | $[7]^2[4][2]$ | $[7]^3[3]$ | $[7]^4$ |
| [5] | [5] | [6][4] | [7][5][3] | $[7]^2[4][2]$ | $[7]^3[3][1]$ | $[7]^4[2]$ | $[7]^5$ |
| [6] | [6] | [7][5] | $[7]^2[4]$ | $[7]^3[3]$ | $[7]^4[2]$ | $[7]^5[1]$ | $[7]^6$ |
| [7] | [7] | $[7]^2$ | $[7]^3$ | $[7]^4$ | $[7]^5$ | $[7]^6$ | $[7]^7$ |

For $p = 11, 13$ we abbreviate the table by omitting the Jordan blocks of length $p$. The general pattern and its symmetries should be apparent from the formulas and these tables.

| p=11 | [1] | [2] | [3] | [4] | [5] | [6] | [7] | [8] | [9] | [10] |
|---|---|---|---|---|---|---|---|---|---|---|
| [1] | [1] | [2] | [3] | [4] | [5] | [6] | [7] | [8] | [9] | [10] |
| [2] | [2] | [3][1] | [4][2] | [5][3] | [6][4] | [7][5] | [8][6] | [9][7] | [10][8] | [9] |
| [3] | [3] | [4][2] | [5][3][1] | [6][4][2] | [7][5][3] | [8][6][4] | [9][7][5] | [10][8][6] | [9][7] | [8] |
| [4] | [4] | [5][3] | [6][4][2] | [7][5][3][1] | [8][6][4][2] | [9][7][5][3] | [10][8][6][4] | [9][7][5] | [8][6] | [7] |
| [5] | [5] | [6][4] | [7][5][3] | [8][6][4][2] | [9][7][5][3][1] | [10][8][6][4][2] | [9][7][5][3] | [8][6][4] | [7][5] | [6] |
| [6] | [6] | [7][5] | [8][6][4] | [9][7][5][3] | [10][8][6][4][2] | [9][7][5][3][1] | [8][6][4][2] | [7][5][3] | [6][4] | [5] |
| [7] | [7] | [8][6] | [9][7][5] | [10][8][6][4] | [9][7][5][3] | [8][6][4][2] | [7][5][3][1] | [6][4][2] | [5][3] | [4] |
| [8] | [8] | [9][7] | [10][8][6] | [9][7][5] | [8][6][4] | [7][5][3] | [6][4][2] | [5][3][1] | [4][2] | [3] |
| [9] | [9] | [10][8] | [9][7] | [8][6] | [7][5] | [6][4] | [5][3] | [4][2] | [3][1] | [2] |
| [10] | [10] | [9] | [8] | [7] | [6] | [5] | [4] | [3] | [2] | [1] |

| p=13 | [1] | [2] | [3] | [4] | [5] | [6] | [7] | [8] | [9] | [10] | [11] | [12] |
|---|---|---|---|---|---|---|---|---|---|---|---|---|
| [1] | [1] | [2] | [3] | [4] | [5] | [6] | [7] | [8] | [9] | [10] | [11] | [12] |
| [2] | [2] | [3][1] | [4][2] | [5][3] | [6][4] | [7][5] | [8][6] | [9][7] | [10][8] | [11][9] | [12][10] | [11] |
| [3] | [3] | [4][2] | [5][3][1] | [6][4][2] | [7][5][3] | [8][6][4] | [9][7][5] | [10][8][6] | [11][9][7] | [12][10][8] | [11][9] | [10] |
| [4] | [4] | [5][3] | [6][4][2] | [7][5][3][1] | [8][6][4][2] | [9][7][5][3] | [10][8][6][4] | [11][9][7][5] | [12][10][8][6] | [11][9][7] | [10][8] | [9] |
| [5] | [5] | [6][4] | [7][5][3] | [8][6][4][2] | [9][7][5][3][1] | [10][8][6][4][2] | [11][9][7][5][3] | [12][10][8][6][4] | [11][9][7][5] | [10][8][6] | [9][7] | [8] |
| [6] | [6] | [7][5] | [8][6][4] | [9][7][5][3] | [10][8][6][4][2] | [11][9][7][5][3][1] | [12][10][8][6][4][2] | [11][9][7][5][3] | [10][8][6][4] | [9][7][5] | [8][6] | [7] |
| [7] | [7] | [8][6] | [9][7][5] | [10][8][6][4] | [11][9][7][5][3] | [12][10][8][6][4][2] | [11][9][7][5][3][1] | [10][8][6][4][2] | [9][7][5][3] | [8][6][4] | [7][5] | [6] |
| [8] | [8] | [9][7] | [10][8][6] | [11][9][7][5] | [12][10][8][6][4] | [11][9][7][5][3] | [10][8][6][4][2] | [9][7][5][3][1] | [8][6][4][2] | [7][5][3] | [6][4] | [5] |
| [9] | [9] | [10][8] | [11][9][7] | [12][10][8][6] | [11][9][7][5] | [10][8][6][4] | [9][7][5][3] | [8][6][4][2] | [7][5][3][1] | [6][4][2] | [5][3] | [4] |
| [10] | [10] | [11][9] | [12][10][8] | [11][9][7] | [10][8][6] | [9][7][5] | [8][6][4] | [7][5][3] | [6][4][2] | [5][3][1] | [4][2] | [3] |
| [11] | [11] | [12][10] | [11][9] | [10][8] | [9][7] | [8][6] | [7][5] | [6][4] | [5][3] | [4][2] | [3][1] | [2] |
| [12] | [12] | [11] | [10] | [9] | [8] | [7] | [6] | [5] | [4] | [3] | [2] | [1] |

Here are some lists of exterior powers $\Lambda^d$ and symmetric powers $S^d$ of modules for $\mathbb{Z}/p$ in characteristic $p$ for $d < p$. These were calculated using the method described in Section 2.8.

$$S^2[1] \cong \Lambda^2[2] \cong [1]$$

$$S^2[2] \cong \Lambda^2[3] \cong [3] \qquad p \geq 3$$

$$S^2[3] \cong \begin{cases} [3]^2 & p = 3 \\ \Lambda^2[4] \cong [5][1] & p \geq 5 \end{cases}$$

$$S^2[4] \cong \Lambda^2[5] \cong \begin{cases} [5]^2 & p = 5 \\ [7][3] & p \geq 7 \end{cases}$$

$$S^2[5] \cong \begin{cases} [5]^3 & p = 5 \\ \Lambda^2[6] \cong \begin{cases} [7]^2[1] & p = 7 \\ [9][5][1] & p \geq 11 \end{cases} \end{cases}$$

$$S^2[6] \cong \Lambda^2[7] \cong \begin{cases} [7]^3 & p = 7 \\ [11][7][3] & p \geq 11 \end{cases}$$

$$S^2[7] \cong \begin{cases} [7]^4 & p = 7 \\ \Lambda^2[8] \cong \begin{cases} [11]^2[5][1] & p = 11 \\ [13][9][5][1] & p \geq 13 \end{cases} \end{cases}$$

$$S^2[n] \cong \Lambda^2[n+1] \cong \begin{cases} [p]^{n-(p-1)/2}[2p-2n-3][2p-2n-7][2p-2n-11]\cdots & p \leq 2n-3 \\ [2n-1][2n-5][2n-9][2n-13]\cdots & p \geq 2n-1 \end{cases}$$

$$S^3[1] \cong \Lambda^3[3] \cong [1] \qquad p \geq 5$$

$$S^3[2] \cong \Lambda^3[4] \cong [4] \qquad p \geq 5$$

$$S^3[3] \cong \Lambda^3[5] \cong \Lambda^2[5] \qquad p \geq 5$$

$$S^3[4] \cong \begin{cases} [5]^4 & p = 5 \\ \Lambda^3[6] \cong \begin{cases} [7]^2[6] & p = 7 \\ [10][6][4] & p \geq 11 \end{cases} \end{cases}$$

$$S^3[5] \cong \begin{cases} [5]^7 & p = 5 \\ \Lambda^3[7] \cong \begin{cases} [7]^5 & p = 7 \\ [11]^2[7][5][1] & p = 11 \\ [13][9][7][5][1] & p \geq 13 \end{cases} \end{cases}$$

$$S^3[6] \cong \begin{cases} [7]^8 & p = 7 \\ \Lambda^3[8] \cong \begin{cases} [11]^4[8][4] & p = 11 \\ [13]^2[12][8][6][4] & p = 13 \\ [16][12][10][8][6][4] & p \geq 17 \end{cases} \end{cases}$$

$$S^3[7] \cong \begin{cases} [7]^{12} & p = 7 \\ \Lambda^3[9] \cong \begin{cases} [11]^7[7] & p = 11 \\ [13]^5[9][7][3] & p = 13 \\ [17]^2[13][11][9][7]^2[3] & p = 17 \\ [19][15][13][11][9][7]^2[3] & p \geq 19 \end{cases} \end{cases}$$

$$S^4[1] \cong \Lambda^4[4] \cong [1] \qquad p \geq 5$$

$$S^4[2] \cong \Lambda^4[5] \cong [5] \qquad p \geq 5$$

$$S^4[3] \cong S^2[5] \qquad p \geq 5$$

$$\Lambda^4[6] \cong S^2[5] \qquad p \geq 7$$

$$S^4[4] \cong S^3[5] \qquad p \geq 5$$

$$\Lambda^4[7] \cong S^3[5] \qquad p \geq 7$$

$$S^4[5] \cong \begin{cases} [5]^{14} & p = 5 \\ [7]^{10} & p = 7 \\ \Lambda^4[8] \cong \begin{cases} [11]^5[9][5][1] & p = 11 \\ [13]^3[11][9][5]^2[1] & p = 13 \\ [17][13][11][9]^2[5]^2[1] & p \geq 17 \end{cases} \end{cases}$$

$$S^4[6] \cong \begin{cases} [7]^{18} & p = 7 \\ \Lambda^4[9] \cong \begin{cases} [11]^{11}[5] & p = 11 \\ [13]^8[9][7][5][1] & p = 13 \\ [17]^3[15][13][11][9]^2[7][5]^2[1] & p = 17 \\ [19]^2[15][13]^2[11][9]^2[7][5]^2[1] & p = 19 \\ [21][17][15][13]^2[11][9]^2[7][5]^2[1] & p \geq 23 \end{cases} \end{cases}$$

$$S^5[1] \cong \Lambda^5[5] \cong [1] \qquad p \geq 7$$

$$S^5[2] \cong \Lambda^5[6] \cong [6] \qquad p \geq 7$$

$$S^5[3] \cong \Lambda^5[7] \cong S^2[6] \qquad p \geq 7$$

$$S^5[4] \cong S^3[6] \qquad p \geq 7$$

$$\Lambda^5[8] \cong S^3[6] \qquad p \geq 11$$

$$S^5[5] \cong S^4[6] \qquad p \geq 7$$

$$\Lambda^5[9] \cong S^4[6] \qquad p \geq 11$$

$$S^5[6] \cong \begin{cases} [7]^{36} & p = 7 \\ \Lambda^5[10] \cong \begin{cases} [11]^{22}[10] & p = 11 \\ [13]^{18}[10][6][2] & p = 13 \\ [17]^{10}[14][12][10]^3[8][6]^2[4][2] & p = 17 \\ [19]^6[18][16][14]^2[12][10]^3[8]^2[6]^2[4][2] & p = 19 \\ [23]^2[22][18]^2[16]^2[14]^2[12]^2[10]^3[8]^2[6]^2[4][2] & p = 23 \\ [26][22][20][18]^2[16]^2[14]^2[12]^2[10]^3[8]^2[6]^2[4][2] & p \geq 29 \end{cases} \end{cases}$$

Next we list the effects of Schur functors $S^\lambda$ on modules for $\mathbb{Z}/p$ in characteristic $p$, where $\lambda$ is a partition of $d < p$. These were calculated using the method described in Section 2.9.

$$S^{[2][1]}[2] \cong [2] \qquad p \geq 5$$

$$S^{[2][1]}[3] \cong [5][3] \qquad p \geq 5$$

$$S^{[2][1]}[4] \cong \begin{cases} [5]^4 & p = 5 \\ [7]^2[4][2] & p = 7 \\ [8][6][4][2] & p \geq 11 \end{cases}$$

$$S^{[2][1]}[5] \cong \begin{cases} [5]^8 & p=5 \\ [7]^5[5] & p=7 \\ [11][9][7][5]^2[3] & p \geq 11 \end{cases}$$

$$S^{[2][1]}[6] \cong \begin{cases} [7]^{10} & p=7 \\ [11]^4[8][6]^2[4][2] & p=11 \\ [13]^2[10][8]^2[6]^2[4][2] & p=13 \\ [14][12][10][8]^2[6]^2[4][2] & p \geq 17 \end{cases}$$

$$S^{[3][1]}[2] \cong S^{[2][1]^2}[3] \cong [3] \qquad p \geq 5$$

$$S^{[3][1]}[3] \cong S^{[2][1]^2}[4] \cong \begin{cases} [5]^3 & p=5 \\ [7][5][3] & p \geq 7 \end{cases}$$

$$S^{[3][1]}[4] \cong S^{[2][1]^2}[5] \cong \begin{cases} [5]^9 & p=5 \\ [7]^6[3] & p=7 \\ [11][9][7]^2[5][3]^2 & p \geq 11 \end{cases}$$

$$S^{[3][1]}[5] \cong \begin{cases} [5]^{21} & p=5 \\ S^{[2][1]^2}[6] \cong \begin{cases} [7]^{15} & p=7 \\ [11]^6[9][7]^2[5]^2[3]^2 & p=11 \\ [13]^3[11][9]^2[7]^3[5]^2[3]^2 & p=13 \\ [15][13][11]^2[9]^2[7]^3[5]^2[3]^2 & p \geq 17 \end{cases} \end{cases}$$

$$S^{[2]^2}[2] \cong [1] \qquad p \geq 5$$

$$S^{[2]^2}[3] \cong [5][1] \qquad p \geq 5$$

$$S^{[2]^2}[4] \cong \begin{cases} [5]^4 & p=5 \\ [7]^2[5][1] & p=7 \\ [9][5]^2[1] & p \geq 11 \end{cases}$$

$$S^{[2]^2}[5] \cong \begin{cases} [5]^{10} & p=5 \\ [7]^7[1] & p=7 \\ [11]^2[9][7][5]^2[1]^2 & p=11 \\ [13][9]^2[7][5]^2[1]^2 & p \geq 13 \end{cases}$$

$$S^{[2]^2}[6] \cong \begin{cases} [7]^{15} & p=7 \\ [11]^7[9][7][5]^2[1]^2 & p=11 \\ [13]^4[11][9]^2[7][5]^3[1]^2 & p=13 \\ [17][13]^2[11][9]^3[7][5]^3[1]^2 & p \geq 17 \end{cases}$$

$$S^{[4][1]}[2] \cong S^{[2][1]^3}[4] \cong [4] \qquad p \geq 7$$

$$S^{[4][1]}[3] \cong S^{[2][1]^3}[5] \cong \begin{cases} [9][7][5][3] & p \geq 11 \\ [7]^3[3] & p = 7 \end{cases}$$

$$S^{[4][1]}[4] \cong S^{[2][1]^3}[6] \cong \begin{cases} [7]^{12} & p = 7 \\ [11]^4[10][8][6]^2[4]^2[2] & p = 11 \\ [13]^2[10]^2[8]^2[6]^2[4]^2[2] & p = 13 \\ [14][12][10]^2[8]^2[6]^2[4]^2[2] & p \geq 17 \end{cases}$$

$$S^{[4][1]}[5] \cong S^{[2][1]^3}[7] \cong \begin{cases} [7]^{32} & p = 7 \\ [11]^{17}[9][7]^2[5]^2[3][1] & p = 11 \\ [13]^{11}[11][9]^3[7]^3[5]^3[3]^2[1] & p = 13 \\ [17]^3[15][13]^3[11]^3[9]^4[7]^4[5]^3[3]^2[1] & p = 17 \\ [19][17][15]^2[13]^3[11]^3[9]^4[7]^4[5]^3[3]^2[1] & p \geq 19 \end{cases}$$

$$S^{[3][2]}[2] \cong [2] \qquad p \geq 7$$

$$S^{[3][2]}[3] \cong [7][5][3] \qquad p \geq 7$$

$$S^{[3][2]}[4] \cong \begin{cases} [7]^8[4] & p = 7 \\ [11]^2[8]^2[6]^2[4]^2[2] & p = 11 \\ [12][10][8]^2[6]^2[4]^2[2] & p \geq 13 \end{cases}$$

$$S^{[3][2]}[5] \cong \begin{cases} [7]^{25} & p = 7 \\ [11]^{11}[9]^2[7]^2[5]^3[3]^2[1] & p = 11 \\ [13]^6[11]^2[9]^3[7]^3[5]^4[3]^2[1] & p = 13 \\ [17][15][13]^2[11]^3[9]^4[7]^3[5]^4[3]^2[1] & p \geq 17 \end{cases}$$

$$S^{[3][1]^2}[3] \cong [5][1] \qquad p \geq 7$$

$$S^{[3][1]^2}[4] \cong \begin{cases} [7]^4[6][2] & p = 7 \\ [10][8][6]^2[4][2] & p \geq 11 \end{cases}$$

$$S^{[2]^2[1]}[3] \cong [3] \qquad p \geq 7$$

$$S^{[2]^2[1]}[4] \cong \begin{cases} [7]^2[4][2] & p = 7 \\ [8][6][4][2] & p \geq 11 \end{cases}$$

$$S^{[2]^2[1]}[5] \cong \begin{cases} [7]^{10}[5] & p = 7 \\ [11]^3[9][7]^2[5]^3[3][1] & p = 11 \\ [13][11][9]^2[7]^2[5]^3[3][1] & p \geq 13 \end{cases}$$

# Appendix B

## Problems

This appendix consists of a selection of open problems connected with the material of this volume. Throughout this chapter, let $p$ be a prime number, let $E = \langle g_1, \ldots, g_r \rangle \cong (\mathbb{Z}/p)^r$ be an elementary abelian $p$-group of rank $r \geq 2$, and let $k$ be an algebraically closed field of characteristic $p$, and let $M$ be a finitely generated $kE$-module. Write $X_i$ for the element $g_i - 1 \in kE$. We begin with a couple of questions of David Eisenbud that are peripherally connected with the main subject of this text.

**Problem 1** Let $M$ be a module of complexity $r$. Then

$$\sum_{n=0}^{\infty} \dim_k \operatorname{Ext}^n_{kE}(M,k)t^n = f(t)/(1-t^2)^r$$

with $f(t)$ a polynomial with integer coefficients and $f(1) > 0$. Eisenbud asks whether it is always true that $f(1) \geq 2^r$.

**Problem 2** Greenlees [139] has constructed a local cohomology spectral sequence

$$H^{**}_{\mathfrak{m}} \widehat{\operatorname{Ext}}^*_{kE}(M,k) \Rightarrow 0.$$

Eisenbud asks whether it can ever happen that the local cohomology of Tate cohomology is non-zero in degree one.

**Problem 3** Given the prime $p$ and the rank $r \geq 2$, what are the possible Jordan types of a $kE$-module of constant Jordan type? A slightly less ambitious but nonetheless extremely difficult problem is to find the possible stable Jordan types.

A great deal of partial information on this problem is given in Chapters 5 and 10. In particular, the simplest stable Jordan type for which we do not know the answer is type [4] [1] with $p = 11$, $r = 2$, see Section 5.14.

Subproblems of Problem 3 include the following.

**Problem 4** Find a proof or a counterexample to Rickard's Conjecture 5.13.1. This states that if a $kE$-module of constant Jordan type has no Jordan blocks of length $j$ then the total number of Jordan blocks of length larger than $j$ is divisible by $p$.

Investigate the stronger form of Rickard's conjecture, stating under the same hypotheses that the number of blocks of length larger than $j$ is divisible by $p^{r-1}$.

The cases $j = 1$ and $j = p - 1$ of Rickard's conjecture are proved in Section 5.13, but the remaining cases are open. The analogue of Rickard's conjecture in characteristic zero is the following, which asks whether Conjecture 5.14.7 is true.

**Problem 5** Let $k$ be an algebraically closed field of characteristic zero, and let $A$, $B$ be commuting $d \times d$ matrices over $k$ with $d \geq 1$, such that for all $\lambda, \mu \in k$, not both zero, $\lambda A + \mu B$ has the same Jordan canonical form (in particular this implies that $A$ and $B$ are nilpotent). Is it true that if there is a Jordan block of some given size then there are also Jordan blocks of all smaller sizes?

In particular, this asks whether there have to be Jordan blocks of size 1. This special case is proved in Theorem 5.14.8.

**Problem 6** Find a proof or a counterexample to Suslin's Conjecture 5.15.1. This states that if a $kE$-module of constant Jordan type has a Jordan block of length $j$ with $2 \leq j \leq p - 1$ then there is either a Jordan block of length $j + 1$ or one of length $j - 1$.

Both Rickard's conjecture and Suslin's conjecture fail for nilvarieties instead of $kE$-modules, so information from the theory of vector bundles and Chern numbers alone cannot hope to prove these conjectures.

**Problem 7** Find a proof or a counterexample to Conjecture 5.15.2 of Carlson, Friedlander and Pevtsova. This states that if $p \geq 5$ and a $kE$-module has stable constant Jordan type $[2]\,[1]^j$ then $j \geq r - 1$.

Corollary 10.4.2 provides partial information for this problem using Chern numbers.

**Problem 8** Let $p \geq 5$ and $E = (\mathbb{Z}/p)^2$. Do there exist $kE$-modules of stable constant Jordan type $[2]\,[1]$ that are not of the form $\Omega^{2n}(kE/J^2(kE))$ or $\Omega^{2n}(\mathsf{Soc}^2(kE))$? More generally, if $p \geq 5$ and $E = (\mathbb{Z}/p)^r$, do there exist $kE$-modules of stable constant Jordan type $[2]\,[1]^{r-1}$ that are not of the form $\Omega^{2n}(kE/J^2(kE))$ or $\Omega^{2n}(\mathsf{Soc}^2(kE))$?

**Problem 9** Suppose that $M$ has constant Jordan type, and let $\phi\colon kE \to kE$ be an automorphism sending each $X_i$ to $X_i$ plus an element of $J^2(kE)$, and let $\phi^*(M)$

be the module obtained by letting $kE$ act on $M$ through $\phi$. Is it true that there is an isomorphism of vector bundles $\mathcal{F}_j(\phi^*(M)) \cong \mathcal{F}_j(M)$ for each $j$ with $1 \le j \le p$?

**Problem 10** Let $M$ be a $kE$-module, and if necessary assume that $M$ has constant Jordan type. Find module theoretic interpretations for expressions such as $H^1(\mathbb{P}^{r-1}, \mathsf{Ker}\,(\theta_M))$, $H^0(\mathbb{P}^{r-1}, \widetilde{M}/\mathsf{Ker}\,(\theta_M))$, $H^i(\mathbb{P}^{r-1}, \mathcal{F}_j(M)(\ell))$, etc.

Note that $H^0(\mathbb{P}^{r-1}, \mathsf{Ker}\,(\theta_M)) = \mathsf{Soc}(M)$.

**Problem 11** Let $p$ be an odd prime. Given a vector bundle $\mathcal{F}$ on $\mathbb{P}^{r-1}$ does there exist a $kE$-module $M$ of constant Jordan type and an integer $i$ with $1 \le i \le p$ such that $\mathcal{F}_i(M) \cong \mathcal{F}$?

For partial information, see the Realisation Theorem 8.1.1.

**Problem 12** Let $p$ be an odd prime, let $r \ge 3$, and let $M$ be a $kE$-module of constant Jordan type. If $\mathcal{F}_i(M) \cong \mathcal{O}(n)$ for some $1 \le i \le p-1$ and some $n \in \mathbb{Z}$, what are the restrictions on the possible values of $i$ and $n$ modulo $p$?

**Problem 13** Let $p = 2$ and $r$ be odd. For what values of $n \ge 0$ do there exist $kE$-modules of constant Jordan type $[2]^n[1]^{r-2}$?

Partial information is given in Lemma 10.5.3. This question is closely related to the question of what Chern numbers can occur for rank $r-2$ vector bundles on the even-dimensional projective space $\mathbb{P}^{r-1}$.

**Problem 14** Let $M$ be a $kE$-module such that for all shifted subgroups $E' \subseteq kE$ of rank two with $X_1 \in kE'$, $M{\downarrow_{E'}}$ is stably isomorphic to a direct sum of copies of $k$. Does it follow that $M$ is stably isomorphic to a direct sum of copies of $k$?

Compare with Theorem 3.2.1 in Okonek, Schneider and Spindler [202].

**Problem 15** Find a proof or a counterexample to the Babylonian Tower Conjecture 10.11.3. This states that an infinitely stably extendable $kE$-module of constant Jordan type with non-projective part of dimension at most $p-2$ has to be a direct sum of syzygies of the trivial module.

**Problem 16** What is the smallest dimension of a $kE$-module with a given rank variety?

A lower bound is given by Carlson's Theorem 12.1.3, and an upper bound is given by taking tensor products of Carlson modules $L_\zeta$, see Section 1.5. These bounds are not at all close to each other.

**Problem 17** For $p = 2$ the lower bound on dimension obtained by Carlson using Bézout's theorem, see Theorem 12.1.3, is often sharp. When $p$ is odd, it seems to be quite far from sharp. What is a good lower bound for $p$ odd?

**Problem 18** Let $\Lambda$ be an exterior algebra on $r$ generators. If $M$ is a $\Lambda$-module of complexity $c$ and Loewy length $r - c + 1$, what are the possibilites for $V_E^{\mathrm{rk}}(M)$?

In Section 12.8 it is shown that $V_E^{\mathrm{rk}}(M)$ must be equidimensional, and that every equidimensional variety expressible as a finite union of set theoretic complete intersections occurs this way, but that leaves a gap in our knowledge.

**Problem 19** Let $\Lambda = \Lambda(X_1, \ldots, X_{2m})$ be an exterior algebra on $2m$ generators. What is the smallest dimension of a $\Lambda$-module whose rank variety is the quadric

$$Y_1Y_2 + \cdots + Y_{2m-1}Y_{2m} = 0\,?$$

There are several ways to produce a module of dimension $2^m$, see Section 12.5. Is this the smallest possible?

**Problem 20** Fløystad [120] investigated monads on projective space of the form

$$\mathcal{O}(-1)^a \to \mathcal{O}^b \to \mathcal{O}(1)^c.$$

Investigate the consequences of his results for modules of Loewy length three for exterior algebras and for elementary abelian $p$-groups.

# References


[1] K. Akin, D. A. Buchsbaum, and J. Weyman, *Schur functors and Schur complexes*, Adv. in Math. **44** (1982), 207–278.

[2] G. Almkvist, *The number of nonfree components in the decomposition of symmetric powers in characteristic p*, Pacific J. Math. **77** (1978), 293–301.

[3] G. Almkvist, *Reciprocity theorems for representations in characteristic p*, Séminaire d'Algèbre Paul Dubreil et Marie-Paule Malliavin, 32ème année (Paris, 1979), Lecture Notes in Mathematics, vol. 795, Springer-Verlag, Berlin/New York, 1980, pp. 1–9.

[4] G. Almkvist, *Representations of $Z/pZ$ in characteristic p and reciprocity theorems*, J. Algebra **68** (1981), 1–27.

[5] G. Almkvist, *Some formulas in invariant theory*, J. Algebra **77** (1982), 338–359.

[6] G. Almkvist and R. Fossum, *Decomposition of exterior and symmetric powers of indecomposable $\mathbb{Z}/p\mathbb{Z}$-modules in characteristic p and relations to invariants*, Séminaire d'Algèbre Paul Dubreil, 30ème année (Paris, 1976–1977), Lecture Notes in Mathematics, vol. 641, Springer-Verlag, Berlin/New York, 1978, pp. 1–111.

[7] J. L. Alperin and L. Evens, *Representations, resolutions, and Quillen's dimension theorem*, J. Pure & Applied Algebra **22** (1981), 1–9.

[8] J. L. Alperin and L. Evens, *Varieties and elementary abelian subgroups*, J. Pure & Applied Algebra **26** (1982), 221–227.

[9] V. Ancona and G. Ottaviani, *Stability of special instanton bundles on $\mathbb{P}^{2n+1}$*, Trans. Amer. Math. Soc. **341** (1994), 677–693.

[10] V. Ancona and G. Ottaviani, *On moduli of instanton bundles on $\mathbb{P}^{2n+1}$*, Pacific J. Math. **171** (1995), 343–351.

[11] V. Ancona and G. Ottaviani, *The Horrocks bundles of rank three on $\mathbb{P}^5$*, J. Reine & Angew. Math. **460** (1995), 69–92.

[12] A. Aramova, L. L. Avramov, and J. Herzog, *Resolutions of monomial ideals and cohomology over exterior algebras*, Trans. Amer. Math. Soc. **352** (1999), no. 2, 579–594.

[13] E. Arrondo, *Schwarzenberger bundles of arbitrary rank on the projective space*, J. London Math. Soc. **82** (2010), 697–716.

[14] M. F. Atiyah, *On the Krull–Schmidt theorem with application to sheaves*, Bull. Soc. Math. France **84** (1956), 307–317.

[15] M. F. Atiyah, V. G. Drinfeld, N. J. Hitchin, and Yu. I. Manin, *Construction of instantons*, Phys. Lett. **65A** (1978), 185–187.

[16] M. F. Atiyah and I. G. Macdonald, *Introduction to commutative algebra*, Addison-Wesley, 1969.

[17] M. F. Atiyah and E. Rees, *Vector bundles on projective* 3*-space*, Invent. Math. **35** (1976), 131–153.

[18] M. F. Atiyah and D. O. Tall, *Group representations, $\lambda$-rings and the $J$-homomorphism*, Topology **8** (1969), 253–297.

[19] M. D. Atkinson and R. Westwick, *Spaces of linear transformations of equal rank*, Linear & Multilinear Algebra **13** (1983), 231–239.

[20] A. B. Aure, *Surfaces on quintic threefolds associated to the Horrocks–Mumford bundle*, Arithmetic of complex manifolds (Erlangen, 1988), Lecture Notes in Mathematics, vol. 1399, Springer-Verlag, Berlin/New York, 1989, pp. 1–9.

[21] M. Auslander and R.-O. Buchweitz, *The homological theory of maximal Cohen–Macaulay approximations*, Mém. Soc. Math. France (N.S.) **38** (1989), 5–37.

[22] L. L. Avramov, R.-O. Buchweitz, S. B. Iyengar, and C. Miller, *Homology of perfect complexes*, Adv. in Math. **223** (2010), 1731–1781.

[23] L. L. Avramov, V. Gasharov, and I. Peeva, *Complete intersection dimension*, Publ. Math. Inst. Hautes Études Sci. **86** (1997), 67–114.

[24] G. S. Avrunin and L. L. Scott, *Quillen stratification for modules*, Invent. Math. **66** (1982), 277–286.

[25] S. Baland, *Modules of constant Jordan type with two non-projective blocks*, J. Algebra **346** (2011), 343–350.

[26] S. Baland, *On the generic kernel filtration for modules of constant Jordan type*, Arch. Math. (Basel) **99** (2012), 305–314.

[27] E. Ballico, *Uniform vector bundles of rank $(n+1)$ on $\mathbb{P}_n$*, Tsukuba J. Math. **7** (1983), 215–226.

[28] E. Ballico, *Vector spaces of matrices of low rank and vector bundles on projective spaces*, Beiträge Algebra Geom. **36** (1995), 119–122.

[29] V. Baranovsky and J. Pecharich, *On equivalences of derived and singular categories*, Cent. Eur. J. Math. **8** (2010), no. 1, 1–14.

[30] M. J. J. Barry, *Decomposing tensor products and exterior and symmetric squares*, J. Group Theory **14** (2011), 59–82.

[31] M. J. J. Barry, *Generators for decompositions of tensor products of modules*, Arch. Math. (Basel) **97** (2011), 503–512.

[32] W. Barth, *Moduli of vector bundles on the projective plane*, Invent. Math. **42** (1977), 63–91.

[33] W. Barth, *Kummer surfaces associated with the Horrocks–Mumford bundle*, Journées de Géométrie Algébrique d'Angers, juillet 1979 (A. Beauville, ed.), Sijthoff and Noordhoff, Alphen aan den Rijn, 1980, pp. 29–48.

[34] W. Barth, *Irreducibility of the space of mathematical instanton bundles with rank* 2 *and $c_2 = 4$*, Math. Ann. **258** (1981), 81–106.

[35] W. Barth, K. Hulek, and R. Moore, *Shioda's modular surface $S(5)$ and the Horrocks–Mumford bundle*, Vector bundles on algebraic varieties (Bombay 1984) (M. F. Atiyah et al., ed.), Oxford University Press, 1987, pp. 35–106.

[36] W. Barth and R. Moore, *Geometry in the space of Horrocks–Mumford surfaces*, Topology **28** (1989), 231–245.

[37] L. B. Beasley, *Spaces of matrices of equal rank*, Linear Algebra Appl. **38** (1981), 227–237.

[38] A. A. Beilinson, *Coherent sheaves on $\mathbb{P}^n$ and problems of linear algebra*, Funct. Anal. Appl. **12** (1978), 214–216.

[39] D. J. Benson, *Representations and cohomology I: Basic representation theory of finite groups and associative algebras*, Cambridge Studies in Advanced Mathematics, vol. 30, Cambridge University Press, 1991, reprinted in paperback, 1998.

[40] D. J. Benson, *Representations and cohomology II: Cohomology of groups and modules*, Cambridge Studies in Advanced Mathematics, vol. 31, Cambridge University Press, 1991, reprinted in paperback, 1998.

[41] D. J. Benson, *Modules for elementary abelian p-groups*, Proceedings of the ICM (Hyderabad 2010), vol. II, 2010, pp. 113–124.

[42] D. J. Benson, *Modules of constant Jordan type with one non-projective block*, Algebras and Representation Theory **13** (2010), 315–318.

[43] D. J. Benson, *Modules of constant Jordan type with small non-projective part*, Algebras and Representation Theory **16** (2013), 29–33.

[44] D. J. Benson, *Modules of constant Jordan type and a conjecture of Rickard*, J. Algebra **398** (2014), 343–349.

[45] D. J. Benson, *Modules for elementary abelian groups and hypersurface singularities*, Commutative Algebra and Noncommutative Algebraic Geometry II, MSRI Publications, vol. 68, Cambridge University Press, 2015, pp. 19–42.

[46] D. J. Benson and J. F. Carlson, *Diagrammatic methods for modular representations and cohomology*, Commun. in Algebra **15** (1987), 53–121.

[47] D. J. Benson and J. F. Carlson, *Projective resolutions and Poincaré duality complexes*, Trans. Amer. Math. Soc. **132** (1994), 447–488.

[48] D. J. Benson, J. F. Carlson, and J. Rickard, *Complexity and varieties for infinitely generated modules, II*, Math. Proc. Camb. Phil. Soc. **120** (1996), 597–615.

[49] D. J. Benson, J. F. Carlson, and J. Rickard, *Thick subcategories of the stable module category*, Fundamenta Mathematicae **153** (1997), 59–80.

[50] D. J. Benson and E. Green, *Nonprincipal blocks with one simple module*, Quarterly Journal of Math (Oxford) **55** (2004), 1–11.

[51] D. J. Benson, S. B. Iyengar, and H. Krause, *Stratifying modular representations of finite groups*, Ann. of Math. **174** (2011), 1643–1684.

[52] D. J. Benson and R. Kessar, *Blocks inequivalent to their Frobenius twists*, J. Algebra **315** (2007), 588–599.

[53] D. J. Benson and J. Pevtsova, *A realization theorem for modules of constant Jordan type and vector bundles*, Trans. Amer. Math. Soc. **364** (2012), 6459–6478.

[54] D. J. Benson and F. Reid, *Modules with small Loewy length*, J. Algebra **414** (2014), 288–299.

[55] I. N. Bernstein, I. M. Gelfand, and S. I. Gelfand, *Algebraic vector bundles on $\mathbb{P}^n$ and problems of linear algebra*, Funktsional. Anal. i Prilozhen. **12** (1978), no. 3, 66–67.

[56] A. I. Bondal and M. M. Kapranov, *Homogeneous bundles*, Helices and Vector Bundles, London Math. Soc. Lecture Note Series, vol. 148, Cambridge University Press, 1990, pp. 45–55.

[57] V. M. Bondarenko and Y. A. Drozd, *Representation type of finite groups*, (transl. from Russian) J. Soviet Math. **20** (1982), 2515–2528.

[58] V. M. Bondarenko and I. V. Lytvynchuk, *The representation type of elementary abelian p-groups with respect to the modules of constant Jordan type*, Algebra and Discrete Mathematics **14** (2012), no. 1, 29–36.
[59] A. Boralevi, *The Horrocks–Mumford bundle restricted to planes*, Collect. Math. **58** (2007), 101–117.
[60] A. Boralevi, D. Faenzi, and E. Mezzetti, *Linear spaces of matrices of constant rank and instanton bundles*, Adv. in Math. **248** (2013), 895–920.
[61] C. Borcea, *On desingularizated Horrocks–Mumford quintics*, J. Reine & Angew. Math. **421** (1991), 23–41.
[62] A. Borel and J.-P. Serre, *Le théorème de Riemann–Roch*, Bull. Soc. Math. France **86** (1958), 97–136.
[63] W. Bosma, J. Cannon, and C. Playoust, *The Magma algebra system, I. The user language*, J. Symbolic Comput. **24** (1997), 235–265.
[64] R. Bott, *Homogeneous vector bundles*, Ann. of Math. **66** (1957), 203–248.
[65] R. M. Bryant and M. Johnson, *Adams operations on the Green ring of a cyclic group of prime-power order*, J. Algebra **323** (2010), 2818–2833.
[66] R.-O. Buchweitz, *Maximal Cohen–Macaulay modules and Tate cohomology over Gorenstein rings*, Unpublished preprint, 1986.
[67] R.-O. Buchweitz, G.-M. Greuel, and F.-O. Schreyer, *Cohen–Macaulay modules on hypersurface singularities II*, Invent. Math. **88** (1987), 165–182.
[68] T. Bühler, *Exact categories*, Expo. Math. **28** (2010), 1–69.
[69] J. Burke and G. Stevenson, *The derived category of a graded Gorenstein ring*, Commutative Algebra and Noncommutative Algebraic Geometry II, MSRI Publications, vol. 68, Cambridge University Press, 2015, pp. 93–123.
[70] J. Burke and M. E. Walker, *Matrix factorizations in higher codimension*, Trans. Amer. Math. Soc. **367** (2015), no. 5, 3323–3370.
[71] J. F. Carlson, *The modular representation ring of a cyclic* 2*-group*, J. London Math. Soc. **11** (1975), 91–92.
[72] J. F. Carlson, *Complexity and Krull dimension*, Representations of Algebras, Puebla, Mexico, 1980, Lecture Notes in Mathematics, vol. 903, Springer-Verlag, Berlin/New York, 1981, pp. 62–67.
[73] J. F. Carlson, *The complexity and varieties of modules*, Integral representations and their applications, Oberwolfach, 1980, Lecture Notes in Mathematics, vol. 882, Springer-Verlag, Berlin/New York, 1981, pp. 415–422.
[74] J. F. Carlson, *The varieties and the cohomology ring of a module*, J. Algebra **85** (1983), 104–143.
[75] J. F. Carlson, *The variety of an indecomposable module is connected*, Invent. Math. **77** (1984), 291–299.
[76] J. F. Carlson, *The cohomology ring of a module*, J. Pure & Applied Algebra **36** (1985), 105–121.
[77] J. F. Carlson, *Varieties for modules*, The Arcata Conference on Representations of Finite Groups (Arcata, California, 1986), Proc. Symp. Pure Math., vol. 47, American Math. Society, 1987, pp. 37–44.
[78] J. F. Carlson, *Varieties and modules of small dimension*, Arch. Math. (Basel) **60** (1993), 425–430.
[79] J. F. Carlson, *Cohomology and induction from elementary abelian subgroups*, Quarterly Journal of Math (Oxford) **51** (2000), 169–181.

[80] J. F. Carlson and E. M. Friedlander, *Exact category of modules of constant Jordan type*, Algebra, arithmetic and geometry: Manin Festschrift, Progr. in Math., vol. 269, Birkhäuser Verlag, Basel, 2009, pp. 259–281.

[81] J. F. Carlson, E. M. Friedlander, and J. Pevtsova, *Modules of constant Jordan type*, J. Reine & Angew. Math. **614** (2008), 191–234.

[82] J. F. Carlson, E. M. Friedlander, and A. A. Suslin, *Modules for $\mathbb{Z}/p \times \mathbb{Z}/p$*, Comment. Math. Helvetici **86** (2011), 609–657.

[83] H. Cartan and S. Eilenberg, *Homological algebra*, Princeton Mathematical Series, no. 19, Princeton Univ. Press, 1956.

[84] P. Cascini, *Weighted Tango bundles on $\mathbb{P}^n$ and their moduli spaces*, Forum Math. **13** (2001), 251–260.

[85] A. Causa, R. Re, and T. Teodorescu, *Some remarks on linear spaces of nilpotent matrices*, Le Matematiche **53 (Suppl.)** (1998), 23–32.

[86] C.-Y. J. Chan, *A correspondence between Hilbert polynomials and Chern polynomials over projective spaces*, Illinois J. Math. **48** (2004), no. 2, 451–462.

[87] X.-W. Chen, *Unifying two results of Orlov on singularity categories*, Abh. Math. Sem. Univ. Hamburg **80** (2010), no. 2, 207–212.

[88] X.-W. Chen, *The singularity category of an algebra with radical square zero*, Doc. Math. **16** (2011), 921–936.

[89] L. Chouinard, *Projectivity and relative projectivity over group rings*, J. Pure & Applied Algebra **7** (1976), 278–302.

[90] W.-L. Chow, *On equivalence classes of cycles in an algebraic variety*, Ann. of Math. **64** (1956), 450–479.

[91] I. Coandă, *On the Bernstein–Gel'fand–Gel'fand correspondence and a result of Eisenbud, Fløystad, and Schreyer*, J. Math. Kyoto Univ. **43** (2003), 429–439.

[92] I. Coandă, *Infinitely stably extendable vector bundles on projective space*, Arch. Math. (Basel) **94** (2010), 539–545.

[93] I. Coandă and G. Trautmann, *The splitting criterion of Kempf and the Babylonian tower theorem*, Commun. in Algebra **34** (2006), 2485–2488.

[94] L. Costa and G. Ottaviani, *Group actions on instanton bundles over $\mathbb{P}^3$*, Math. Nachr. **246–247** (2002), 31–46.

[95] L. Costa and G. Ottaviani, *Nondegenerate multidimensional matrices and instanton bundles*, Trans. Amer. Math. Soc. **355** (2003), 49–55.

[96] W. W. Crawley-Boevey, *On tame algebras and bocses*, Proc. London Math. Soc. **56** (1988), 451–483.

[97] E. C. Dade, *Endo-permutation modules over p-groups, I*, Ann. of Math. **107** (1978), 459–494.

[98] E. C. Dade, *Endo-permutation modules over p-groups, II*, Ann. of Math. **108** (1978), 317–346.

[99] W. Decker, *Das Horrocks–Mumford-Bündel und das Modul-Schema für stabile 2-Vektorbündel über $\mathbb{P}_4$ mit $c_1 = -1$, $c_2 = 4$*, Math. Zeit. **188** (1984), 101–110.

[100] W. Decker, *Stable rank 2 vector bundles with Chern-classes $c_1 = -1$, $c_2 = 4$*, Math. Ann. **275** (1986), 481–500.

[101] W. Decker, N. Manolache, and F.-O. Schreyer, *Geometry of the Horrocks bundle on $\mathbb{P}^5$*, Complex projective geometry (G. Elligsrud et al., ed.), London Math. Soc. Lecture Note Series, vol. 179, Cambridge University Press, 1992, pp. 128–148.

[102] W. Decker and F.-O. Schreyer, *On the uniqueness of the Horrocks–Mumford bundle*, Math. Ann. **273** (1986), 415–443.

[103] W. Decker and F.-O. Schreyer, *Pullbacks of the Horrocks–Mumford bundle*, J. Reine & Angew. Math. **382** (1987), 215–220.

[104] I. Dolgachev and M. Kapranov, *Arrangements of hyperplanes and vector bundles on $P^n$*, Duke Math. J. **71** (1993), 633–664.

[105] S. R. Doty, *The submodule structure of certain Weyl modules for groups of type $A_n$*, J. Algebra **95** (1985), 373–383.

[106] P. Dowbor and H. Meltzer, *Classical vector bundles and representations of quivers*, Bol. Soc. Mat. Mexicana **11** (2005), no. 2, 205–220.

[107] Y. A. Drozd, *Tame and wild matrix problems*, Representation Theory II, Lecture Notes in Mathematics, vol. 832, Springer-Verlag, Berlin/New York, 1980, pp. 242–258.

[108] L. Ein, *Stable vector bundles on projective spaces in Char $p > 0$*, Math. Ann. **254** (1980), 53–72.

[109] L. Ein, *Some stable vector bundles on $\mathbb{P}^4$ and $\mathbb{P}^5$*, J. Reine & Angew. Math. **337** (1982), 142–153.

[110] D. Eisenbud, *Homological algebra on a complete intersection, with an application to group representations*, Trans. Amer. Math. Soc. **260** (1980), 35–64.

[111] D. Eisenbud, G. Fløystad, and F.-O. Schreyer, *Sheaf cohomology and free resolutions over exterior algebras*, Trans. Amer. Math. Soc. **355** (2003), 4397–4426.

[112] D. Eisenbud and J. Harris, *Vector spaces of matrices of low rank*, Adv. in Math. **70** (1988), 135–155.

[113] P. Ellia, *Sur les fibrés uniformes de rang $(n+1)$ sur $\mathbb{P}^n$*, Mém. Soc. Math. France (N.S.), vol. 7, Gauthier–Villars, 1982.

[114] G. Ellingsrud and A. Strømme, *Stable rank-2 vector bundles on $\mathbb{P}^3$ with $c_1 = 0$ and $c_2 = 3$*, Math. Ann. **255** (1981), 123–135.

[115] L. Evens, *The cohomology ring of a finite group*, Trans. Amer. Math. Soc. **101** (1961), 224–239.

[116] D. Faenzi, *A geometric construction of Tango bundle on $\mathbb{P}^5$*, Kodai Math. J. **27** (2004), 1–6.

[117] M. L. Fania and E. Mezzetti, *Vector spaces of skew-symmetric matrices of constant rank*, Linear Algebra Appl. **434** (2011), 2383–2403.

[118] R. Farnsteiner, *Jordan types for indecomposable modules of finite group schemes*, J. Eur. Math. Soc. **16** (2014), 925–989.

[119] W. Feit, *The representation theory of finite groups*, North Holland, Amsterdam, 1982.

[120] G. Fløystad, *Monads on projective spaces*, Commun. in Algebra **28** (2000), no. 12, 5503–5516.

[121] G. Fløystad, *Describing coherent sheaves on projective space*, arXiv:math/0012263, 2001.

[122] R. M. Fossum, *Invariants and Schur functors in characteristic $p > 0$*, Young tableaux and Schur functors in algebra and geometry (Toruń, 1980), Astérisque, vol. 87, Soc. Math. France, 1981, pp. 85–95.

[123] R. M. Fossum, *Decompositions revisited*, Séminaire d'Algèbre Paul Dubreil and Marie-Paule Malliavin, 34ème Année (Paris, 1981), Lecture Notes in Mathematics, vol. 924, Springer-Verlag, Berlin/New York, 1982, pp. 260–295.

[124] E. M. Friedlander and J. Pevtsova, *Representation theoretic support spaces for finite group schemes*, Amer. J. Math. **127** (2005), 379–420, correction: AJM **128** (2006), 1067–1068.
[125] E. M. Friedlander and J. Pevtsova, *Π-supports for modules for finite groups schemes*, Duke Math. J. **139** (2007), 317–368.
[126] E. M. Friedlander and J. Pevtsova, *Generalized support varieties for finite group schemes*, Documenta Math. **Extra Volume Suslin** (2010), 197–222.
[127] E. M. Friedlander and J. Pevtsova, *Constructions for infinitesimal group schemes*, Trans. Amer. Math. Soc. **363** (2011), 6007–6061.
[128] E. M. Friedlander, J. Pevtsova, and A. Suslin, *Generic and maximal Jordan types*, Invent. Math. **168** (2007), 485–522.
[129] W. Fulton, *Intersection theory*, Ergebnisse der Mathematik und ihrer Grenzgebiete, Folge 3, Band 2, Springer-Verlag, Berlin/New York, 1984.
[130] M. Gerstenhaber, *On nilalgebras and linear varieties of nilpotent matrices I*, Amer. J. Math. **80** (1958), 614–622.
[131] M. Gerstenhaber, *On nilalgebras and linear varieties of nilpotent matrices III*, Ann. of Math. **70** (1959), 167–205.
[132] M. Gerstenhaber, *On nilalgebras and linear varieties of nilpotent matrices II*, Duke Math. J. **27** (1960), 21–31.
[133] M. Gerstenhaber, *On dominance and varieties of commuting matrices*, Ann. of Math. **73** (1961), 324–348.
[134] M. Gerstenhaber, *On nilalgebras and linear varieties of nilpotent matrices IV*, Ann. of Math. **75** (1962), 382–418.
[135] S. P. Glasby, C. E. Praeger, and B. Xia, *Decomposing modular tensor products: 'Jordan partitions', their parts and p-parts*, Israel J. Math. **209** (2015), 215–233.
[136] E. S. Golod, *The cohomology ring of a finite p-group*, Dokl. Akad. Nauk. SSSR **125** (1959), 703–706, (Russian).
[137] A. L. Gorodentsev and A. N. Rudakov, *Exceptional vector bundles on projective spaces*, Duke Math. J. **54** (1987), 115–130.
[138] J. A. Green, *The modular representation algebra of a finite group*, Illinois J. Math. **6** (1962), 607–619.
[139] J. P. C. Greenlees, *Commutative algebra in group cohomology*, J. Pure & Applied Algebra **98** (1995), 151–162.
[140] A. Grothendieck, *Sur la classification des fibrés holomorphes sur la sphère de Riemann*, Amer. J. Math. **79** (1957), 121–138.
[141] A. Grothendieck, *Éléments de géométrie algébrique I*, Publ. Math. IHES (1960), no. 4, 1–228.
[142] A. Grothendieck, *Éléments de géométrie algébrique II*, Publ. Math. IHES (1961), no. 8, 1–222.
[143] T. A. Hannula, T. G. Ralley, and I. Reiner, *Modular representation algebras*, Bull. Amer. Math. Soc. **73** (1967), 100–101.
[144] D. Happel, *Triangulated categories in the representation theory of finite dimensional algebras*, London Math. Soc. Lecture Note Series, vol. 119, Cambridge University Press, 1988.
[145] R. Hartshorne, *Complete intersections and connectedness*, Amer. J. Math. **84** (1962), 497–508.

[146] R. Hartshorne, *Residues and duality*, Lecture Notes in Mathematics, vol. 20, Springer-Verlag, Berlin/New York, 1966.

[147] R. Hartshorne, *Algebraic geometry*, Graduate Texts in Mathematics, vol. 52, Springer-Verlag, Berlin/New York, 1977.

[148] R. Hartshorne, *Stable vector bundles and instantons*, Comm. Math. Phys. **59** (1978), 1–15.

[149] R. Hartshorne, *Stable vector bundles of rank* 2 *on* $\mathbb{P}^3$, Math. Ann. **238** (1978), 229–280.

[150] R. Hartshorne, *Algebraic vector bundles on projective spaces: a problem list*, Topology **18** (1979), 117–128.

[151] J. Herzog and D. Popescu, *Thom–Sebastiani problems for maximal Cohen–Macaulay modules*, Math. Ann. **309** (1997), 677–700.

[152] J. Herzog, B. Ulrich, and J. Backelin, *Linear maximal Cohen–Macaulay modules over strict complete intersections*, J. Pure & Applied Algebra **71** (1991), 187–202.

[153] W. Hesselink, *Singularities in the nilpotent scheme of a classical group*, Trans. Amer. Math. Soc. **222** (1976), 1–32.

[154] L. Hille, *Homogeneous vector bundles and Koszul algebras*, Math. Nachr. **191** (1998), 189–195.

[155] F. Himstedt and P. Symonds, *Exterior and symmetric powers of modules for cyclic 2-groups*, J. Algebra **410** (2014), 393–420.

[156] F. Hirzebruch, *Topological methods in algebraic geometry*, Grundlehren der mathematischen Wissenschaften, vol. 131, Springer-Verlag, Berlin/New York, 1978.

[157] N. Hoffmann, *Independent parameters for special instanton bundles on* $\mathbb{P}^{2n+1}$, Journal of Geometry and Physics **61** (2011), 2321–2330.

[158] M. Holloway and R. Kessar, *Quantum complete rings and blocks with one simple module*, Quarterly Journal of Math (Oxford) **56** (2005), 209–221.

[159] G. Horrocks, *A construction of locally free sheaves*, Topology **7** (1968), 117–120.

[160] G. Horrocks, *Examples of rank three vector bundles on five dimensional projective space*, J. London Math. Soc. **18** (1978), 15–27.

[161] G. Horrocks and D. Mumford, *A rank* 2 *vector bundle on* $\mathbb{P}^4$ *with* $15,000$ *symmetries*, Topology **12** (1973), 63–81.

[162] I. Hughes and G. Kemper, *Symmetric powers of modular representations, Hilbert series and degree bounds*, Commun. in Algebra **28** (2000), 2059–2088.

[163] K. Hulek, *Geometry of the Horrocks–Mumford bundle*, Algebraic geometry (Bowdoin 1985) (S. J. Bloch, ed.), Proc. Symp. Pure Math., vol. 46, Part II, American Math. Society, 1987, pp. 69–85.

[164] K. Hulek, *The Horrocks–Mumford bundle*, Vector bundles in algebraic geometry (Durham 1993), London Math. Soc. Lecture Note Series, vol. 208, Cambridge University Press, 1995, pp. 139–177.

[165] K. Hulek and A. Van de Ven, *The Horrocks–Mumford bundle and the Ferrand construction*, Manuscripta Math. **50** (1985), 313–335.

[166] B. Ilic and J. M. Landsberg, *On symmetric degeneracy loci, spaces of symmetric matrices of constant rank and dual varieties*, Math. Ann. **314** (1999), 159–174.

[167] S. Iyengar (ed.), *Twenty-four hours of local cohomology*, Graduate Studies in Mathematics, American Math. Society, 2008.

[168] S. Iyengar and R. Takahashi, *Annihilation of cohomology and strong generation of module categories*, Int. Math. Res. Not. **2016** (2016), no. 2, 499–535.

[169] S. Jackowski and J. E. McClure, *Homotopy decomposition of classifying spaces via elementary abelian subgroups*, Topology **31** (1992), 113–132.

[170] K. Jaczewski, M. Szurek, and J. Wiśniewski, *Geometry of the Tango bundle*, Proceedings of the conference on algebraic geometry (Berlin, 1985) (Leipzig), Teubner-Texte Math., vol. 92, Teubner, 1986, pp. 177–185.

[171] G. D. James and A. Kerber, *The representation theory of the symmetric group*, Cambridge University Press, 1981.

[172] H. Kaji, *Example of $\sigma$-transition matrices defining the Horrocks–Mumford bundle*, Tokyo J. Math. **12** (1989), 21–32.

[173] S. O. Kaptanoğlu, *Structure and detection theorems for $k[C_2 \times C_2]$-modules*, Rend. Sem. Mat. Univ. Padova **123** (2010), 169–189.

[174] S. O. Kaptanoğlu, *$p$-power points and modules of constant $p$-power Jordan type*, Commun. in Algebra **39** (2011), 3781–3800.

[175] B. Keller, *Chain complexes and stable categories*, Manuscripta Math. **67** (1990), 379–417.

[176] B. Keller and D. Vossieck, *Sous les catégories dérivées*, Comptes Rendus Acad. Sci. Paris, Série I **305** (1987), 225–228.

[177] G. Kempf, *A criterion for the splitting of a vector bundle*, Forum Math. **2** (1990), 477–480.

[178] H. Knörrer, *Cohen–Macaulay modules on hypersurface singularities I*, Invent. Math. **80** (1987), 153–164.

[179] F. M. Kouwenhoven, *The $\lambda$-structure of the Green ring of cyclic $p$-groups*, The Arcata Conference on the Representation Theory of Finite Groups (P. Fong, ed.), Proc. Symp. Pure Math., vol. 47, 1987, pp. 451–466.

[180] D. Kraines, *Massey higher products*, Trans. Amer. Math. Soc. **124** (1966), 431–449.

[181] O. Kroll, *Complexity and elementary abelian $p$-groups*, J. Algebra **88** (1984), 155–172.

[182] N. M. Kumar, *Construction of rank two vector bundles on $\mathbb{P}^4$ in positive characteristic*, Invent. Math. **130** (1997), 277–286.

[183] N. M. Kumar, C. Peterson, and A. P. Rao, *Construction of low rank vector bundles on $\mathbb{P}^4$ and $\mathbb{P}^5$*, J. Alg. Geometry **11** (2002), 203–217.

[184] N. M. Kumar, C. Peterson, and A. P. Rao, *Degenerating families of rank two bundles*, Proc. Amer. Math. Soc. **131** (2003), 3681–3688.

[185] N. M. Kumar, C. Peterson, and A. P. Rao, *Monads on projective spaces*, Manuscripta Math. **112** (2003), 183–189.

[186] E. Lee, *A modular quintic Calabi–Yau threefold of level* 55, Canadian J. Math. **63** (2011), 616–633.

[187] K. J. Lim, *The varieties for some Specht modules*, J. Algebra **321** (2009), 2287–2301.

[188] M. Linckelmann, *Quillen stratification for block varieties*, J. Pure & Applied Algebra **172** (2002), 257–270.

[189] I. G. Macdonald, *Symmetric functions and Hall polynomials*, Oxford University Press, 1995, 2ed.

[190] S. MacLane, *Categories for the working mathematician*, Graduate Texts in Mathematics, vol. 5, Springer-Verlag, Berlin/New York, 1971.

[191] L. Manivel and E. Mezzetti, *On linear spaces of skew-symmetric matrices of constant rank*, Manuscripta Math. **117** (2005), 319–331.

[192] J. D. McFall, *How to compute the elementary divisors of the tensor product of two matrices*, Linear & Multilinear Algebra **7** (1979), no. 3, 193–201.

[193] J. D. McFall, *On elementary divisors of the tensor product of two matrices*, Linear Algebra Appl. **33** (1980), 67–86.

[194] R. M. Miro-Roig and J. A. Orus-Lacort, *On the smoothness of the moduli space of mathematical instanton bundles*, Compositio Math. **105** (1997), 109–119.

[195] T. Muir and W. H. Metzler, *A treatise on the theory of determinants*, 2ed, Dover, 1960.

[196] S.-Y. Nam and Y.-T. Oh, *$\lambda$-ring structure of the Green ring of a cyclic $p$-group*, J. Algebra **338** (2011), 92–113.

[197] S.-Y. Nam and Y.-T. Oh, *Symmetric powers of the $p+1$-dimensional indecomposable module of a cyclic $p$-group and the $\lambda$-structure of its Green ring*, J. Algebra **368** (2012), 75–91.

[198] A. Neeman, *Algebraic and analytic geometry*, Cambridge University Press, 2007.

[199] C. W. Norman, *On the Jordan form of the tensor product over fields of prime characteristic*, Linear & Multilinear Algebra **38** (1995), 351–371.

[200] C. W. Norman, *On Jordan bases for the tensor product and Kronecker sum and their elementary divisors over fields of prime characteristic*, Linear & Multilinear Algebra **56** (2008), 415–451.

[201] J. O'Halloran, *A simple proof of the Gerstenhaber–Hesselink theorem for nilpotent matrices*, Commun. in Algebra **15** (1987), 2017–2023.

[202] C. Okonek, M. Schneider, and H. Spindler, *Vector bundles on complex projective spaces*, Birkhäuser Verlag, Basel, 1980; 3ed. 2011.

[203] C. Okonek and H. Spindler, *Mathematical instanton bundles on $\mathbb{P}^{2n+1}$*, J. Reine & Angew. Math. **364** (1986), 35–50.

[204] D. O. Orlov, *Triangulated categories of singularities and equivalences between Landau–Ginzburg models*, Sbornik Math **197** (2006), 1827–1840.

[205] G. Ottaviani and E. Rubei, *Quivers and the cohomology of homogeneous vector bundles*, Duke Math. J. **132** (2006), 459–508.

[206] D. G. Quillen, *The spectrum of an equivariant cohomology ring, I*, Ann. of Math. **94** (1971), 549–572.

[207] D. G. Quillen, *The spectrum of an equivariant cohomology ring, II*, Ann. of Math. **94** (1971), 573–602.

[208] D. G. Quillen, *Higher algebraic $K$-theory I*, Algebraic $K$-theory I: Higher $K$-theories (H. Bass, ed.), Lecture Notes in Mathematics, vol. 341, Springer-Verlag, Berlin/New York, 1973, pp. 85–147.

[209] D. G. Quillen, *Projective modules over polynomial rings*, Invent. Math. **36** (1976), 167–171.

[210] D. G. Quillen and B. B. Venkov, *Cohomology of finite groups and elementary abelian subgroups*, Topology **11** (1972), 317–318.

[211] T. Ralley, *Decomposition of products of modular representations*, J. London Math. Soc. **44** (1969), 480–484.

[212] A. P. Rao, *A note on cohomology modules of rank two bundles*, J. Algebra **86** (1984), 23–34.

[213] A. P. Rao, *Mathematical instantons in characteristic two*, Compositio Math. **119** (1999), 169–184.

[214] F. Reid, *Varieties for modules of small dimension*, Ph.D. thesis, University of Aberdeen, 2014.

[215] J.-C. Renaud, *The characters and structure of a class of modular representation algebras of cyclic p-groups*, J. Austral. Math. Soc. **26** (1978), 410–418.

[216] J.-C. Renaud, *The decomposition of products in the modular representation ring of a cyclic group of prime power order*, J. Algebra **58** (1979), 1–11.

[217] J.-C. Renaud, *On modular representation algebras and a class of matrix algebras*, J. Austral. Math. Soc. **33** (1982), 351–355.

[218] J.-C. Renaud, *Recurrence relations in a modular representation algebra*, Bull. Austral. Math. Soc. **26** (1982), 215–219.

[219] J. Rickard, *Derived categories and stable equivalence*, J. Pure & Applied Algebra **61** (1989), 303–317.

[220] C. M. Ringel, *The representation type of local algebras*, Representations of Algebras, Lecture Notes in Mathematics, vol. 488, Springer-Verlag, Berlin/New York, 1974.

[221] A. N. Rudakov, *The Markov numbers and exceptional bundles on* $\mathbb{P}^2$, Izvestiya **32** (1989), 99–112.

[222] E. Sato, *On the decomposability of infinitely extendable vector bundles on projective spaces and Grassmann varieties*, J. Math. Kyoto Univ. **17** (1977), 127–150.

[223] H. Schoutens, *Projective dimension and the singular locus*, Commun. in Algebra **31** (2003), no. 1, 217–239.

[224] J.-P. Serre, *Faisceaux algébriques cohérents*, Ann. of Math. **61** (1955), no. 2, 197–278.

[225] J.-P. Serre, *Géométrie algébrique et géométrie analytique*, Ann. Inst. Fourier (Grenoble) **6** (1955–1956), 1–42.

[226] J.-P. Serre, *Modules projectifs et espaces fibrés à fibre vectorielle*, Séminaire P. Dubreil, M.-L. Dubreil-Jacotin et C. Pisot (1958), Fasc. 2, Exp. 23.

[227] I. R. Shafarevich, *Basic algebraic geometry*, Springer-Verlag, Berlin/New York, 1977.

[228] B. Srinivasan, *The modular representation ring of a cyclic p-group*, Proc. London Math. Soc. **14** (1964), 677–688.

[229] Robert Steinberg, *Générateurs, relations et revêtements de groupes algébriques*, Colloq. Théorie des Groupes Algébriques (Bruxelles, 1962), Librairie Universitaire, Louvain, 1962, pp. 113–127.

[230] A. A. Suslin, *Projective modules over polynomial rings are free (Russian)*, Dokl. Akad. Nauk SSSR **229** (1976), 1063–1066.

[231] J. Sylvester, *On the dimension of spaces of linear transformations satisfying rank conditions*, Linear Algebra Appl. **78** (1986), 1–10.

[232] P. Symonds, *The complexity of a module and elementary abelian subgroups: a geometric approach*, Proc. Amer. Math. Soc. **113** (1991), 27–29.

[233] P. Symonds, *Cyclic group actions on polynomial rings*, Bull. London Math. Soc. **39** (2007), 181–188.

[234] H. Tango, *An example of indecomposable vector bundle of rank* $n-1$ *on* $\mathbb{P}^n$, J. Math. Kyoto Univ. **16** (1976), 137–141.

[235] H. Tango, *On morphisms from projective space* $\mathbb{P}^n$ *to the Grassmann variety* $\mathrm{Gr}(n, d)$, J. Math. Kyoto Univ. **16** (1976), 201–207.

[236] H. Tango, *On vector bundles on* $\mathbb{P}^n$ *which have* $\sigma$*-transition matrices*, Tokyo J. Math. **16** (1993), 1–29.

[237] J. Tate, *Homology of Noetherian rings and local rings*, Illinois J. Math. **1** (1957), 14–27.

[238] A. N. Tyurin, *Finite dimensional vector bundles over infinite varieties*, Math. USSR Izv. **10** (1976), 1187–1204.

[239] A. N. Tyurin, *Collected works I: Vector bundles*, Universitätsverlag Göttingen, 2008, F. Bogomolov et al., eds.

[240] V. K. Vedernikov, *A family of rank-2 mathematical instanton bundles on* $P_3$, Publ. RIMS, Kyoto Univ. **33** (1997), 573–598.

[241] A. Van de Ven, *On uniform vector bundles*, Math. Ann. **195** (1972), 245–248.

[242] B. B. Venkov, *Cohomology algebras for some classifying spaces*, Dokl. Akad. Nauk. SSSR **127** (1959), 943–944.

[243] P. J. Webb, *Bounding the ranks of* $\mathbb{Z}G$*-modules by their restrictions to elementary abelian subgroups*, J. Pure & Applied Algebra **23** (1982), 311–318.

[244] D. L. Wehlau, *Invariants for the modular cyclic group of prime order via classical invariant theory*, J. Eur. Math. Soc. **15** (2013), 775–803.

[245] R. Westwick, *Spaces of linear transformations of equal rank*, Linear Algebra Appl. **5** (1972), 49–64.

[246] R. Westwick, *Spaces of matrices of fixed rank*, Linear & Multilinear Algebra **20** (1987), 171–174.

[247] R. Westwick, *Examples of constant rank spaces*, Linear & Multilinear Algebra **28** (1990), 155–174.

[248] R. Westwick, *Spaces of matrices of fixed rank II*, Linear Algebra Appl. **235** (1996), 163–169.

[249] K. Wolffhardt, *The Hochschild homology of complete intersections*, Trans. Amer. Math. Soc. **171** (1972), 51–66.

[250] J. Worch, *Categories of modules for elementary abelian p-groups and generalized Beilinson algebras*, J. London Math. Soc. **88** (2013), 649–668.

[251] Y. Yoshino, *Cohen–Macaulay modules over Cohen–Macaulay rings*, London Math. Soc. Lecture Note Series, vol. 146, Cambridge University Press, 1990.

# Index

For EU product safety concerns, contact us at Calle de José Abascal, 56–1°,
28003 Madrid, Spain or eugpsr@cambridge.org.

www.ingramcontent.com/pod-product-compliance
Lightning Source LLC
LaVergne TN
LVHW101458200926
848483LV00001B/156

* 9 7 8 1 1 0 7 1 7 4 1 7 7 *